NEW HAVEN
CLOCKS & WATCHES

With a Special Section on New Haven Movements

Tran Duy Ly

Arlington Book Company, Inc.
1997

NEW HAVEN CLOCKS & WATCHES

With a Special Section on New Haven Movements

Tran Duy Ly

HISTORY by **CHRIS H. BAILEY**

INDEX by **SONYA L. SPITTLER**

INTRODUCTION by **THOMAS J. SPITTLER**

First Edition

First Published in USA 1997 by Arlington Book Company, Inc.
2706 Elsmore Street, Fairfax, Virginia 22031-1409 USA

ISBN 0-930163-75-3

The Arlington Book Company proudly presents its newest and long-awaited publication, "**New Haven Clocks & Watches**" by Tran Duy Ly.

The New Haven Clock Company has a fascinating history producing clocks for over 100 years. This is the first book to combine most New Haven catalogs and many photographs of actual New Haven clocks and movements into one definitive source combined with an excellent price guide. You can be certain that the clocks shown contain their original hands, pendulums, glasses, finials, and all other case trim. Meticulous detail, resulting from exhaustive horological research, is presented in this volume in the same easy to use format that has made Tran Duy Ly's other publications so popular.

The supplementary, current price guide has been compiled by America's top horologists and provides professional estimates for clocks in original mint condition. This price guide will be kept current and always made available by the Arlington Book Company.

With this outstanding New Haven reference book at your finger tips you have the best advice from experts throughout America on identifying, dating, and collecting these historical clocks. We are sure that it will be an invaluable guide in assisting your decision to buy a particular clock as well as helping you obtain the best possible pricing advice.

ARLINGTON BOOK COMPANY, INC.

INTRODUCTION

I began to seriously collect clocks while living in England in 1974 and most of the clocks I found were American clocks by the New Haven Clock Company and its associated British sales firm Jerome & Co. Many of these clocks had an interesting label showing a factory identified as "The New Haven Clock Company formerly the Jerome Manufacturing Company." I wish I had had a book such as ***New Haven Clocks & Watches*** by Tran Duy Ly to help me better understand the history of the New Haven Clock Company and identify the many clocks and watches they produced.

Tran Duy Ly has produced ***New Haven Clocks & Watches*** to complement his other books on Seth Thomas clocks, Waterbury clocks, Ansonia clocks and Welch clocks. In my opinion, these five firms combined to produce over 75% of the clocks manufactured in America. This exciting new volume on New Haven clocks is produced to the usual high standards of the Arlington Book Company. It contains an amazing 38 chapters and over 500 pages on different types of New Haven clocks and a supplementary price guide. The many views provided will help identify correct dials, hands, pendulums, weights, gongs, movements and trim.

What intrigues me most about this book is the 11 page history and chronology written by the noted horologist Chris Bailey. Chris clearly develops the story of the giant of American clockmaking, Chauncey Jerome, and his relationship with Hiram Camp, the first president of the New Haven Clock Company. I have always felt that if Camp had named the firm the Chauncey Jerome Clock Company he would have captured its true identity in the way the Seth Thomas Clock Company did. But this did not happen. This caused problems for the fledgling New Haven Clock Company when it tried to sell its products in England. The British identified the name "Jerome" with quality American clocks and were willing to accept no substitutes. To overcome the situation New Haven used the name of their independent British sales firm, Jerome & Co., and New Haven clocks with the Jerome & Co. label were successfully sold in England from 1856 to 1904. Even today the name "Jerome" in England is bigger than Seth Thomas and E. N. Welch combined.

So it is fitting that the Arlington Book Company has filled a void by producing this long awaited book on the New Haven Clock Company. Tran Duy Ly, more than any other author of horological books in America, has repeatedly produced books that are right on target for clock collectors around the world. Each new book builds on the solid foundation he has established in the past. ***New Haven Clocks & Watches***, like his other books, is well written and easy to use. It contains over 2,200 quality illustrations and photographs. The section on kitchen clocks contains the most models ever seen in print. In addition to hundreds of pages of clocks there are even 24 pages of New Haven watches.

I feel the 50 pages of movements and parts showing extremely clear and detailed illustrations of 61 New Haven movements are outstanding. Most of these pages show a front and rear view of the movements and for the first time I can recall, each movement part is clearly identified. For the repairman, these 50 pages alone are worth the price of the book.

In summary, I am truly excited about Tran Duy Ly's new book ***New Haven Clocks & Watches***. If you own a New Haven clock and wish to find out more about it, or if you deal in clocks and need to know up-to-date prices, or if you repair clocks and could use wonderful illustrations of New Haven movements, this is the book for you.

TOM SPITTLER
New Carlisle, Ohio

ACKNOWLEDGEMENTS

I would like to take this opportunity to express my sincere appreciation to the following friends who provided me with valuable material, information and up-to-date prices without which this publication could not have been accomplished.

Tran Duy Ly

ALLEN, J. C.
Carthage, MO.

AMERICAN CLOCK AND WATCH MUSEUM
Bristol, CT.

BAILEY, Chris H.
FNAWCC*
Bristol, CT.

BERGER, Steven & Marsha
FNAWCC
Fountain Hills, AZ

BOXENHORN, Jerold
S. Bellmore, N. Y.

BURLINGAME, Roy
East Moline, IL.

COWAN, Doug
FNAWCC
Cincinnati, OH.

D'AMICO, Joe
Bergenfield, N. J.

DYE, Merrill, D.
Hagaman, N. Y.

EHRHARDT, Roy
Umatilla, FL.

FUNK, Larry
Northfield, IL.

GORRELL, David
Annapolis, MD.

HAYENGA, Lubertus H.
Alameda, CA.

JUNCK, Russel
Des Moines, IA

KRAUSE, James
Hungtington Beach, CA.

LEE, J. David
Delanson, N. Y.

LOOFT, Terry & Carole
Wilmington, OH.

LOWE, Michael G.
E. Peoria, IL.

MALTZ, Millie & Jerry
New Rochelle, N. Y.

MARKLEY, Kenneth, A.
Camp Hill, PA.

MATHER, William K.
FNAWCC
Randolph Center, VT.

MICHAEL'S ANTIQUE CLOCKS
Oklahoma City, OK.

MILLER, Denzil
Miamiburg, OH.

MITTELSTADT, William
Sheboygan, WI.

MORGAN, David
FNAWCC*
Whitehouse, N. J.

NAWCC., INC.
Columbia, PA.

O'BRIEN, Gene
St. Louis, MO.

PEER, Mark
Mark of Time
Sarasota, FL.

PETRUCELLI, Steve
Cranbury, N. J.

RAMBO, Chuck
Appleton, WI.

RAPPAPORT, Larry
Great Neck, N. Y.

REICHEL, Robert
FNAWCC
Kent, WA.

RICHEY, Earl
Alexandria, VA.

RICHMOND, Leon
Leasburg, N. C.

SADOWSKI, Steve
Maspeth, N. Y.

SCHOLI, Mike
Springfield, PA.

SERNAK, Joseph
Stirling, N. J.

SHANNON, David
Alexandria, VA.

SIMMONS, J. W.
FNAWCC*
Eagle Rock, VA.

SPITTLER, Tom & Sonya
FNAWCC
New Carlisle, OH.

TANNER, John
Upland, CA.

TAYLOR, Morgan Jay
Alameda, CA.

WARNER, Jane & David
FNAWCC
St. Louis, MO.

WILKEY, Kim & Marty
Norfolk, England

WINGATE, Robert M.
Dallas, TX.

WONG, Dennis
Falls River, MA.

WYER, Jack
S. Setauket, N. Y.

ZINGSHEIM, Roy
Fond-du-Lac, WI.

* Star Fellow

SPECIAL THANKS

From the Arlington Book Company to Bill Simmons who so generously donated his collection of original Clock and Watch Catalogs.

Bill Simmons was NAWCC President from 1979 to 1981.

J. W. Simmons
Box 56
Eagle Rock, VA 24085

July 25, 1995

Tran Duy Ly
Arlington Book Co.
2706 Elsmore St.
Fairfax, VA 22031

Dear Tran,

For many years I have collected original catalogs of clocks and watches. They have given me much enjoyment in both finding them and then using the vast amount of original pictures and descriptions of clocks and watches they contain. Over the years these rare catalogs have often proved to be the only source of information available on a particular clock and I was only able to identify it and date it from one of these catalogs.

I have tremendous respect for the work you have done over the past 25 years at the Arlington Book Co. in republishing many of these original catalogs and also in your authoring the magnificent books on Seth Thomas, Ansonia, Waterbury, and Welch as well as the two volumes on American clocks, Calendar clocks and your Longcase clock book. These books have quickly become the standards for the horological community worldwide. They have reached tens of thousands of individuals from beginning collectors to the most advanced horologists.

For this reason I have decided to give you my entire collection enclosed herewith all original catalogs knowing they will then be seen by tens of thousands in your hands and the horological community will be much better off because of that.

Sincerely,

Bill Simmons

CONTENTS

Page		
	4	**Introduction**
	5	**Acknowledgements**
	7	**Contents**
	8	**Important Tips**
	14	**History**
	21	**Chronology**
	506	**Index**

Page		
	25	Advertising Clocks
	28	Alarm Clocks
	49	Automobile Clocks
	59	Banjo Clocks
	65	Calendar Clocks
	87	Carriage Clocks
	94	Chime Clocks
	102	Crystal Regulators
	108	Dials
	110	Electric Clocks
	116	Figure Clocks
	131	Hanging Clocks
	182	Inkstand Clocks
	184	Leather Cases
	185	Longcase Clocks & Standing Regulators
	192	Mantel Clocks - **Cabinet Clocks**
	217	Mantel Clocks - **Enameled Imitation Cases**
	223	Mantel Clocks - **Enameled Iron Cases**
	237	Mantel Clocks - **Enameled Wood Cases**
	247	Mantel Clocks - **Marble Cases**
	264	Mantel Clocks - **Marbloid Cases**
	266	Mantel Clocks - **Metal Cases**
	282	Mantal Clocks - **Musical Clocks**
	283	Mantel Clocks - **Tambour Clocks**
	292	Mantel Clocks -**Tile Cases**
	294	Mantel Clocks - **Wood Cases**
	350	Mission Clocks
	357	Movement Illustrations
	401	Movement Parts
	408	Novelty Clocks
	453	Novelty Clocks - **Wood Cases**
	461	Plush Clocks
	466	Porcelain Clocks
	475	Statuettes and Candlesticks
	479	Watchman's Clocks
	482	Watches
	495	Watches - Advertising
	497	Wrist Watches

IMPORTANT TIPS

The New Haven Clock Company made and sold clocks for over 100 years and it is assured that within these pages you will find exact representations of the company's best work . The most complete selection of New Haven clocks are presented in this edition which provides a comprehensive guide to a vast and thorough exhibition of antique clocks. Years of exhaustive research and compilation of numerous clock catalogs have resulted in clear and accurate renderings of these fine timepieces. The use of these catalog illustrations has also permitted the display of modifications and changes in the appearances of many models, showing clearly that the changes were made by the company and are not the result of modifications or restoration occurring after the clock left the factory. In addition photographs of actual clocks are included when they show additional detail not available from catalog illustrations.

Clocks of all types may be found in this book, including: Advertising, Alarm, Automobile, Banjo, Calendar, Carriage, Chime, Crystal Regulators, Electric, Figure, Inkstand, Hanging, Longcase, Mantel, Mission, Novelty, Plush, Porcelain, Statuette, Candlestick, Watchman's Clocks, Watches and Wrist Watches. In addition, a wide variety of movements are also included.

- All models are completely and professionally indexed.
- Clocks of the same type are grouped together.
- Clocks with similar appearance are grouped next to each other.
- Beside each clock is the year of the catalog from which the illustration was taken and this represents the date of production. However, some popular clocks may have been produced a few years earlier or later.
- Most clocks also have their original selling prices, descriptions, specifications and dimensions.
- A separate price list has been compiled from the top expert horologists in the field and provides **professional estimates** for antique clocks in **mint original condition**. This list will be updated annually and available from the Arlington Book Company.

This book is designed to provide you with a great deal of the information you will need to make wise choices as the marketplace demands knowledgeable and fast decision making.

Remember that clock prices are subjective. If the clock is in original mint condition and brings you joy, don't spend too much time bargaining. A friend recently paid $2,000.00 for a New Haven Flying Pendulum No. 2 like the one shown on the cover of this book. The clock was in original mint condition and filled a gap in his collection of flying pendulum clocks. However, the catalog price for this clock is $1,500.00. Did he pay too much? I say a fervent "NO!" A clock worth $1,000.00 to you may be worth $1,500.00 to the man behind you. My major rule has always been, "if you like it... buy it...," it's likely that you won't live long enough to find another like it. This is especially true with rare clocks.

As you may know, there are now many more members of the National Association of Watch and Clock Collectors competing for what appears to be fewer clocks. This means that you might have to exhibit some patience and leniency in deciding upon a new purchase. You might now consider a repainted dial or a replaced finial on a rare clock. It's harder to find what you want and you have to exercise prudent but open minded caution.

It is good to establish relationships with specific clock dealers who you can contact from time to time to make them more aware of your clock needs. "Out of sight, out of mind..." is the old expression. In fact, it might be a good idea to contact these dealers prior to a local or regional meeting to arrange to have them bring clocks for your acceptance or refusal prior to showing to everyone.

With the exception of "opportunity" clocks, it is usually most advantageous to not collect more than one specimen of a specific clock because if you attempt to sell at a future date, it becomes a little more difficult when you have duplicate items; some current examples of "opportunity" clocks are: Seth Thomas Regulators Nº 4, 5, 7, 10, 10 Double Time, 12, 14, 15, 16, 19, 60, and 63, or some other manufacturers such as Ansonia Regulator No. 9, Gale fancy case Astronomical Gallery calendar; Howard Regulator Nº 6, 36, 60; Ithaca Nº 1 sweep second Calendar, or some rare, or early American and European clocks, or clocks that do something besides tell time, such as Annular, Conical, Industrial, Mystery, and rare swinging clocks.

It is also advisable to consider staying within certain areas of specialty so that you know what to look for and where to find it, and to who you can sell it, when necessary. An example would be the collector - who specializes in American Weight Driven Regulator, Electric, Grandfather, Novelty or unusual French clocks.

Learn to make quick decisions... the person next to you might walk off with your prize. The guidelines in this book will help you avoid expensive mistakes, and will help you throughout your collecting career. Whether hobbyist, investor or serious collector, these "useful tips" should always be a part of your decision making process. Above all, remember that clock prices fluctuate according to price trends and economic conditions. Clocks purchased during a peak period may not have comparable value later.

The factors that determine the price of antique clocks are:

- Rarity
- Desirability
- Condition

Rarity is defined in terms of supply or availability. A "rare" clock is usually expensive and is usually available only in limited numbers, and chances are, will never lose its value.

Desirability is a key factor affecting the antique clock market. Some clocks are decorative and well suited for use in any home or office. Examples of clocks in this category are Calendar, Flying Pendulum, Crystal Regulator, Hanging Regulator, and Longcase Clocks.

It used to be the rule that Grandfather or Hanging Clocks under eight feet were the most desirable and practical. However with the many new houses that have cathedral ceiling, very large clocks are much in demand and the bigger the better. Some manufacturers produced these clocks in two styles: one to stand on the floor, and the other to hang on a wall. Of the clocks designed with two styles, the hanging style is considered to be the most valuable.

Large clocks are generally more expensive than small models, except in the case of marble clocks where the small sizes are the most valuable (they weigh less and are easier to display in the average home).

Clocks under glass domes are very attractive - but, these glass domes are mostly nonstandard and almost impossible to find. They can, however, be custom-made at considerable cost. The beveled glass on the Crystal Regulator or Carriage Clock, if cracked or chipped, can be replaced without difficulty at reasonable cost. In addition to the style or design, the movement is another important consideration in determining value. Eight day movements with weight driven or striking mechanisms are the most desirable. Other factors determining value and desirability of antique clocks include:

- Historical Significance

- Horological Importance
- Regional Factors
- Brand name recognition

Condition: The condition of an antique is another major consideration in determining a realistic market price. Whether you are buying, selling, trading, collecting, investing or speculating in antique clocks, the condition is a critical factor.

Only clocks in *original mint* condition and in good running order will command top prices. A replaced or damaged dial; missing parts; dull finish; worn-off gold gilt; chipped beveled glass; tablets which are either cracked, flaking, or refinished; or missing trim will reduce prices accordingly.

Clocks that have been overly restored or repaired are clearly controversial. The greater the amount of restoration the more controversy. The following criteria may be applied when analyzing an antique clock:

- Dial & Hands
- Movement
- Maker's Label
- Case
- Tablet or Decorative Glass (If any)

The **dial** is most important and should be studied carefully as the initial step in analyzing the condition of a clock.

Look for these important signals:

- Is the dial indigenous to the style and period of the clock?
- Does the dial fit the case properly without signs of extra holes or tampering?
- Are the dial holes in perfect alignment with the winding arbor(s) and center arbor?
- Are there any signs of restoration, repainting, or replacement?
- Are there any markings on the dial? (signature, maker's name, etc.) Are they original?
- Are the **hands** original?

Clocks that are "MINT ORIGINAL" always have a dial that is indigenous to the style and period of the piece. The dial is properly fastened to the case in perfect alignment with the movement. The dial retains its original finish with small signs of wear, very minor hairline cracks, and slightly faded numerals.

The mechanism or **movement** in a clock is the next logical place to proceed with the investigation. The key points to cover when examining a clock movement are:

- Does the movement fit the case (backboard) without signs of extra holes or tampering?
- Are there any missing pieces? Broken teeth? Signs of major repairs?
- Is the movement signed or stamped with a maker's name? If so, is the name consistent with the name on the dial and/or clock label?

The clock **label** may provide the only means of determining the origin of the clock. It often also contains the directions for its care and use. In some instances there are dates corresponding to a maker's patent or improvement in the mechanism. The questions with regard to the label are:

- Is the label intact and legible with only minor signs of shrinkage, wear, fading, etc.?
- Are there signs of replacement with a reproduction label?

Depending upon the style of the clock, manufacturers used a variety of materials to construct clock **cases**:

- Wood (Hardwood & Veneers)
- Metal (Iron, Tin, Brass, Nickel, Copper)
- Marble & Ceramics (China, Porcelain)

The important points to consider in examining a **wooden** clock case are:

- Does the door fit the case properly?
- Is the backboard intact and original?
- Is there any significant structural damage (cracking, splitting)?
- Is there any missing or replaced ornamentation (finials, hardware, etc.)?
- Has the case been refinished?

To **refinish** a wooden case or not has been

a philosophical question that has been debated for years. Does one leave the finish as found and discount any damage and dirt as badges of antiquity or does one go for the original "look" of the clock?

In this quandary we are often at odds with others in our family in that many clocks are prominent in size and make a statement in any room they are placed. Hence there is always a temptation to make the **"Antique"** look as nice as possible; especially if one paid a considerable sum to own such a treasure. Concurrently, such questions arise such as, "Should I polish the brass?" We all know that we should be in agreement with the sages who advise "Leave the clock alone or you destroy its value" but are really troubled by the drab appearance and the inability to see what just might be beautiful wood. Of course there is wisdom in that sage advice of a refined sort, "Do nothing that is not reversible." Let's approach the dilemma of restoration in stages from complete tolerance (no restoration) to complete restoration.

Often wood cases are dirty from oils and heating and cooking gasses which give them a dull appearance to say the least. On items such as a British longcase clock from the eighteenth century a simple waxing with a quality paste wax will remove the surface dirt without removing the patina (that is the good dirt, wax and oils that gives true antiques a subtle shading and age). On more modern items with the same dirt on the surface one might try a gentle application of Murphy Oil Soap or a mild soap cleaner. Be sure to check that the finish is not shellac as shellac often whitens when a water based cleaner is used. More resistant dirt may be attacked with a product like Cotton Cleanser or mechanic's hand cleaner applied with a towel and buffed. The clock should be allowed to dry for a few weeks and then waxed.

Commonly we encounter a shellac finish that has alligatored or checkered, or a clock that has been repeatedly French polished with more shellac over and over again. This condition may be addressed by carefully rubbing with medium steel wool saturated with alcohol, a solvent for shellac. When the layers of loose shellac have been loosened and put into solution immediately wipe off while wet with a fresh paper towel. Do not try to remove all the finish, but simple remove all the bad shellac on the top. In this instance immediately wax with a quality paste wax. A beautiful shellac finish will result. I have found that other solvents such as acetone or lacquer thinner work the same way for lacquer finish, but laquer is not as fast drying as shellac and must be allowed to cure before waxing.

Finally, if the finish is absolutely fully destroyed, painted over, or already removed it must be refinished. Don't give up too quickly on an old finish as the procedures mentioned above are usually adequate enough to renew any old finish and the results are usually better than refinishing. When stripping always try to use the least abusive stripper to remove as little patina as possible and also to avoid removing the filler from the grain of the wood. Now is the time to replace missing trim pieces and fill the nail holes that seem to appear in moldings.

Putting finish back on your clock takes time and is dirty work. Fill open grained woods such as walnut and mahogany. Don't use stains that are too light as most antiques were on the dark side by current standards. Finally multiple coats of finish must be applied and rubbed out with fine steel wool. This is where most refinishing jobs fail. The first coat of finish looks good and that is all that is done. Wrong! Use many coats. After the final coat of finish rub with the finest steel wool and wax.

Most **white metal** and some **brass** ornamental clocks' cases are gilt or silver plated. Great care must be taken when cleaning these surfaces. The thin layer of gilt or silver plating will be easily removed by cleaning solutions and strong rubbing. If you believe cleaning is necessary, hire the services of a professional and avoid the expense of re-gilding or re-plating.

In the case of **Porcelain, China** and **Marble** clocks, condition is the most important factor. If such a clock has been broken and glued back together it is severely reduced in value. Color is the second most important factor in estimating value. Bright reds, blues and frosted pinks are usually considered to be the most desirable. In order to receive top prices, clocks must be in ORIGINAL MINT condition and in good running order. Damaged dials, or chipped porcelain will **substantially** reduce prices.

Tablets or decorative glass panels were incorporated into the designs of many shelf and wall clocks. In this type of decoration, paint is applied to the rear side of the glass. In addition, the sequence of steps is the reverse of the procedure normally used on opaque surfaces such as wood or metal. Hence the name "REVERSE GLASS PAINTING" is derived from this decorative art form. There are three points to note in examining a glass tablet:

- Does the tablet appear to be original?
- Are there any signs of restoration?
- Are there any cracks or flaking?

All of these factors determine the condition of an antique clock. A clock that meets most or all of these test standards is deemed "MINT ORIGINAL."

What about those clocks that didn't stand the test of time, lower quality clocks? It is hard to say what these clocks are really worth. There is a market for these lower grade clocks but any controversy surrounding a clock decreases its desirability. Many books that are written on this subject suggest that clocks are in very short supply, that all clocks are highly collectible, and that one should consider the clocks in poorer condition by simply adjusting the market price slightly lower.

I do not share in this view, nor do I recommend that you use this guide in that way. There are many more clocks offered for sale each year than are buyers for them. Most of these are the lower quality variety and are rarely sold to knowledgeable buyers. I urge you to stand at the loading area at the end of a MART-sale and witness the large percentage of unsold clocks being returned to their homes. Most of these are lower quality clocks!

Of course, some mint condition clocks also leave the Mart-sale unsold simply because the owners would rather take them home than accept less than their too high asking price.

Concern yourself only with clocks in "MINT ORIGINAL" condition. Don't settle for anything less than the best. If you have patience to wait for that perfect piece, and the resources to acquire it, you will be rewarded in the satisfaction of knowing you own the best.

BUYING, SELLING, INVESTING, COLLECTING, SPECULATING.

There are three basic personalities associated with every clock enthusiast: Collector, Speculator, Investor. This may sound strange as you most probably are interested in antique clocks as a hobby. Whether you consider yourself to be a hobbyist or a professional, the antique clock market of the 2000's must be approached in these terms.

COLLECTOR - The collector is an individual who has adopted antique clocks as a hobby. He acquires clocks for pleasure and personal enjoyment. Factors such as aesthetic appeal, historic, or technical aspects of antique clocks are very important considerations to him. The ownership of an antique clock often stimulates peripheral interests while providing a decorative example of a period in history. In his approach to the antique clock market, the collector expects to pay a fair price for a fine example.

SPECULATOR- The speculator purchases a clock for the purpose of resale. The motives behind speculation are clearly profit oriented. Speculators sometimes achieve folk hero status in "rags to riches" stories of buying and selling antique clocks. There is always a chance that you will uncover a "great find" at a garage sale and turn a handsome profit upon selling the clock, but overall, speculating in antique clocks is too risky for the average person.

INVESTOR- The investor approaches the antique clock market in a businesslike manner. His important considerations are capital appreciation, safety, and liquidity.

When investing for capital appreciation, ask:

- What is your cost basis?
- What would you do with the money as an alternative investment?

If you purchase a clock at or above the market price, you can not expect to achieve capital appreciation in the short term. The investment must be made at a price that allows room for growth. Compare and contrast alternative investment opportunities at the time of purchase. Factors such as money market rates, inflation, stock and bond prices, etc. should be considered prior to investing in clocks.

The investment "safety" of clocks is another important consideration. The market for antique clocks should continue to grow at slow steady rates, but liquidity depends solely upon market demands. As in any other market, the selling price of your clock will depend on how long you are willing to wait for the profit margin you are willing to accept.

AUCTIONS - Auctions are an important vehicle in the antique clock market. There are numerous public and private auctions held across the country each year. They attract a great deal of publicity and warrant special consideration. Auctions are "happenings" that are attended by all types of collectors, investors, and speculators. Find a reputable auction gallery and attend frequently. They are often an open forum for information exchange and price guidelines.

Exercise great care in bidding on antique clocks at an auction. An important point to keep in mind is that there will be only one winning (highest) bid. If and when your bid is accepted you must acknowledge the fact that you are the high bidder. This means that none of the other participants felt that the piece was worth as much or more than you did! I remind you of the phrase "Caveat Emptor": buyer beware. The necessary ingredients for success at an auction are **knowledge**, **experience**, and **money**. A deficiency in any of these areas will produce unsatisfactory results.

BOOKS are the cornerstone of knowledge. There are hundreds of titles written on the subject of clocks and watches, and reading horological books will prepare you to search for the antique originals. Building a large and complete clock library can be very helpful. In addition, expanding your horological library is always a good investment. Quite often, the books themselves become rare treasures when copies are not available because they are out of print.

My experience over the years has taught me that these few simple guidelines, if put to proper use, will save time, money and possible disappointment. This book is designed to be your partner...a point of reference for all your collecting needs. It is my wish that it will help make your antique clock collecting experience enjoyable and profitable.

Be the Early Bird ...

... Time waits for no man

Tran Duy Ly

THE NEW HAVEN CLOCK COMPANY

© by Chris H. Bailey

One cannot fully discuss the history of the New Haven Clock Company without recounting the activities of Hiram Camp (1811-1892) and his uncle, Chauncey Jerome (1793-1868). Camp sought his father, Samuel Camp's, permission to work at the clock business for his uncle and on his 18th birthday, in 1829, he left his home in Plymouth, Connecticut for nearby Bristol. Jerome was then a partner in the firm of Jeromes & Darrow who were involved in the manufacture of wooden movement shelf clocks.

About 1833, the firm of Jeromes & Darrow was dissolved and was succeeded by a new firm, C. & N. Jeromes. When the financial recession of 1837 came to New England, the Jerome firm found itself in difficulty and on May 10, 1837 filed for bankruptcy.

During the fall of 1837, Chauncey Jerome went to Richmond, Virginia to attempt to collect money owed their firm and there he formulated ideas for the production of a brass movement clock that could be made cheaper than ever before. His cases, hands, dials and other appurtenances would remain little changed, so he needed only a cheaply produced one-day brass movement. His brother, Noble Jerome (1800-1861) designed this movement which could be made from stampings of rolled sheet brass and it was patented on June 27, 1839. This clock became a major production item for years and brought new prosperity to Jerome and to the Connecticut clock manufacturing industry.

About 1842, Chauncey Jerome went to New Haven and purchased the carriage factory of Isaac Mix & Sons who had failed and there made cases, keeping his movement manufacturing in Bristol. Movements were shipped from Bristol to New Haven, via the Farmington canal, and installed in cases and shipped to many areas of the United States and some foreign countries. The operation was interrupted on April 23, 1845 when a fire destroyed the major movement factory in Bristol of which Hiram Camp was then superintendent. Some tools and lathes were saved, but all of the presses and dies went through the fire and losses were estimated at $75,000.

After the fire, Jerome moved all his manufacturing facilities to New Haven and there Hiram Camp supervised the building of a new movement factory which became operational during the summer of 1845. About 1847, it was proposed that Camp work independently and deliver completed clocks, boxed and ready for shipment by Jerome. During 1850, Jerome formed a new firm which he called the Jerome Manufacturing Company and the following year Camp built a wooden building on the west side of Hamilton Street, near the Jerome facility, where he made movements and supplied them to Jerome.

Demand for Camp's movements became more acute in succeeding years and it was decided that a company should be formed. On February 7th, 1853**, the New Haven Clock Company was incorporated with $20,000 capital "to manufacture, sell and deal in clocks and timekeepers of every description..." Hiram Camp was chosen

---

***Previously published sources, including those by the company, incorrectly record this date as February 17th. Official minutes of Director's meetings prove the incorporation actually took place on February 7th.*

president, treasurer and agent for the company and was to receive $1,000 per year for his services.

Of the 800 shares made available at $25 each, Hiram Camp purchased 200. Another 200 shares were purchased by the firm of English & Welch. One partner in this firm, James Edward English (1812-1890), was the great grandson of the clockmaker, Isaac Doolittle (1721-1800) of New Haven. He entered into partnership with Harmanus Madison Welch (1813-1889) about 1848 and they ran a prosperous lumber and shipping business for more than 40 years. They were large investors in the Jerome and New Haven enterprises. H. M. Welch, the younger brother of Elisha N. Welch, had been involved with clockmaking since about 1845 when he established H. M. Welch & Company at Plainville, Conn. In 1860, H. M. Welch became the Mayor of the city of New Haven and served until 1863, the same year he formed the First National Bank of New Haven of which he was president until his death.

Substantial growth was shown the first several years of New Haven's production and the first dividend of $2.50 per share of stock was voted January 31, 1855. Business became uncertain shortly thereafter with the failure of the Jerome Manufacturing Company, their major customer. Speaking of the failure in a brief manuscript history written in 1892, Camp recalled,

" Mr. Jerome himself not being posted up as to the standing of the Co. but thinking the Co. was making maney engaged in building a Church on Wooster Square, New Haven, but things began to go hard, and in their extremity the Secretary, Mr. S. B. Jerome, made an agreement with P. T. Barnum the great showman, who had been interested in the clock business with one Theadore [sic] Terry in Bridgeport where Barnum resided. Terry had sold out to Barnum a refuse lot of almost useless stuff. The plan now was to merge the two companies into one. Barnum said that their Co. owed about $20,000 which was afterwards proved to be over $70,000, which Jerome Co. had to assume after the merging of the two companies."

The result of this merger and other financial burdens was that the Jerome Manufacturing Company filed for bankruptcy on February 14, 1856.

Jerome's creditors numbered more than two hundred and fifty and included among them were Hiram Camp for $1,519.09; English & Welch for $47,804.24; Elisha Welch for $21,889.34. Additionally, the firm of English & Welch in partnership with Elisha Welch was owed $6,389.41. Of the $568,000 worth of debts against the Jerome Manufacturing Company, little more than $52,000 was raised and paid to creditors. Hiram Camp recorded,

"At the time of the failure of the Jerome Co. the New Haven Clock Co. had quite an amount charged up for movts. etc. Mr. Camp had given his name individually to prevent the failure, to the amount of $6,000 in order to save the New Haven Co. He had received as collateral $16,000 [worth of] Barnum's acceptances, and also a large lot of Marine Movts. and clock cases, this was in order to be prepared to keep up the supply for the English trade. Mr. Camp was now keeping the books of the New Haven Co. He credited the things tangible for the benefit of the New Haven Clock Co. and kept Barnum's acceptances for which he received 8 cents on the dollar. The New Haven Clock Co. thus losing but a little while Mr. Camp lost about $15,000, sold 4/5 of his clock stock, bank stock, Building Association stock, etc."

With the assets of the Jerome Manufacturing Company up for sale, it was decided that the New Haven Clock Company should attempt to buy them. On April 24, 1856, the firm of English & Welch put up an additional $20,000 and the capital stock of the company was raised to $40,000. The Jerome company was purchased and on Wednesday, January 28, 1857, the Annual Meeting of New Haven Clock Company "was held at the office of the late Jerome Manufacturing Company." On that date, Elisha N. Welch, clock manufacturer of Bristol, Conn. was elected a Director. He then owned 67 shares of stock, but had sold his stock by the following year's meeting and was not reelected a Director.

Hiram Camp wrote the following letter, dated April 17, 1857 and was presented to the company's Directors on that date:

"To the directors in the
New Haven Clock Co.

Gentlemen,

Being desirous of making a final settlement of the claim I have made against your Company on account of my individual endorsements of the paper of the late Jerome Manufacturing Company, I hereby propose in consideration of three thousand dollars, to be paid me by said New Haven Clock Compy, in their three notes for the sum of one thousand dollars each. One note four months, one six months, and one eight months, from date - all of said notes to be dated April 20th, 1857. Said notes to be in full satisfaction for all claims I have or ought to have against said company, and in consideration of the giving and paying of said notes, I hereby agree to devote all my time and services in the employment of said Clock Company, according to the best of my abilities, for the term of three years from the first of January, 1858, for the annual payment of Twelve hundred dollars. The same to be paid monthly, so long as I shall remain in the employment of the said New Haven Clock Company."

(signed) Hiram Camp

It was voted to execute the notes spoken of above and Camp was retained as President for the three years and in fact, for many thereafter.

During 1859, the company merged sales efforts with E. N. Welch as noted in their Director's minutes, "That this Company will join Elisha N. Welch in the business of selling clocks in the City of New York, in the manner and form as is shown in a written agreement entered into between the said Welch and us bearing date, New York, January 12, 1859 ..." The agreement itself has apparently not survived and no further mention is made in the records. This was, however, the company's entry into the American Clock Company of New York who acted as their sales agents for many Connecticut clock manufacturers until 1879.

During 1860, the capital of the company was declared insufficient and was raised from $40,000 to $150,000. At that time, the company was employing 300 men and 15 women and paying out monthly wages of over $10,000. Annually, they were using 255,000 lbs. of brass and 800,000 feet of lumber. The $147,000 worth of materials used produced 168,000 clocks valued at $325,000. With the outbreak of the War between States in 1861, business slowed, but picked up the next year. A fire occurred in April, 1866, which Camp said burned "the old shop". This apparently had little effect on the company as it was not even mentioned in the company's official records.

James E. English, who had been serving as secretary of the company since January, 1856, felt the pressures of political office and resigned July 1, 1865. Mr. English had been a member of the state legislature 1855-56, a state Senator 1856-59 and a member of the U.S. Congress from 1861-65. After his resignation, he was elected Governor of Connecticut which office he held 1867-71, followed by service as U.S. Senator from 1875-77.

Edward Stevens, Chauncey Jerome's son-in-law, was chosen as English's replacement as secretary. Stevens came to New Haven from Liverpool, England where he had been selling clocks under the firm name of Jerome & Company. Stevens was permitted to purchase 800 shares of stock in January 1865, when the capital of the company was raised from $150,000 to $200,000. Business slowed in 1866 and 1867 but picked up thereafter. In 1870, the company was using two 200 hp steam engines for power and employing 174 men, 21 women and 31 children. They paid out $155,000 in wages during the year. Their production included 46,032 movements valued at $60,000 and 178,086 complete clocks valued at $500,000.

New Haven's records noted January 23, 1878: "Notice was given by the Secretary that the company's largest customer, Jerome & Co. was now a limited company, that Edward Steven was no longer responsible personally, but that Jerome & Co. Limited had large capital and was worthy of credit. The continuance of the same terms to Jeromc & Co. firm was established in Great Britain during Chauncey Jerome's career to sell his clocks to the English market. With the dissolution of Jerome's business in America in 1856, the English sales firm Jerome & Co. continued to sell American products.

During the 1870's the New Haven Clock Company carried a general line of clocks, particularly ogee, cottage, steeple, beehive and some wall clocks, particularly those with lever escapements. Their clocks were illustrated in the catalogs of the American Clock Company and they are not known to have issued separate catalogs during that time. When the business of the American Clock Company was broken up in 1879, New Haven began to issue their own catalogs.

Sales offices were opened in 1879 in Chicago, New York and about that time in San Francisco. When the 1880 catalog was issued, more than half of the clocks manufactured by the company were new additions to the line since 1878. By 1887, the company had a huge line of items and issued a catalog of 206 pages. Edward Ingraham (1887-1972), former president of the E. Ingraham Company noted,"As far back as 1910, I recall my father and Mr. Stockton discussing New Haven and remarking that no company can make money in the clock business with such an extensive line as New Haven showed in their catalog."

From June, 1879 until June, 1880 the Company was employing 460 men, 52 women and 88 children. They worked 10 hours a day, six days a week. The factory ran full time 12 months a year. An average day's wages for skilled workmen was $2.00 and $1.00 per day was paid an ordinary worker. More than $223,000 worth of materials were utilized to produce $471,875 worth of clocks. Pocket watches were first added to the line in 1880. At that time small backwinding clock movements were put in watch cases one inch thick and three inches in diameter.

During April, 1883, a fire destroyed a full third of the case shop, but it was rebuilt and running again thirty days later. No dividends were paid stockholders that year and this was the first hiatus in 16 years. Business picked up in 1884 and 1885, but again became slow in 1886. Thereafter finances were in a critical state for more than a decade and one half.

On January 27, 1886, the president was "authorized to treat with the Jerome & Co. to acquire such parts of their plant as may be deemed for the best interests on the New Haven Clock Co. ..." The Trademark "Jerome & Co." was used on some New Haven clocks as early as 1880 and this practice continued until after 1900. On October 20, 1904, the Directors voted "that this company purchase the good will and business of Jerome & Company Ltd. for the sum of $12,000 from Pierce N. Welch ..." Pierce N. Welch, then owner of the British firm, was the son of Harmanus M. Welch.

After several years of slow business, the capital of the company was raised in May, 1890 from $200,000 to $500,000 in hopes the new money would put life into the firm. This did not happen and on September 1, 1891, "A letter was read from Hiram Camp accompanying his letter of resignation." This was referred to an executive committee of Directors who accepted his resignation, effective January 1, 1892. The following January 29th, Samuel Arthur Galpin (1846-1902) was chosen to replace Camp and was paid $4,000 per year.

On March 8, 1894, it was voted to issue notes up to $50,000 for funds for the business and to authorize the mortgage of the land, buildings and machinery of the company. Finances were so critical at this time that six days later at a special meeting, the Directors voted, "that when the funds which are now being raised to the amount of $50,000 are secured and exhausted and the necessity for a new loan shall arise, the affairs of New Haven Clock Company shall be immediately placed in the hands of a Receiver, to be wound up as soon as possible."

During March of 1896, additional money was needed, but the company did not declare bankruptcy. Another $50,000 was raised and the factory and machinery were again mortgaged. On March 11, 1897, the company was reorganized and on that day, 7,600 shares of stock were cancelled and the capital of the company reduced from $500,000 to $110,000 in hopes of a rehabilitation of credit. It was also agreed that if the business was not in better financial condition at the end of 1897, it would be liquidated.

Business did improve under Galpin's leadership and on January 24, 1901 he gave an oral report which, in part, stated that the business' increase in its volume "had rendered possible this year a better showing than for 20 years past." Galpin was an able leader and was spoken of as "the man who saved the Company" and "built bricks with straws." Galpin did not live to see the company's full return to a profitable business as he died suddenly in London, England, August 5, 1902. In October, the company assumed the payment of almost £250 relating to his illness, death and return of his remains to New Haven.

Walter Chauncey Camp (1859-1925) was chosen successor to Galpin in 1902 and the following year a 10% dividend was paid to stockholders, the first since 1885! During 1904, plans were underway to expand and modernize the watch manufacturing department and machines were added which allowed them to be produced more cheaply. Wrist watches were first added to the line of products in 1915 and received their first boost in popularity from soldiers during World War I. Walter Camp, unrelated to the former president Hiram Camp, was a well known athlete and often called the "Father of Football at Yale." During World War I, he also introduced his "Daily Dozen Exercises" which were widely popular.

Camp was president of the company until February 19, 1923 when he was succeeded by Edwin Park Root. Root held office until February 24, 1929 and was succeeded by Richard Henry Whitehead. When Whitehead took over the company, it again had serious financial problems. When the Great Depression began a few months thereafter, the situation became even more grave. In writing about Whitehead, former New Haven Director Philip H. English stated: "Mr. Whitehead's work for the company extended from 1922 until August 1946, and resulted in continuous progress in methods and financial control. He was an able administrator and made enormous effort during the War years in close cooperation with the United States Naval Laboratory at Washington."

During 1938, the company began to use a quartz crystal oscillator for timing all balance wheels for accuracy in their watches. The "Elm City" model pocket watch was being produced so inexpensively at that time that it sold to the trade for 56¢ each.

Near the time of the beginning of World War II, the firm was employing 1,500 workers and normally producing about three million timepieces and clocks a year. As a result of war work, the alarm clock business was terminated in 1941, automobile clocks were last produced in January, 1942, and watches discontinued in 1942 except for a few special orders for Army and Navy use.

During 1938, the company had first begun engineering work on fuzes and the need for these became critical after the war started. Horological manufacture dwindled from 96% in 1940 to 62% in 1942, none in 1943, and only 1% in 1944. During 1944 and 1945 the company produced fuzes for 3,200,000 bombs and parts for 40 million anti-aircraft fuzes and controls for Navy mines.

During July of 1945, newspapers announced that Paul V. Eisner and Max Taussig, co-owners of the Rensie Watch Company of New York had been elected to the board of Directors. The Rensie Watch Company represented prominent Swiss watch manufacturing concerns. During 1946 clock and watch production resumed, but on March 6th of that year, the company was reorganized and after 93 years, the name changed to the New Haven Clock and Watch Company. Richard H. Whitehead, who had headed the company successfully for many years, resigned. Speaking of this time, Philip H. English wrote, "The main difficulties arose from an attempt to extend sales from six to nine million dollars annually, without adequate working capital for such a very large increase. Attempts by the former management to restrain such a rapid expansion were unheeded."

Former president Richard H. Whitehead wrote in 1957: "It took many years to ruin the company after I left because it was in such a strong position. ... When I quit in 1946 I realized that the Company had fallen into unfavorable hands & there was little I could do. ... So after a decade of service to the Company I quit rather than be a front for people I found I did not care to associate with."

Once the company was in the hands of Swiss watchmaking interests, its financial deterioration progressed steadily. Paul V. Eisner took the Presidency in 1946 and was succeeded by Lawrence Robbins in 1949, James Heitzman in 1950, Max Geller in 1951 and Seymour Ziff in 1956. On December 6, 1956, the company filed a petition for corporate reorganization under Chapter X of the Bankruptcy Act.

During this period, activities were limited to the production on a minor scale of condensers, power pack supplies (and other related items in the "Condenser Products Division") and of pocket watches and pedometers. In 1957, stock hit a low of 12 ½ ¢ a share.

Though the company emerged from the reorganization in February, 1958 with a deficit of only $457,511, during December, 1959, the management decided that it was impossible to continue any profitable manufacturing operations other than those of the Condenser Products Division. These had accounted for a small portion of the company's production since 1945. By 1959, the company had not paid a dividend to stockholders since 1946 and there was none in the forseeable future. Operating losses during 1959 were $1,170,393 with a total deficit by December of that year of $3,178,121.00.

On March 22-24, 1960, all of the manufacturing facilities of the New Haven Clock and Watch Company were sold at public auction and private negotiation. This included more than 500 machines, tools, dies, jigs and fixtures and work in progress. Total sales of the facilities amounted to half a million dollars. The company's buildings on Hamilton Street in New Haven, totalling more than a quarter of a million square feet of floor space were partially rented until they could be sold.

The Condenser Products Division was moved to Brooksville, Florida where the remains of the Company proposed to enter a real estate development venture. In March, 1969, Edward Ingraham wrote: "A successor company known as Haven Industries (headquarters in Florida) amalgamating in 1969 with a bankrupt sugar company. They issue a terrible statement and the amalgamated company starts off close to receivership." Mr. Ingraham later commented that "no vestige remains" of the New Haven Clock Company. However, the factory complex at New Haven still stands even though a portion was badly damaged by a fire about 1988.

HIRAM CAMP, PRESIDENT OF THE NEW HAVEN CLOCK COMPANY 1853-1892

REFERENCES

All materials below are from the Edward Ingraham Library of the American Clock & Watch Museum, Bristol, Conn., unless otherwise noted.

1. Official Minutes Book of the New Haven Clock Company, 1853-1905.

2. Sketch of the Clockmaking Business, by Hiram Camp. (Original handwritten copy by Camp, 1892).

3. Auction Sale Notice of New Haven Clock & Watch Company business, 1960.

5. Memoirs of Edward Ingrahams, Clockmaker.

6. Handwritten correspondence:
   a. Richard H. Whitehead to Philip H. English, 1957.
   b. Richard Ingraham to Lockwood Barr, February 7, 1957.
   c. Edwards Ingraham to E. Ingraham Company salesmen, April 7, 1949
   d. Philip H. English to Lockwood Barr, March 4, 1957.

7. The Story of the New Haven Clock Company. Published by the company, ca. 1945.

8. Manuscript history compiled on the New Haven Clock Company by Lockwood Barr, (deceased), 1957.

9. History of the Class of 1880. Yale College, pp. 103-5.

10. History of the City of New Haven, edited by Edward E. Atwater. New York, 1887. pp. 577-81. (Bristol Public Library, Bristol, Conn.)

11. Commemorative Biographical Record of New Haven County, Conn. Chicago, 1902. pp. 12-16. (Connecticut State Library, Hartford)

12. Descendants of James Welch, By Charles B. Welch. 1946. pp. 35-6. (Connecticut State Library)

13. 1860, 1870, 1880 Industrial Census records of New Haven Connecticut. (Connecticut State Library)

14. Commerce Manufactures and Resource of the City of New Haven and Environs. 1881. p. 101.

NEW HAVEN CHRONOLOGY

1793	June 10th. Chauncey Jerome born at Canaan, CT.
1804	Lyman Jerome dies leaving young family in which Chauncey is only 11.
1806	Winter. Chauncey Jerome apprenticed to Col. Allyn Wells at Plymouth to learn trade as house carpenter.
1811	Hiram Camp born.
1816	Winter. While employed by Eli Terry as a casemaker, Jerome makes first pillar & scroll clock cases.
1821	Chauncey Jerome moves from Plymouth to Bristol, CT and establishes a clock casemaking business.
1829	At age 18, Hiram Camp begins working for his uncle in the firm of Jeromes & Darrow at Bristol, CT.
1838	30-hour brass clock introduced by Jerome.
1844	October. By this date Hiram Camp is superintendent of Jerome's clock movement shop at Bristol.
1845	April 23rd. Jerome's movement factory burns.
1845	Summer. Hiram Camp supervises construction of a new Jerome movement factory at New Haven, CT.
c.1847	Camp begins independent operation supplying clocks to Jerome.
1850	February 14th. Jerome Manufacturing Company incorporated.
1851	Camp constructed wooden clock factory to manufacture brass movements for Jerome Manufacturing Company.
1853	February 7th. New Haven Clock Company incorporated with $20,000 capital. Hiram Camp chosen president.
1855	English sales firm called "Jerome & Company" formed to supply English market.
1856	February 14th. Jerome Manufacturing Company files for bankruptcy because of bad business dealings.

1856	April 24th. Stockholders of New Haven Clock Company attempt to raised money to purchase bankrupt Jerome Manufacturing Co.
1856	May 3rd. New Haven newspapers announce the purchase of the bankrupt Jerome firm by the New Haven Clock Company.
1859	January 12th. New Haven Clock Company becomes involved with American Clock Company at New York City as sales agents for their clocks.
1860	June. 300 men and 15 women employed. 168,000 clocks valued at $325,000 produced in previous 12 months.
1861	Business slows because of Civil War but improves the following year.
1865	American Clock Company begins to issue trade catalogs which include New Haven Clocks.
1866	April. New Haven's "old shop" burns. New factory built. Business slow in this and following year.
1870	June. 174 men, 21 women and 31 children employed. Production during last 12 months includes 46,023 clock movements valued at $60,000 and 178,086 complete clocks valued at $500,000.
1879	New sales offices opened in Chicago, New York and San Francisco.
1879	October. American Clock Company, New York sales agent for New Haven and other firms, ceases business.
1880	New Haven Clock Company issues separate trade catalog of 72 pages of which nearly half are new models introduced since 1878. Firm's catalog also offers clocks made by E. Howard & Co., E. Ingraham & Co., and Itchaca calendar clocks, but are dropped after a few years. Non-jewelled pocket watches added to line.
1880	June. Company employs 460 men, 52 women and 88 children. Clock produced for 12 months previous valued at $471,875.
1883	April. Fire destroys 1/3 of the case shop. Is rebuilt in 30 days. Marble clocks with French movements introduced into line along with imported cuckoo and novelty clocks. Watchman's electric clock and Estel's patent program clocks introduced. Plush cased and iron cased clocks introduced.
1884	Cabinet clocks and "art tile" clocks introduced. Hall clocks added to line, some models being offered until 1927.
1886	New Haven's business slow and financial situation does not improve for about 15 years.
1889	Mechanical dial alarm clocks introduced as well as some luminous standard dial alarm clocks.

1890	May. Capital of New Haven raised to half a million dollars in hopes of improving financial situation, but is in vain.
1891	September 1st. Hiram Camp resigns as President of the firm, effective January 1, 1892.
1892	January 29th. Samuel A. Galpin chosen new President.
1894	March 8th. New Haven's Directors vote to raise $50,000, nearing bankruptcy.
1895	Oak cased parlor clocks appear in line.
1897	March 11th. Company is reorganized and capital reduced for $110,000. Business to be liquidated if not improved by the end of year.
1901	January 24th. President reports best in 20 years for firm.
1902	August 5, 1902. Samuel A. Galpin dies at London, England.
1902	Walter C. Camp "father of football at Yale" chosen to succeed Galpin as President.
1903	10% dividend paid to stockholders, the first since 1885.
1904	Watch manufacturing department expanded and modernized.
1904	October 20th. Entire business of English firm of Jerome & Company Ltd., is purchased. "Mission" oak clocks added to line about this year.
1907	Automobile clocks/watches offered by this time.
1911	Banjo and Tambour clocks appears in line.
1913-1914	8-day alarm clocks being offered by this year.
1915	Wrist watches added to New Haven product line. Old fashioned 30-hour OG clocks still being offered.
1920	German-style wall clocks added to the line.
1923	February 19th. Edwin P. Root succeeds Walter Camp as President of firm.
1927	Many banjo clocks offered, but all other wall clocks dropped.
1929	February 24th. Richard H. Whitehead succeeds Edwin Root as President of firm. Great Depression with stock crash in November.
1941	New Haven firm employing about 1,500 and manufacturing about 3 million clocks and watches a year. Alarm clocks business terminated.
1942	Clock and watch production reduces to 62% and the following year to 0 % as production is

	converted to war products.
1945	July. Swiss watch company buys interest in New Haven firm and soon gains control.
1946	Production resumes without adequate capital for increase in volume. Richard Whitehead resigns as president.
1946	March 6th. Company reorganized and name changed to New Haven Clock and Watch Company.
1946-1956	Succession of short-lived Presidents as firm deteriorates.
1956	December 6th. Nearly bankrupt, firm files for reorganization. Clock and watch manufacture ceases and only condenser business remains, which was moved to Florida.
1960	March 22-24. All manufacturing facilities, machine, tools, etc. and work in progress sold at public auction.

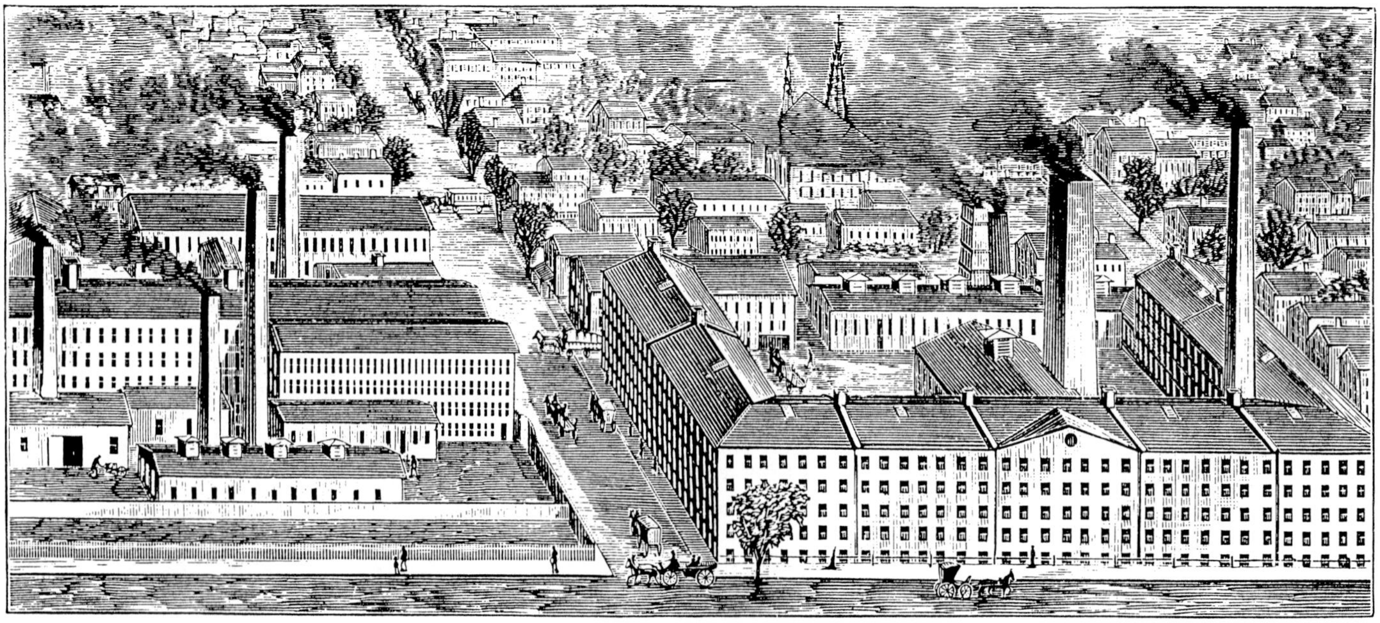

FACTORIES, NEW HAVEN, CONN., U. S. A.

ADVERTISING CLOCKS

1

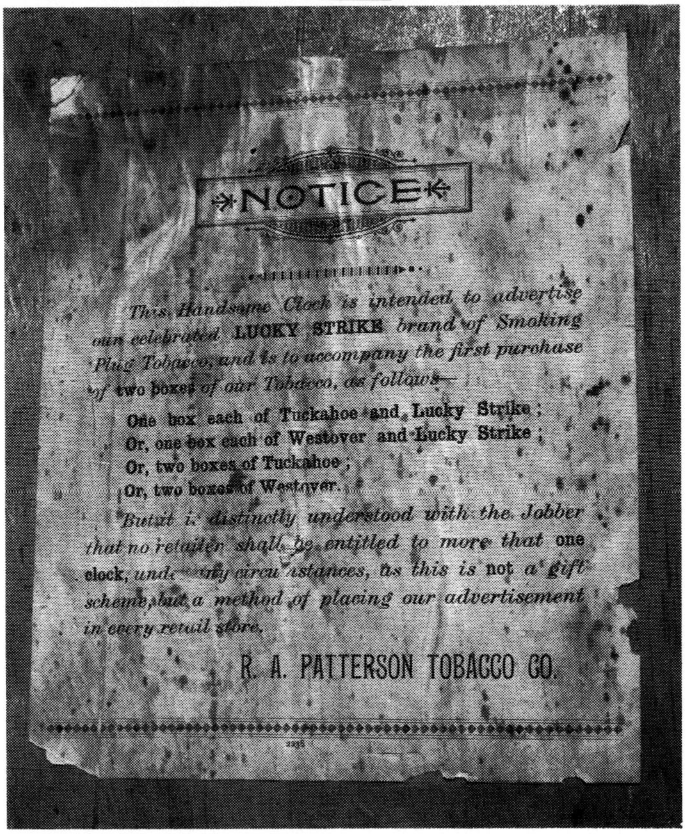

NOTICE

This Handsome Clock is intended to advertise our celebrated LUCKY STRIKE brand of Smoking Plug Tobacco, and is to accompany the first purchase of two boxes of our Tobacco, as follows—

One box each of Tuckahoe and Lucky Strike;
Or, one box each of Westover and Lucky Strike;
Or, two boxes of Tuckahoe;
Or, two boxes of Westover.

But it is distinctly understood with the Jobber that no retailer shall be entitled to more than one clock, under any circumstances, as this is not a gift scheme, but a method of placing our advertisement in every retail store.

R. A. PATTERSON TOBACCO CO.

Lucky Strike important label states:
"No retailer shall be entitled to more than one clock."
{The Millie & Jerry Maltz Collection}
{Photo courtesy of Jack Wyer}

2

Dial removed showing
Lucky Strike 8-Day Time Movement.
{The Millie & Jerry Maltz Collection}
{Photo courtesy of Jack Wyer}

3

Close-up of **Lucky Strike** lower door with gold letters on black. Reverse painting on glass reads: **"BRAGG, CUMMINGS & CO. BANGOR, ME."**
{The Millie & Jerry Maltz Collection}
{Photo courtesy of Jack Wyer}

4

LUCKY STRIKE
Ca. 1910. Rosewood veneer case. 8-Day Time. 6 o'clock wind. Height, 25 inches. **"R. A. PATTERSON TOB. CO. TOBACCO"** printed on 12 inch paper dial.
{The Millie & Jerry Maltz Collection}
{Photo courtesy of Jack Wyer}

ADVERTISING CLOCKS

5

DUFFY'S PURE MALT
Ca. 1910. 8-Day Time. Height, 32 inches. Width, 18 inches. 12 inch paper dial. Figure eight style case. Doors made of wood, probably Pine - Not of Paper Mache. Advertising reads:
DUFFY'S PURE MALT
WHISKEY
Absolutely Pure & Unadulterated
TRADEMARK
{The Millie & Jerry Maltz Collection}
{Photo courtesy of Jack Wyer}

6

Close-up of **Duffy's** 8-Day Time.
Movement showing winding at 6 o'clock.
{The Millie & Jerry Maltz Collection}
{Photo courtesy of Jack Wyer}

7

Reverse painting on glass on lower door of **Duffy's Pure Malt Whiskey** with company trademark. Advertising letters are carved onto door.
{The Millie & Jerry Maltz Collection}
{Photo courtesy of Jack Wyer}

ADVERTISING CLOCKS

8

SNOWBALL

SNOWBALL
Ca. 1910. Height, 24½ inches. Width, 16½ inches. Dial 12 inches. Movement identical to number 2 page 25. Advertising reads:
SNOWBALL BAKING POWDER
PURE WHOLESOME
{The Millie & Jerry Maltz Collection}

9

HARVARD BREWING CO.
Ca. 1880. Height, 28 inches. Width, 20 inches. The case is a form of picture frame with gold gilt on wood. Reverse painting on glass reads:
HARVARD PURE MALT BEVERAGES EXCLUSIVELY
LOWELL, MASS. U.S.A.
{The Millie & Jerry Maltz Collection}

10

SATURN ELECTRIC

A solid Oak case with Natural Wood finish, panel space for name or advertising card 6¾ inches by 12 inches. Height, 35 inches; width, 15½ inches; 12-inch dial. Equipped with a slow speed (200 r. p. m.) self-starting sub-synchronous Westinghouse motor.

Retail, **$25.00**

11

REFEREE ELECTRIC

A commercial clock in Oak or Mahogany finish with a large panel 11⅞ inches by 10¾ inches for name or advertising card. Height, 35½ inches; width, 16 inches; 12-inch dial. Equipped with a slow speed (200 r. p. m.) self-starting sub-synchronous Westinghouse motor.

Retail, **$22.50**

12

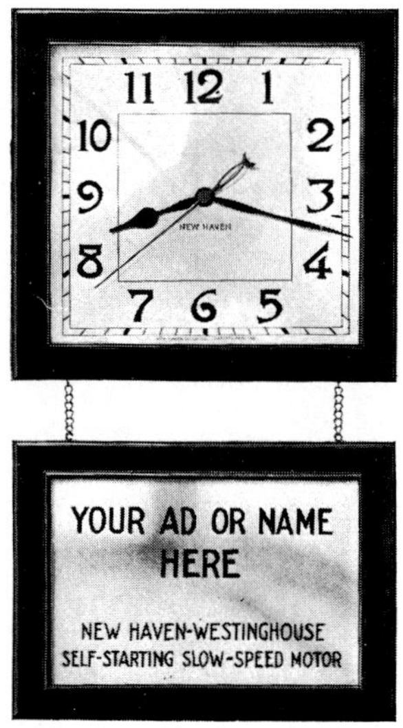

BALCONY ELECTRIC

A clock for office or commercial use finished in Oak or Mahogany. The hanging panel has display space for name or advertisement. 11¾ inches by 8 inches. Clock frame is 14½ inches square with 12-inch dial. Equipped with a slow speed (200 r. p. m.) self-starting sub-synchronous Westinghouse motor.

Retail, **$20.00**

ALARM CLOCKS

13

1900

14

1886

15

1886

ALERT - ALARM

NICKEL

Height, 7 inches. Dial 3 inches.
1 Day Time and Alarm,......$2.80

ALERT - ALARM T. P.

NICKEL

Height, 6 ½ inches.
1 Day Time and Alarm, Nickel....$2.95
1 Day Time and Alarm, Plush......$3.45

ALERT - ALARM

EXTRA CASE

Height, 6 ½ inches.
1 Day Time and Alarm....$3.25

16

1883

EGYPTIAN.

NICKEL OR GILT.

Height, 8 inches. Dial, 4 inches. 1 Day Time.
1 Day Time, Calendar. Also with Alarm.
Finished with Sunk Back and Cast Brass Polished Fronts.

17

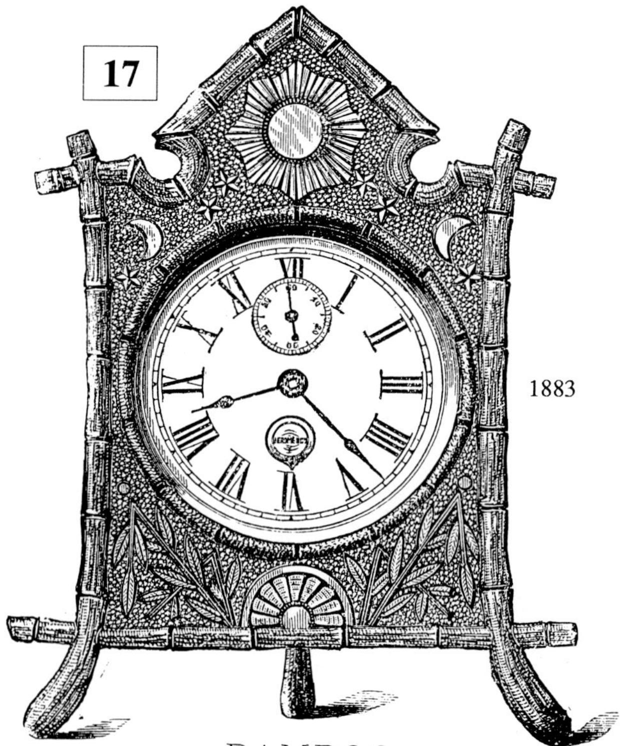

1883

BAMBOO.

NICKEL OR GILT.

Height, 8¾ inches. Dial, 4 inches.
1 Day Time. 1 Day Time, Calendar. Also with Alarm.
Finished with Sunk Back and Cast Brass Fronts.

ALARM CLOCKS

18

1911

BRUMMEL

HEIGHT ....10½ INCHES
WIDTH ....8¼ INCHES
DIAL ....5 INCHES
FINISH, OXIDIZED COPPER
MOVEMENT, 1 DAY LONG ALARM
PRICE ....LIST, $5.25

19

1911

THE PARAL ALARM

HEIGHT ....5½ INCHES
WIDTH ....3⅜ INCHES
DIAL, PORCELAIN ......2 INCHES
FINISH, GILT, SILVER, GUN METAL OR OLD BRASS
PRICE ....LIST, $5.45

20

1911

SENTRY ALARM

HEIGHT ....10¾ INCHES
WIDTH ....6¾ INCHES
DIAL ....4⅛ INCHES
FINISH, ENAMELED RED, BLUE, OR GREEN
MOVEMENT, 1 DAY INTERMITTENT ALARM
PRICE ....LIST, $4.35
PRICE, WITH BRONZE, NICKEL OR
GILT FINISH ....LIST EXTRA, .45

21

1883

MOONSHINE

WOOD CASE WITH NICKEL OR BRASS FRONT.
Height, 6 ½ inches.
1 Day Time and Alarm.

22

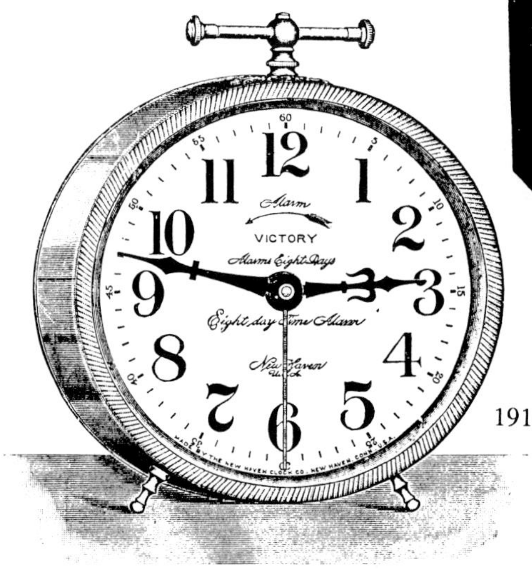

1911

VICTORY

HEIGHT ....5½ INCHES
WIDTH ....4½ INCHES
DIAL ....4⅛ INCHES
CASE, 4½-INCH, SEAMLESS BRASS, KNURLED FRONT, WITH 4-INCH BELL METAL GONG ON BACK
FINISH, NICKEL PLATED
MOVEMENT, 8 DAY TIME ALARM, RINGS (AT HOUR SET) UNTIL STOPPED BY PUSHING IN ON BUTTON ON BACK OF CLOCK MARKED "STOP ALARM." THIS SWITCH IS AUTOMATICALLY OPERATED SO THAT CLOCK RINGS AT SAME HOUR NEXT MORNING. ALSO HAS "ALARM OR SILENT" SWITCH FOR USE WHEN TIME ONLY IS WANTED
PRICE..LIST, $4.80

23

1913

CASCADE

TIMEPIECE . . . . LIST, $2.65
INTERMITTENT ALARM . . . LIST, 3.10
HEIGHT, 4⅜ INCHES. WIDTH, 3¼ INCHES.
DIAL, 2 INCHES.
CASE, ORNAMENTAL BRASS, OLD BRASS FINISH.

ALARM CLOCKS

24

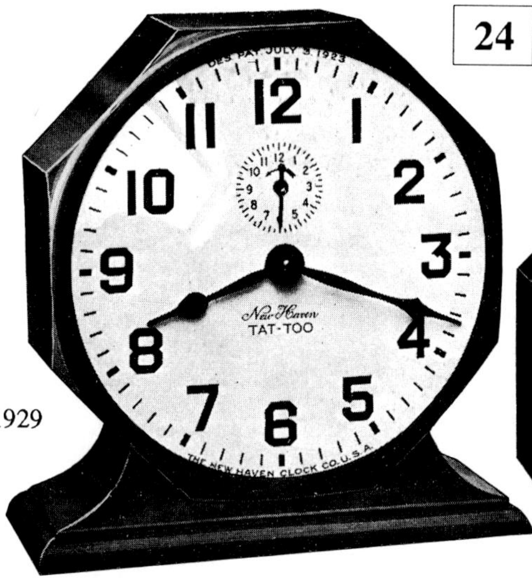

1929

"TAT-TOO" ARTLARM

Height, 5 inches; 4½-inch Dial. Dependable 40-hour Movement. Back Bell Alarm with Shut-off. Seamless Metal Case with Moulded Base to match. Convex Krack-Proof Krystal. Green Pastel Finish.

	White Dial
Broken Lots, each	$4.26
Dozen Lots, each	4.16
Case Lots of 24, each	4.06

25

1929

"TAT-TOO" ARTLARM, Radium

Height, 5 inches; 4½-inch Dial. Dependable 40-hour Movement, Back Bell Alarm with Shut-off. Seamless Metal Case with Moulded Base to match. Convex Krack-Proof Krystal. Green Pastel Finish.

	Radium Dial
Broken Lots, each	$5.62
Dozen Lots, each	5.52
Case Lots of 24, each	5.42

26

1886

PADLOCK (Alarm).

NICKEL OR POLISHED BRASS.

Height, 6 inches. Dial, 3 inches.

1 Day Time, Nickel or Brass ..$6.50
1 " " Alarm 7.50

27

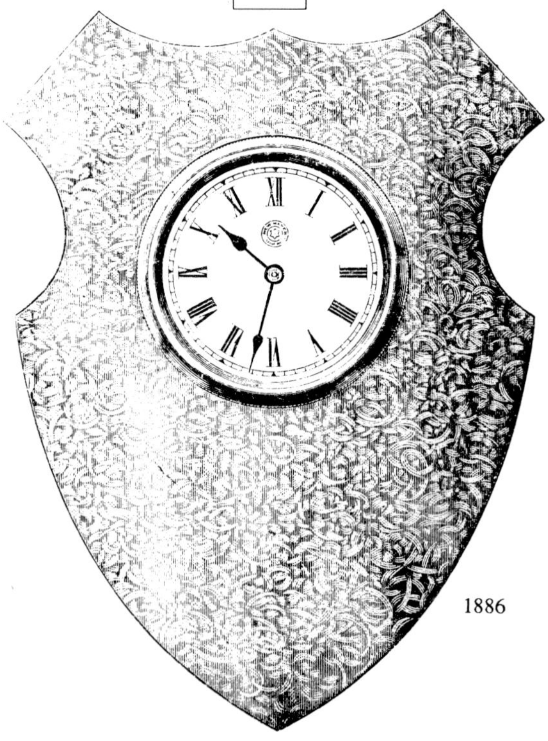

1886

SHIELD N0. 1, T. P. ALARM

Height, 6 inches. Dial, 2 inches. Made of Figured Brass with Gilt Finised.
1 Day Time Calendar. 1 Day Time, Alarm.
1 Day Time, Alarm........$4.10

28

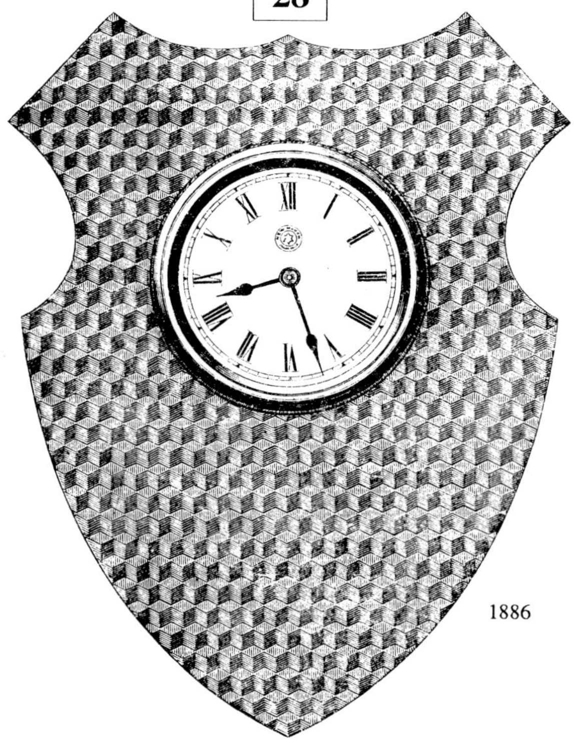

1886

SHIELD N0. 2, T. P. ALARM

Height, 6 inches. Dial, 2 inches. Made of Figured Brass with Gilt Finised.
1 Day Time Calendar. 1 Day Time, Alarm.
1 Day Time, Alarm........$4.10

ALARM CLOCKS

The year entered beside each clock is the year of the catalog from which the illustration was taken. Some clocks may have been produced a few years earlier.

29

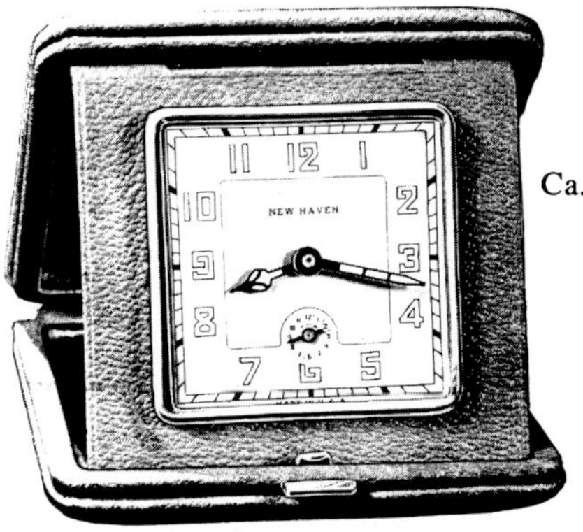

Ca. 1938

TOURIST No. 2 Alarm

A handsome, traveling alarm clock with genuine leather folding case in Tan, Brown, Navy or Black. Two-tone etched, gold-color dial with RADIUM numerals and hands. Polished brass hands and bezel. Unbreakable crystal. 30-Hour alarm movement. Size when closed: 4⅛ inches wide, 4⅛ inches deep, 1½ inches high. Packed in lots of 24. Shipping weight: 19 lbs.

30

CEYLON

1911

HEIGHT ....8 INCHES
WIDTH ....6¾ INCHES
DIAL, PORCELAIN ....4⅛ INCHES
CASE, SOLID MAHOGANY
MOVEMENT, 8 DAY TIME ALARM, RINGS (AT HOUR SET) UNTIL STOPPED BY PUSHING IN ON BUTTON ON BACK OF CLOCK MARKED "STOP ALARM." THIS SWITCH IS AUTOMATICALLY OPERATED SO THAT CLOCK RINGS AT SAME HOUR NEXT MORNING

31

Ca. 1937

CAPRI

Cast metal case finished in Rose, Green, Black or Ivory. Two-tone, gold-color dial; polished brass bezel. Convex glass. Full-toned alarm bell with top shut-off. 30-Hour movement. Height, 6¼ inches. Width, 5⅞ inches. Packed in lots of 12. Shipping weight: 28 lbs.

RETAIL $1.50

32

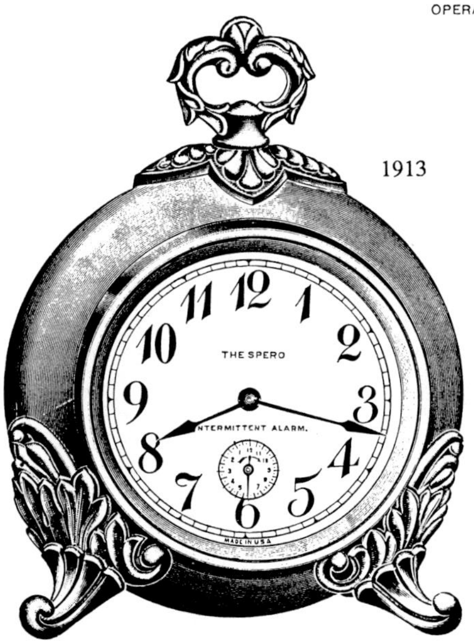

1913

SPERO

INTERMITTENT ALARM. LIST, $3.65
HEIGHT, 5¼ INCHES. WIDTH, 4 INCHES.
DIAL, 2½ INCHES.
CASE, ORNAMENTAL BRASS, OLD BRASS FINISH.

33

1900

BEACON ALARM—Nickel or Enamel.

Height, 6½ inches. Width, 4¼ inches. Dial, 4 inches.

List price, One-day Alarm, Nickel, $1.60
" " One-day Alarm, Enamel, 1.65

34

CAMPUS

1913

TIME ALARM. LIST, $1.60
HEIGHT, 6 INCHES. WIDTH, 4⅛ INCHES.
DIAL, 4 INCHES.
CASE AND ALL TRIMMINGS NICKEL PLATED.

ALARM CLOCKS

36

35

1900

CHAMPION—Long Alarm.

Dial, 4½ inches.

List price, One-day Time, Alarm, $2.50

1900

TATTOO—Intermittent Alarm.

Dial, 4½ inches.

List price, One-day Time, Alarm, $2.65

37

1911

TATTOO ALARM

(BLACK DIAL, LUMINOUS FIGURES AND HANDS)

HEIGHT ....6¼ INCHES
WIDTH ....4½ INCHES
DIAL, BLACK ....4 INCHES, WITH LUMINOUS FIGURES AND HANDS
CASE, 4½-INCH, SEAMLESS BRASS, KNURLED FRONT, WITH 4-INCH BELL METAL GONG ON BACK
FINISH, NICKEL PLATED
MOVEMENT, 1 DAY ALARM, RINGS INTERMITTENTLY EVERY 20 SECONDS FOR 15 MINUTES, FITTED WITH "ALARM AND SILENT" SWITCH
PRICE, WITH LUMINOUS DIAL ....LIST, $2.95

38

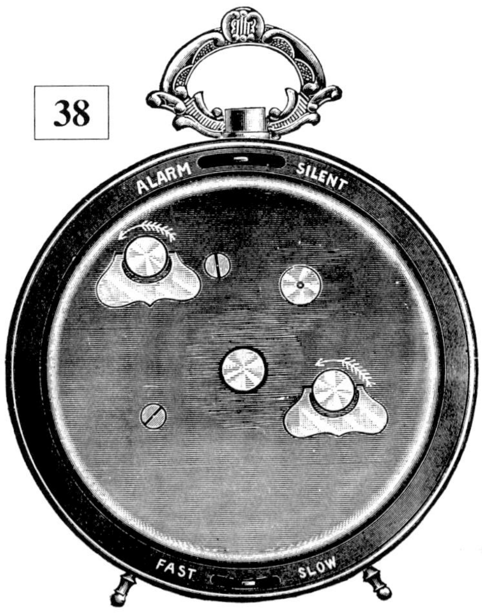

1911

BACK VIEW

Showing how Bell is attached on Tattoo and Champion Alarms.

ALARM CLOCKS

39

1900

BEACON ALARM — Luminous Dial, Nickel.

Height, 6½ inches. Width, 4¼ inches. Dial, 4 inches.

List price, One-day Alarm, Luminous, $1.85

40

1900

BEACON—Alarm.

Dial, 4 inches.

List price, One-day Time, Alarm, $1.75

41

1900

42

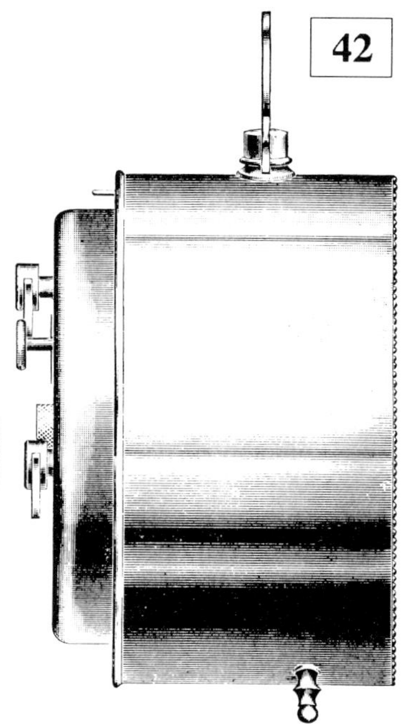

1900

43

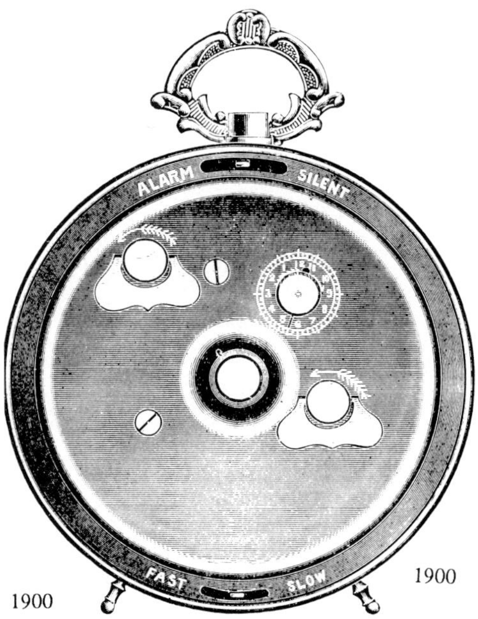

1900

CHAMPION ALARM CLOCK

A loud, continuous alarm. Fitted with a 4½-inch Dial and a 4-inch Bell-Metal Gong on the back. The Alarm rings, unless shut off, for five minutes.

List price, One-day Time, Alarm, $2.35

ALARM CLOCKS

44

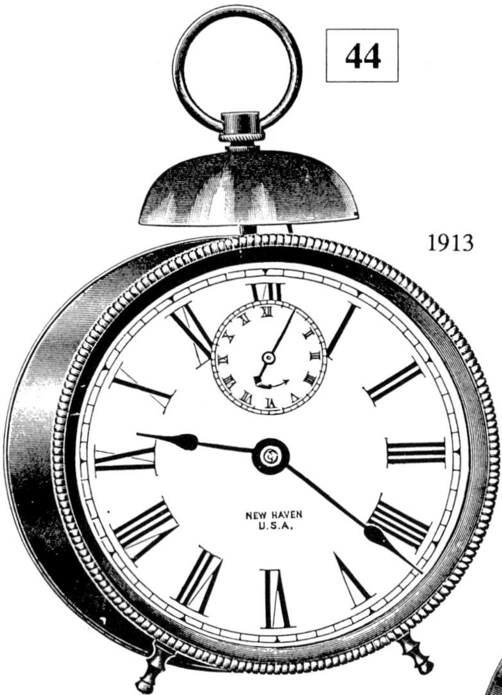

1913

SPRITE

TIME ALARM. LIST, $2.05
HEIGHT, 7¼ INCHES. WIDTH, 4½ INCHES.
DIAL, 4⅛ INCHES.
FITTED ALARM ONLY.

45

1889

SULTANA
HEIGHT, 11 INCHES.
DIAL, 9 INCHES.
1 Day Time, Alarm........ $6.50.
1 Day Strike.................... $6.50.

46

1889

STING REPEATING STRIKE

NICKEL.

By touching the button on the side, this clock repeats the hour last struck until ten minutes of the next.

Code Word.		Price.
(Abrade)	1 Day Repeating Strike..	$4.00

47

1900

TOCSIN Long Alarm.

Dial, 4½ inches.

The Alarm, unless shut off, rings continuously for five minutes.

List price, One-day Time, Alarm, $2.10

48

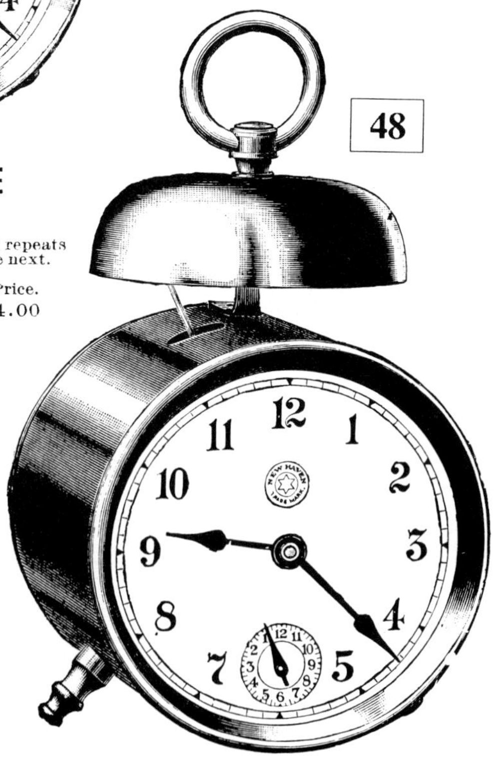

1886

STING ALARM

1 Day Time, Alarm........... $3.25
1 Day Time, Alarm, Gilt...... 3.35

ALARM CLOCKS

The year entered beside each clock is the year of the catalog from which the illustration was taken. Some clocks may have been produced a few years earlier.

49

WHISTLER. 1883

NICKEL, ROMAN DIAL.
Dial, 4 inches.
1 Day Time. 1 Day Time, Alarm.

50

WHISTLER ALARM.

NICKEL. FANCY DIAL. 1883
Dial, 4 inches.
1 Day Time. 1 Day Time, Alarm.

51

DAY DREAM. 1883

NICKEL
1 Day Time, Alarm.
One Spring for both Time and Alarm.

52

GLOBE. 1883

NICKEL
1 Day Time Calendar. 1 Day Time, Alarm.
1 Day Time Calendar, Alarm. 1 Day Strike.

53

1883

ACORN.

NICKEL
Dial, 3 inches.
1 Day Time. 1 Day Time, Alarm.

54

JIGGER 1911

HEIGHT ....6½ INCHES
WIDTH ....4⅛ INCHES
DIAL ....4 INCHES
CASE, 4⅛-INCH, SEAMLESS BRASS, WITH 2¼-INCH BELL
FINISH, NICKEL PLATED
MOVEMENT, 1 DAY ALARM, RINGS INTERRUPTEDLY WITH SILENT INTERVALS OF 2 SECONDS (HAS SWITCH FOR STOPPING ALARM)
PRICE ....LIST, $1.90
PRICE, WITH LUMINOUS DIAL ....LIST EXTRA, .30

ALARM CLOCKS

55

1913

LARK

(FRONT VIEW)

56

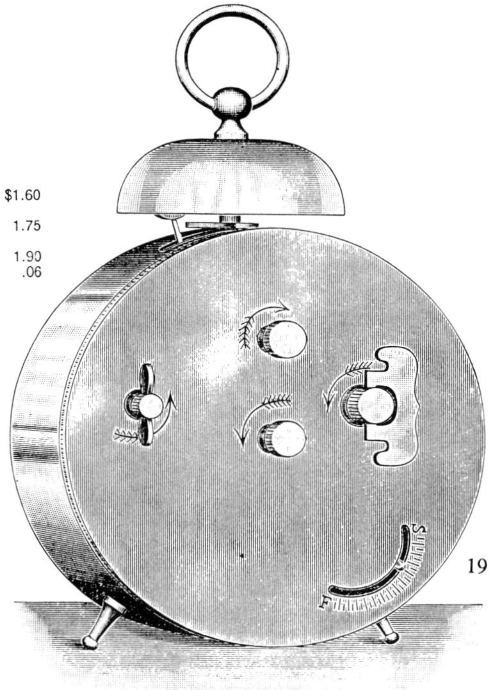

1913

LARK

(BACK VIEW)

LARK SWITCHLESS

HEIGHT ....6½ INCHES
WIDTH ....4⅛ INCHES
DIAL ....4 INCHES
CASE, 4⅛-INCH, SEAMLESS BRASS, WITH KNURLED FRONT AND FLUSH BACK WITH KNURLED EDGE. HAS 2¼-INCH BELL
FINISH, NICKEL PLATED CASE, BELL, MAT, BACK AND ALL TRIMMINGS

PRICE, 1 DAY TIME PIECE, ALARM ....LIST,	$1.60
PRICE, 1 DAY TIME PIECE, ALARM CALENDAR ....LIST,	1.75
PRICE, 1 DAY TIME PIECE, ALARM LUMINOUS ....LIST,	1.90
PRICE, WITH ENAMEL FINISH ....LIST EXTRA,	.06

1911

57

1886

LARK T. P. (Alarm).

NICKEL.

Dial, 3 inches.

1 Day Time, Alarm .$3.10

58

1886

LARK REPEATING STRIKE.

NICKEL.

By touching the button on the side, this clock repeats the hour last struck until ten minutes of the next.

Code Word.		Price.
(Abyss)	1 Day Repeating Strike..	$3.60

59

1886

LEATHER TRAVELING CASE.

LARK STRIKE AND ALARM.

Code Word.		Price.
(Alas)	1 Day Repeating Strike........	$6.10
(Ache)	1 " Alarm	5.10

ALARM CLOCKS

60

1895

START ALARM.
NICKEL.
1 Day Time, Alarm
Dial, 4 inches. . . . $2.35

61

1900

FLY ALARM—Nickel.

Dial, 4 inches.

List price, One-day Time Alarm, $1.55

62

1900

FLASH.

LUMINOUS DIAL. ALARM.

Dial, 4 inches.

1 Day Time. $2.75
1 Day Time, Alarm. 3.25

63

1886

HERALD.

WOOD CASE. WITH NICKEL OR BRASS FRONT.

Height, 6½ inches. Dial, 4 inches.
One Spring only.

Code Word.		Price.
(Rascal)	1 Day Time, Alarm.	$3.10

ALARM CLOCKS

64

1906

AUTO ALARM

Four-inch Seamless Case, with Knurled Front,
and Bell on the back.
Fitted with Shut-off Switch.

One-day Time, Alarm . . . . List Price, $1.65

65

1906

AUTO BACK VIEW

Showing Bell attached.

66

1913

IN CLOCKS FITTED WITH INTERMITTENT ALARM MOVEMENT, THE ALARM RINGS AND STOPS IN 20-SECOND INTERVALS FOR 15 MINUTES. ALL CLOCKS HAVE "ALARM OR SILENT" SWITCH.

67

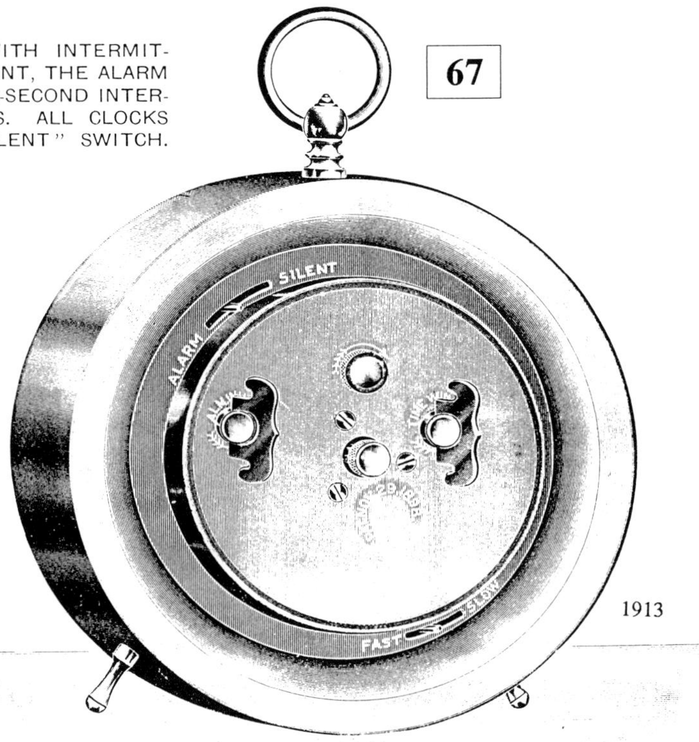

1913

GIANT
(FRONT VIEW)

GIANT

INTERMITTENT ALARM . . LIST, $4.10
WITH LUMINOUS DIAL . . . . LIST, 4.40
HEIGHT, 7¼ INCHES. WIDTH, 6⅛ INCHES.
DIAL, 5 INCHES.

GIANT
(BACK VIEW)

ALARM CLOCKS

68

1913

RAPID FIRE

INTERMITTENT ALARM. LIST, $2.20
HEIGHT, 6⅛ INCHES. WIDTH, 4½ INCHES.
DIAL, 4⅛ INCHES.

69

1913

STING

NICKEL FINISH . . . . LIST, $1.70
GILT FINISH . . . . LIST, 1.85
HEIGHT, 3 INCHES. WIDTH, 2¼ INCHES.
DIAL, WHITE, 2 INCHES.

70

1913

CIRCLE

TIMEPIECE . . . . LIST, $2.20
ALARM . . . . LIST, 2.50
HEIGHT, 4¼ INCHES. WIDTH, 3¼ INCHES.
DIAL, 2 INCHES.

71

1913

THE COMMUTER

CONTINUOUS AND INTERMITTENT ALARM. LIST, $4.20
HEIGHT, 7⅛ INCHES. WIDTH, 5¼ INCHES.
DIAL, 4½ INCHES.

72

1900

ELFIN—Nickel.

Height, 8 inches. Dial, 6 inches.

List price, One-day Time, Alarm, $2.75
" " One-day Strike, 2.75

ALARM CLOCKS

73

1913

EIGHT-DAY AUTOMATIC
(Black Dial, Luminous Figures and Hands.)

RUNS AND ALARMS 8 DAYS. LIST, $5.55
HEIGHT, 7¼ inches. WIDTH, 6 inches.
DIAL, 4¼ inches.

75

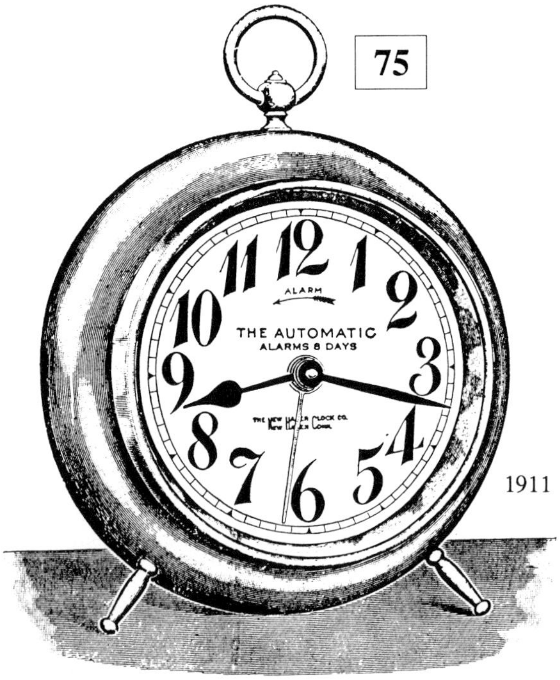

1911

EIGHT DAY AUTOMATIC

HEIGHT ....7¼ inches
WIDTH ....6 inches
DIAL ....4¼ inches
CASE, 6-inch, Seamless Brass, with 4-inch Bell Metal Gong on Back
FINISH, Nickel Plated
MOVEMENT, 8 day Time Alarm, Rings (at hour set) until Stopped by Pushing Button on Back of Clock marked "Stop Alarm." This Switch is automatically operated so that Clock rings at same hour next morning. Also has "Alarm or Silent" Switch for use when Time only is wanted
PRICE ....List. $5.25
PRICE, with Luminous Dial ....List extra .30

74

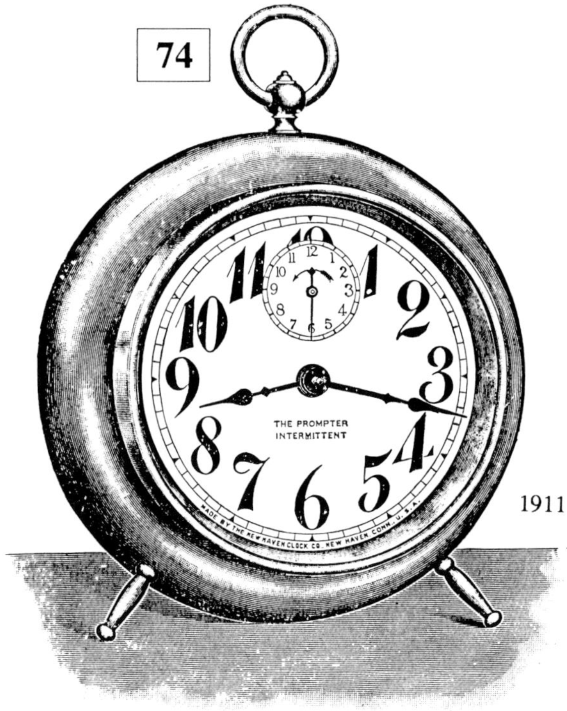

1911

PROMPTER

HEIGHT ....7¼ inches
WIDTH ....6 inches
DIAL ....4¼ inches
CASE, 6-inch, Seamless Brass, with 4-inch Bell Metal Gong on Back
FINISH, Nickel Plated
MOVEMENT, 1 day Alarm, Rings Intermittently every 20 seconds for 15 minutes. Fitted with "Alarm or Silent" Switch
PRICE ....List, $3.50
PRICE, with Luminous Dial ....list extra, .30

76

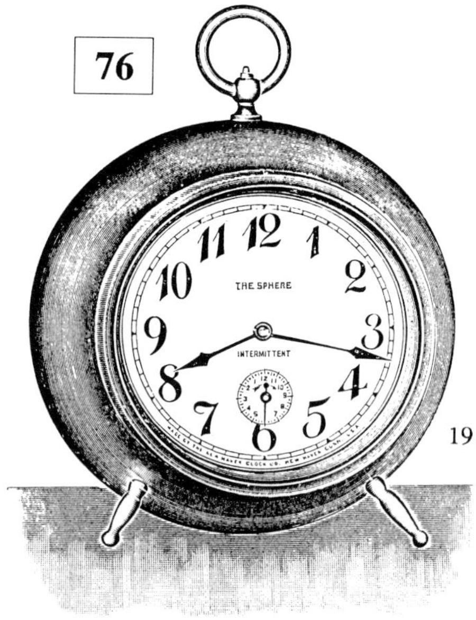

1911

SPHERE

HEIGHT ....5 inches
WIDTH ....4 inches
DIAL ....2½ inches
CASE 4 inch Seamless Brass with 3 inch Bell Metal Gong on Back
FINISH, Nickel Plated
MOVEMENT, 1 day Alarm, Rings Intermittently every 20 seconds for 5 minutes. Fitted with "Alarm or Silent" Switch
PRICE..List, $3.05

ALARM CLOCKS

77

1906

THE JUNIOR TATTOO ALARM

Finished in Nickel or Gilt.
Two-inch Dial.
Fitted with an Intermittent Alarm Movement that rings every few seconds for five minutes.
Fitted with switch to shut off alarm.

Nickel Finish . . . . List Price, $2.60
Gilt Finish . . . . " " 2.85

78

1906

Back View
JUNIOR TATTOO
Showing Bell attached.

79

1906

JUNIOR TATTOO IN LEATHER CASE

Height, 3¼ inches. Width, 3 inches.
Two-inch Paper Dial.
Fitted with an Intermittent Alarm Movement.

Clock . . . . List Price, $2.60
Leather Case . . . . " " 1.60
Complete . . . . " " 4.20

81

1913

WANDA

INTERMITTENT ALARM. LIST, $5.80
HEIGHT, 3¾ INCHES. WIDTH, 3¼ INCHES.
DIAL, PORCELAIN, 2 INCHES.

80

PADUA ALARM
(Also called **ALCOVE TATTOO**)

1911

HEIGHT ....4 INCHES
WIDTH ....3¼ INCHES
DIAL, PORCELAIN ......2 INCHES
FINISH, GILT, SILVER, GUN METAL OR OLD BRASS
PRICE ....LIST, $4.40

82

EIGHT DAY SELF-SETTER

1911

HEIGHT ....8⅛ INCHES
WIDTH ....6⅛ INCHES
DIAL ....5 INCHES
CASE, 6⅛-INCH SEAMLESS BRASS, KNURLED FRONT WITH 4-INCH BELL METAL GONG ON BACK
FINISH, NICKEL PLATED
MOVEMENT, 8 DAY TIME ALARM, RINGS (AT HOUR SET) UNTIL STOPPED BY PUSHING BUTTON ON BACK OF CLOCK MARKED "STOP ALARM." THIS SWITCH IS AUTOMATICALLY OPERATED SO THAT CLOCK RINGS AT SAME HOUR NEXT MORNING. ALSO HAS "ALARM OR SILENT" SWITCH FOR USE WHEN TIME ONLY IS WANTED
PRICE ....LIST. $5.25

83

THE MACON ALARM

1911

HEIGHT ....4¼ INCHES
WIDTH ....3⅛ INCHES
DIAL, PORCELAIN ......2 INCHES
CASE, HEAVY BRASS
FINISH, GILT, SILVER, GUN METAL OR OLD BRASS
PRICE ....LIST, $5.15

ALARM CLOCKS

84

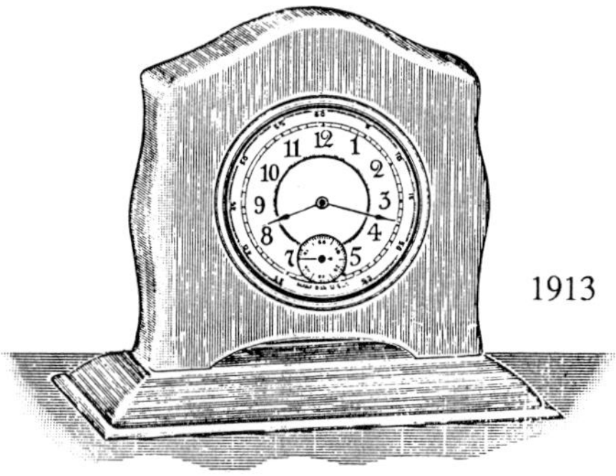

1913

KEITH

TIMEPIECE. LIST, $4.65
HEIGHT, 3½ INCHES. WIDTH, 4 INCHES.
DIAL, WHITE, 1¾ INCHES.

85

1913

KNOX

TIMEPIECE. LIST, $4.65
HEIGHT, 3½ INCHES. WIDTH, 3¼ INCHES.
DIAL, WHITE, 1¾ INCHES.

86

1913

KEATING

TIMEPIECE. LIST, $4.65
HEIGHT, 3¾ INCHES. WIDTH, 2¾ INCHES.
DIAL, WHITE, 1¾ INCHES.

87

1913

WALDO

INTERMITTENT ALARM. LIST, $6.10
HEIGHT, 3½ INCHES. WIDTH, 3 INCHES.
DIAL, PORCELAIN, 2 INCHES.

88

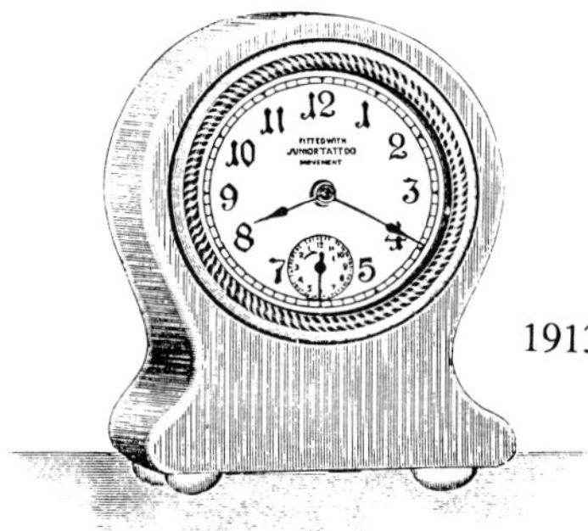

1913

WATSON

INTERMITTENT ALARM. LIST, $6.10
HEIGHT, 3¾ INCHES. WIDTH, 3 INCHES.
DIAL, PORCELAIN, 2 INCHES.

89

1913

KALON

TIMEPIECE. LIST, $4.65
HEIGHT, 3½ INCHES. WIDTH, 3 INCHES.
DIAL, WHITE, 1¾ INCHES.

90

1913

KELVIN

TIMEPIECE. LIST, $4.65
HEIGHT, 3⅝ INCHES. WIDTH, 4 INCHES.
DIAL, WHITE, 1¾ INCHES.

91

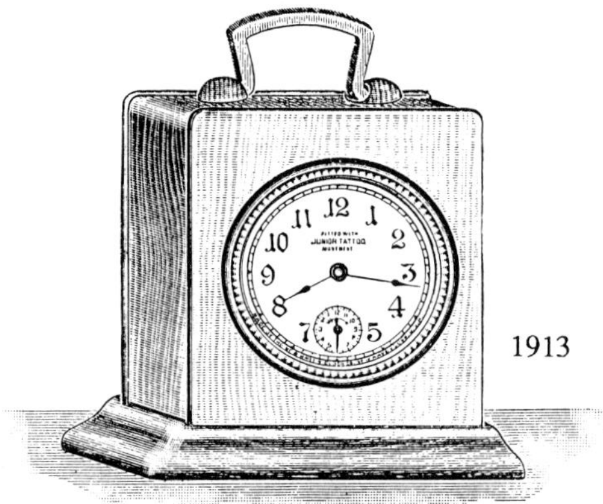

1913

WESTON

INTERMITTENT ALARM. LIST, $5.80
HEIGHT, 5 INCHES. WIDTH, 3¾ INCHES.
DIAL, PORCELAIN, 2 INCHES.

92

1913

NEMO

ONE-DAY INTERMITTENT ALARM. LIST, $3.65
HEIGHT, 4¾ INCHES. WIDTH, 4¼ INCHES.
DIAL, WHITE, 2 INCHES.

ALARM CLOCKS

93

1929

EASEL ARTLARM

Gothic cathedral design. Walnut finished, Moulded Easel Case, Enameled Dial. Krack Proof Krystal. 40-hour dependable movement, shaped bell on back, long steady alarm with shut-off. Size: height, 4¾ inches.

	Plain Dial	Radium Dial
Broken Lots, each........	$7.00	$8.32
Dozen Lots, each..........	6.90	8.22
Case Lots of 24, each....	6.80	8.12

94

1913

MORITA

LIST, $10.90

HEIGHT, 9¾ INCHES. WIDTH, 6⅜ INCHES.

95

NEGUS

ONE-DAY TIMEPIECE. LIST, $2.20

HEIGHT, 4¾ INCHES. WIDTH, 4½ INCHES.

DIAL, WHITE, 2 INCHES.

1929

96

1913

CHICO

ONE-DAY ALARM . . . . . . LIST, $2.90

EIGHT-DAY ALARM . . . . . LIST, 5.40

HEIGHT, 8¾ INCHES. WIDTH, 6¼ INCHES.

DIAL, 3¾ INCHES.

ONE-DAY, FANCY TWO-COLOR DIAL.

EIGHT-DAY, WHITE ARABIC DIAL.

97

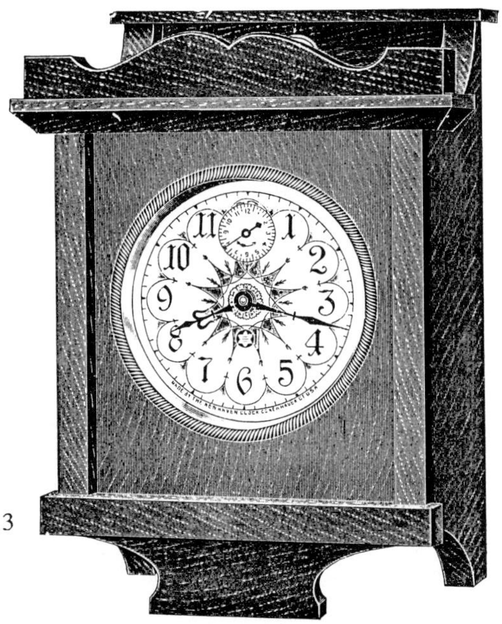

1913

CAIRO

(HANGING AND MANTEL.)

ONE-DAY ALARM. LIST, $2.90

HEIGHT, 9 INCHES. WIDTH, 6⅜ INCHES.

DIAL, FANCY, TWO-COLOR, 3¾ INCHES.

ALARM CLOCKS

98

Ca. 1936

SQUARE TELL-TALE, Plain Dial
3¾ inches Square, Krack-proof Krystal, 3½ inch Dial, Russet Bronze Case, 40-hour Movement, Back Bell Alarm with Shut-off.

	Plain Dial
Broken Lots, each	$2.92
Dozen Lots, each	2.82
Case Lots of 50, each	2.72

99

SQUARE TELL-TALE, Radium
3¾ inches Square, Krack-proof Krystal, 3½ inch Dial, Russet Bronze Case, 40-hour Movement, Back Bell Alarm with Shut-off.

	Radium Dial
Broken Lots, each	$4 28
Dozen Lots, each	4.18
Case Lots of 50, each	4.08

100

Ca. 1937

TICK TOCK

Metal case finished in Rose, Green, Black or Ivory. Two-tone, gold-color dial; polished brass bezel. Convex glass. Full-toned alarm bell with top shut-off. 30-Hour movement. Height, 5½ inches. Width, 5 inches. Packed in lots of 12. Shipping weight: 20 lbs.

RETAIL $1.50

101

1913

TRAVELER TATTOO

LIST, $4.65

HEIGHT, 3 INCHES. WIDTH, 3 INCHES.

CASE, GENUINE LEATHER.

102

1913

STUDY TATTOO

LIST, $4.40

HEIGHT, 4⅛ INCHES. WIDTH, 3 INCHES.

103

Ca. 1936

TOM-TOM, Plain Dial
Height, 5½ inches. Width, 4⅞ inches. 4¼ inch Cubist Dial. Krack-proof Krystal. Nickel Plated Octagon Case. 40 Hour Movement, Back Bell Intermittent Alarm with Shut-off. Removable Springs.

	White Dial
Broken Lots, each	$4.60
Dozen Lots, each	4.48
Case Lots of 24, each	4.38

104

Ca. 1936

TOM-TOM, Radium
Height, 5½ inches. Width, 4⅞ inches. 4¼ inch Cubist Dial. Krack-proof Krystal. Nickel Plated Octagon Case. 40 Hour Movement, Back Bell Intermittent Alarm with Shut-off. Removable Springs.

	Radium Dial
Broken Lots, each	$5.94
Dozen Lots, each	5.84
Case Lots of 24, each	5.74

105

LIBRARY TATTOO

1911

HEIGHT ....4 INCHES
WIDTH ....2¾ INCHES
DIAL, PORCELAIN ......2 INCHES
CASE, HEAVY BRASS
FINISH, GILT, SILVER, GUN METAL OR OLD BRASS
PRICE ....LIST, $5.45

ALARM CLOCKS

The year entered beside each clock is the year of the catalog from which the illustration was taken. Some clocks may have been produced a few years earlier.

EIGHT-DAY TIME ALARM CLOCKS IN SOLID MAHOGANY CASES.
THEY RUN AND ALARM EIGHT DAYS. DIALS, PORCELAIN, 4⅛ INCHES.
BELL METAL GONG ON BACK. FITTED WITH "ALARM OR SILENT" SWITCH.

106

1913

CECIL
LIST, $10.15
HEIGHT, 7½ INCHES. WIDTH, 8¼ INCHES.

107

1913

CISCO
LIST, $10.15
HEIGHT, 6¾ INCHES. WIDTH, 7¼ INCHES.

108

1913

GUEST TATTOO
LIST, $5.85
HEIGHT, 6½ INCHES. WIDTH, 4½ INCHES.
CASE, SOLID MAHOGANY WITH INLAID MARQUETRY.

109

1913

CERES
LIST, $10.15
HEIGHT, 8¼ INCHES. WIDTH, 7⅜ INCHES.

110

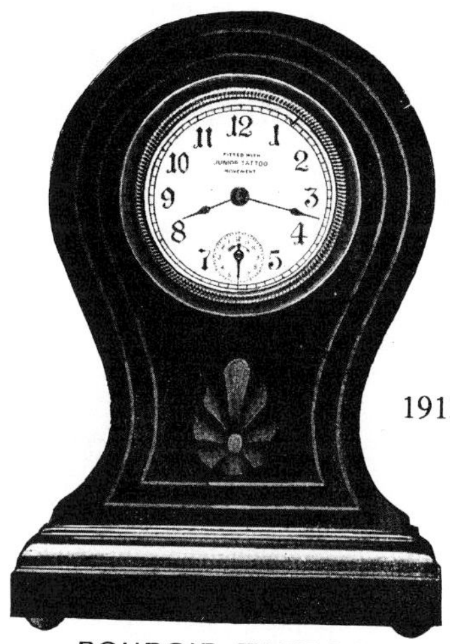

1913

BOUDOIR TATTOO
LIST, $5.85
HEIGHT, 6½ INCHES. WIDTH, 4½ INCHES.
CASE, SOLID MAHOGANY WITH INLAID MARQUETRY.

ALARM CLOCKS

111

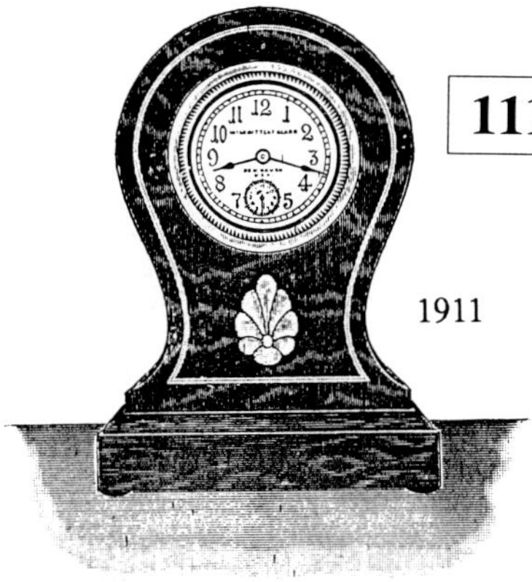

1911

THE EUNICE ALARM

HEIGHT ....6½ INCHES
WIDTH ....4½ INCHES
DIAL, PORCELAIN ......2 INCHES
CASE, SOLID MAHOGANY WITH INLAID MARQUETRY DECORATIONS
PRICE ....LIST, $5.85

112

1911

THE MARVEL ALARM

HEIGHT ....3⅞ INCHES
WIDTH ....2⅞ INCHES
DIAL, PORCELAIN ......2 INCHES
CASE, HEAVY BRASS
FINISH, GILT, SILVER, GUN METAL OR OLD BRASS
PRICE ....LIST, $5.15

113

1911

THE IRMA ALARM

HEIGHT ....3½ INCHES
WIDTH ....4⅛ INCHES
DIAL, PORCELAIN ......2 INCHES
FINISH, GILT, SILVER, GUN METAL OR OLD BRASS
PRICE ....LIST, $5.45

114

1913

WELBY

INTERMITTENT ALARM. LIST, $6.50
HEIGHT, 6 INCHES. WIDTH, 4¼ INCHES.
DIAL, PORCELAIN, 2 INCHES.

115

1911

CASTRO

HEIGHT ....6½ INCHES
WIDTH ....6½ INCHES
DIAL, FANCY, TWO COLOR ....3¾ INCHES
CASE, SOLID OAK FRAME, FLEMISH FINISH, HOLDING AN AUTO ALARM, THE BEZEL AND CASE OF WHICH ARE FINISHED IN OXIDIZED BRONZE
MOVEMENT, 1 DAY ALARM WITH "ALARM OR SILENT" SWITCH
PRICE ....LIST, $2.65

116

1929

117

1929

"TAT-TOO JR." ARTLARM

Height, 2¾ inches; 2¼-inch Silvered Dial. Dependable 40-hour Movement. Back Bell Alarm with Shut-off. Seamless Metal Case with Moulded Base to match. Krack-Proof Krystal. Pastel Colors—Rose, Green, Yellow or Blue.

	Silvered Dial	Radium Dial
Broken Lots, each	$4.26	$5.62
Dozen Lots, each	4.16	5.52
Case Lots of 24, each	4.06	5.42

ALARM CLOCKS

118

1929

1929

119

120

1929

DORIS ARTLARM

Gothic Design, Metal Case, White Dial, Radium Numerals, Artistic Hands. Krack-Proof Krystal. 40-hour Dependable Movement, Long Steady Alarm with Shut Off. Pastel Colors, Rose, Green or Blue with black trim. **Height, 8 [illegible] inches**

	Radium Dial
Broken Lots, each	$7.00
Dozen Lots, each	6.90
Case Lots of 24, each	6.80

TIDY-TOT

Height, 3⅜ inches; Width, 2¾ inches; 2¼-inch Dial. Krack-Proof Krystal. Nickel-plated Octagon Case. 40-hour Movement. Back Bell, Intermittent Alarm with Shut-off.

	Silvered Dial	Radium Dial
Broken Lots, each	$4.60	$5.94
Dozen Lots, each	4.48	5.84
Case Lots of 24, each	4.38	5.74

121

122

1929

123

124

1929

ABBEY JR. ARTLARM

Mahogany finished moulded case, gothic design. Krack-Proof Krystal. 40-hour dependable movement, shaped bell on back, long steady alarm with shut-off. Height, 3½ inches.

	Plain Dial	Radium Dial
Broken Lots, each	$4.90	$6.24
Dozen Lots, each	4.80	6.14
Case Lots of 24, each	4.70	6.04

SQUARE ARTLARM

Mahogany finished moulded case. Krack-Proof Krystal. 40-hour dependable movement, shaped bell on back, long steady alarm with shut-off. Height, 4¼ inches.

	Plain Dial	Radium Dial
Broken Lots, each	$4.90	$6.24
Dozen Lots, each	4.80	6.14
Case Lots of 24, each	4.70	6.04

125

1929

1929

126

TICK-TOCK

Height, 4¼ inches; 3¾-inch Dial. Dependable 40-hour Movement. Back Bell Alarm with Shut-off. Seamless Metal Case with Moulded Base to match. Convex Krack-Proof Krystal. Pastel Colors—Blue, Green, Yellow or Rose.

	White Dial	Radium Dial
Broken Lots, each	$3.54	$4.90
Dozen Lots, each	3.44	4.80
Case Lots of 24, each	3.34	4.70

ALARM CLOCKS

127

1929

"TIP-TOP TRAVELER"
Height, 1⅞ inches. Silvered Radium Dial. Second Hand. Pull-out Set, Stem Wind. Nickel-plated Octagon Case. Bowed Crystal **$4.76**

128

Ca. 1937

SOFT-TICK
A quiet-running alarm clock with top shut-off, pearl-gray metal case, modernistic dial and a 30 hour movement. The concealed alarm bell has a clear, musical tone. Height, 4½ inches. Width, 4 inches.

129

Ca. 1932

POPEYE
The famous Popeye attractively reproduced in full colors on the case and on the dial. The tick of this alarm clock is unusually soft. Ivory-colored metal case with black trim; and full-toned, concealed alarm with top shut-off. Height, 4½ inches. Width, 4 inches.

130

Ca. 1937

CHROME-SQUARE ALARM
A smart, small size alarm clock with chromium case; new, two-tone, index dial with applied numerals and ball feet. Width, 2½ inches. Height, 2⅞ inches.

131

Ca. 1937

SLUMBER STOPPER
A low priced alarm clock with a subdued tick and top shut-off. The case is of metal and is available in a choice of colors with attractively designed dials in colors to match the case. Height, 4½ inches. Width, 4½ inches.
Slumber Stopper No. 1, Green and Gold; No. 2, Green and Silver; No. 3, Pearl Gray and Black; No. 10, Pearl Gray.

132

FUTURIST - Musical Alarm
A real musical alarm with two tuneful concealed bells - pleasant as a chime. Case is cast metal in black with gold trim. Dial is cream enamel with black cubist numerals and attractive hands. Height, 6 inches. Width, 4 inches

Ca. 1932

133

1885

Height, 5 inches. Dial, 4 inches.

SNIPE. (Alarm.)

NICKEL.

1 Day Time, Alarm.......... .. .... 5 75

Ca. 1932

134

NELSON No. 2 (30 Hour Alarm)
A modernistic boudoir clock with alarm fitting, brown suede finished case with polished gilt trimings. Two-tone, index dial with latest type applied hour indicators. Full-toned alarm bell. Height, 4¾ inches. Width, 6¼ inches.

AUTOMOBILE CLOCKS

The year entered beside each clock is the year of the catalog from which the illustration was taken. Some clocks may have been produced a few years earlier.

135

1911

MINEOLA

HEIGHT ....4¾ INCHES
WIDTH ....4 INCHES
DIAL, PORCELAIN ......3 INCHES
FINISHED IN POLISHED BRASS, NICKEL OR BLACK ENAMEL
MOVEMENT, FINE QUICK-BEAT 8 DAY TIME PIECE WITH SECOND HAND

136

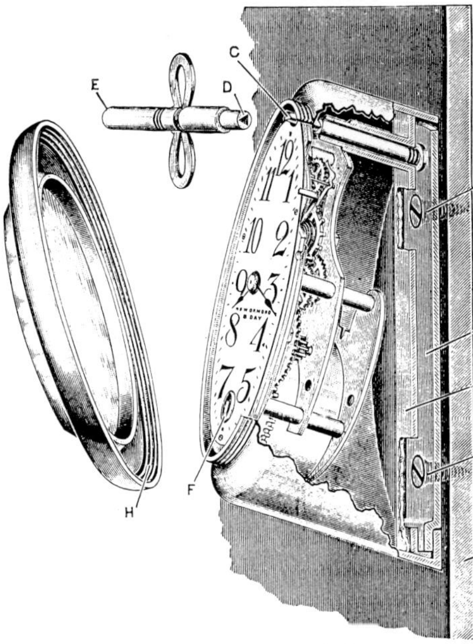

1911

SECTIONAL VIEW

SHOWING SPECIAL FEATURES

A. LOCKING PLATE WHICH SCREWS OR BOLTS TO DASHBOARD AND INTO WHICH CLOCK LOCKS
B. SCREWS OR BOLTS, HOLDING LOCKING PLATE TO DASHBOARD
C. BOLT WITH TRIANGULAR HEAD (REQUIRING SPECIAL KEY) WHICH LOCKS CLOCK TO LOCKING PLATE
D. END OF KEY TO LOCK CLOCK TO LOCKING PLATE
E. END OF KEY FOR WINDING CLOCK
F. WINDING ARBOR
G. BACK OF CLOCK WHICH LOCKS INTO LOCKING PLATE
H. SCREW BEZEL WHICH MAKES CLOCK DUST- AND WATER-PROOF
I. DASHBOARD

137

1911

SAVANNAH

HEIGHT ....4½ INCHES
WIDTH ....4⅛ INCHES
DIAL, PORCELAIN ......3 INCHES
FINISHED IN POLISHED BRASS, NICKEL OR BLACK ENAMEL
MOVEMENT, FINE QUICK-BEAT 8 DAY TIME PIECE WITH SECOND HAND

138

1911

WESTBURY

HEIGHT ....4⅛ INCHES
WIDTH ....4 INCHES
DIAL, PORCELAIN ......3 INCHES
FINISHED IN POLISHED BRASS, NICKEL OR BLACK ENAMEL
MOVEMENT, FINE QUICK-BEAT 8 DAY TIME PIECE WITH SECOND HAND

AUTOMOBILE CLOCKS

The year entered beside each clock is the year of the catalog from which the illustration was taken. Some clocks may have been produced a few years earlier.

139

1911

NEW ORMOND

HEIGHT ....4 INCHES
WIDTH ....4 INCHES
DIAL, PORCELAIN ......3 INCHES
FINISHED IN POLISHED BRASS, NICKEL OR BLACK ENAMEL
MOVEMENT, FINE QUICK-BEAT 8 DAY TIME PIECE WITH SECOND HAND

140

1906

ORMOND

This clock, designed for Motor-car and Motor-boat use, has a heavy polished brass off-set case, 4½ inches in diameter, and is dust and water proof. The dial, which is 2¾ inches in diameter, is of porcelain with black Arabic figures.

The winding, setting and regulating is all done from the outside, making it unnecessary to remove the mechanism from the case.

The body of clock locks onto the back flange enabling user to attach flange permanently to dashboard and yet be able to unlock his clock and carry it to a place of safety.

Clock runs 8 days with one winding.

Our patents cover a "Time Recording" device.
The Bezel is fitted to turn friction tight on the case, and a red pointer is painted on the glass to enable the user to set pointer at hour of starting on a run. At end of run he can readily see the time elapsed.

List Price, $9.05

141

1911

142

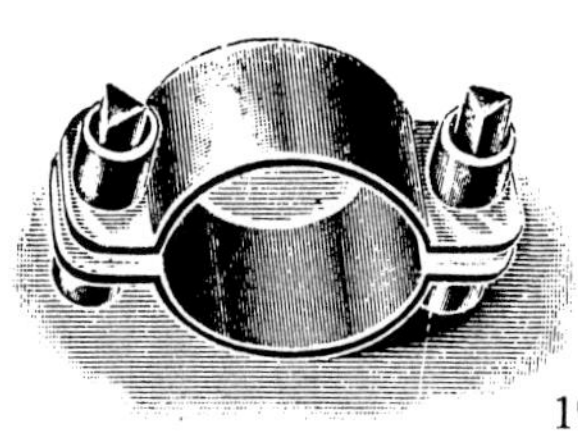

1911

SHOWING HOW ATTACHED

143

1911

THE FIAT MOTOR CYCLE CLOCK

HEIGHT ....2⅝ INCHES
WIDTH ....2⅛ INCHES
DIAL, WHITE ....1½ INCHES
CASE, HEAVY BRASS, HOLDING A COMPLETE WATCH. HAS A BRACKET WITH DEVICE FOR LOCKING TO HANDLE BAR OF BICYCLE OR MOTOR CYCLE
FINISH, POLISHED BRASS, NICKEL, OR BLACK ENAMEL
MOVEMENT, 1 DAY TIME PIECE
PRICE..LIST, $3.65

THE NEBO MOTOR CYCLE CLOCK

HEIGHT....................3 INCHES
WIDTH..................2⅝ INCHES
DIAL, WHITE.........1½ INCHES
CASE, CAST BRASS, HOLDING A COMPLETE WATCH. HAS A BRACKET WITH DEVICE FOR LOCKING TO HANDLE BAR OF BICYCLE OR MOTOR CYCLE
FINISH, POLISHED BRASS, NICKEL, OR BLACK ENAMEL
MOVEMENT, 1 DAY TIME PIECE
PRICE ....LIST, $3.65

AUTOMOBILE CLOCKS

144

1911

THE DUANE AUTO CLOCK

HEIGHT ....3½ INCHES
WIDTH..................2⅞ INCHES
DIAL, WHITE.........1½ INCHES
CASE, HEAVY BRASS WITH OFFSET HOLDING A COMPLETE WATCH
FINISH, POLISHED BRASS, NICKEL, OR BLACK ENAMEL
MOVEMENT, 1 DAY TIME PIECE
PRICE... ....LIST, $2.90
CLOCK IS SCREWED TO DASHBOARD

145

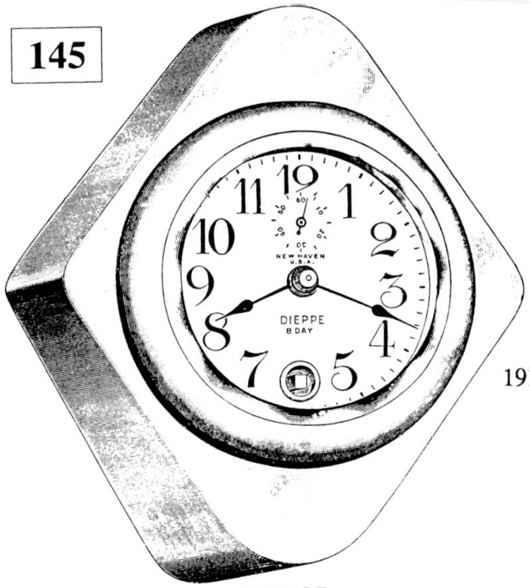

1911

DIEPPE

HEIGHT ....5⅝ INCHES
WIDTH ....4½ INCHES
DIAL, PORCELAIN ......3 INCHES
FINISHED IN POLISHED BRASS, NICKEL OR BLACK ENAMEL
MOVEMENT, FINE QUICK-BEAT 8 DAY TIME PIECE WITH SECOND HAND

8 DAY MOTOR CLOCKS

146

1911

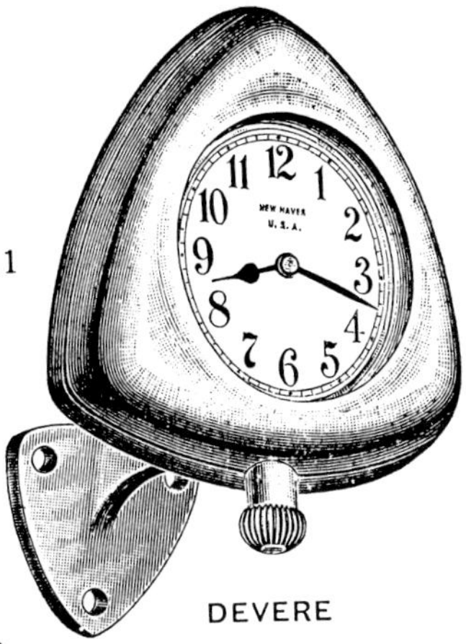

DEVERE

THE DEVERE AUTO CLOCK

HEIGHT................3½ INCHES
WIDTH..................2⅞ INCHES
DIAL, WHITE..........1⅛ INCHES
CASE, HEAVY BRASS WITH OFFSET HOLDING A COMPLETE WATCH
FINISH, POLISHED BRASS, NICKEL OR BLACK ENAMEL
MOVEMENT, 1 DAY TIME PIECE
PRICE...LIST, $3.65
CLOCK IS SCREWED TO DASHBOARD

147

1906

DAYTONA

This clock designed for Motor-car and Motor-boat use, has a polished brass off-set case, 3 inches in diameter. The clock has a 1½-inch paper dial with heavy black numerals. The clock can be wound and set from the outside, and runs thirty hours with one winding.

One-day Time ....List Price, $ 2.80

148

1911

JERICHO

HEIGHT ....4⅞ INCHES
WIDTH ....4⅞ INCHES
DIAL, PORCELAIN ......3 INCHES
FINISHED IN POLISHED BRASS, NICKEL OR BLACK ENAMEL
MOVEMENT, FINE QUICK-BEAT 8 DAY TIME PIECE WITH SECOND HAND

AUTOMOBILE CLOCKS

149

1937

No. 165 AV—8-Day Pull Wind movement with back set. Black dial with white numerals and radium hands. Selected, non-glare gunmetal plate glass mirror with beveled edges. Size, 2½" x 7".

150

1917

COWL
8-Day Time
Height, 3½ Inches Width, 3½ Inches
Dial, Silver, 2¼ Inches
Is Attached to Cowl-Board with Three Screws. Projects only ¾-inch from Board

151

No. 5000 MIRROR CLOCK COMBINATION 1932

Is easily installed in the present bracket and offers unusual visibility from all parts of the car. The high grade 8-day movement, built to withstand the vibrations of motoring, has two main springs, a dust proof case and conveniently placed winding and setting stem.

Mirrors are made of the finest plate glass, polished and beveled, with a choice of non-glare gunmetal, gold, or clear glass. Size 2½ x 7 inches.

Plain Dial List Price.....$9.50 Radium Dial List Price.....$10.50

152

1932

JEROME

A quality clock at popular prices, finished in rich French Bronze with an 8-day, two main spring movement and dust proof case. Stem wind and set. Can be attached to dash or over the windshield. Case is 2½ inches in diameter.

List Price...............$9.00

153

1932

MODERNE

A handsome case in Butler Satin finish, unusually thin with an 8-day jeweled movement, two main springs, radium numerals and hands and stem wind and set with back regulator. Easily installed on dash or over the windshield. Size 3½ x 2⅞ inches. Radium Dial.

Radium Dial List Price......$12.50

AUTOMOBILE CLOCKS

154

1937

No. 95 AV—30-Hour Pull Wind movement with back set. Orange and white dial of hexagonal design photographed on mirror. White enameled hands. Selected plate glass mirror with beveled edges. Size, 2½" x 7".

156

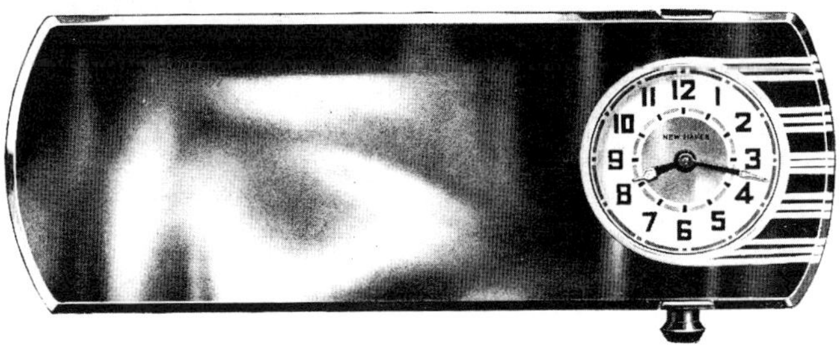

1937

No. 99 AV—30-Hour Pull Wind movement with back set. White and grey dial with black numerals. Black hands with red tips. Decorative, selected plate glass mirror with beveled edges. Size, 2½" x 7".

158

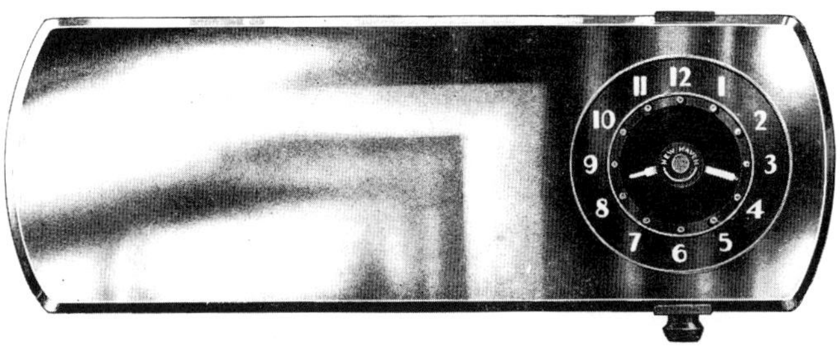

1937

No. 92 AV—30-Hour Pull Wind movement with back set. Attractive orange and white dial photographed on mirror. White enameled hands. Selected plate glass mirror with beveled edges. Size, 2½" x 7".

159

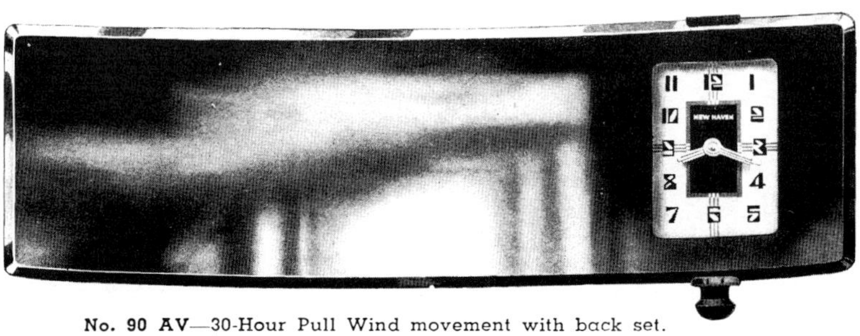

1937

No. 90 AV—30-Hour Pull Wind movement with back set. White and grey dial with black center panel and numerals. Selected plate glass, Convex mirror with beveled edges. Size, 2½" x 8".

155

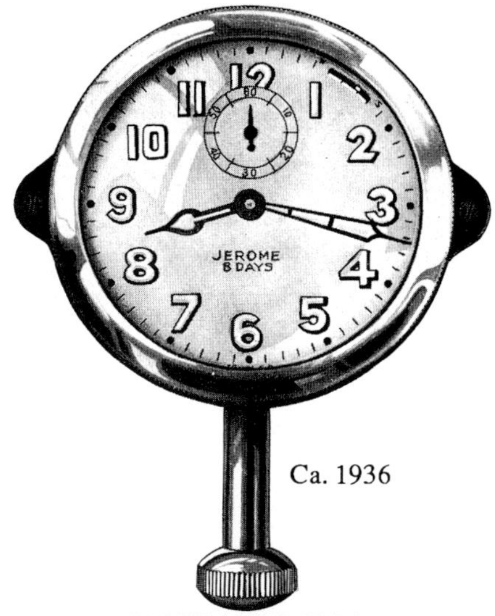

Ca. 1936

JEROME AUTO CLOCK

8 Day, Stem Wind, Dependable Movement, Two Springs, Thin Model, Visible Regulator, Gun Metal Case, Nickeled Sash and Stem. For use over Windshield or on Instrument Board, easy to attach, no holes to cut.

Plain Dial $12.50
Radium Dial 14.00

157

1937

JEROME—Designed for installation on the dash or over the windshield. 8-Day movement—Stem Wind and Set. Silvered and black dial with white enameled hands. Size, 2½" in diameter.

159-A

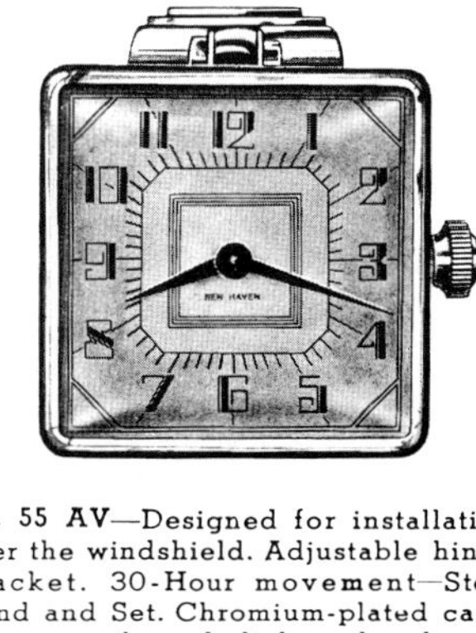

1937

No. 55 AV—Designed for installation over the windshield. Adjustable hinge bracket. 30-Hour movement—Stem Wind and Set. Chromium-plated case; two-tone silvered dial; and unbreakable crystal. Size, 2½" square. Plain or radium dial.

AUTOMOBILE CLOCKS

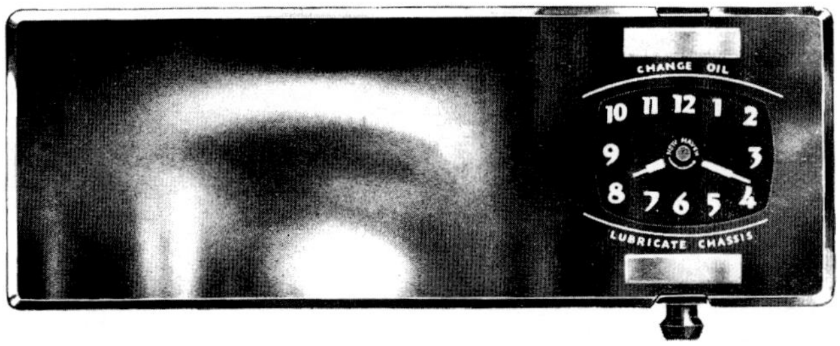

160

1937

No. 96 *AV*—30-Hour Pull Wind movement with back set. Black dial with white numerals and hands. Frosted panels on mirror for record of oil changing and chassis lubrication. Size, 2½" x 7".

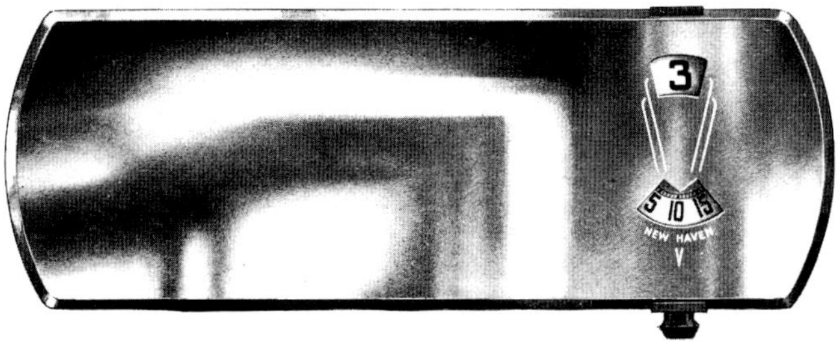

162

1937

No. 81 AV—30-Hour, Jump Hour, Pull Wind movement with back-set. Attractive. modernistic design in white on mirror. Selected plate glass mirror with beveled edges. Size of mirror, 2½" x 7".

163

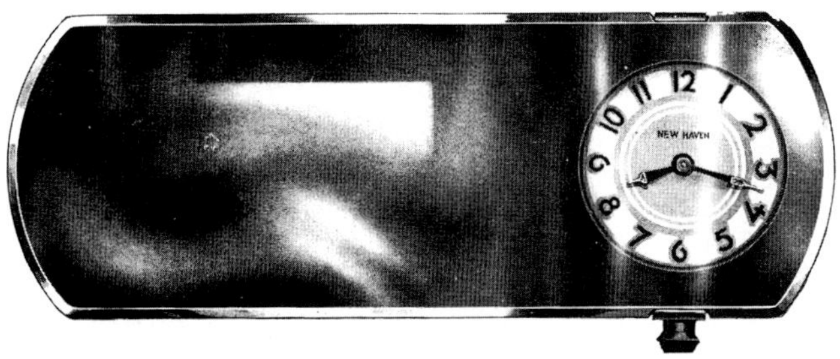

1937

No. 93 *AV*—30-Hour Pull Wind movement with back set. White and grey dial with black numerals. Black hands with red tips. Selected plate glass mirror with beveled edges. Size, 2½" x 7". ,

165

1937

No. 17 AV—30-Hour movement—Stem Wind and set. Black dial with white numerals and white enameled hands. Beveled-edge, plate glass mirror in non-glare gunmetal or clear. Size, 2½" x 7".

Mirrors are regularly furnished in non-glare gunmetal finish.

161

1937

No. 79 AV—A 30-Hour header-bar clock with adjustable hinge bracket for installation over the windshield. Smart, orange and white dial of hexagonal design photographed on mirror. White enameled hands. Selected plate glass mirror with beveled edges. Size, 2½" square.

164

1937

No. 73 *AV*—A 30-Hour header-bar clock with adjustable hinge bracket for installation over the windshield. Attractive, orange and white dial photographed on mirror. White enameled hands. Selected plate glass mirror with beveled edges. Size, 2½" square.

Mirror Brackets Available on Request.

(Specify make and year of cars when ordering.)

AUTOMOBILE CLOCKS

166

No. 55AV - 6A (Plain) List $2.45

30-Hour movement—Stem Wind and set. Adjustable hinge bracket for installation over the windshield. Chromium-plated case; two-tone silvered dial with black numerals and hands. Unbreakable crystal. Size: 2½" square. Packed singly. Shipping weight: 11 oz.

No. 55AV - 7A (Radium) List $3.00

Same as the above but with radium numerals and hands.

Ca. 1927

167

Ca. 1932

No. 73 AV-1 - 30-Hour Pull Wind clock with adjustable hinge bracket for center winding strip or over-head installation. Orange and white photographic dial. White enameled hands. Selected, non-glare Gunmetal plate glass mirror with beveled edges. Size: 2½ inches square.

168

Ca. 1927

No. 2AV (Gunmetal or Champagne) List $3.75

30-Hour Pull Wind movement with back-set. Black numerals on a white track. White hands with red tips. Frosted panels on mirror for record of oil changing and chassis lubrication. Selected, non-glare Gunmetal or Champagne plate glass mirror with beveled edges. Size: 2½" x 7". Packed singly. Shipping weight: 15 oz.

169

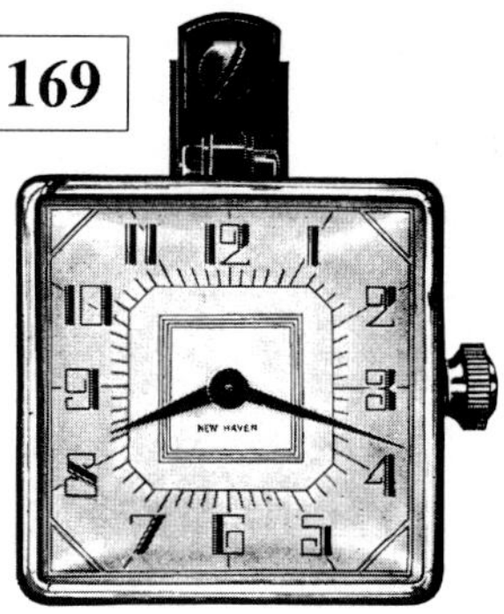

Ca. 1932

No. 55 AV-8A - 30-Hour movement—Stem Wind and set. Adjustable hinge bracket for center windshield strip or over head installation. Chromium-plated case; two-tone silvered dial with black numerals and hands. Unbreakable crystal. Size: 2½ inches square.

170

1911

THE AUTO WATCH

SIZE, No. 16
FINISH, NICKEL PLATED
CRYSTAL, HEAVY AND BEVELED
DIAL, WHITE ARABIC, WITH HEAVY FIGURES AND HEAVY HANDS. DIAL CAN BE INVERTED IF SO ORDERED
MOVEMENT, STEM WIND AND STEM SET
PRICE..LIST, $1.75

FOR USE IN TOILET CASES AS FITTED INTO CLOSED CARRIAGES AND LIMOUSINE AUTOMOBILES

171

1917

BELMONT
8-Day Time
Height, 4⅜ Inches   Width, 4⅜ Inches
Dial, Silver, 2¾ Inches

AUTOMOBILE CLOCKS

30-Hour Automobile Clocks
Compensated for Temperature changes.

172

Ca. 1932

No. 1001 MIRROR CLOCK COMBINATION

A popularly priced mirror clock with a 30-hour movement and artistic dial and hands. The plate glass is first quality, polished and beveled. Available in non-glare gunmetal, gold, or clear glass. Size 2½ x 7 Inches.

	List Price
GOLD	$2.95
GUNMETAL	2.75
CLEAR	2.75

173

Ca. 1932

ASCOT

Plain Dial List Price...$2.35 Radium Dial List Price ..$3.25

An unusually attractive 30-hour clock with a chromium trimmed case, unbreakable crystal, novel dial design of unusual visibility and convenient stem wind and set. 2½ inch square case.

All New Haven Automobile Clocks are made of the best available materials and are guaranteed against defects in material and workmanship.

174

Ca. 1927

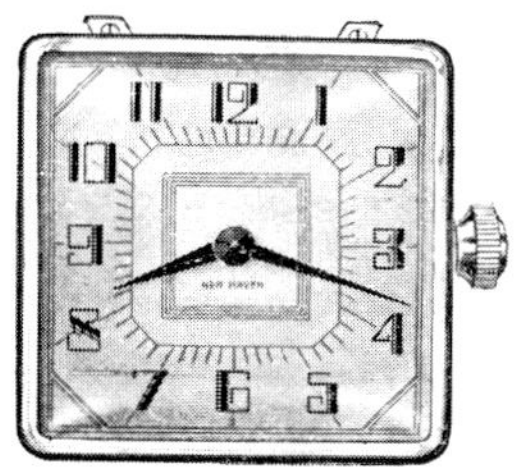

Ascot - A popular model for installation on dash or over the windshield, with 30-hour movement, two-tone silvered dial, chromium case and unbreakable crystal Size, 2½ inches square Plain or Radium Dial.

175

1932

176

1932

AUTO Radium Dial

List Price $2.75

Black dial with radium numerals and hands.

AUTO Plain Dial

List Price $1.90

Enamel dial with black numerals and hands.

These clocks are equipped with unbreakable crystals, dull finished, chromium plated cases, and 30-hour movements with stem wind and set. They can be installed either on the dash or over the windshield. Only a screwdriver is required for installation.

AUTOMOBILE CLOCKS

177

8-DAY
Movements

30-HOUR
Movements

Ca. 1927

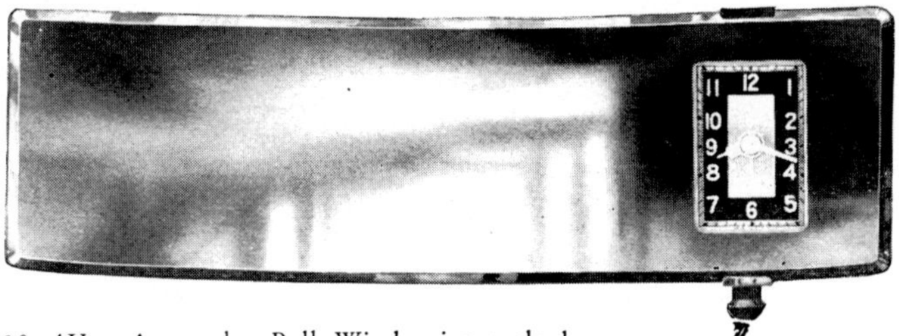
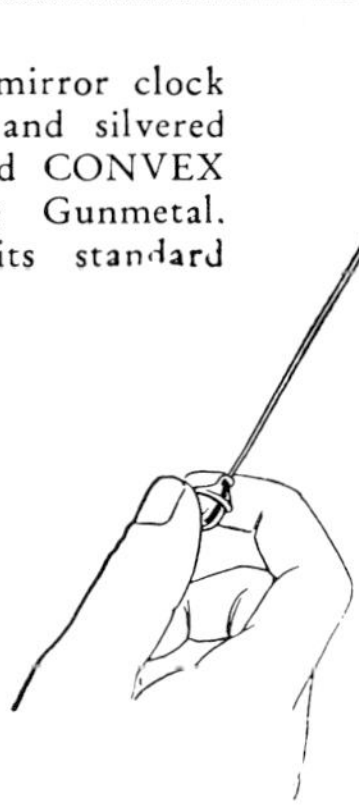

30 AV - A popular Pull Wind mirror clock with 30-hour movement, black and silvered dial, radium hands, back set, and CONVEX mirror. Available in non-glare Gunmetal. only. Size, 2½ x 8 inches. Fits standard brackets.

178

31 AV - A modern Jump Hour, Pull Wind mirror clock combination with 30-hour movement, back-set, and beveled, plate glass mirror. Available in non-glare Gunmetal, Gold or Clear Glass. Size, 2½ x 7 inches.

179

61 AV - A new 8-day, Jump Hour, Pull Wind mirror clock combination, with a high grade, accurate movement, back-set, and plate glass, beveled edge mirror. Available in Clear, non-glare Gunmetal, or Gold. Size, 2½ x 7 inches.

180

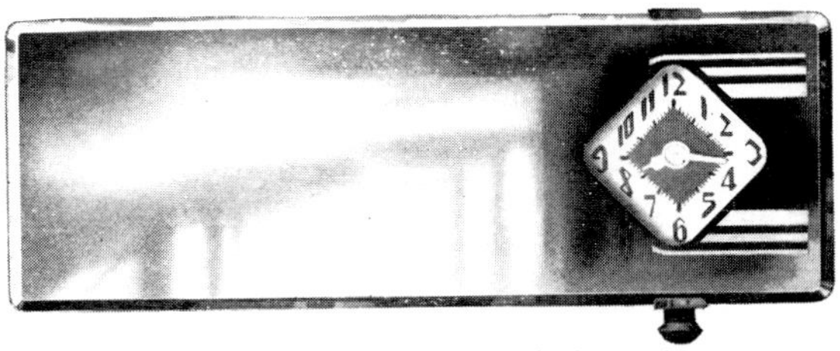

32 AV - A Pull Wind mirror clock combination with 30-hour movement, radium hands and bronze dial, back-set, and beveled, plate glass mirror. Available in non-glare Gunmetal, Gold or Clear Glass. Size, 2½ x 7 inches.

181

62 AV - A high grade, 8-day, pull wind movement with two-tone silvered and black dial, radium hands, and selected plate glass, beveled edge mirror. Available in Clear, non-glare Gunmetal, or Gold. Size, 2½ x 7 inches.

182

33 AV - A Pull Wind mirror clock as illustrated with 30-hour movement, radium hands and black and silvered dial. First quality plate glass, beveled edges. Available in non-glare Gunmetal, Gold or Clear Glass. Size, 2½ x 7 inches.

34 AV - Same as the No. 33 AV but with conventional Stem Wind, and white hands.

183

63 AV - An 8-day, Pull Wind mirror clock with black and bronze dial, radium hands, back set, and CONVEX mirror. Available in non-glare Gunmetal, only. Size, 2½ x 8 inches. Fits standard brackets.

AUTOMOBILE CLOCKS

184

No. 7001
New 8-Day Model
With Aviation Dial
Size 2½ x 8 inches

List Price $11.00

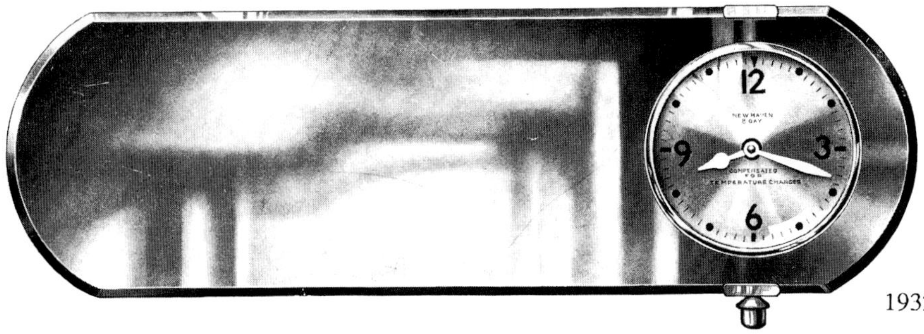

1932

Sized to fit the present bracket

185

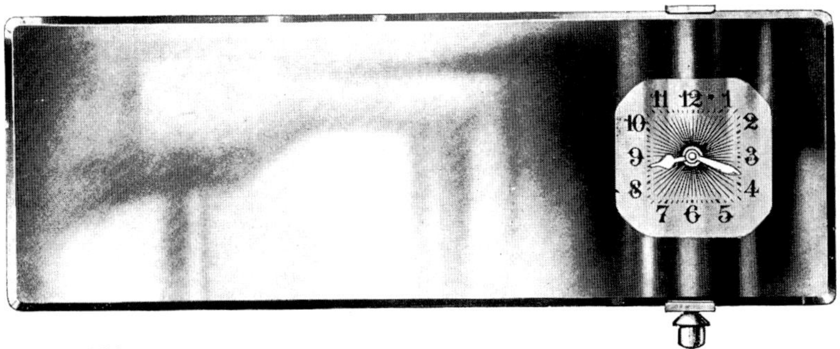

1932

No. 2001
New 30-Hour Model
Back Set, Size 2½ x 7 inches

List Price Plain Dial $4.00
" " Radium " $5.00

Bicycle clock complete with slotted bracket attaches clock to bicycle see #2117 page 490

Bicycle Clock

186

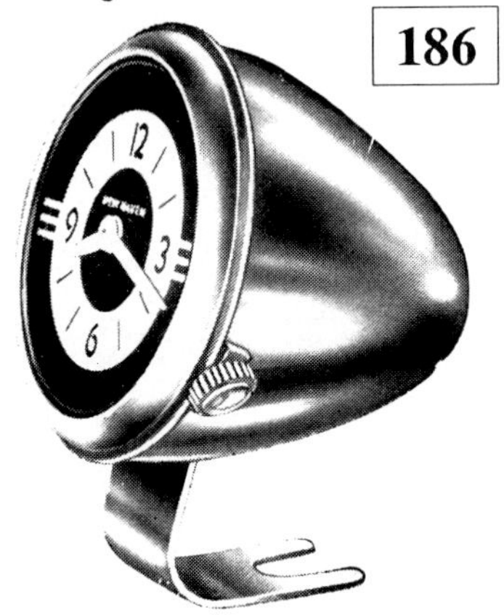

Ca. 1927

BICYCLE CLOCK

Slotted bracket attaches clock to bicycle by slipping under handle-bar post bolt. Chrome, streamlined case. Visible aviation type dial. Unbreakable crystal. Stem wind, stem set 30-hour movement. Size: Width, 2 inches. Height, 2½ inches. Depth, 2⅛ inches. Packed in lots of 24. Shipping weight: 8 lbs.

RETAIL $1.95

A Remarkable ..... Convenience

A revolutionary change in winding methods that eliminates the bothersome winding stem and simplifies a tedious job. A few pulls on the small cable complete the winding operation.

The movement is the same dependable time-piece, compensated for temperature changes and built to withstand the vibrations of motoring.

All mirrors are made from the Finest Plat Glass and are available in Non-Glare Gun Metal, Gold or clear Glass.

187

Ca. 1927

No. 77AV List $3.35

30-Hour Pull Wind header-bar clock in heavy plate glass. Adjustable hinge bracket for installation over the wind-shield. Black numerals on a rose-tinted grey glass background. Grey hands. Size: 2½" square. Packed singly. Shipping weight: 13 oz.

BANJO CLOCKS

188

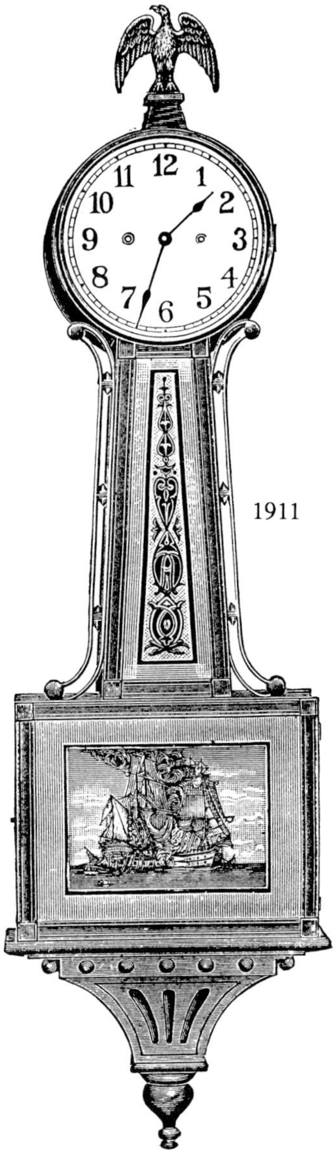

1911

WILLARD

HEIGHT ....44½ INCHES
WIDTH ....12½ INCHES
DIAL ....8 INCHES, SILVER FINISH
CASE, SOLID MAHOGANY, WITH CAST GOLD PLATED TRIMMINGS
MOVEMENT, 30 DAY TIME PIECE
PRICE..LIST, $48.00

This clock was designed and made by Edw. K. Willard, a skilled clock-maker in London, England, in 1801. Its beauty and practical worth attracted the attention of an American clock manufacturer, George Worthington, who was in London at the time, and securing the patent rights for the United States, he introduced the Willard to the American trade in 1812. We are now manufacturing this clock in limited numbers for special trade.

189

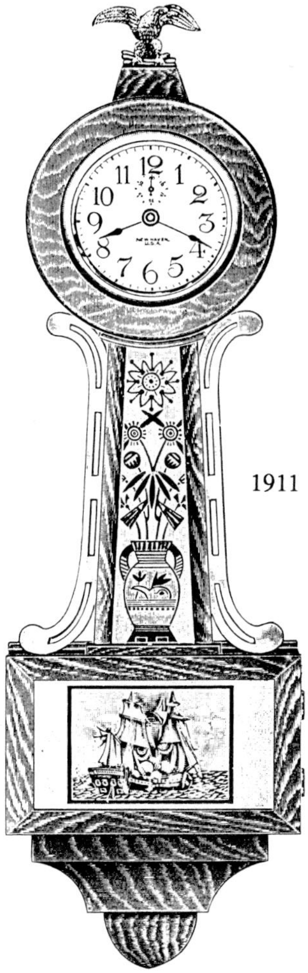

1911

WILLIS

HEIGHT...........16⅞ INCHES
WIDTH..................5 INCHES
DIAL, PORCELAIN...3 INCHES
CASE, SOLID MAHOGANY, WITH CAST BRASS SIDE ORNAMENTS AND DECORATED GLASS PANEL
MOVEMENT, FINE QUICK-BEAT 8 DAY TIME PIECE
PRICE...........................LIST, $10.15

190

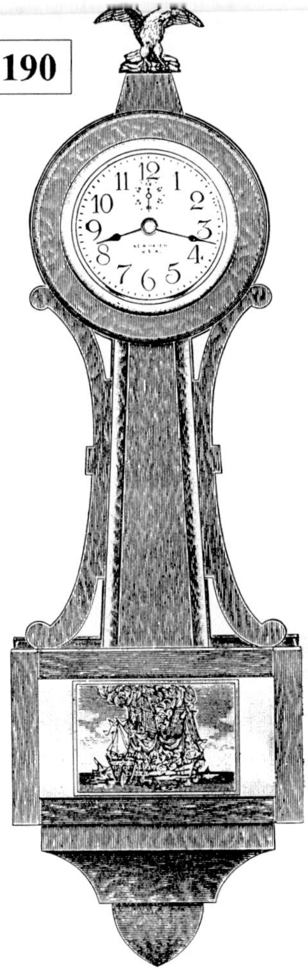

1911

WILLMAR

HEIGHT ....16⅞ INCHES
WIDTH ....5 INCHES
DIAL, PORCELAIN..3 INCHES
CASE, SOLID MAHOGANY
MOVEMENT, FINE QUICK-BEAT 8 DAY TIME PIECE
PRICE ....LIST, $8.75

191

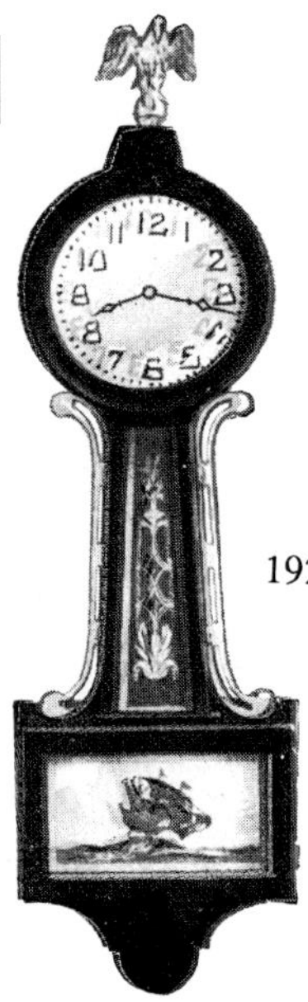

1928

WILLIS

Height, 17⅞ inches. Width, 5½ inches. Convex Glass, Gold Ornaments. Solarchrome Panels, Solid Mahogany Case.
8 Day Jeweled with Raised Gold Numerals $24.00
8 Day, with 3½ inch Silver Dial 20.60

WILLIS

BANJO CLOCKS

192

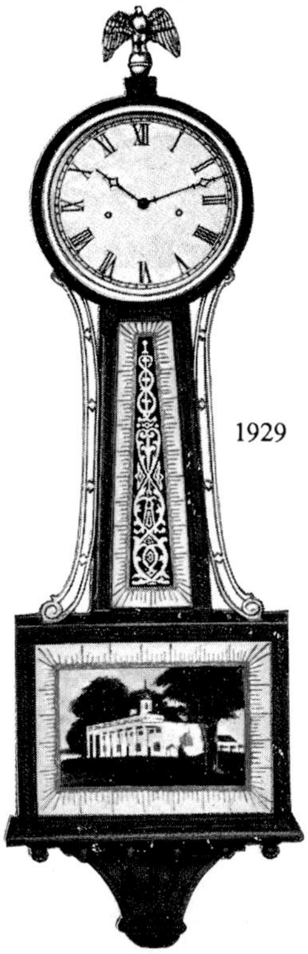

1929

WILLARD

Height, 41 inches. Width, 12¼ inches. 8-inch Antique Dial. Solid Mahogany Case, Convex Crystal, Attractive Side and Top Ornaments, Decorated Panels.

Strike, 8-day Pendulum, List $59.10

Timepiece, 8-day Weight, List $72.80

193

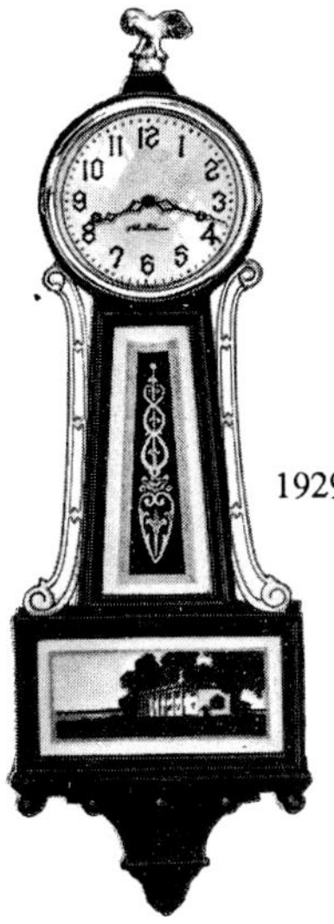

1929

"WINSOME" Time

Height, 27 inches. Width, 8½ inches. Dial 5½ inches. Solid Mahogany Case, Antique Finish, Convex Crystal. Attractive Side and Top Ornaments. Decorated Panels. 8-day Pendulum Timepiece.

List, Silvered Dial $33.20

List, Raised Numerals . .$36.40

194

1929

"WINSOME" Strike

Height, 27 inches. Width, 8½ inches. Dial 5½ inches. Solid Mahogany Case, Antique Finish, Convex Crystal, Attractive Side and Top Ornaments. Decorated Panels. 8-day Pendulum Hour and Half-hour Strike.

List $37.80

195

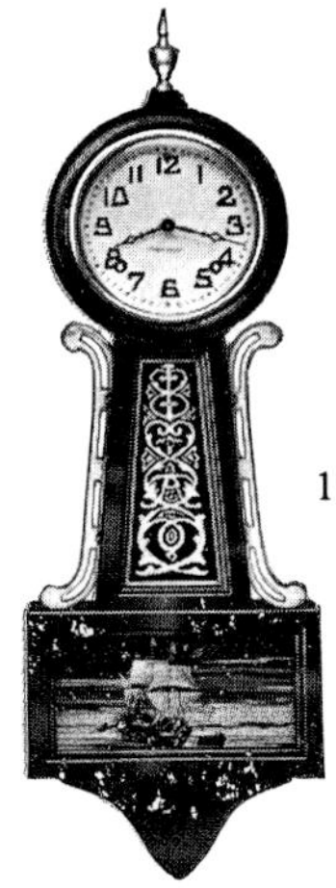

1929

WINETKA

Height, 18½ inches. Width, 6 inches. 3½-inch Silvered Dial, with raised Numerals. Attractive Side Ornaments and Top Piece. Decorated Panels, Bowed Crystal. Solid Mahogany Case, 8-day Pendulum Movement with Hour and Half-hour Strike. Colors, Rose, Green, Blue, Yellow or Mahogany.

List $28.00

BANJO CLOCKS OF SOLID MAHOGANY

Hand-Rubbed Duco Finish

196

1929

MYRAN

Height, 12½ inches. Width, 7½ inches. 5½-inch Antique Dial. Polished Sash, Solid Mahogany Case with Burled Maple Panel. 8-day Jeweled Lever Timepiece.

List $30.00

197

1929

WILLIS

Height, 18 inches. Width, 5 inches. 3½-inch Silvered Dial. Solid Mahogany Case, Decorated Panels, Attractive Side and Top Ornaments. Colors, Green, Blue or Mahogany. 8-day Jeweled Lever Timepiece.

List, Raised Numerals . . .$24.00

List, Plain Numerals $21.00

198

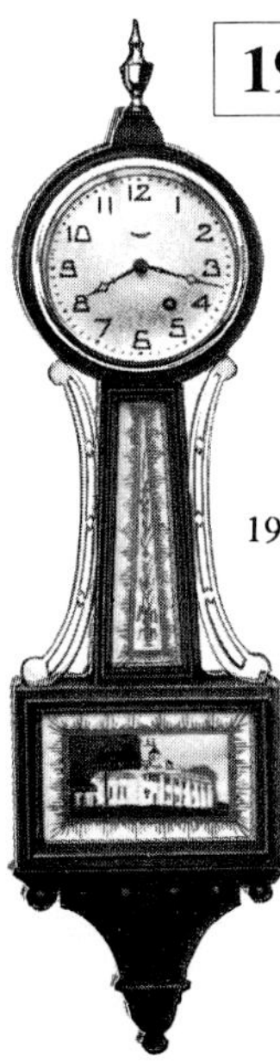

1929

WILMOT

Height, 20 inches. Width, 5 inches. 3¾-inch Antique Dial. Cast Sash, Attractive Side Ornaments and Top Piece. Decorated Gold Leaf Panels, Solid Mahogany Case, 8-day Jeweled Lever Timepiece.

List $36.40

199

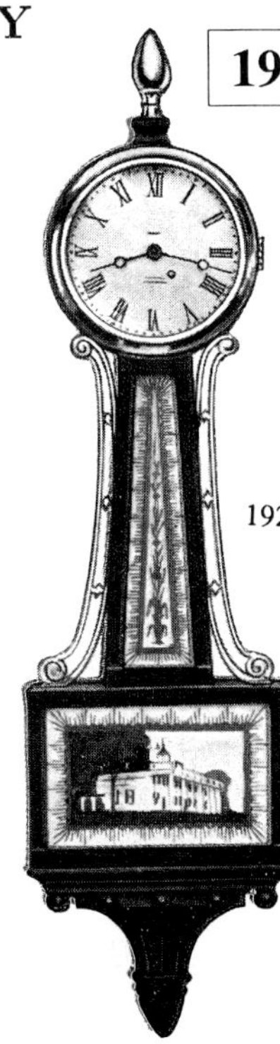

1929

WHALLEY

Height, 32¼ inches. Width, 8¼ inches. 6-inch Antique Dial. Heavy Cast Sash, Attractive Side Ornaments and Top Piece, Decorated Gold Leaf Panels, Solid Mahogany Case, 8-day Jeweled Lever Timepiece.

List $59.10

BANJO CLOCKS

201

200

WINDROW

LIST, $19.65
HEIGHT, 23½ INCHES.
WIDTH, 7¾ INCHES.
CAST BRASS SIDE ORNAMENTS.
DECORATED PANEL.

1913

WALDRON

LIST, $19.65
HEIGHT, 24¾ INCHES.
WIDTH, 8½ INCHES.
CAST BRASS SIDE ORNAMENTS.
DECORATED PANEL.

1913

204

203

202

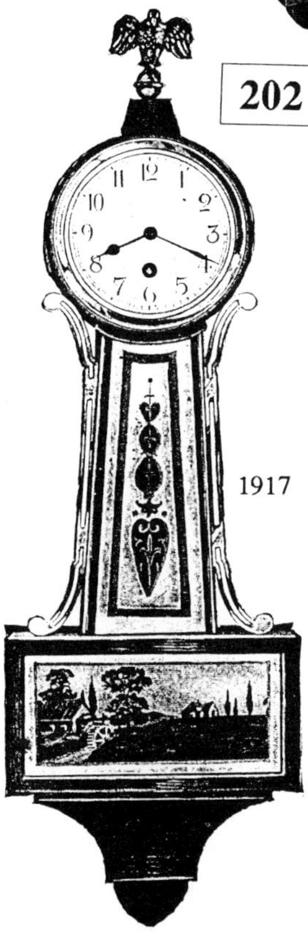

1917

WINDROW, No. 2
8-Day Lever Timepiece
Height, 23½ in. Width, 7¾ in.
Solid Mahogany Case with Cast Metal Side Ornaments and Decorated Glass Panels.
Dial, 5-inch Porcelain with French Cast Sash and Beveled Bowed Glass.

1913

WALTON

LIST, $18.90
HEIGHT, 24¾ INCHES.
WIDTH, 8½ INCHES.

WINSTON

LIST, $18.90
HEIGHT, 23½ INCHES.
WIDTH, 7¾ INCHES.

1913

BANJO CLOCKS

EIGHT-DAY CLOCKS. FINE, QUICK-BEAT SPRING TIMEPIECES.
4-INCH PORCELAIN DIALS. FRENCH CAST SASH, BEVELED GLASS.
DECORATED GLASS PANELS. SOLID MAHOGANY CASES

205

1913

WARDROP

LIST, $19.65
HEIGHT, 24¼ INCHES.
WIDTH, 9¼ INCHES.
DECORATED PANEL.

208

1913

WARREN

LIST, $18.90
HEIGHT, 24¼ INCHES.
WIDTH, 9¼ INCHES.

206

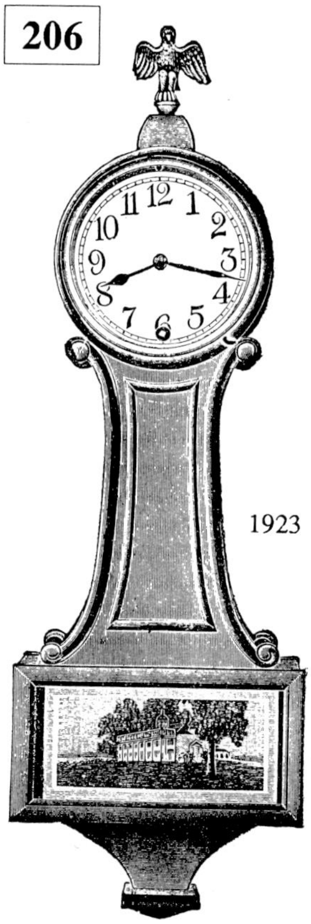

1923

WELTON

Height, 25½ inches; width, 8¼ inches. 6-inch silver dial. Solid mahogany. 8-day pendulum time.
No. 9765 Each.................$34.80

207

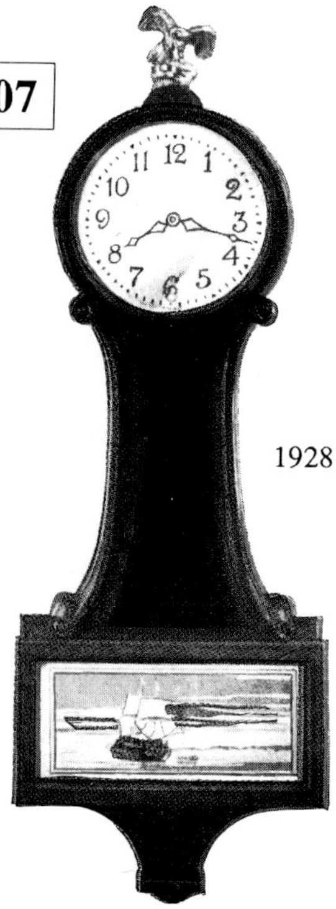

1928

WELTON

Height, 25¾ inches. Width, 8⅛ inches. Dial, 6 inches. Beveled Glass, Gilt Eagle. Antique Mahogany, Black Panels.
8 Day Pendulum Time......$33.20

209

1923

WARING

Height, 38 inches; width, 11½ inches. 8-inch silver finish dial case, mahogany finish. Decorated glass panels. Movement, 8-day half-hour strike cathedral gong, or 8-day time.
No. 9761 Time, each............$27.50
No. 9762 Cathedral Gong, each.. 29.70

BANJO CLOCKS

210

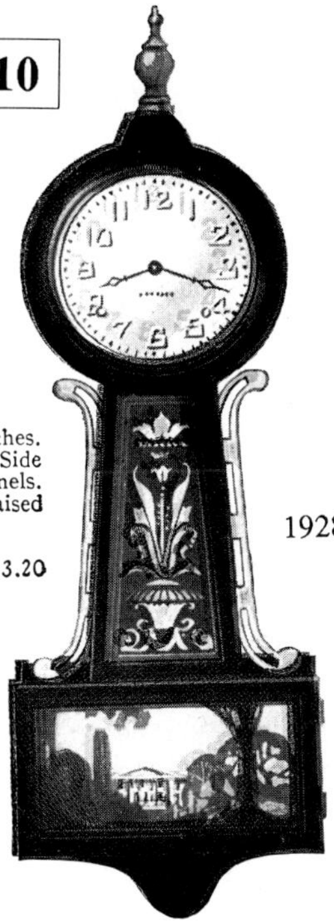

1928

WINETKA
Height, 18 inches. Width, 6 inches. Solid Mahogany Case. Gilt Side Ornaments. Solarchrome Panels. 3½ inch Silver Dial with Raised Gold Numerals. Bowed Glass.
8 Day Pendulum, Hour and Half-Hour Strike $33.20

211

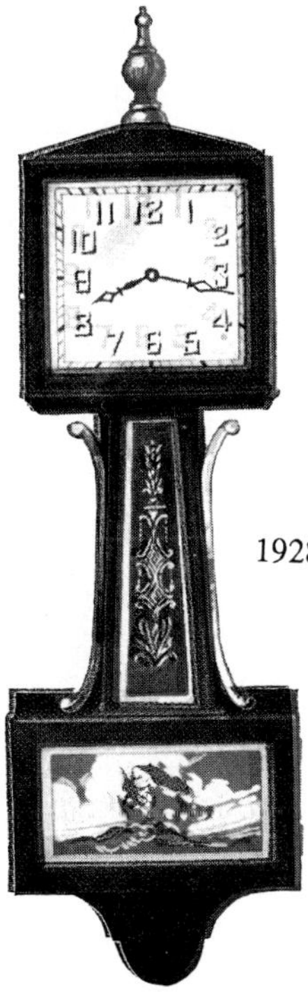

1928

WINONA
Height, 17½ inches. Width, 5 inches. 3½ inch Square Silver Dial with Raised Gold Numerals. Bowed Glass, Solid Mahogany Case, Solarchrome Panels. Gilt Side Ornaments.
8 Day Jeweled Timepiece......$24.75

212

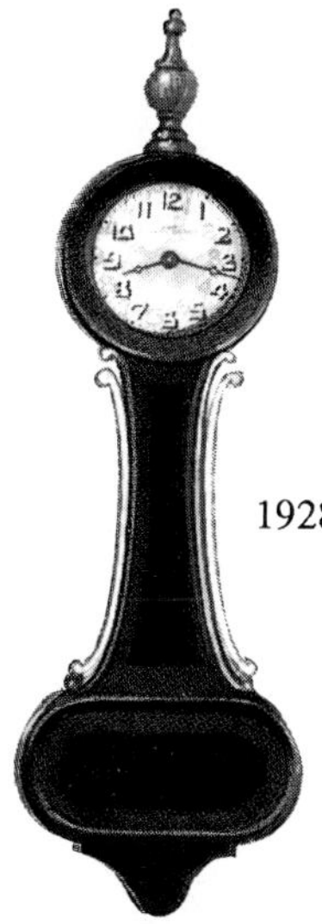

1928

WINNER
Height, 13½ inches. Width 4 inches. Bowed Crystal. Antique Mahogany.
8 Day Jeweled, with Raised Gold Dial $18.20
8 Day, with 2½ inch Silver Dial 15.25

213

1920

WILSON
30-Day Spring Timepiece
Height, 40½ Inches Width, 13¼ Inches
8-inch Silver Finish Dial
Decorated Glass Panels
Ornamental Cast Metal Side Pieces, Gold Plated
Solid Mahogany Case

214

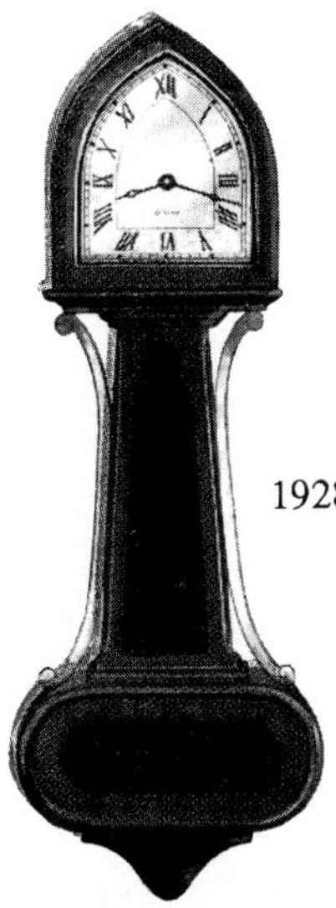

1928

WAVERLY
Height, 12¾ inches. Width 4 inches. Gold Dial, 2½ inches. Bowed Crystal. Antique Mahogany, Black Panels.
8 Day Lever Timepiece........$16.60

215

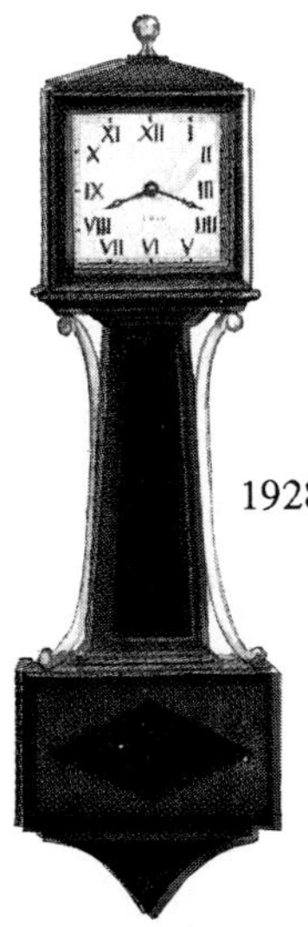

1928

WORTHY
Height, 13⅞ inches. Width, 3¾ inches. Square Gold Dial, 2½ inches. Bowed Crystal. Antique Mahogany, Black Panels.
8 Day Lever Timepiece........$16.60

216

1928

RICHELIEU
HEIGHT, 21¼ inches
WIDTH, 6¼ inches
CASE, Antique Finish
DIAL, 3½ inch Silver
12 DAY Movement

BANJO CLOCKS

217

Ca. 1932

WHITNEY

A really handsome clock in a mahogany finished case, 30½ inches high, 9½ inches wide, with artistic panel designs and ornaments. Eight-day pendulum movement with 6-inch silvered dial. Available with either chime or No. 2 duo-rod strike movement.

218

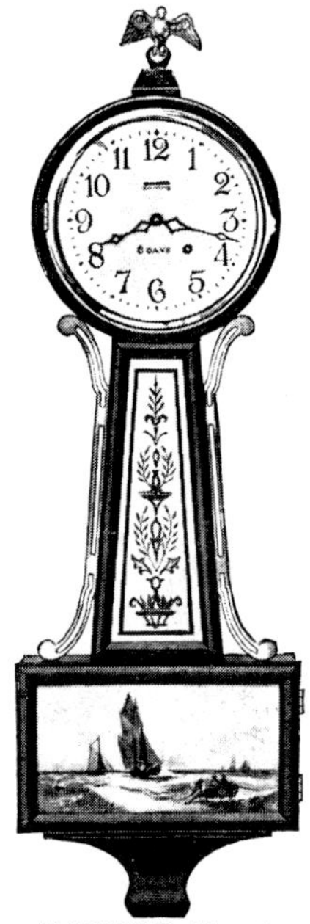

Ca. 1932

***WARWICK* Timepiece**

A gracefully modern clock, beautifully proportioned. Mahogany finished case with decorative panels and side ornaments. Eight-day lever timepiece movement. 24½ inches high, 7⅜ inches wide, 5½-inch cream colored dial.

219

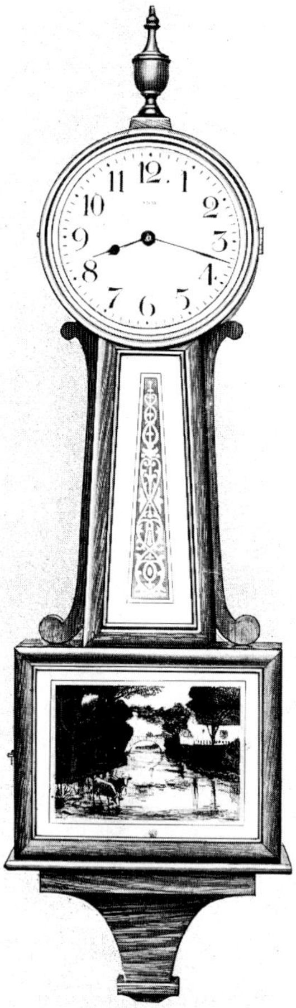

1923

WARING

HEIGHT, 40 inches WIDTH, 11½ inches
8 inch Silver Dial
Mahogany Finish Case
Decorated Glass Panels
MOVEMENT, 8 Day Half-hour Strike, Cathedral Gong, or 8 Day Time.

220

1923

WAYLAND

8 Day Pendulum Timepiece
HEIGHT, 26 inches WIDTH, 9½ inches
6 inch Silver Dial with Beveled Convex Glass
Solid Mahogany Case

221

1923

WASHINGTON

30 Day Pendulum Timepiece
HEIGHT, 41 inches WIDTH, 12¼ inches
8 inch Silver Dial
Hand Painted Glass Panels
Ornamental Cast Metal Side Pieces, Gold Plated. Solid Mahogany Case

JEROME & COMPANY

New Haven, Conn.

"Jerome & Company" was a trademark adopted by the New Haven Clock Company in the 1870's and was used by them on some clock models until at least 1900. This firm's double dial calendar clocks often carried this trademark.

222

1885

PLOVER - PERPETUAL CALENDAR.

Height, 18 inches. Dial, 6 Inches. Calendar Dial, 4 inches.
8 Day Strike, Cathedral Gong....$20.25

CALENDAR CLOCKS

NEW HAVEN
Jerome & Co.

223

1885

CABINET No. 4
WITH PERPETUAL CALENDAR.
ASH OR MAHOGANY. WHITE DIAL.
Height, 18 ¾ inches. Dial, 5 Inches.

8 Day Strike, Ash	$11 25
8 " " Alarm, Ash	11 85
8 " " Cathedral Gong, Ash	11 75
8 " " " " Alarm, Ash	12 35
8 " " Mahogany	11 65
8 " " Alarm, Mahogany	12 25
8 " " Cathedral Gong, Mahogany	12 15
8 " " " " Alarm, Mahogany	12 75

224

1885

CABINET No. 5
WITH PERPETUAL CALENDAR.
OAK OR MAHOGANY. WHITE DIAL.
Height, 19 ½ inches. Dial, 5 Inches.

8 Day Strike, Oak	$12 00
8 " " Alarm, Oak	12 60
8 " " Cathedral Gong, Oak	12 50
8 " " " " Alarm, Oak	13 10
8 " " Mahogany	12 40
8 " " Alarm, Mahogany	13 00
8 " " Cathedral Gong Mahogany	12 90
8 " " " " Alarm, Mahogany	13 50

CALENDAR CLOCKS

NEW HAVEN
Jerome & Co.

225

PLEASE NOTE:

Clocks marked "Jerome & Co." or having this trademark were manufactured by the New Haven Clock Company, not by the earlier firm of Chauncey Jerome. "Jerome & Co." was a trademark on many New Haven models in the 1880's.

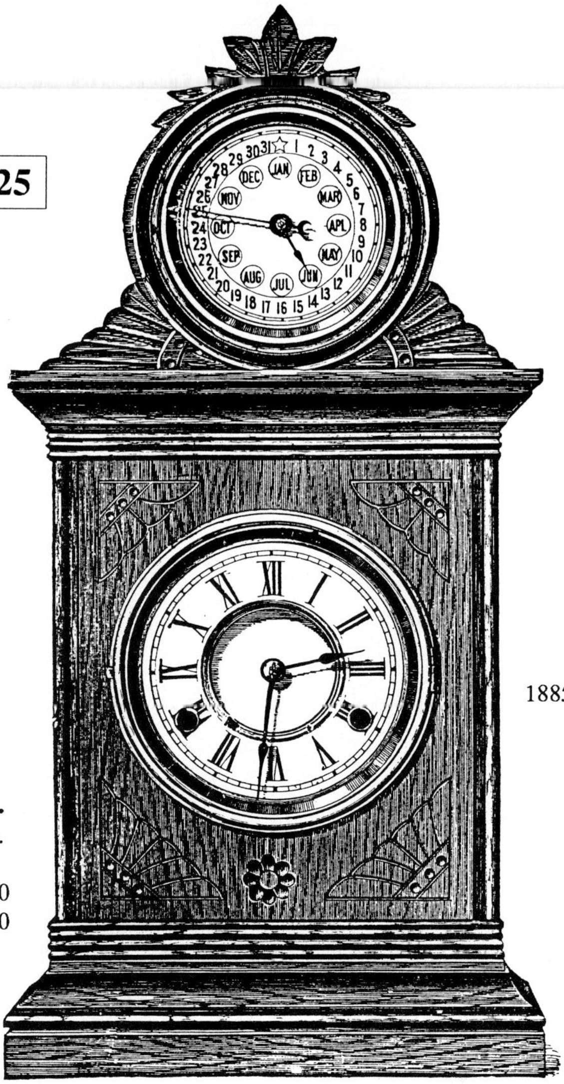

1885

CURLEW

CURLEW.
PERPETUAL CALENDAR.
Height, 18 ½ inches. Dial, 5 Inches.
Calendar Dial, 4 inches.
8 Day Strike..............................$17.70
8 Day Strike, Cathedral Gong..$18.20

226

1885

MALLARD.

MALLARD.
PERPETUAL CALENDAR.
OAK OR MAHOGANY.
Height, 13 inches. Width, 14 ½ inches.
Time and Calendar Dial, 4 inches.
8 Day Time, Calendar....$15.00

CALENDAR CLOCKS

NEW HAVEN

Jerome & Co.

227

1885

CABINET No. 7
WITH PERPETUAL CALENDAR.

ASH.

Height, 21 ⅛ inches. Dial, 5 Inches.

8 Day Strike	$11 70
8 " " Alarm	12 30
8 " " Cathedral Gong	12 20
8 " " " " Alarm	12 80

228

CABINET - WITH PERPETUAL CALENDAR.

Oak. 8-Day Time & Strike **New Haven** movement. Height, 21 ¾ inches. Width, 10 ¾ inches. Time dial 4 ½ inches. Calendar dial 3 ⅜ inches. **Jerome** calendar movement.

{Roy Burlingame Collection}

{Photo courtesy of Michael G. Lowe}

229

Close-up of **Jerome** perpetual calendar mechanism.

{Roy Burlingame Collection}

{Photo courtesy of Michael Lowe}

CALENDAR CLOCKS

NEW HAVEN
Jerome & Co.

230

Side view of Cabinet No 6 Calendar.
{Larry B. Funk Collection}

231

CABINET No. 6
WITH PERPETUAL CALENDAR.
(Cabinet No. 6 without Calendar see #736 page 195)
Oak case. Eight Day Time and Strike. Height, 21⅜ inches. Base width, 12¾ inches. Side to side no molding, 9¼ inches, with molding 10¾ inches. Top width, 11⅛ inches. Time Dial, 5 inches. Calendar Dial, 4 inches.
{Larry B. Funk Collection}

232

Rear view of Cabinet No. 6 Calendar.
{Larry B. Funk Collection}

233

Close up of New Haven clock movement.
{Larry B. Funk Collection}

CALENDAR CLOCKS

NEW HAVEN

Jerome & Co.

The year entered beside each clock is the year of the catalog from which the illustration was taken. Some clocks may have been produced a few years earlier.

234

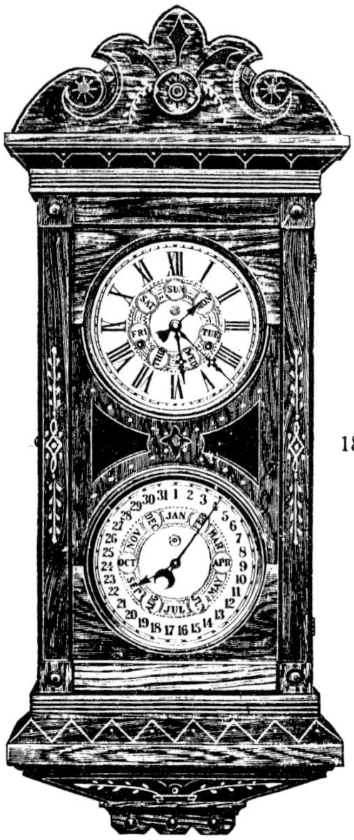

1886

DNEISTER.

WALNUT, CALENDAR.

This Base can be removed for mantel use.

Height, 37½ inches. Dials, 8 inches.

235

1886

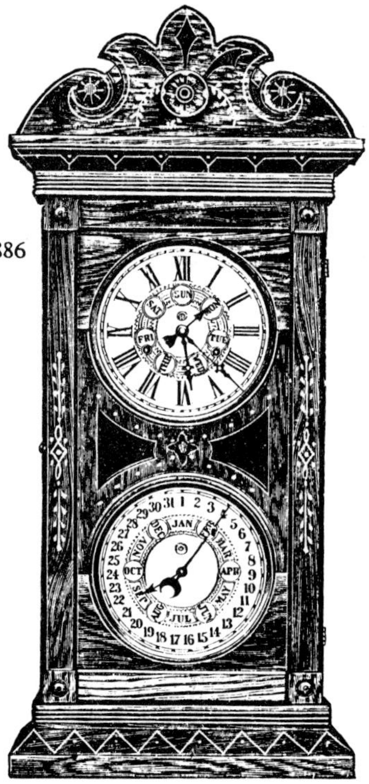

DNEISTER.

WALNUT, CALENDAR.

Showing Clock with Base off.

Height, 34 inches. Dials, 8 inches.

Code Word.		Price.
(Pad)	8 Day Time	$18.75
(Paradise)	8 " " Spanish Dial	18.75
(Pagan)	8 " Strike	20.30
(Plumb)	8 " " Spanish Dial	20.30

This Clock is furnished with an Adjustable Base,
which can be removed for mantel use,
or put on when desired as a Hanging Clock.
We will fit any of the Black Walnut Clocks
with the Wire Bell, if so ordered.

236

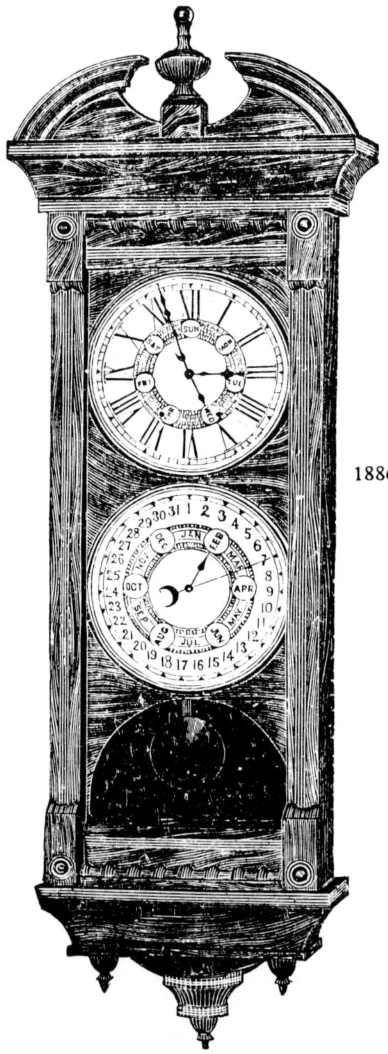

1886

ROKEBY (Spring).
Perpetual Calendar.

WALNUT.

Height, 63 inches. Dials, 12 inches.

Code Word.		Price.
(Punch)	8 Day Time, Spring	$39.00
(Pup)	8 " " Weight	44.00
(Milk)	8 " " Perpetual Calendar	45.00
(Mink)	8 " Strike, " "	46.50

CALENDAR CLOCKS

NEW HAVEN
Jerome & Co.

237

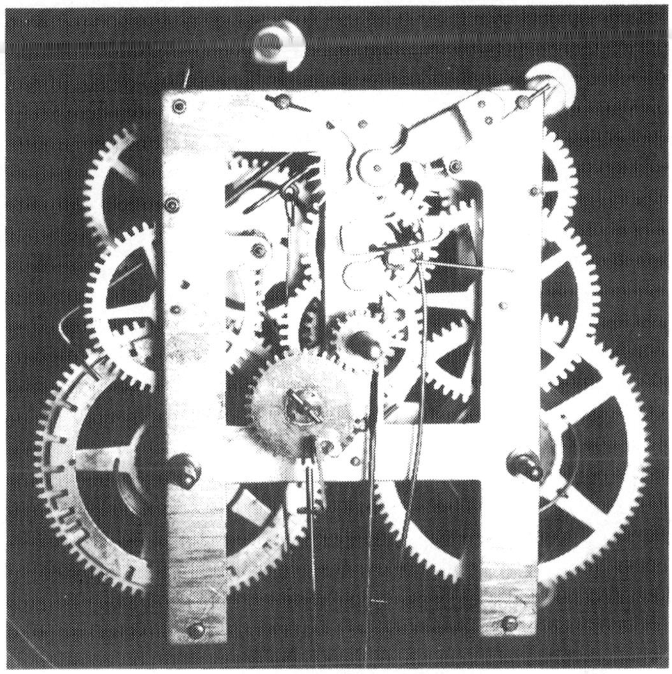

8-Day Time & Strike movement of
Register Mantel Calendar figure #239
{Larry B. Funk Collection}

238

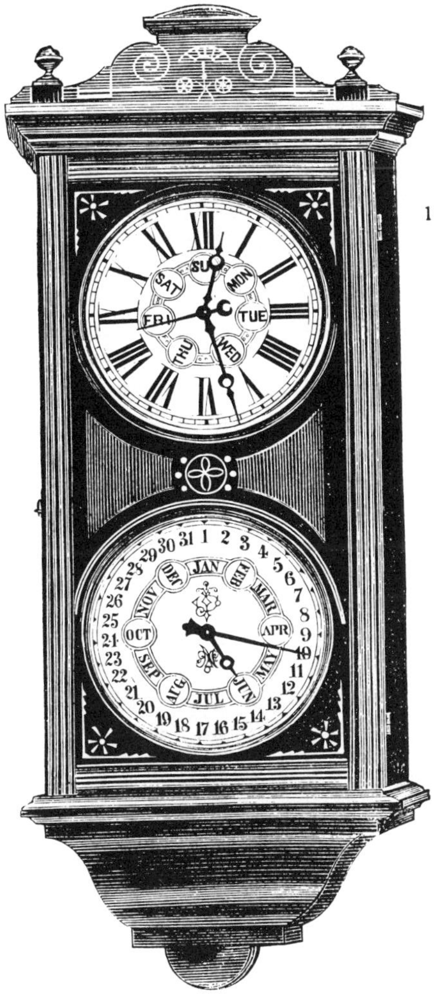

1880

REGISTER.

Hanging Calendar (Spring).

Walnut.

Height, 31 inches.
Dial, 8 inches.
8 Day Time.
8 Day Strike.
With Perpetual Calendar.

REGISTER MANTEL CALENDAR
Walnut. Eight day time. Eight day strike.
Height, 28 inches. Width, 14 inches.
Perpetual Calendar.

239

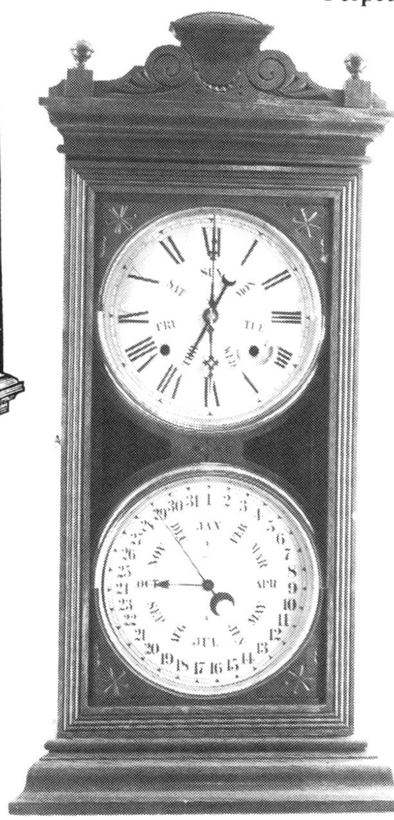

REGISTER
{Larry B. Funk Collection}

240

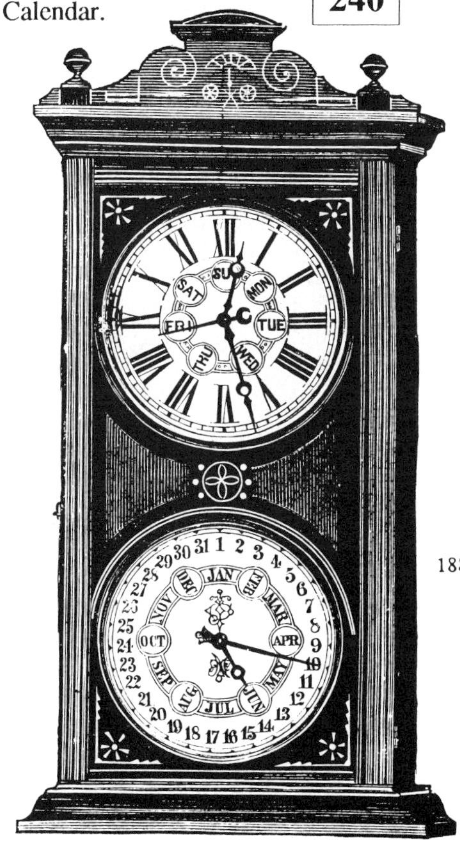

1880

REGISTER
From 1880 Catalog.

CALENDAR CLOCKS

NEW HAVEN

Jerome & Co.

PLEASE NOTE:

Clocks marked "Jerome & Co." or having this trademark were manufactured by the New Haven Clock Company, not by the earlier firm of Chauncey Jerome. "Jerome & Co." was a trademark on many New Haven models in the 1880's.

242

LEADER

Walnut. 8-Day Time & Strike. Height, 34 ½ inches. Depth, 8 ½ inches. Width, 19 ½ inches.

{Private Collection}

241

1886

HERON.

WALNUT VENEER (SOLID WALNUT CIRCLE).

Height, 31 inches. Dial, 12 inches.

Code Word.		Price.
(Tabular)	8 Day Time, Perpetual Calendar	$21.00
(Tacit)	8 " Strike, " " Gong	23.00

243

Dial removed showing movement of **Leader Calendar Clock.** {Private Collection}

CALENDAR CLOCKS

NATIONAL CALENDAR CLOCK COMPANY

(no place given)

Clocks bearing this label have very similar cases to the common "Fashion" models manufactured by the Southern Calendar Clock Company. Movements were manufactured by the New Haven Clock Company about 1890.

245

Fashion Calendar in walnut case. Height, 32 inches. Width, 16½ inches. 7 inch black dials. Eight-day time & strike New Haven clock movement with simple calendar mechanism attached to bottom dial. Letters "**FASHION**" on glass door in flat gold.
{Ben Gravolet Collection}

244

CROWN
Walnut. 8-Day Time & Strike.
Height, 34 ½ inches. Depth, 8 ½ inches.
Width, 19 ½ inches. {Private Collection}

246

Close-up of **Crown** calendar mechanism.
{Private Collection}

247

Close-up of Fashion Calendar dials with simple calendar mechanism. Top & bottom center circle reads: "**MANUFACTURED FOR NATIONAL CALENDAR CLOCK CO.**"
{Ben Gravolet Collection}

CALENDAR CLOCKS

NEW HAVEN

National Calendar Clock Co.

248

O.G. CALENDAR CLOCK. Rosewood Veneer. Spring. 8-Day Time & Strike. Height, 26". Width, 15 ¼". Short pendulum. 7-inch black dials with gold numbers. Center circle reads; **"Manufactured For National Calendar Clock Co."**. {Private Collection}

249

COLUMN CALENDAR CLOCK.
Veneer Imitation Graining.
8-Day Time & Strike.
Height, 27". Width, 16".
Center circle reads: **"Manufactured For National Calendar Clock Co."**.
{Private Collection}

250

Upper Movement of **COLUMN CALENDAR CLOCK.** {Private Collection}

251

Close-up of **Column** calendar mechanism. The simple calendar mechanism is attached to the dial. {Private Collection}

CALENDAR CLOCKS

The year entered beside each clock is the year of the catalog from which the illustration was taken. Some clocks may have been produced a few years earlier.

253

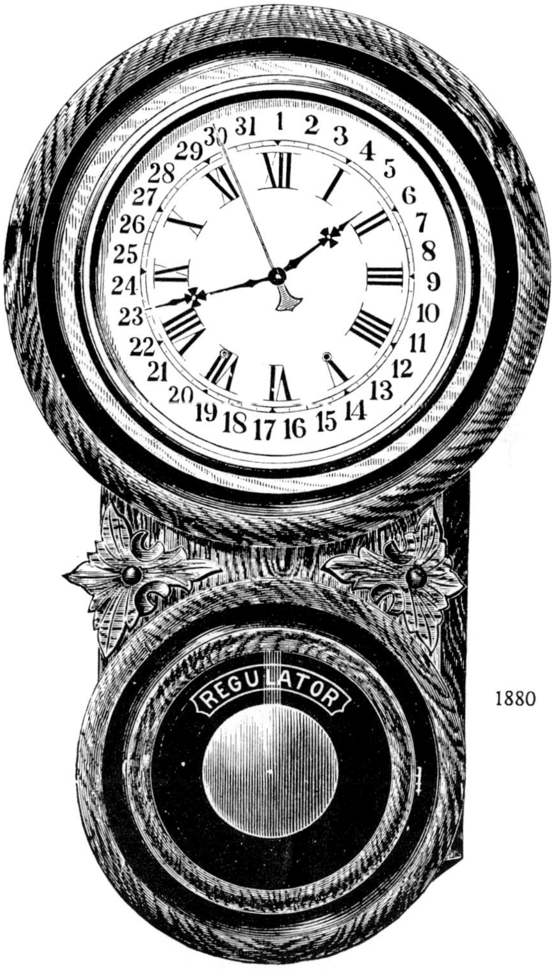

1880

REGULATOR, D. R. CALENDAR.

Spring.

Height, 31 inches. 8 Day Time.
Dial, 12 inches. 8 Day Strike.

252

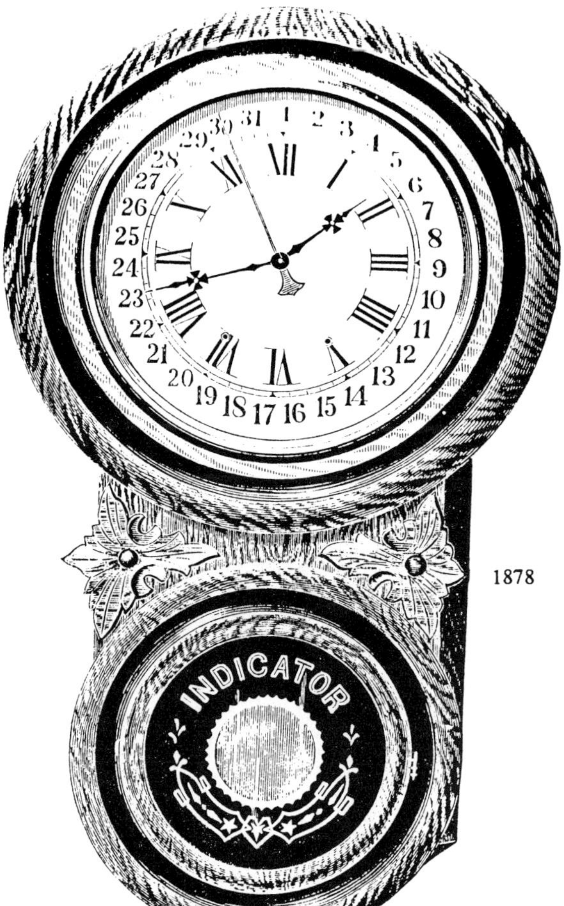

1878

INDICATOR, D. R.

With or without Calendar.

Height, 30 ½ inches. Dial, 12 Inches.
8 Day, Time, Spring.
8 Day, Strike, Spring.

CALENDAR CLOCKS

254

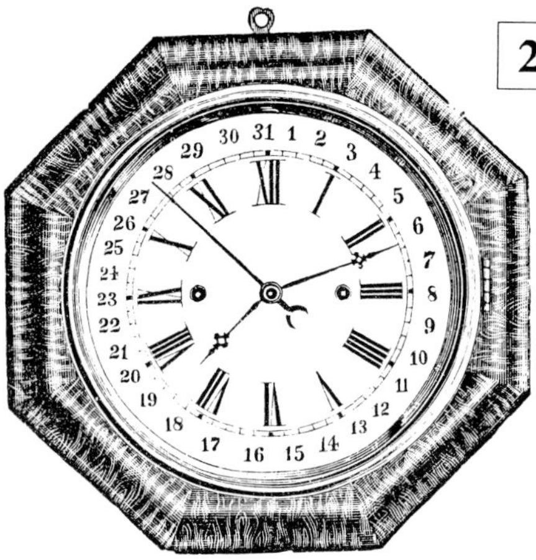

1900

WOOD LEVER Calendar

Polished Veneered.

List price, One-day, 8-inch, Time, Calendar, $4.50
" " One-day, 10-inch, Time, Calendar, 5.25

255

1878

ROUND CALENDAR LEVER.

8 Day. Time. 8 inch.

256

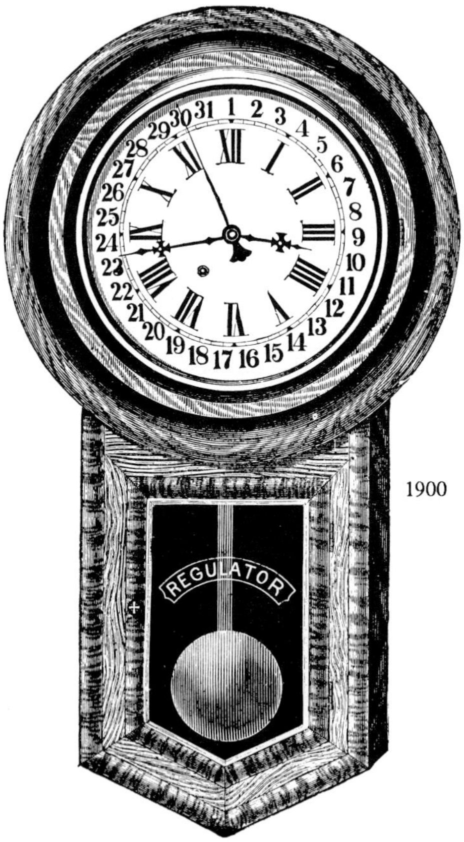

1900

REGULATOR CALENDAR

Oak or Walnut Veneer, Solid Oak or Walnut Circle.
Height, 32 inches. Dial, 12 inches.
List price, Eight-day Time, Calendar, $ 9.55
" " Eight-day Strike, Calendar, 10.55

257

1878

INDICATOR.

CALENDAR.

Height, 32 inches.
Dial, 12 Inches.
8 Day, Time, Spring.
8 Day, Strike, Spring.

CALENDAR CLOCKS

259

Close-up of **No. 1 Regulator** Label.
{The Millie & Jerry Maltz Collection}
{Photo courtesy of Jack Wyer}

258

No. 1 REGULATOR
Walnut veneer case. Ca. 1895. 8-Day Time and Calendar. Height, 32 inches. Width, 18 inches.
{The Millie & Jerry Maltz Collection}
{Photo courtesy of Jack Wyer}

260

Close-up of **No. 1 Regulator**
8-Day Time and Calendar Movement.
{The Millie & Jerry Maltz Collection}
{Photo courtesy of Jack Wyer}

CALENDAR CLOCKS

For more Calendar clocks see pages 182-183

261

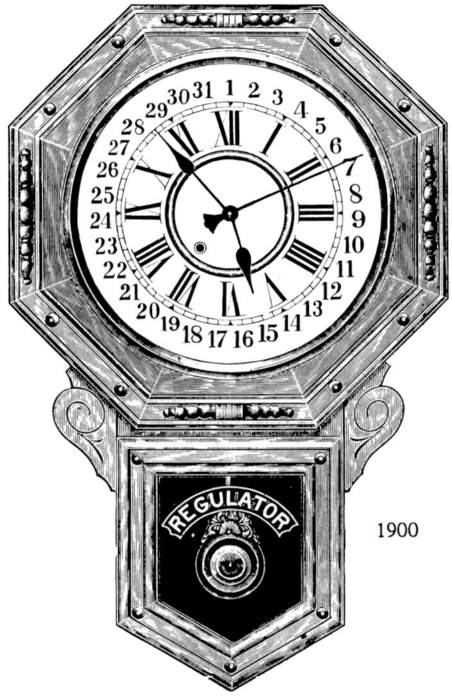

1900

BRADDOCK, CALENDAR—Solid Oak.

Height, 27 inches. Dial, 12 inches.

List price, Eight-day Time, Calendar, $6.50
" " Eight-day Strike, Calendar, 7.50

262

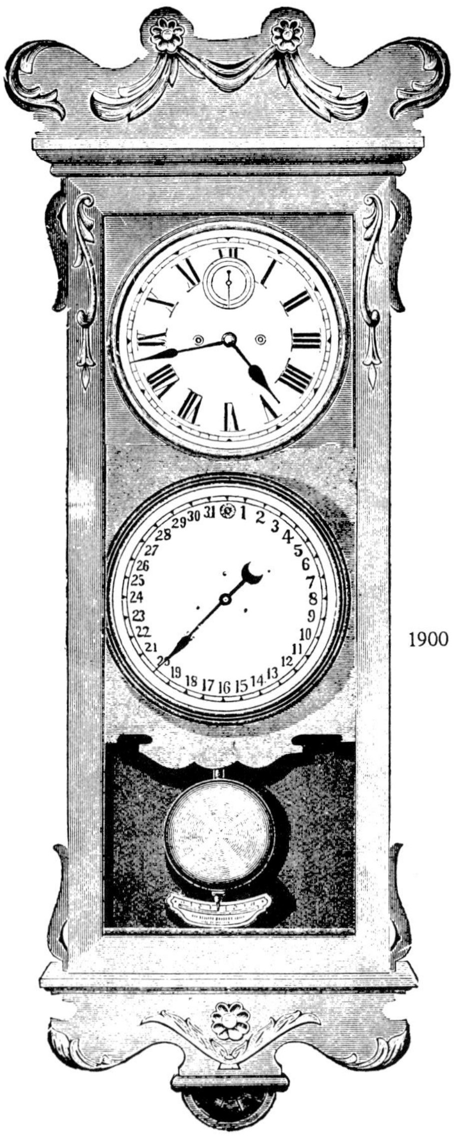

1900

AUSTRIAN CALENDAR—Cherry.

Height, 48¾ inches. Width, 19 inches. Dial, 10 inches.

List price, Eight-day Strike, Gong, Calendar, with Double Dial, $21.00

263

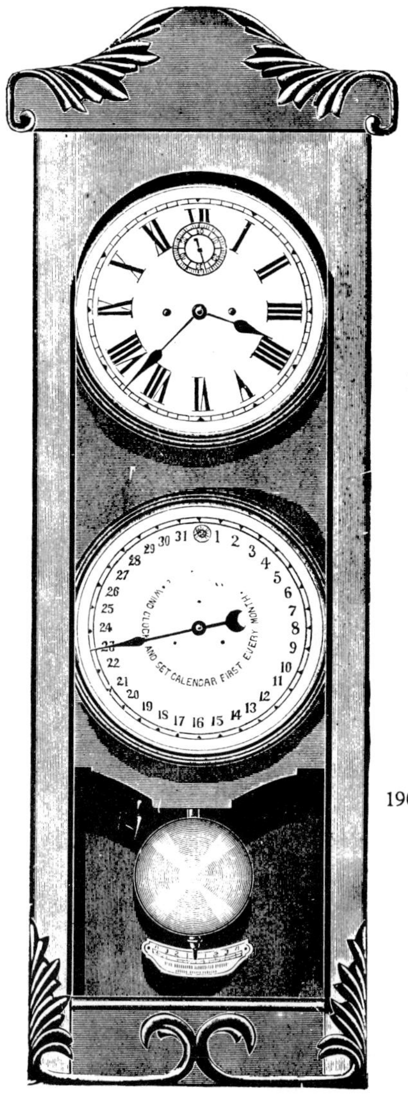

1900

GLORIANA, CALENDAR—Solid Oak.

Height, 42 inches. Dial, 10 inches.

List price, Eight-day Time, Calendar, $23.65

264

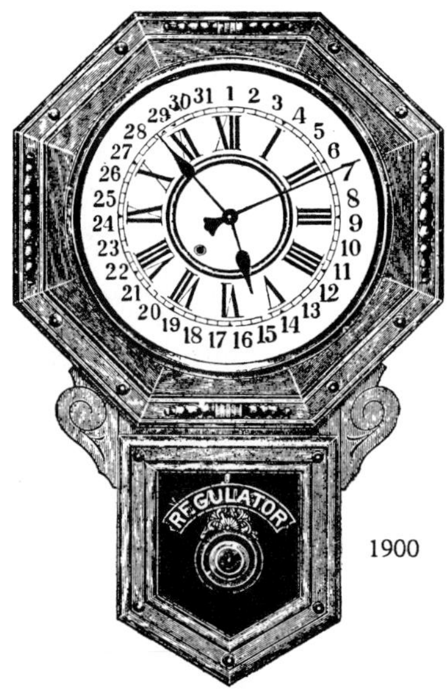

1900

BRADDOCK

Solid Oak.

Height, 21 inches. Dial, 10 inches.

List price, Eight-day Time, . $5.75
" " Eight-day Strike, . 6.75
" " Eight-day Time, Calendar, 6.25
" " Eight-day Strike, Calendar, 7.25

CALENDAR CLOCKS

265

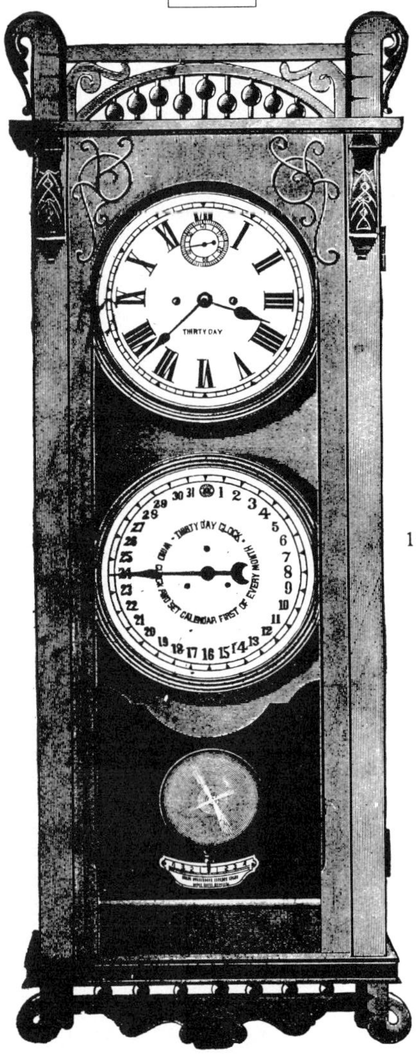

1894

Intrepid.
Height, 49 inches; Dial, 10 inches.
30 Day Time, Calendar. $30.00
{Photo courtesy of Roy Ehrhardt}

266

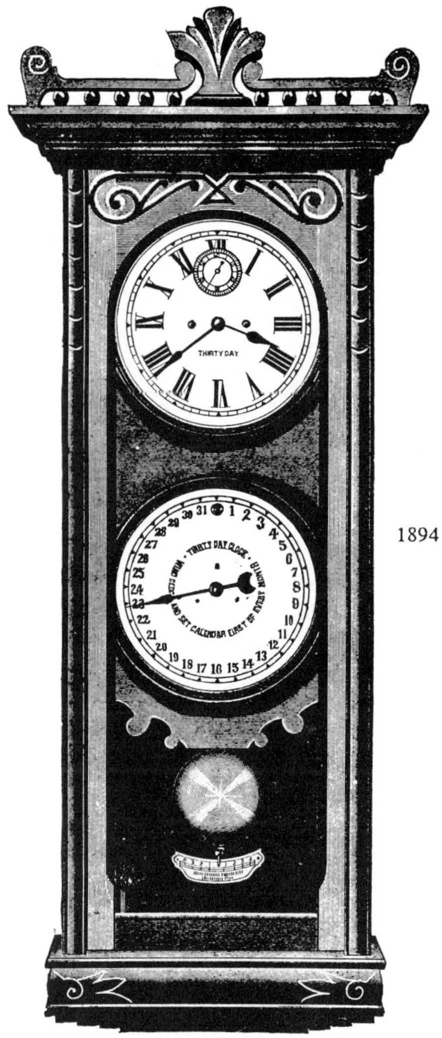

1894

Barbara.
Height, 47 inches; Dial, 10 inches.
30 Day Time, Calendar. $26.00
30 Day Time, $22.50
{Photo courtesy of Roy Ehrhardt}

CALENDAR CLOCKS

For more Calendar clocks see pages 182-183

267

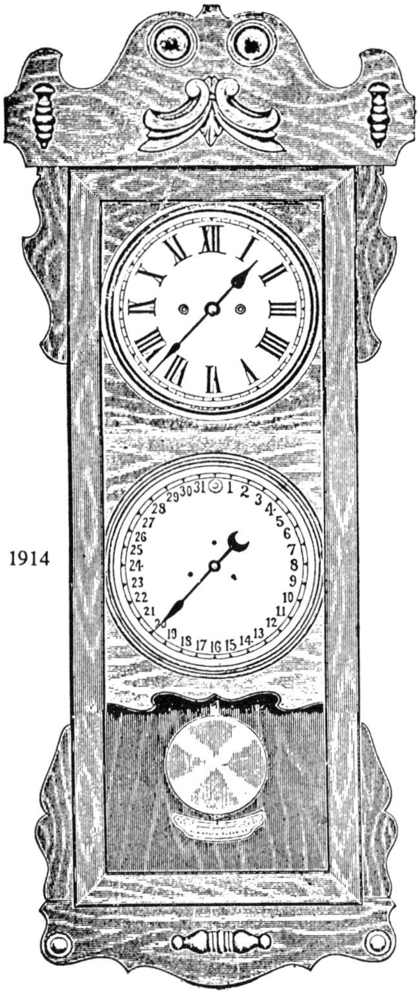

1914

RUTLAND
(SHOWING 30-DAY CALENDAR FITTING.)

EIGHT-DAY SPRING TIME PIECE. LIST, $14.50
HEIGHT, 48 INCHES. WIDTH, 19 1/2 INCHES.
DIAL, 10 INCHES.
CASE, SOLID OAK OR MAHOGANY.
(ALSO FITTED WITH WEIGHT MOVEMENT.)

	EXTRA LIST
FOR 8-DAY STRIKE, CATHEDRAL GONG	$1.50
FOR 30-DAY SPRING TIMEPIECE	2.20
FOR 8-DAY CALENDAR	.45
FOR 30-DAY CALENDAR	2.20
FOR 8-DAY WEIGHT (TIMEPIECE ONLY)	3.65

268

RUTLAND
{Leon Richmond Collection}

269

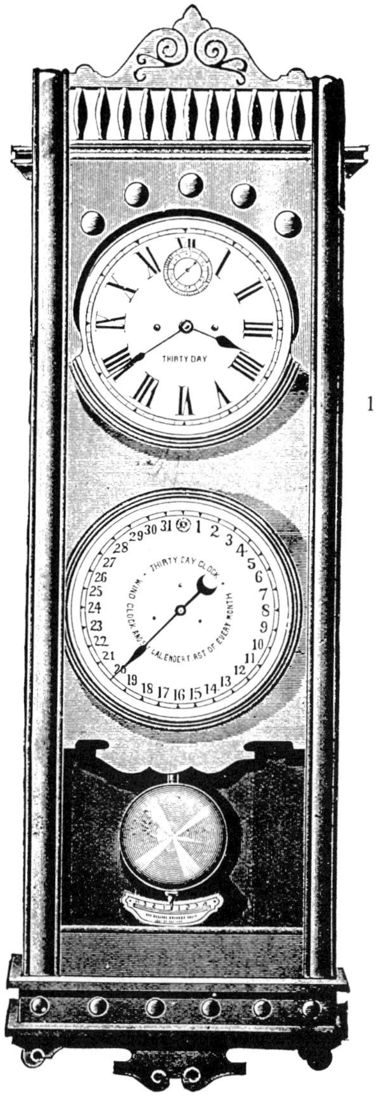

1895

HEBE. Calendar.
Solid Oak.
Height, 49 inches; Dials, 10 inches.

30 Day Time,	List, $22.50
30 Day Time, Calendar,	List, 25.80

CALENDAR CLOCKS

270

1880

DROP OCTAGON CALENDAR, GILT.

Height, 24 inches. Dial, 12 inches.
8 Day Time. Also made in R. C.
8 Day Strike. See Price List.

271

1900

BLAKE, CALENDAR — Solid Oak.

Height, 27 inches. Dial, 12 inches.

List price, Eight-day Time, Calendar, $6.25
" " Eight-day Strike, Calendar, 7 25

272

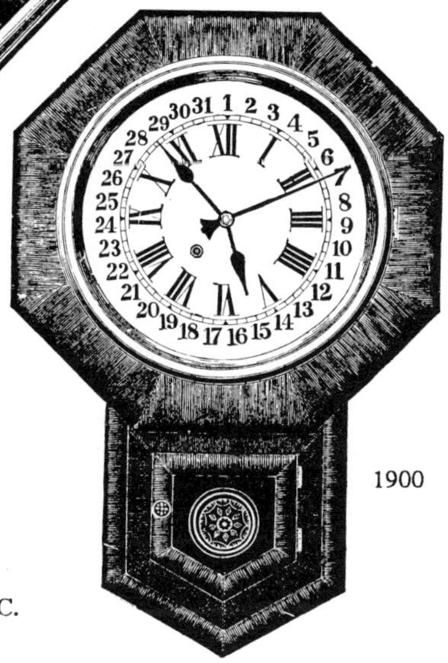

1900

TEN-INCH DROP OCTAGON—Calendar.

Polished Veneered Rosewood or Zebra or Solid Oak.

Height, 21 inches. Dial, 10 inches.

List price, Eight-day Time, Calendar, $6.00
" " Eight-day Strike, Calendar, 7.00

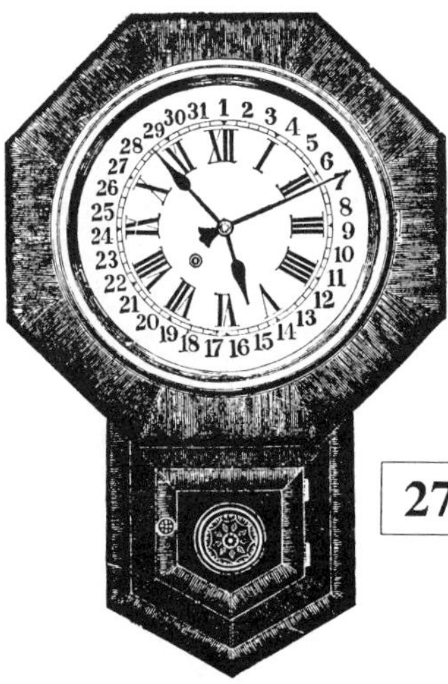

1900

273

EIGHT-INCH DROP OCTAGON—Calendar.

Polished Veneered Rosewood or Zebra or Solid Oak.

Height, 18¼ inches.

List price, Eight-day Time, Calendar, $5.75
" " Eight-day Strike, Calender, 6.75

274

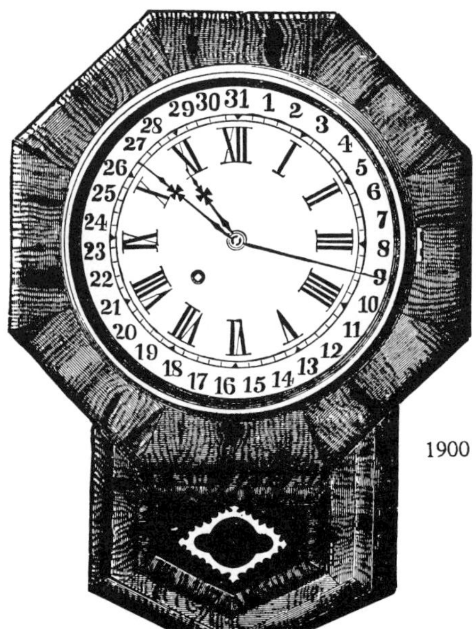

1900

12-INCH DROP OCTAGON, CALENDAR

Rosewood, Zebra or Oak.
Height, 24 inches. Dial, 12 inches.
List price, Eight-day Time, Calendar, $6.25
" " Eight-day Strike, Calendar, 7.25

CALENDAR CLOCKS

275

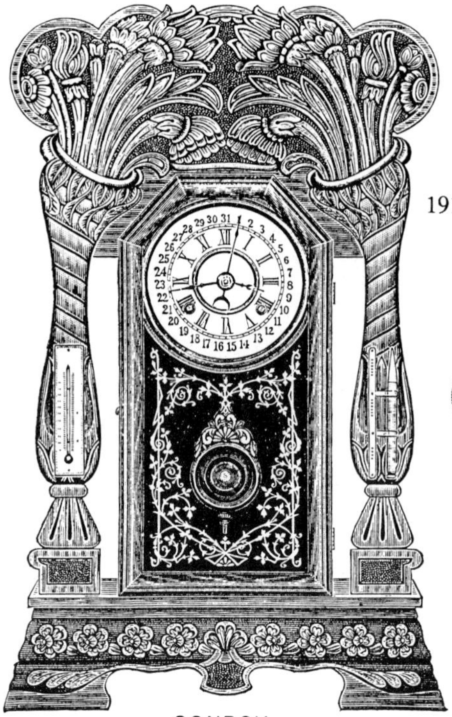

1911

CONROY

HEIGHT ....25 INCHES
WIDTH .....15½ INCHES
DIAL ....6 INCHES.
CASE, OAK WITH THERMOMETER AND BAROMETER
PACKED, 6 IN A BOX
MOVEMENT, 8 DAY HALF-HOUR STRIKE WITH CALENDAR
PRICE......................................LIST, $5.55

276

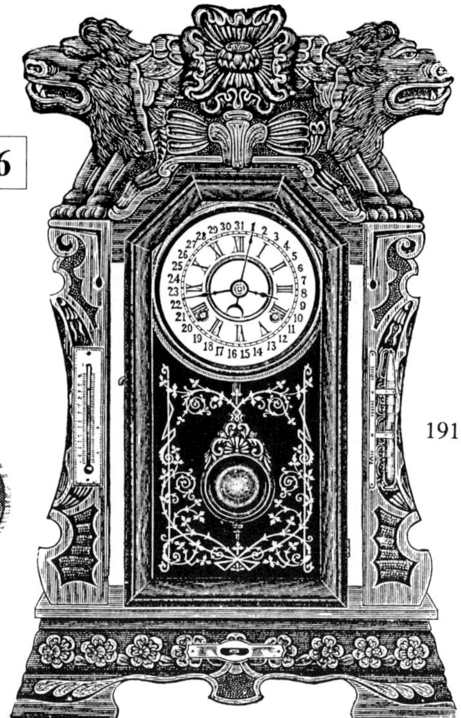

1911

SAMPSON

HEIGHT ....25 INCHES
WIDTH ......15½ INCHES
DIAL ....6 INCHES
CASE, OAK WITH THERMOMETER, BAROMETER AND LEVEL
PACKED, 6 IN A BOX
MOVEMENT, 8 DAY HALF-HOUR STRIKE WITH CALENDAR
PRICE.......................................LIST, $5.85

277

BABSON

OAK.
Calendar, Thermometer, Barometer and Level. 8-Day Time & Strike.
Height, 25 inches.
Width, 15 ½ inches.
(William Mittlestadt Collection)

278

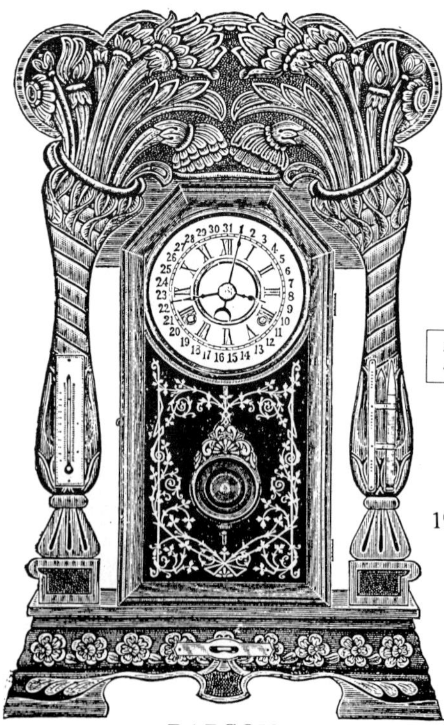

1911

BABSON

BABSON

HEIGHT ....25 INCHES
WIDTH ......15½ INCHES
DIAL ....6 INCHES
CASE, OAK WITH THERMOMETER, BAROMETER AND LEVEL
PACKED, 6 IN A BOX
MOVEMENT, 8 DAY HALF-HOUR STRIKE WITH CALENDAR
PRICE...LIST, $5.85

279

1911

GIBSON

HEIGHT ....25 INCHES
WIDTH ....15½ INCHES
DIAL ....6 INCHES
CASE, OAK WITH THERMOMETER AND BAROMETER
PACKED, 6 IN A BOX
MOVEMENT, 8 DAY HALF-HOUR STRIKE WITH CALENDAR
PRICE.......................................LIST, $5.55

CALENDAR CLOCKS

280

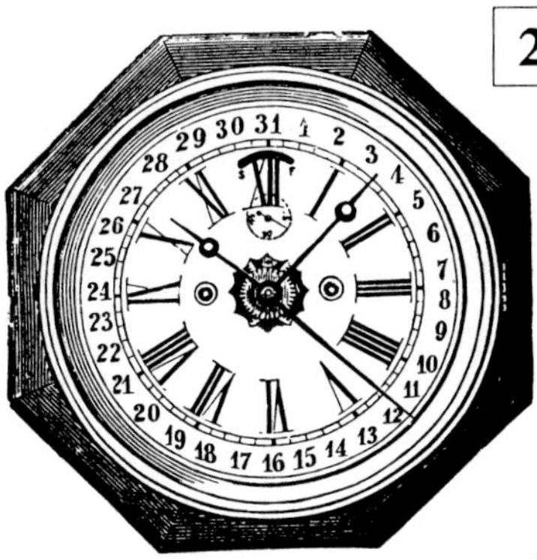

1880

ROUND CORNER CALENDAR LEVER.

1 Day Strike. 10 inch Dial.

8 Day Time. 8 inch Dial.

281

1911

TAMPA

(SHOWING 8 DAY CALENDAR FITTING)

HEIGHT ....37¾ INCHES
WIDTH ....15½ INCHES
DIAL ....12 INCHES
CASE, SOLID OAK
PRICE, 8 DAY TIME ....LIST, $7.40
PRICE, 8 DAY STRIKE, CATHEDRAL GONG LIST, 8.90
PRICE, WITH CALENDAR ....LIST EXTRA, .45

TAMPA

282

1886

CALENDAR.

PLUSH, ASSORTED COLORS.

No. of Clocks in a Case.	Net Weight.	Gross Weight.	Cubic Meas't.
12	6 lbs.	13 lbs.	1 ft. 0 in.

Code Word.		Price.
(Abib)	1 Day Time, Calendar	$6.25

This movement has cut and polished steel pinions, double roller club tooth escapement, straight line lever, screw pillars.

283

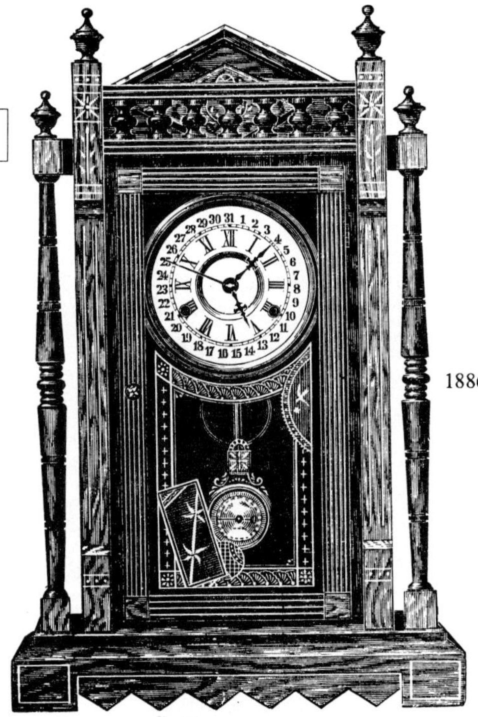

1886

VISTULA.

BLACK WALNUT.

Height, 23½ inches. Dial, 6 inches.

Code Word.		Price.
(Ostrich)	8 Day Strike	$8.50
(Plaque)	8 " " Alarm	9.10
(Ounce)	8 " " Cathedral Gong	9.00
(Witch)	8 " " " " Alarm	9.60
(Kyloes)	8 " " with Calendar	10.00

CALENDAR CLOCKS

284

1880

SPRITE CALENDAR.

Nickel.

4 inch Dial.

1 Day Time, Calendar.

285

1880

SPRITE CALENDAR ALARM.

Nickel.

4 inch Dial.

1 Day Time, Calendar, Alarm.

286

1886

START (T. P. Cal. Alarm).

Dial, 4 inches.

Code Word.		Price.
(Tame)	1 Day Time, Calendar, Alarm..	$2.40

287

1886

START (Calendar T. P.).

NICKEL.

Dial, 4 inches.

Code Word.		Price.
(Mountain)	1 Day Time, Calendar	$1.95

CALENDAR CLOCKS

For more Calendar clocks see pages 182-183

289

1900

FLY ALARM CALENDAR—Nickel.

Dial, 4 inches.

List price One-day Alarm Calendar, $1.80

288

1900

BEACON CALENDAR—Nickel or Enamel.

Height, 6½ inches. Width, 4¼ inches. Dial, 4 inches.

List price, One-day Time, Calendar, $1.75
" " One-day Alarm, Calendar, 1.90

290

1900

FLY TIME CALENDAR

Dial, 4 inches.

List price, One-day Time Calendar, $1.65

291

1880

GLOBE CALENDAR.

Nickel.

Height, 6 inches. Dial, 4 inch.
1 Day Time, Calendar.
1 Day Time, Alarm and Calendar.
8 Day Time, Calendar.

CALENDAR CLOCKS

292

1884

ECHO CALENDAR

NICKEL FINISHED. DIAL, 4½ INCHES.

Time Figures in Black. Calendar Figures in Red.

1 Day Time, Calendar ... $2.20

293

1900

SAMPSON—Oak.

Height, 23¾ inches. Width, 15 inches.
Dial, 6 inches.
Fitted with Thermometer, Barometer and Spirit Level.
Eight-day, Half-hour Strike, Calendar.

List price, $6.05

294

1884

CALENDAR, ALARM.

NICKEL FINISHED WITH SUNK BACK.

1 Day Time, Calendar, Alarm $2.90

295

1884

WHISTLER

NICKEL FINISHED.
DIAL, 4 INCHES.

1 Day Time, Calendar. $2.20

296

1884

WHISTLER CALENDAR

NICKEL FINISHED. DIAL, 4 INCHES.

1 Day Time, Calendar, Alarm. $2.90

CARRIAGE CLOCKS

297

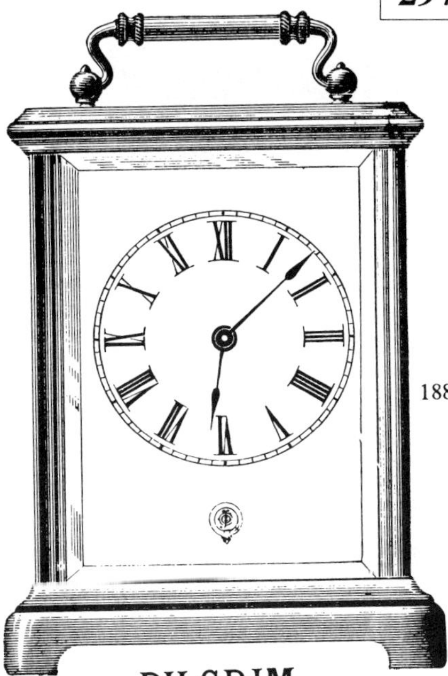

1880

PILGRIM.

Nickel.

Height, 6 inches.

1 Day Time, Alarm.

Beveled glass front.

298

1885

PILGRIM

SINGLE SPRING.

NICKEL WITH GILT MAT.

Height, 7 inches. Dial, 2 ¾ inches.

1 Day Time, Alarm................$5.45

299

1885

PILGRIM

NICKEL.

Height, 7 inches. Dial, 2 ¾ inches.

1 Day Strike.............................. $5.60
1 Day Strike, Gilt Mat................ $5.60
1 Day Strike, fancy Dial..............$5.60
1 Day Strike, Alarm.....................$6.35
1 Day Strike, Alarm, Gilt Mat.... $6.35
1 Day Strike, Alarm, fancy Dial. $6.35

300

1881

301

1881

PILGRIM

NICKEL. BEVELED EDGE, PLATE GLASS FRONT, GLASS SIDES AND DOOR. ALSO MADE WITH FANCY DIAL, IN GILT OR NICKEL MAT.

Height, 6 inches. 1 Day Time, Alarm. 1 Day Strike. 1 Day Strike, Alarm.

CARRIAGE CLOCKS

302

1885

PEACOCK.

EXTRA GOLD PLATED.

Height, 6½ inches. Dial, 2½ inches.

8 Day Time, Open Dial..........$18 75

8 " " Closed Dial.......... 18 75

303

1886

PEACOCK.

EXTRA GOLD PLATED.

Height, 6½ inches. Dial, 2½ inches.

8 Day Time........$14.50

304

1880

ALERT—PLUSH.

Height, 7 inches.
1 Day Time.
1 Day Time, Alarm.
Made in a variety of colors.

305

1889

ACME.

SOLID CAST BRASS, HAND FINISHED, ENAMEL DIAL.

1 Day Time..........$7.50

CARRIAGE CLOCKS

306

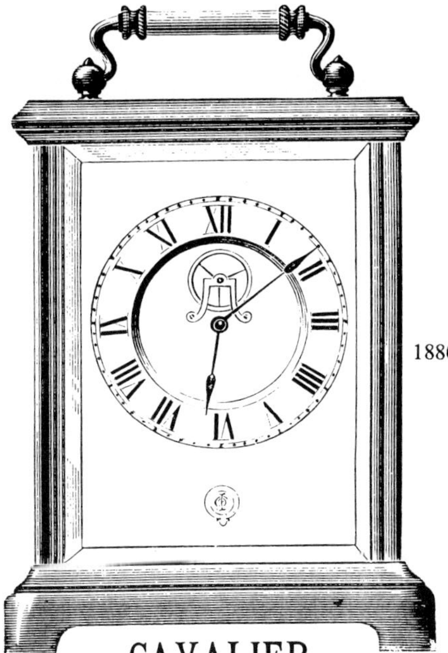

1880

CAVALIER.

Nickel.

Height, 6 inches.

1 Day Time, Alarm.

Beveled glass front.

307

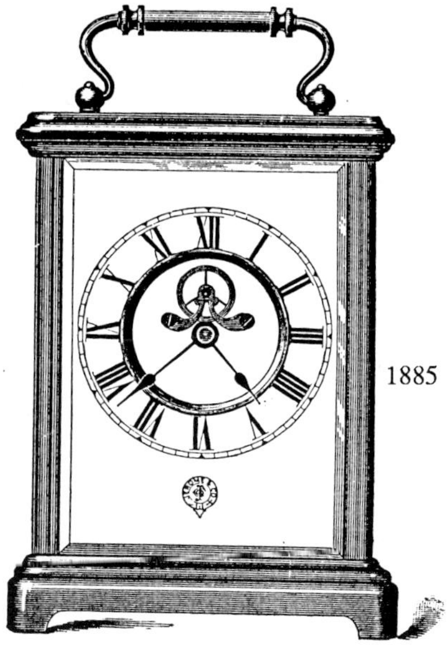

1885

CAVALIER.

NICKEL WITH VISIBLE BALANCE.
Height, 7 inches. Dial, 2 ¾ inches.
1 Day Time, Alarm........................ $5.95
1 Day Time, Alarm,
Black and Gilt Dial........................ $6.15

308

1885

PIGEON.

EXTRA NICKEL CASE.

Height, 5 inches. Dial, 2¼ inches.

1 Day Time, Open Dial............$7 15
1 " " Closed Dial........... 7 15

309

1885

TITANIA.

OAK, MAPLE OR MAHOGANY.

Gilt Ornaments.

Height, 6 inches. Dial, 3 inches.

Code Word.		Price.
(Sable)	1 Day Time, Alarm.................	$5 20

310

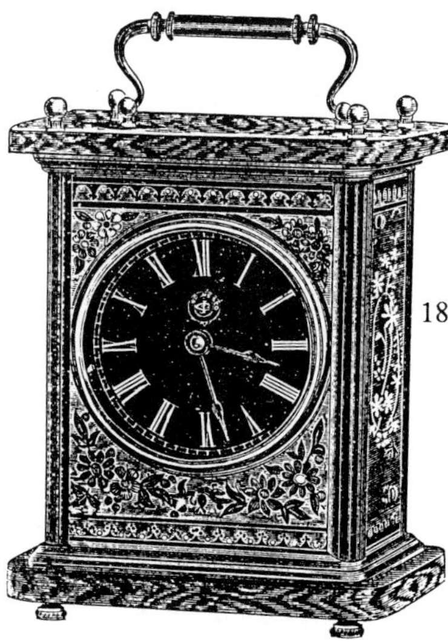

1885

STARLING.

OAK OR MAHOGANY TOP AND BASE.
ENAMELED METAL FRONT AND
SIDE PLATES.

Height, 7 inches.

1 Day Time, Alarm....$5 95

311

1885

OBERON.

OAK, MAPLE OR MAHOGANY.

Visible Balance.

Height, 6 inches. Dial, 3 inches.

1 Day Time, Alarm,........$6 00

CARRIAGE CLOCKS

312

1900

TRURO

French Satin Gold Finish.
Height, 5 inches. Width, 2¾ inches.
Eight-day Time with Enameled Porcelain Dial.

List price, $11.25

313

PANZA

1911

HEIGHT ....4 INCHES
WIDTH ....2¾ INCHES
DIAL, PORCELAIN ......2 INCHES
CASE, SOLID BRASS
FINISH, GILT, SILVER OR GUN METAL
PRICE, 1 DAY TIME PIECE ...LIST, $3.95
PRICE, WITH 1 DAY INTERMITTENT ALARM ....LIST EXTRA, 1.50

314

1900

CARRIAGE ALARM

Height, 7¼ inches. Width, 4½ inches.
Dial, 2¾ inches.
Nickel-plated Case, with Fancy Gilt Front and Handle.
Alarms intermittently every twenty seconds for fifteen minutes.
One-day Time, Alarm ....List Price, $4.25

315

1886

PADLOCK REPEATING STRIKE.

NICKEL OR POLISHED BRASS.

Height, 6 inches. Dial, 3 inches.

By touching a button on the side, this clock can be made to repeat the hour up to ten minutes of the next.

Code Word. Price.
(Agirse) 1 Day Repeating Strike $8.50

316

1886

ADELA.

SOLID CAST BRASS, HAND FINISHED.

Height, 8¾ inches.
Dial, 2½ inches.

This Movement has Square Frame, with Three Plates, allowing the Main Spring to be removed without disturbing the Escapement; Steel Polished Pinions, Screw Pillars, Straight Line Lever, Double Roller Club Tooth Escapement and Jeweled Balance; equal to a fine Watch Escapement in finish and time keeping qualities.

8 Day Time. ....**Price, $20.00**

CARRIAGE CLOCKS

317

1911

MESSENGER

HEIGHT ....7¼ INCHES
WIDTH ....4¼ INCHES
DIAL ....2¾ INCHES
CASE, NICKEL PLATED, WITH FANCY GILT FRONT AND HANDLE, GLASS IN SIDES
PRICE, 1 DAY TIME ALARM ....LIST, $4.35
PRICE, 1 DAY STRIKE ....LIST, 4.35

318

1900

SAMOA

French Satin Gold or Bronze Finish.
Height, 7¼ inches. Width, 4 inches.
One or Eight-day Time.
Two-inch Ivorized Porcelain Dial.

List price, One-day Time, Satin Gold, $7.00
" " One-day Time, Bronze, . 6.30
" " Eight-day Time, Satin Gold, 8.50
" " Eight-day Time, Bronze, 7.80

Fitted to order only with Jeweled Sash in Brilliant, Ruby, Emerald, Amethyst, Sapphire, Pearl and Turquoise or alternate stones.
Extra list, $2.10

319

1906

NANA

Height, 5½ inches. Width, 3½ inches.
Two-inch Porcelain Dial.
Ormolu Gold Plated Case.
One-day Time ....List Price, $4.25
One-day Time, Alarm .. " " 5.75

320

1889

BADGER.

REPEATING ALARM. Height, 7¼ inches.

1 Day Repeating Alarm...............$5.50

No Chance to Sleep Over.—The mechanism in this Clock is so adjusted that the alarm will repeat its signal at intervals of three minutes until it has alarmed ten times. Any one who has become accustomed to the ordinary alarm, and repeatedly slept through its warning, will appreciate the persistency of this Clock in getting him up.

321

1880

SUTHERLAND.

NICKEL.

Height, 8 inches.
1 Day Time. 1 Day Time, Calendar.
1 Day Time, Alarm. 1 Day Time, Calendar, Alarm.
1 Day Strike. 8 Day Time.
Dial can either have Arabic or Roman figures.
Solid or Glass Sides. Made also with Brass Mat.

CARRIAGE CLOCKS

322

1911

KENWICK

HEIGHT ....4¼ INCHES
WIDTH ....2⅞ INCHES
DIAL, PORCELAIN ..1¾ INCHES
CASE, SOLID BRASS
FINISH, GILT, SILVER, GUN METAL OR OLD BRASS
PRICE, 1 DAY TIME PIECE ....LIST, $5.25
PRICE, 1 DAY TIME PIECE ALARM..LIST EXTRA, 1.50
PRICE, 1 DAY STRIKE ....LIST EXTRA, 1.50

323

1920

EDMOND

Eight-Day Half-Hour Repeating Strike
Height, 5¾ Inches Width, 3½ Inches
Dial, Porcelain, 2⅛ Inches
Case, Polished Brass
Edmond in Leather Case
Height, 5½ Inches Width, 4¼ Inches

324

1911

KENNETH

HEIGHT ....4⅜ INCHES
WIDTH ....2⅞ INCHES
DIAL, PORCELAIN ..1¾ INCHES
CASE, SOLID BRASS
FINISH, GILT, SILVER, GUN METAL OR OLD BRASS
PRICE, 1 DAY TIME PIECE ....LIST, $5.25
PRICE, 1 DAY TIME PIECE ALARM..LIST EXTRA, 1.50
PRICE, 1 DAY STRIKE ....LIST EXTRA, 1.50

325

NUGGET In Leather Case
Height, 3 Inches Width, 2⅝ Inches

1920

326

1911

NUGGET

HEIGHT ....3 INCHES
WIDTH ....2⅛ INCHES
DIAL, PORCELAIN ..1¼ INCHES
CASE, HEAVY BRASS
FINISH, GILT, SILVER, GUN METAL OR OLD BRASS
MOVEMENT, 1 DAY TIME PIECE
PRICE ....LIST, $4.35

CARRIAGE CLOCKS

327

1913

PARAL
LIST, $3.95
HEIGHT, 5½ INCHES. WIDTH, 3⅜ INCHES.

328

1913

MARVEL
LIST, $3.65
HEIGHT, 3⅞ INCHES. WIDTH, 2⅞ INCHES.

329

1913

CHOPIN
LIST, $3.95
HEIGHT, 5¼ INCHES. WIDTH, 3¼ INCHES.

330

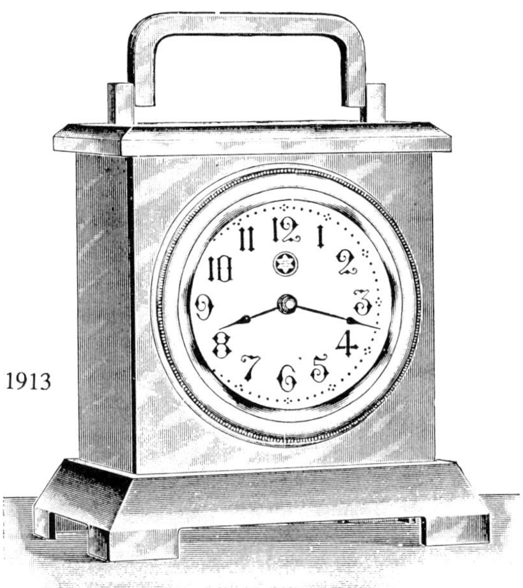

1913

PARIS
LIST, $5.50
HEIGHT, 4¾ INCHES. WIDTH, 3½ INCHES.

331

LEROY 1900

French Satin Gold Finish.
Height, 3¾ inches. Width, 2¼ inches.
One-day Time with Enameled Porcelain Dial.

List price, $4.90

332

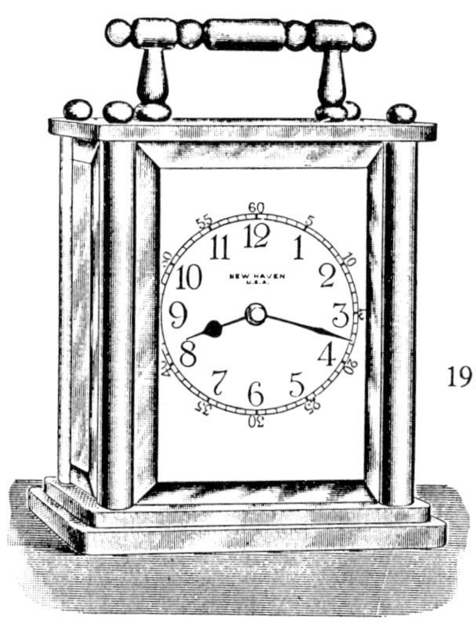

1913

NADIR 1911

HEIGHT ....3 INCHES
WIDTH ....2⅛ INCHES
DIAL, PORCELAIN ..1¼ INCHES
CASE, HEAVY BRASS
FINISH, GILT, SILVER, GUN METAL OR OLD BRASS
MOVEMENT, 1 DAY TIME PIECE
PRICE ....LIST, $4.35

CHIME CLOCKS

333

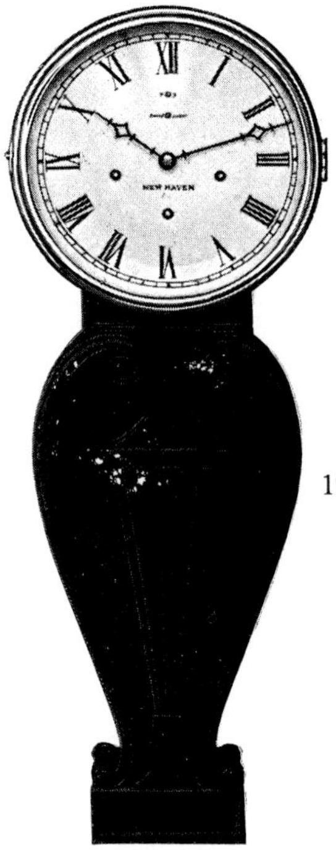

1929

"SHERATON"

Height, 24 inches. Width, 9 inches. 8-inch Antique Dial, Polished Sash. Solid Mahogany, Hand - Carved Case, Antique Finish. 8-Day Pendulum Westminster Chime with Chorded Strike.
List $88.25

334

1929

"ART MODERNE"

Height, 19 inches. Width, 11½ inches. 8-inch Antique Dial, Polished Sash. Solid Mahogany, Hand Rubbed, Gilt Ornament, Bird's Eye Maple Panel. 8-Day Pendulum Westminster Chime with Chorded Strike.
List $88.25

335

1929

"CHIPPENDALE"

Height, 24½ inches. Width, 9 inches. 8-inch Antique Dial, Polished Sash. Solid Mahogany, Hand - Carved Case, Antique Finish. 8-Day Pendulum Westminster Chime with Chorded Strike.
List $88.25

336

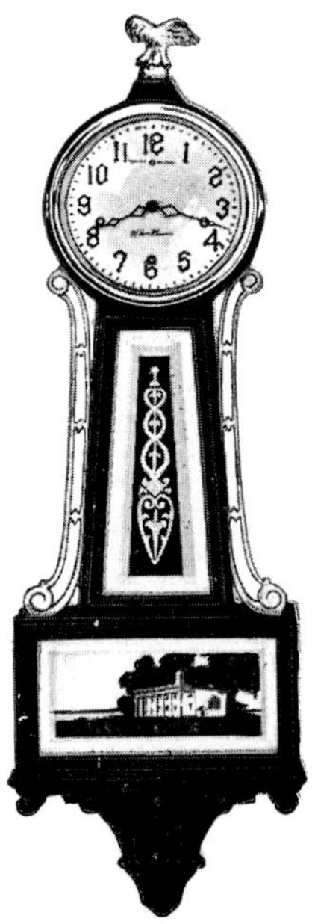

1929

"WINSOME" Banjo Chime

Height, ., inches. Width, 8½ inches. 8-Day Pendulum Chime Movement with Chorded Strike. Solid Mahogany, Antique Finish, Decorated Panels. 5⅝ inch Silvered Dial with Raised Numerals.
List $74.40

337

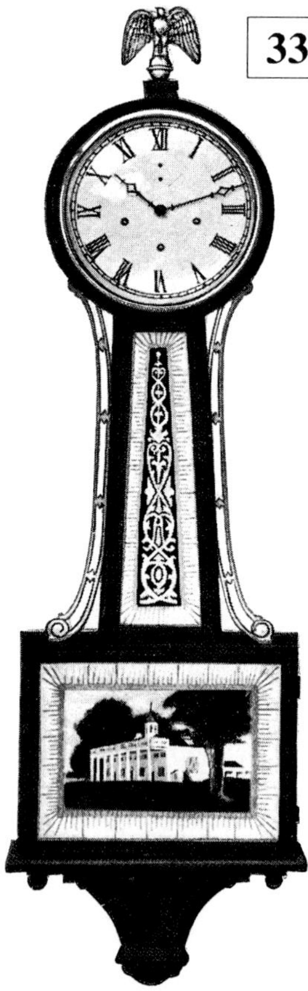

1929

"WILLARD" Banjo Chime

Height, 42 inches. Width, 12 inches. 8-inch Antique Dial. Solid Mahogany, Decorated Panels. 8 - Day Pendulum Chime Movement with Chorded Strike.
List $98.00

338

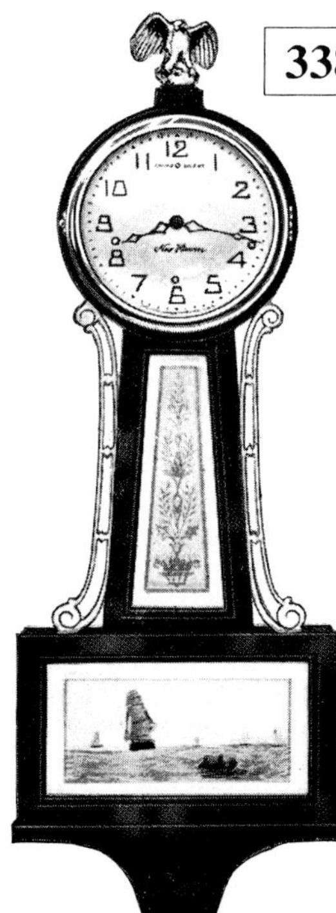

1929

"WHITNEY" CHIME

Height, 30½ inches. Width, 9½ inches. 6-inch Antique Dial, Polished Sash. Mahogany Finish, Brass Side Ornaments, Eagle Top, Decorated Panels. 8 - Day Pendulum Westminster Chime with Chorded Strike.
List $56.00

CHIME CLOCKS

339

CHIME No. 4
Height, 20½ inches. Width, 12¼ inches. Depth, 8½ inches. Dial, 5½ inches. Heavy brass face. Separate chime movement. Side wound. Five wire gongs. Movement mounted with iron frame. 8-day with Westminster chime. Willcock patent Canada 1836.
{Terry & Carole Looft Collection}

340

Back view of **Chime No. 4** movement.
{Terry & Carole Looft Collection}

341

Front view of **Chime No. 4** movement.
{Terry & Carole Looft Collection}

342

Side view of **Chime No. 4** showing the chime-silent lever.
{Terry & Carole Looft Collection}

CHIME CLOCKS

Movement illustrations for this clock on page 400

343

1895

EIGHT-BELL CHIME No. I.

Case of Solid Mahogany with best French Satin Gilt Trimmings, or of Oak with Antique Silver Trimmings.

Height, 18 inches.

Eight-day Movement with Westminster Chime of Eight Bells, chiming every quarter. The hour is struck on a rich Cathedral Gong. Seven-inch Metal Dial with Etched Rim, raised Numerals and Ornaments to match Trimmings on Case.

1900 List price for either finish, $41.75

344

1900

FOUR-BELL WESTMINSTER CHIME, No. 3.

Height, 18½ inches. Width, 12 inches.
Light Oak with Oxidized Silver or Bronze Trimmings; or Mahogany with French Satin Gold Trimmings.
Six-inch Metal Dial with Fancy Etched Center, finished to match Trimmings.

List price, $55.75

345

1911

CHIME No. 4

CHIME No. 4

HEIGHT ....20 INCHES
WIDTH ....12¼ INCHES
DIAL, ETCHED SILVER.............6 INCHES, WITH CAST BRASS ORNAMENTS
CASE, SOLID MAHOGANY
MOVEMENT, 8 DAY HOUR STRIKE AND QUARTER-HOUR CHIME ON CATHEDRAL GONGS
PRICE .... LIST, $34.80

CHIME CLOCKS

WESTMINSTER CHIMES

These Clocks are fitted with an Eight-Day Strike movement of the Standard Cabinet Pattern and a Chime movement which plays upon four Cathedral Gongs the same tune as that of the Chimes in Westminster Abbey. The music for each quarter is shown on this page. These Gongs are tuned with great care and are played by mechanism made under Willcock patent of August 28, 1894. The richness of tone is secured by the Gunther Resonator, patented March 27, 1896. The hour is struck on a fifth gong.

346

1911

CHIME No. 6

HEIGHT ....15¼ INCHES
WIDTH ....11½ INCHES
DIAL, SILVER ....8 INCHES
CASE, SOLID ANTIQUE OAK WITH SILVER ORNAMENTS
MOVEMENT, 8 DAY HOUR STRIKE AND QUARTER HOUR CHIME ON CATHEDRAL GONGS
PRICE ....LIST, $54.3

347

1911

CHIME No. 7.

HEIGHT ....17½ INCHES
WIDTH ....13 INCHES
DIAL, SILVER ....8 INCHES
CASE, SOLID ANTIQUE OAK WITH SILVER ORNAMENTS
MOVEMENT, 8 DAY HOUR STRIKE AND QUARTER-HOUR CHIME ON CATHEDRAL GONGS
PRICE ....LIST, $65.25

348

1900

FOUR-BELL WESTMINSTER CHIME, No. 5.

Height, 22¼ inches. Width, 12¼ inches.
Light Oak with Oxidized Silver or Bronze Trimmings; or Mahogany with French Satin Gold Trimmings.
Six-inch Metal Dial with Fancy Etched Center, finished to match Trimmings.

List price, $36.25

CHIME CLOCKS

349

1929

INGLEWOOD ✤ WESTMINSTER CHIME

HEIGHT . . 12½ inches WIDTH . . 9½ inches DIAL . . 7 inches, Silver Finish
CASE, Solid Mahogany, Rich Dark Brown Cast Sash, Beveled Convex Glass
MOVEMENT, New 8 Day Pendulum; Rod Chime and Strike

350

Ca. 1937

DURHAM Chime

SELF-STARTING ELECTRIC or 8-DAY PENDULUM

Richly carved mahogany case; modern, two-tone metal dial, 5¼ inches wide. Gold-plated hands and raised, gold-plated, applied hour indicators and numerals. Polished brass sash with convex crystal. Full Westminster, self-adjusting, quarter-hour chimes and hour strike. Height, 11⅞ inches. Width, 8¾ inches. Packed singly. Shipping weight: 10 lbs.

RETAIL $36.75

351

1929

ABBEY ✤ WESTMINSTER CHIME

HEIGHT . . 15½ inches WIDTH . . 10½ inches DIAL . . 7 inches, Silver Finish
CASE, Solid Mahogany, Rich Dark Brown Cast Sash, Beveled Convex Glass
MOVEMENT, New 8 Day Pendulum; Rod Chime and Strike

352

Ca. 1937

CLOISTER Chime

SELF-STARTING ELECTRIC or 8-DAY PENDULUM

Selected mahogany case with redwood burl veneer raised panels. Modern 5¼-inch, two-tone metal dial. Raised, gold-plated, applied hour indicators and numerals. Gold-plated hands. Polisned brass sash with convex crystal. Full Westminster, self-adjusting, quarter-hour chimes and hour strike. Height, 13 inches. Width, 8½ inches. Packed singly. Shipping weight: 12 lbs.

RETAIL $44.50

CHIME CLOCKS

353

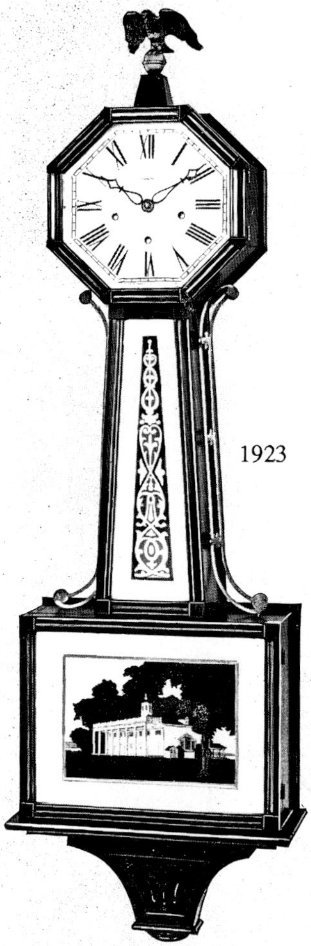

1923

WASHINGTON ✵ CHIME
(Patent Applied For)
LENGTH, 37 inches WIDTH, 12¼ inches
DIAL, 7 inches, Silver
Solid Mahogany Case

354

1929

INGOMAR ⚜ WESTMINSTER CHIME
HEIGHT . . 10 inches WIDTH . . 20½ inches DIAL . . 7 inches, Silver Finish
CASE, Solid Mahogany, Rich Dark Brown Cast Sash, Beveled Convex Glass
MOVEMENT, New 8 Day Pendulum; Rod Chime and Strike

355

1929

GRANVILLE ⚜ WESTMINSTER CHIME
HEIGHT . . 9½ inches WIDTH . . 20 inches DIAL . . 7 inches, Silver Finish
CASE, Solid Mahogany, Rich Dark Brown Cast Sash, Beveled Convex Glass
MOVEMENT, New 8 Day Pendulum; Rod Chime and Strike

356

Ca. 1937

DUNA Chime

SELF-STARTING ELECTRIC or 8-DAY PENDULUM

Mahogany plywood, tambour model. Two-tone metal dial, 5¼ inches wide. Black hands with raised, gold-plated, applied numerals. Polished brass sash with convex crystal. Full Westminster, self-adjusting, quarter-hour chimes and hour strike. Height, 8⅛ inches. Width, 19¼ inches. Packed singly. Shipping weight: 11 lbs.

RETAIL $27.75

357

Ca. 1938

PEERLESS Chime

SELF-STARTING ELECTRIC or 8-DAY PENDULUM

Solid walnut case with claro-swirl walnut veneer front and sides. Genuine marquetry inlay in front panel. Two-tone metal dial, 5¼ inches wide. Gold-plated hands and raised, gold-plated, applied numerals. Polished brass sash with convex crystal. Full Westminster, self-adjusting, quarter-hour chimes and hour strike. Height, 11 inches. Width, 7¾ inches. Packed singly. Shipping weight: 11 lbs. 3 oz.

RETAIL $30.00

CHIME CLOCKS

FULL WESTMINSTER CHIMES WITH CHORDED STRIKE

Chimes on silver-toned rods at the quarter, half-hour, third quarter and hour. Eight-day, self-adjusting chime and strike with easily removable main springs. The flexible steel-feather control is an exclusive feature found only in New Haven chimes. Hand-rubbed cases of solid mahogany. Duco Finish. Requires six-inch mantel.

358

1929

"TOLEDO" Westminster Chime

Height, 9 inches. Base, 19 inches. Dial, 5½ inches. 8-Day Pendulum Chime Movement, Silvered Dial with Raised Numerals, Convex Crystal, Polished Sash. Solid Mahogany, Antique Finish with Dark Panels and Base.
List $44.50

359

1929

"ORLEANS" Westminster Chime

Height, 9¾ inches. Base, 20¼ inches. 8-Day Pendulum Chime Movement, Silvered Dial with Raised Numerals, Beveled Convex Crystal, Cast Sash. Solid Mahogany, Antique Finish with Dark Panels and Base.
List $71.00

360

1929

"LINCOLN" Westminster Chime

Height, 9¾ inches. Base, 21 inches. Dial, 6 inches. 8-Day Pendulum Chime Movement, Silvered Dial with Raised Numerals, Beveled Convex Crystal, Cast Sash. Solid Mahogany Case with Dark Panels.
List $53.40

361

1929

"HARKNESS" Westminster Chime

Height, 10 inches. Base, 21 inches. Dial, 6 inches. 8-Day Pendulum Chime Movement, Silvered Dial with Raised Numerals, Beveled Convex Crystal, Cast Sash. Solid Mahogany Case with Dark Panels.
List $53.40

362

1929

"CHESTER" Westminster Chime

Height, 10½ inches. Base, 22½ inches. 8-Day Pendulum Chime Movement, 7-inch Silvered Dial with Raised Numerals, Beveled Convex Crystal, Cast Sash. Solid Mahogany, Antique Finish with Dark Panels and Base.
List $74.00

CHIME CLOCKS

FULL WESTMINSTER CHIMES

Full Westminster Chime on Silver-Toned Rods with Self-Adjusting Chorded Strike. Solid Mahogany, Hand Rubbed, Durable Duco Finish.

The year entered beside each clock is the year of the catalog from which the illustration was taken. Some clocks may have been produced a few years earlier.

363

1929

"AMIENS" Westminster Chime

Height, 12 inches. Width, 10 inches. 6-inch Silvered Dial with Raised Numerals, Cast Sash, Beveled Convex Crystal. Solid Mahogany, Antique Finish with Two-Tone Panels. 8-Day Pendulum Chime Movement.
List $88.25

364

1929

"GAMMA" Westminster Chime

Height, 9½ inches. Base, 21 inches. 5½-inch Silvered Dial with Raised Numerals. 8-Day Pendulum Chime Movement. Convex Crystal. Polished Sash. Mahogany-Finished Case, Solid Mahogany Front with Modernistic Design.
List $41.40

365

1929

"THETA" Westminster Chime

Height, 9½ inches. Base, 21 inches. 5½-inch Silvered Dial with Raised Numerals. 8-Day Pendulum Chime Movement. Convex Crystal. Polished Sash. Mahogany-Finished Case, Solid Mahogany Front with Modernistic Design.
List $41.40

366

1929

"SIGMA' Westminster Chime

Height, 9½ inches. Base, 21 inches. 5½-inch Silvered Dial with Raised Numerals. 8-Day Pendulum Chime Movement. Convex Crystal. Polished Sash. Mahogany-Finished Case, Solid Mahogany Front with Modernistic Design.
List $41.40

CRYSTAL REGULATORS

367

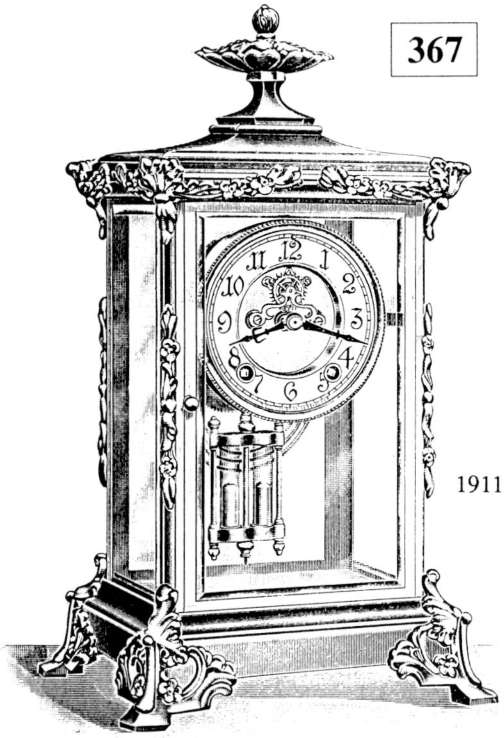

1911

TECUMSEH

HEIGHT ....15¼ INCHES
WIDTH ....8½ INCHES
DIAL, PORCELAIN...........4 INCHES, VISIBLE ESCAPEMENT
CASE, HEAVY METAL, FINISHED IN POLISHED GOLD PLATE, FITTED WITH BEVELED FRENCH PLATE GLASS FRONT, SIDES AND BACK
FINISH, ON TRIMMINGS, ORMOLU GOLD PLATE OR FRENCH BRONZE
MOVEMENT, 8 DAY HALF-HOUR STRIKE, CATHEDRAL GONG
PRICE ....LIST, $42.00

368

1911

THAMES

HEIGHT ....16 INCHES
WIDTH ....9 INCHES
DIAL, PORCELAIN..............4 INCHES, VISIBLE ESCAPEMENT
CASE, HEAVY METAL, FINISHED IN POLISHED GOLD PLATE, FITTED WITH BEVELED FRENCH PLATE GLASS FRONT, SIDES AND BACK
FINISH, ON TRIMMINGS, ORMOLU GOLD PLATE OR FRENCH BRONZE
MOVEMENT, 8 DAY HALF-HOUR STRIKE, CATHEDRAL GONG
PRICE ....LIST, $43.50

369

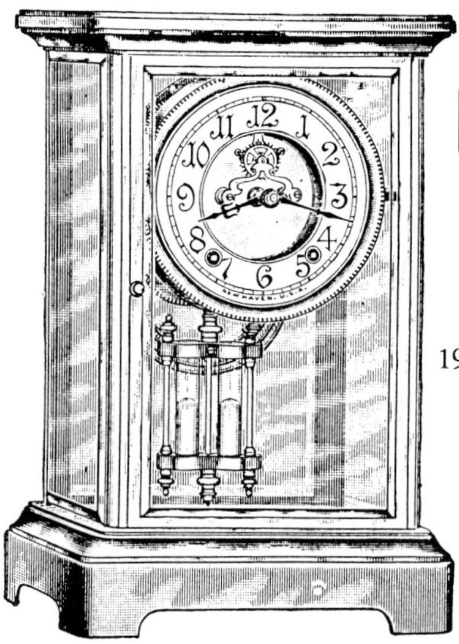

1911

THERON

HEIGHT.......................11 INCHES
WIDTH ....7⅛ INCHES
DIAL, PORCELAIN ....4 INCHES, VISIBLE ESCAPEMENT
CASE, HEAVY METAL WITH BEVELED FRENCH PLATE GLASS FRONT, SIDES AND BACK
FINISH, GOLD PLATED
MOVEMENT, 8 DAY HALF-HOUR STRIKE, CATHEDRAL GONG
PRICE ....LIST, $32.00

370

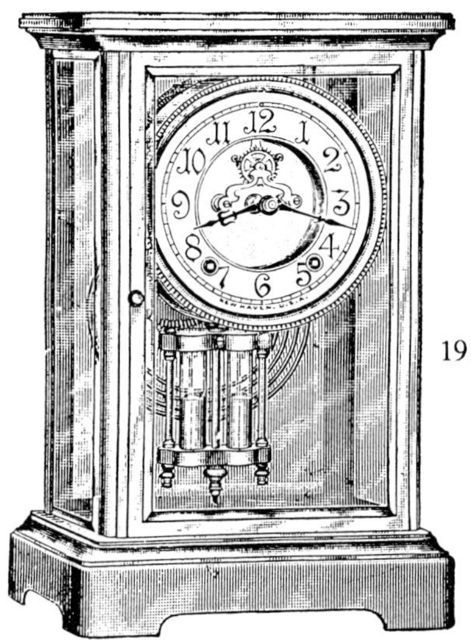

1911

THESPIA

HEIGHT ....10 INCHES
WIDTH ....6¼ INCHES
DIAL, PORCELAIN.......... 4 INCHES, VISIBLE ESCAPEMENT
CASE, HEAVY METAL WITH BEVELED FRENCH PLATE GLASS FRONT, SIDES AND BACK
FINISH, GOLD PLATED
MOVEMENT, 8 DAY HALF-HOUR STRIKE, CATHEDRAL GONG
PRICE ....LIST, $29.00

CRYSTAL REGULATORS

The year entered beside each clock is the year of the catalog from which the illustration was taken. Some clocks may have been produced a few years earlier.

371

1911

TOLLAND

HEIGHT ....15½ INCHES
WIDTH ....8½ INCHES
DIAL, PORCELAIN...........4 INCHES, VISIBLE ESCAPEMENT
CASE, HEAVY METAL, FINISHED IN POLISHED GOLD PLATE, FITTED WITH BEVELED FRENCH PLATE GLASS FRONT, SIDES AND BACK
FINISH, ON TRIMMINGS, ORMOLU GOLD PLATE OR FRENCH BRONZE
MOVEMENT, 8 DAY HALF-HOUR STRIKE, CATHEDRAL GONG
PRICE ....LIST, $42.00

372

1911

TIOGA

HEIGHT ....12¼ INCHES
WIDTH ....9 INCHES
DIAL, PORCELAIN...........4 INCHES, VISIBLE ESCAPEMENT
CASE, HEAVY METAL, FINISHED IN POLISHED GOLD PLATE, FITTED WITH BEVELED FRENCH PLATE GLASS FRONT, SIDES AND BACK
FINISH, ON TRIMMINGS, ORMOLU GOLD PLATE OR FRENCH BRONZE
MOVEMENT, 8 DAY HALF-HOUR STRIKE, CATHEDRAL GONG
PRICE ....LIST, $40.50

373

1911

THOREAU

THOREAU

HEIGHT ....15¼ INCHES
WIDTH ....8½ INCHES
DIAL, PORCELAIN...........4 INCHES, VISIBLE ESCAPEMENT
CASE, HEAVY METAL, FINISHED IN POLISHED GOLD PLATE, FITTED WITH BEVELED FRENCH PLATE GLASS FRONT, SIDES AND BACK
FINISH, ON TRIMMINGS, ORMOLU GOLD PLATE OR FRENCH BRONZE
MOVEMENT, 8 DAY HALF-HOUR STRIKE, CATHEDRAL GONG
PRICE ....LIST, $32.00

CRYSTAL REGULATORS

375

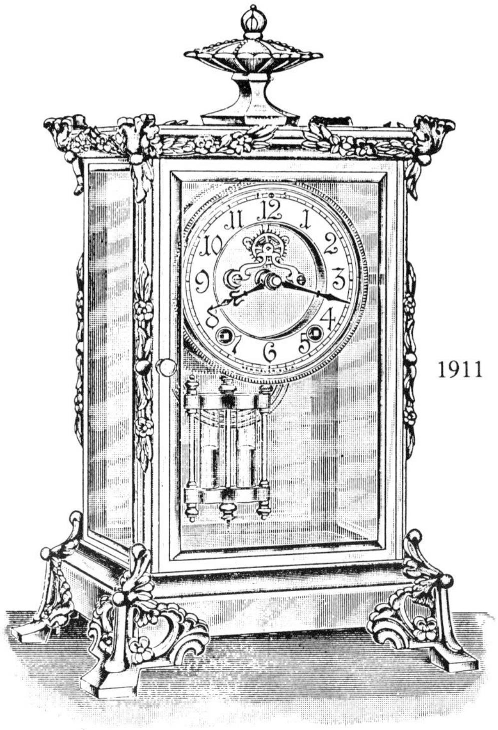

1911

TYCOON

HEIGHT....................14¼ INCHES
WIDTH ....8½ INCHES
DIAL, PORCELAIN..........4 INCHES, VISIBLE ESCAPEMENT
CASE, HEAVY METAL, FINISHED IN POLISHED GOLD PLATE, FITTED WITH BEVELED FRENCH PLATE GLASS FRONT, SIDES AND BACK
FINISH, ON TRIMMINGS, ORMOLU GOLD PLATE OR FRENCH BRONZE
MOVEMENT, 8 DAY HALF-HOUR STRIKE, CATHEDRAL GONG
PRICE ....LIST, $40.50

374

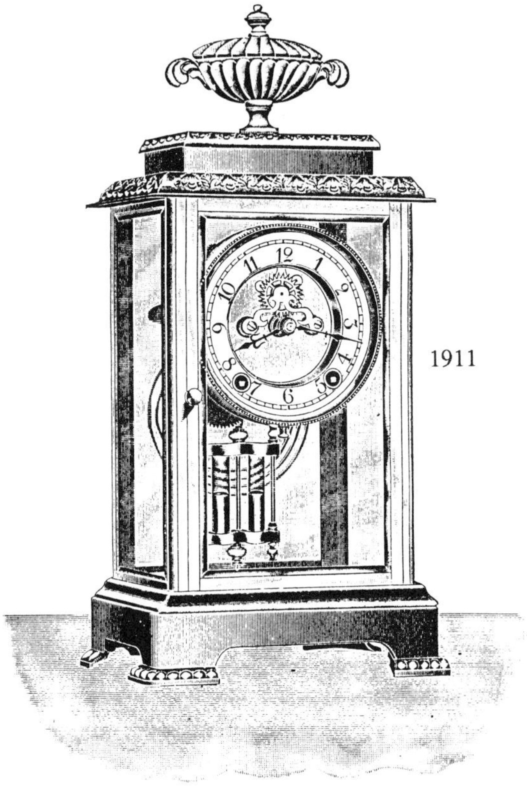

1911

TUCSON

HEIGHT ....14¼ INCHES
WIDTH..... ....7 INCHES
DIAL, PORCELAIN...........4 INCHES, VISIBLE ESCAPEMENT
CASE, HEAVY METAL, FINISHED IN POLISHED GOLD PLATE, FITTED WITH BEVELED FRENCH PLATE GLASS FRONT, SIDES AND BACK
FINISH, ON TRIMMINGS, ORMOLU GOLD PLATE OR FRENCH BRONZE
MOVEMENT, 8 DAY HALF-HOUR STRIKE, CATHEDRAL GONG
PRICE ....LIST, $39.25

376

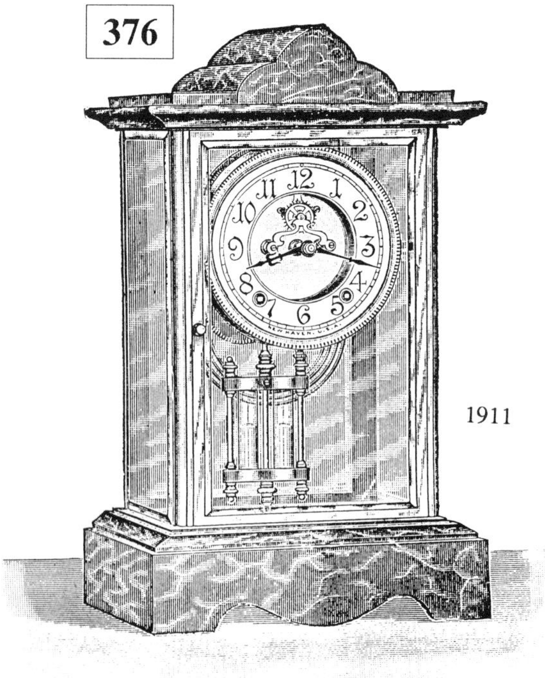

1911

THOPAS

THOPAS

HEIGHT ....13 INCHES
WIDTH.......................7½ INCHES
DIAL, PORCELAIN...........4 INCHES, VISIBLE ESCAPEMENT
CASE, HEAVY METAL WITH BEVELED FRENCH PLATE GLASS FRONT, SIDES AND BACK
FINISH, GOLD PLATED WITH IMPORTED BRAZILIAN ONYX TOP AND BOTTOM
MOVEMENT, 8 DAY HALF-HOUR STRIKE, CATHEDRAL GONG
PRICE ....LIST $48.50

CRYSTAL REGULATORS

377

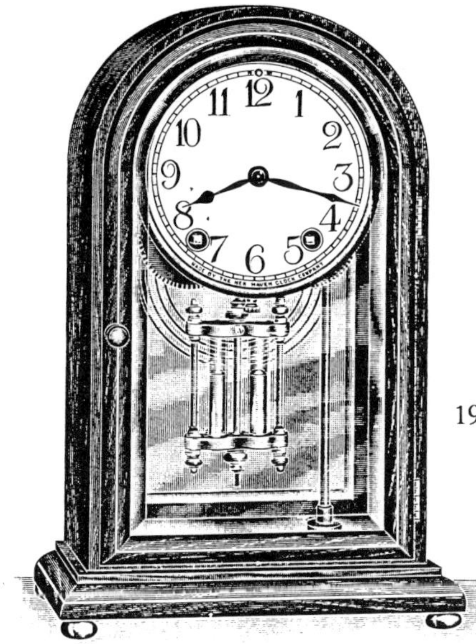

1917

THURSTON
Height, 11¼ Inches Width, 7 Inches
Dial, Porcelain, 4 Inches
Case, Solid Mahogany
Beveled Plate Glass Front and Back

378

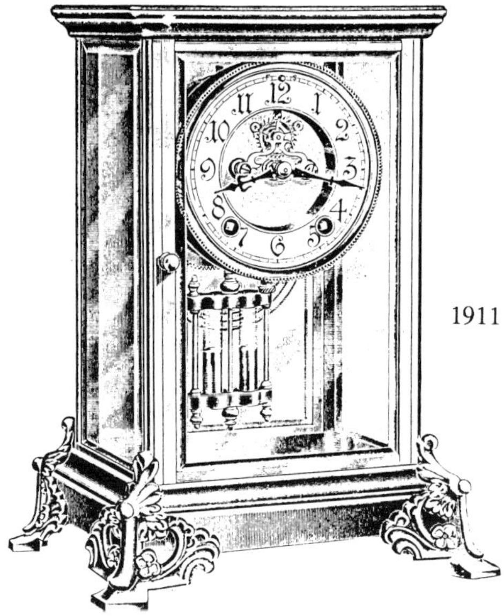

1911

TOLUSO

HEIGHT ....11½ INCHES
WIDTH ....8½ INCHES
DIAL, PORCELAIN...........4 INCHES, VISIBLE ESCAPEMENT
CASE, HEAVY METAL, FINISHED IN POLISHED GOLD PLATE, FITTED WITH BEVELED FRENCH PLATE GLASS FRONT, SIDES AND BACK
FINISH, ON TRIMMINGS, ORMOLU GOLD PLATE OR FRENCH BRONZE
MOVEMENT, 8 DAY HALF-HOUR STRIKE, CATHEDRAL GONG
PRICE ....LIST, $34.75

379

1911

TORLA

HEIGHT ....12 INCHES
WIDTH ....8½ INCHES
DIAL, PORCELAIN...........4 INCHES, VISIBLE ESCAPEMENT
CASE, HEAVY METAL, FINISHED IN POLISHED GOLD PLATE, FITTED WITH BEVELED FRENCH PLATE GLASS FRONT, SIDES AND BACK
FINISH, ON TRIMMINGS, ORMOLU GOLD PLATE OR FRENCH BRONZE
MOVEMENT, 8 DAY HALF-HOUR STRIKE, CATHEDRAL GONG
PRICE ....LIST, $39.25

380

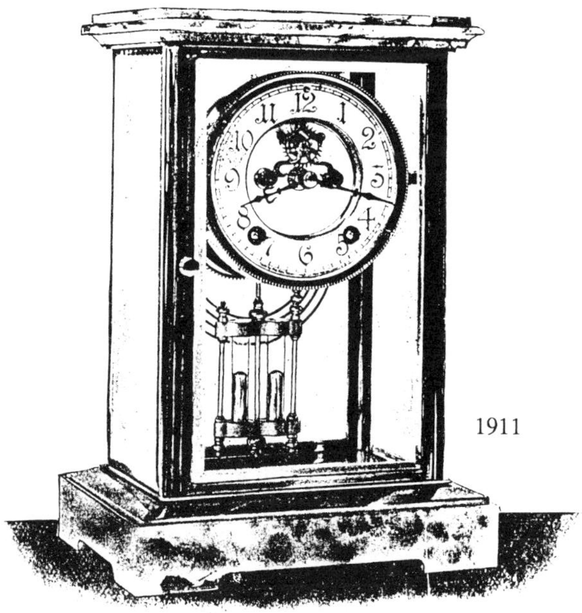

1911

THRACE

HEIGHT ....11 INCHES
WIDTH ....7½ INCHES
DIAL, PORCELAIN...........4 INCHES, VISIBLE ESCAPEMENT
CASE, HEAVY METAL WITH BEVELED FRENCH PLATE GLASS FRONT, SIDES AND BACK
FINISH, GOLD PLATED WITH IMPORTED BRAZILIAN ONYX TOP AND BOTTOM
MOVEMENT, 8 DAY HALF-HOUR STRIKE, CATHEDRAL GONG
PRICE ....LIST $45.00

CRYSTAL REGULATORS

Movements are Rack and Snail, 8-Day Half Hour Strike, Cathedral Gong. Plates and Wheels Polished

381

1917

THEONE
Height, $11\frac{1}{4}$ Inches Width, 7 Inches
Dial, Porcelain, 4 Inches
Case, Solid Mahogany
Beveled Plate Glass Front and Back

382

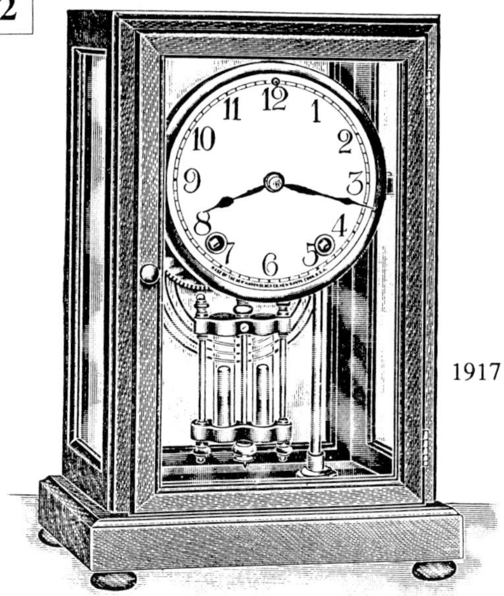

1917

TUXFORD
Height, $10\frac{1}{2}$ Inches Width, $6\frac{7}{8}$ Inches
Dial, Porcelain, 4 Inches
Case, Solid Mahogany
Beveled Plate Glass Front, Back and Sides

383

1917

THETA
Height, 13 Inches Width, 9 Inches
Dial, Porcelain, 7 Inches
Case, Solid Mahogany
Beveled Plate Glass Front and Back

384

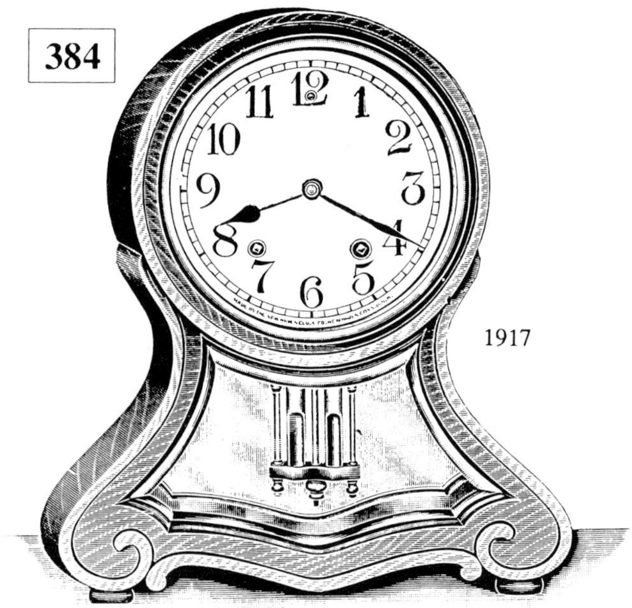

1917

TAUNTON

Movement Rack and Snail, 8-Day Half-hour Strike, Cathedral Gong. Plates and Wheels Polished.

Height, $11\frac{1}{2}$ Inches Width, 10 Inches
Dial, Silver Finish, 7 Inches

Case, Solid Mahogany. Beveled Plate Glass Front and Back.

CRYSTAL REGULATORS

385

Back view of **Theta**.
{Terry & Carole Looft Collection}

386

THETA

Height, 15¼ inches. Width, 9¾ inches. Depth, 6 ¼ inches. Dial 5 ¼ inches. All beveled glass. Four sides. Mercury pendulum. Porcelain dial. Inlaid caning around the top. 8-day time and strike.
{Terry & Carole Looft Collection}

387

Close-up back view of **Theta** movement.
{Terry & Carole Looft Collection}

388

Side view of **Theta** showing glass side and canning on the top.
{Terry & Carole Looft Collection}

DIALS

389

1895

Empire Sash, Convex Glass, Porcelain Dial.

No. 221.................................6 inch.
" 228.................................5 "

390

1895

Empire Sash, Convex Glass, Silver Dial, with Gilt Overlay.

No. 219.................................6 inch.
" 222.................................5 "

391

1895

Empire Sash, Convex Glass, Porcelain Dial, Visible Escapement.

No. 223.................................6 inch.
" 224.................................5 "

392

1895

Porcelain Dial, French Sash.

No. 225.................................6 inch.
" 229.................................5 inch.

393

1895

American Gilt.

No. 238.....................6 inch.
" 242.....................6 " —New

394

1895

Porcelain Dial, French Sash, Visible Escapement.

No. 227.................................6 inch.
" 223.................................5 "

395

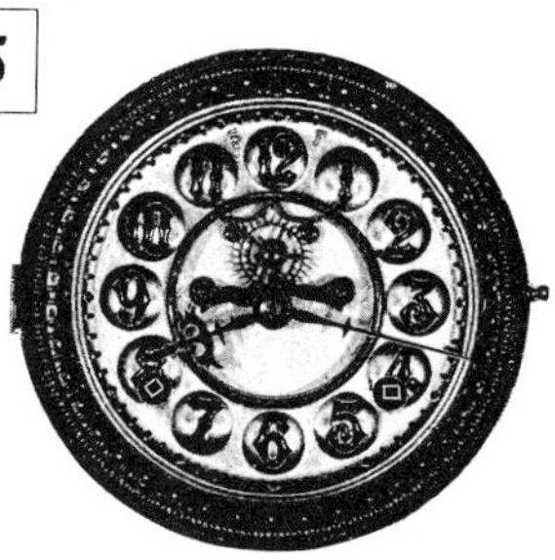

1895

American Gilt, Visible Escapement.

No. 241.................................6 inch.

396

1895

American White, Visible Escapement.

No. 237.................6 inch—Roman.
" 235.................6 " —Arabic.

397

1895

American White.

No. 236.................6 inch—Roman.
" 234.................6 " —Arabic.

398

1895

Empire Sash, Convex Glass, Silver Dial with Gilt Overlay.

No. 239 ....4 inch.

399

1895

Empire Sash, Convex Glass, Silver Dial, Gilt Center.

No. 240.................................4 inch.

400

1895

American Gilt.

No. 231 ....4 inch.

DIALS

401

1900

Decorated Ivorized Porcelain Dial,
with Cast Sash.
Made in 4 in. and 3 in.

402

1900

Pearl Dial, with Cast Sash.
Made in 6 in., 5 in., 4 in. and 3 in.
Also made for Visible Escapement
in 6 in. and 5 in.

403

1900

Pearl Dial, with Struck Sash.
Made in 6 in., 5 in., 4 in. and 3 in.
Also made for Visible Escapement
in 6 in. and 5 in.

404

1900

Ivorine Dial, with Cast Sash.
Made in 6 in., 5 in., 4 in. and 3 in.
Also made for Visible Escapement
in 6 in. and 5 in.

405

1900

Ivorine Dial, with Struck Sash.
Made in 6 in., 5 in., 4 in. and 3 in.
Also made for Visible Escapement
in 6 in. and 5 in.

406

1900

Plain Dial, with Struck Sash.
Made in 6 in., 5 in., 4 in. and 3 in.
Also made for Visible Escapement
in 6 in. and 5 in.

407

1900

Gilt Dial, with Struck Sash.
Made in 6 in. and 5 in.
Also made for Visible Escapement
in 6 in. and 5 in.

408

1900

Ivorized Porcelain Dial,
with Cast Sash.
Made in 6 in., 5 in., 4 in. and 3 in.

409

1900

Ivorized Porcelain Dial,
Visible Escapement, Cast Sash.
Made in 6 in. and 5 in.

410

1900

Ivorized Porcelain Dial,
Cast Sash, with Jewels.
Made in 5 in., 4 in. and 3 in.

ELECTRIC CLOCKS

411

1929

"WHITNEY" ELECTRIC
Height, 30½ inches. Width, 9½ inches. 6-inch Silvered Dial. Mahogany Finished Case. Attractive Ornaments and Decorated Panels Large Sweep Second Hand, Front Starting Lever and Sealed Motor.
Keystone List.........$33.20

412

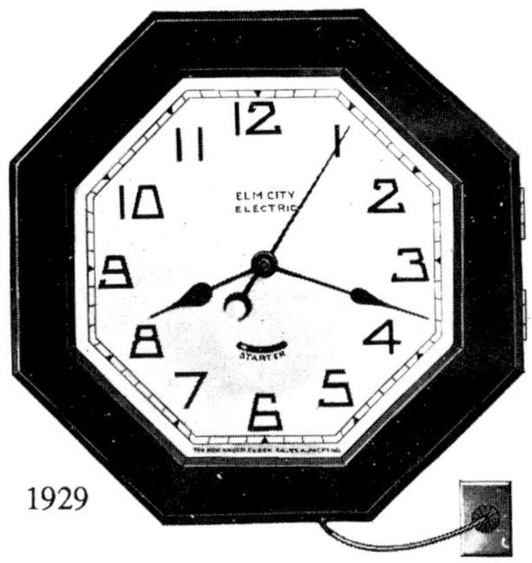

1929

"GALLERY" ELECTRIC
Width, 14½ inches. 12-inch Dial. Mahogany Finished Case of Octagon Design. Large Sweep Second Hand, Front Starting Lever and Sealed Motor.
Keystone List.........$21.00

413

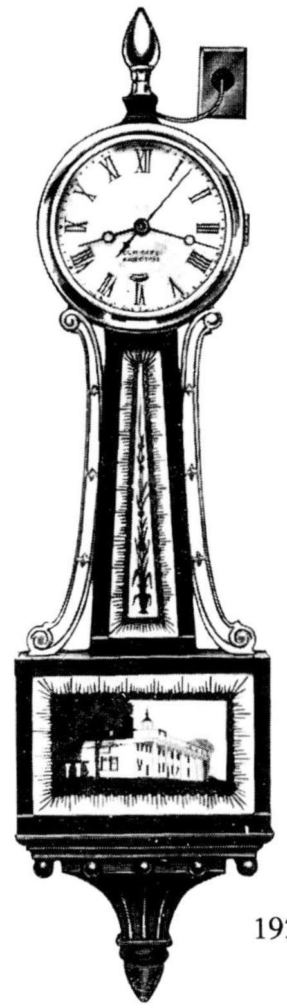

1929

"WHALLEY" ELECTRIC
Height, 32 inches. Width, 8 inches. 6-inch Antique Dial, Cast Sash, Attractive side and top Ornaments. Solid Mahogany Case with Decorated Panels in Gold Leaf. Large Sweep Second Hand, Front Starting Lever and Sealed Motor.
Keystone List.........$59.10

414

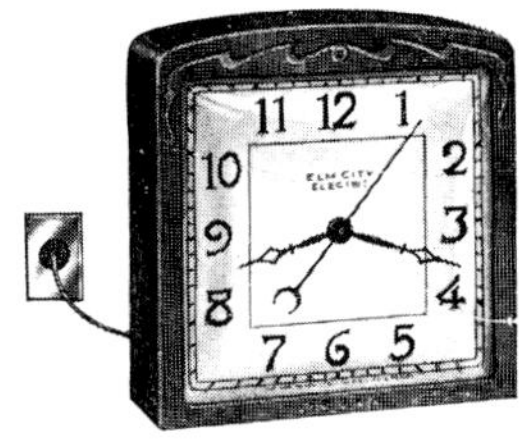

1929

"SQUARE" ELECTRIC
Height, 4¼ inches. Mahogany Finished Moulded Case. Krack-Proof Krystal. Square Enamel Dial. Large Sweep Second Hand, Starting Lever on Back and Sealed Motor.
Keystone List.....$13.30

415

1929

"LARINA" ELECTRIC
Width, 8½ inches. 5½ inch Dial. Convex Crystal, Colors, Rose, Yellow, Blue, Green or White. Large Sweep Second Hand, Front Starting Lever and Sealed Motor.
Keystone List........................$13.30

416

1929

"12-INCH WALL" ELECTRIC
Width, 14½ inches. 12-inch Dial. Mahogany Finish. Large Sweep Second Hand, Front Starting Lever and Sealed Motor.
Keystone List...............$21.00

"8-INCH WALL" ELECTRIC
Width 9¾ inches. 8-inch Dial. White Enamel. Otherwise same as above.
Keystone List................$18.20

417

1929

"MARAVIA" ELECTRIC
Height, 27½ inches. Width, 16 inches. 12-inch Antique Dial, Ornate Hands, Solid Mahogany Hand Carved Case with Maple Panel. Large Sweep Second Hand, Front Starting Lever and Sealed Motor.
Keystone List........................$59.10

"MADENA" ELECTRIC
Height 18½ inches. Width 11½ inches. 8-inch Dial. Otherwise similar to above
Keystone List.........................$51.80

418

1929

"WARWICK" ELECTRIC
Height, 22½ inches. Width, 7⅜ inches. 6-inch Dial, Convex Crystal, Decorated Panels, Attractive side and top Ornaments. Mahogany Finish. Large Sweep Second Hand, Front Starting Lever and Sealed Motor.
Keystone List.........$21.00

ELECTRIC CLOCKS

Elm City *Electric* Clocks

For use on alternating current only, regularly furnished for 110-120 volts, 60 cycle.

New Haven Electrics made in the early 1930's using Westinghouse motors are often marked New Haven-Westinghouse.

419

1929

"AMIENS" ELECTRIC

Height, 12 inches. Width, 10 inches. 6-inch Silvered Dial with Raised Numerals. Beveled Convex Crystal. Solid Mahogany Antique Finish with Two-Tone Panels. Large Sweep Second Hand, Front Starting Lever and Sealed Motor.

Keystone List............$72.80

420

1929

"GOTHIC" ELECTRIC

Height 7½ inches. Width, 5½ inches. 3½ inch Silvered Dial. Convex Crystal. Mahogany Finished Moulded Case. Large Sweep Second Hand, Front Starting Lever and Sealed Motor.

Keystone List......$18.20

421

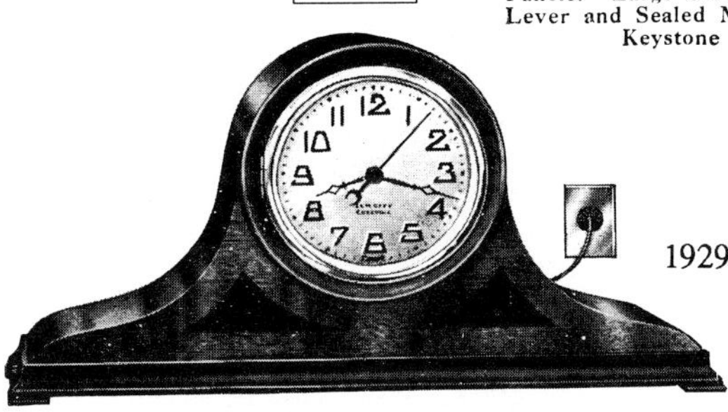

1929

"PETITE" ELECTRIC

Height, 6¾ inches. Width, 13 inches. 3½ inch Silvered Dial with Raised Numerals. Solid Mahogany Case, Antique Finish with Dark Panels. Large Sweep Second Hand, Front Starting Lever and Sealed Motor.

Keystone List................$21.00

422

1929

"TAMBOUR NO. 90" ELECTRIC

Height, 9¼ inches. Width, 20 inches. 5½-inch Silvered Dial. Convex Crystal. Mahogany Finished Case. Large Sweep Second Hand, Front Starting Lever and Sealed Motor.

Keystone List........................$21.00

423

1929

"HARMONIOUS" ELECTRIC

Height, 9¼ iches. Width, 20 inches. 5½ inch Silvered Dial with Raised Numerals. Beveled Convex Crystal, Mahogany Finished Case. Large Sweep Second Hand, Front Starting Lever and Sealed Motor.

Keystone List......................$25.90

424

1929

"TOLEDO" ELECTRIC

Height, 9 inches. Base 19 inches. 5½ inch Silvered Dial with Raised Numerals, Convex Crystal, Solid Mahogany Case with Dark Panels and Base. Large Sweep Second Hand, Front Starting Lever and Sealed Motor.

Keystone List.........................$42.00

ELECTRIC CLOCKS

425

Ca. 1937

LURAY Electric
NHW 611-53T

An all-metal case finished in Black, Ivory or Green with heavy nickel sash and flush back. Attractive 5-inch ivory-tinted dial with modernistic hands and bowed glass crystal. Diameter of case 6⅞ inches. Depth of case 1¾ inches.

426

Ca. 1937

FARMINGTON Strike NHW 611-85S

An early American design with Mt. Vernon panel scene, mahogany case in antique finish, hour and half-hour strike, sweep second hand, Roman numeral dial and rubber cushion feet. Width, 8¼ inches. Height, 14 inches.

427

Ca. 1937

SHELLEY Strike NHW 611 76S

A distinctive upright mahogany case in antique finish with raised figured panel, hour and half-hour strike, etched numerals, 5⅝-inch silvered dial, bowed crystal, sweep second hand and rubber cushion feet. Height, 10½ inches. Width, 9¼ inches.

428

Ca. 1937

TURIN Chime or Strike
SELF-STARTING ELECTRIC or 8-DAY PENDULUM

Case of mahogany plywood with raised panel. Two-tone metal dial, 5¼ inches wide. Black hands with raised, gold-plated, applied numerals. Polished brass sash with convex crystal. Full Westminster, self-adjusting, quarter-hour chimes and hour strike; or hour and half-hour strike. Height, 11⅝ inches. Width, 10 inches. Packed singly. Shipping weight: 11 lbs.

RETAIL $25.00 (Chime)
RETAIL $16.75 (Strike)

429

Ca. 1937

TAMBOUR No. 8 Strike NHW 611-73S

A mahogany tambour case in antique finish with raised dark panels, beaded base design, 4⅞ inch silvered dial, hour and half-hour strike, sweep second hand, polished brass sash, convex crystal and rubber cushion feet. Base, 18 inches. Height, 8 inches.

430

Ca. 1937

WHITNEY Electric Strike EC 611-37S

An attractive banjo clock with decorative panels and ornaments, mahogany finish; hour and half-hour strike; 5⅝-inch silvered dial, raised numerals, with sweep second hand; convex crystal with polished brass sash; rubber cushion wall rests. Height, 29¼ inches. Width, 9⅝ inches

431

Ca. 1937

GALLERY ELECTRIC
EC 611-30T

A commercial office timepiece in octagon case, mahogany finish; dependable timepiece movement; 12-inch white dial with large prominent hands and sweep second hand; rubber cushion wall rests. Width, 14½ inches.

ELECTRIC CLOCKS

432

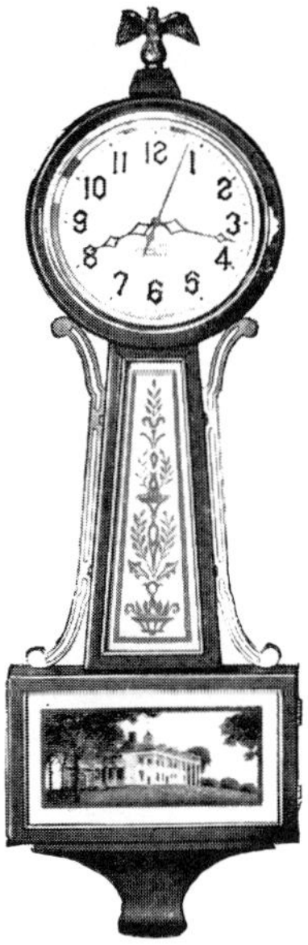

Ca. 1937

BANJO No. 3 **NHW 611-131T**

A banjo timepiece in a light mahogany finish with picture panels; 4⅞-inch silvered dial, raised numerals, sweep second hand, convex crystal, polished brass sash; rubber cushion wall rests, opening in back for wall plug. Height, 24¼ inches. Width, 7½ inches.

433

Ca. 1937

EASEL

Available with either timepiece or alarm movement. The moulded case is a mahogany shade of Gothic design mounted on an easel stand with white-enamel dial, buffed brass sash and unbreakable crystal. Height, 4⅝ inches. Width of stand, 4⅝ inches.

Easel No. 1 Timepiece
Easel No. 2 Alarm

434

Ca. 1937

CAMPUS ALARM

A moderately priced electric alarm clock in cast metal case with full-toned concealed alarm bell and white legible dial. Available in Green, Blue or Rose with Nickel trim. Also made in Black with Nickel trim. Width, 5 inches. Height, 6¼ inches.

435

Ca. 1937

EUREKA ALARM

A charming continental design in a nickel and black metal case, 3⅛-inch ivory-tinted dial, nickel-plated ball feet with rubber rests and full-toned alarm bell. Width, 3¾ inches. Height, 4¼ inches.

436

Ca. 1937

FUTURIST Electric Alarm

An attractively styled, all metal case in Black and Gold; cream enamel dial; black cubist numerals and modernistic hands. The concealed alarm bell is full-toned, with a pleasant rhythmic beat Width, 4 inches. Height, 6 inches.

New Haven Electrics made in the early 1930's using Westinghouse motors are often marked New Haven-Westinghouse.

437

Ca. 1937

14-INCH WALL **EC 611-31T**

A large commercial timepiece in a case of imported selected mahogany, antique finish, corner panels in relief; 14-inch white dial, large prominent hands and numerals, sweep second hand, rubber cushion wall rests, two wall hangers, Width, 16¼ inches.

438

Ca. 1937

INCA **NHW 611-134T**

A desk timepiece of selected mahogany in a light antique finish, 3½ inch silvered dial, polished brass sash, beveled crystal, brass feet. Height, 7¾ inches. Width, 5¼ inches.

439

Ca. 1937

ELKO Timepiece **2 TB**

A low priced timepiece for mantel or radio, in mahogany finish with a decorated panel, fluted sides, 4⅞-inch raised numeral dial, buffed brass sash and rubber cushion feet. Height, 7¾ inches. Width, 6⅞ inches.

ELECTRIC CLOCKS

440

Ca. 1937

ELWOOD Strike

SELF-STARTING ELECTRIC

Imported selected mahogany case. 5¼-inch, two-tone metal dial. Convex crystal with polished brass sash. Hour and half-hour strike. Height, 11¼ inches. Width, 7½ inches. Packed singly. Shipping weight: 8 lbs.

RETAIL $11.25

441

Ca. 1937

ELECTRIC TATTOO Alarm

A handsome metal case in nickel and black with a 3¼-inch metal dial, heavy cast base with rubber rests and full toned alarm bell. Width at base, 4 inches. Height, 4⅜ inches.

442

Ca. 1937

COMMUTER ALARM

An attractive design with a nickel trimmed black metal case, heavy cast base with rubber rests, 3¼-inch ivory tinted dial and full-toned alarm bell. Width at base, 4 inches. Height, 4⅜ inches.

444

Ca. 1937

BANTRY Alarm

SELF-STARTING ELECTRIC

Metal case available in black and chrome or ivory and polished brass. 3½ inch, two-tone dial. Patented bell alarm NOT OF THE ORDINARY BUZZER TYPE. Ball feet with rubber rests. Height, 4¼ inches. Width, 3¾ inches. Depth, 2 inches. Packed singly. Shipping weight: 1 lb. 12 oz.

RETAIL $3.95

443

STYLIS Timepiece Ca. 1937

SELF-STARTING ELECTRIC

This timepiece tells time at a glance with large, legible numerals which flip over like the pages of a book. Case of natural mahogany with white holly trim. Brown numerals on ivory-color tabs to harmonize with the case. Height, 3¾ inches. Width, 7¾ inches. Depth, 3¾ inches. Packed singly. Shipping weight: 3 lbs. 10 oz.

RETAIL $9.95

445

Ca. 1937

OCTET ALARM

A modernistic style in black metal with nickel trim, 3¼-inch metal dial, heavy cast base with rubber rests and full-toned alarm bell. Width at base, 4 inches. Height, 4¼ inches.

446

Ca. 1937

MARS Alarm

SELF-STARTING ELECTRIC

Small moulded clock in Walnut shade, beautifully grained. Inlaid brass numerals and hour indicators. Metal dial. Patented bell alarm NOT OF THE ORDINARY BUZZER TYPE. Height, 5⅛ inches. Width, 4½ inches. Depth, 2½ inches. Packed singly. Shipping weight: 2 lbs. 2 oz.

RETAIL $3.95

447

ONYX No. 2 Timepiece Ca. 1937

SELF-STARTING ELECTRIC

Beautifully mottled onyx case. Dark brown enameled numeral band with polished brass numerals. Two-tone metal dial. Polished brass sash. Adjustable easel support. Height, 6 inches. Width, 6 inches. Depth, 2 inches.

RETAIL $9.95

448

Ca. 1937

DEVERE Strike

SELF-STARTING ELECTRIC

Modernized Gothic, solid mahogany case combined with plywood. African-scented veneer front. Dark brown enameled numeral band—5⅛ inches in diameter. Two-tone metal dial with polished brass sash. Hour and half-hour strike. Height, 9¼ inches. Width, 7 inches. Packed singly. Shipping weight: 3 lbs. 7 oz.

RETAIL $15.00

ELECTRIC CLOCKS

449

Ca. 1937

WINSLOW EASEL Alarm
SELF-STARTING ELECTRIC

Chromium plated case of simple linear design. 3½-inch square, two-tone silvered dial. Self-adjusting easel. Patented bell alarm NOT OF THE ORDINARY BUZZER TYPE. 4½ inches square. 2 inches deep. Packed singly. Shipping weight: 2 lbs.

RETAIL $5.95

WINSLOW EASEL Timepiece

—Similar to the above but with round, two-tone bronze finish dial; and self-starting electric timepiece movement.

RETAIL $4.95

450

Ca. 1937

VILNA Timepiece
SELF-STARTING ELECTRIC

Solid mahogany case with redwood buri veneer front. 3½-inch, two-tone bronze finish dial with polished brass sash. Height, 6⅛ inches. Width, 6 inches. Depth, 2⅜ inches. Packed singly. Shipping weight: 2 lbs. 7 oz.

RETAIL $6.95

451

Ca. 1937

RENFREW Timepiece
SELF-STARTING ELECTRIC

Solid mahogany case with African-scented veneer front and white holly trim. 3½-inch, two-tone silvered dial with etched numerals and polished brass sash. Height, 5¾ inches. Width, 5¾ inches. Depth, 2¾ inches. Packed singly. Shipping weight: 2 lbs. 7 oz.

RETAIL $6.95

452

Ca. 1937

CLAREMONT Timepiece
SELF-STARTING ELECTRIC

Natural finish solid mahogany case enriched with maple and red gum ply. 3½-inch, two-tone bronze finish dial with polished brass sash. Height, 6 inches. Width, 5⅝ inches. Depth, 2¼ inches. Packed singly. Shipping weight: 2 lbs. 8 oz.

RETAIL $6.95

453

Ca. 1937

MONTROSE Timepiece
SELF-STARTING ELECTRIC

Beautifully grained solid mahogany and plywood case. 3½-inch silvered dial with etched numerals. Polished brass sash. Width, 12¼ inches. Height, 5⅝ inches. Depth, 2¾ inches. Packed singly. Shipping weight: 3 lbs. 2 oz.

RETAIL $6.95

454

Ca. 1937

ALFORD Timepiece
SELF-STARTING ELECTRIC

Case of solid mahogany with white holly trim. Ebony-finished base. 3½-inch, two-tone bronze finish dial with polished brass sash. Height, 5⅛ inches. Width, 4½ inches. Depth, 2 inches. Packed singly. Shipping weight: 1 lb. 15 oz.

RETAIL $6.95

ALFORD Electric Alarm

Same as above but with patented bell alarm NOT OF THE ORDINARY BUZZER TYPE.

RETAIL $7.95

455

Ca. 1937

DEVON Timepiece
SELF-STARTING ELECTRIC

Case of plywood herring-bone, matched walnut veneer. 3½-inch, two-tone bronze finish dial with polished brass sash. Height, 5½ inches. Width, 5 inches. Depth, 3½ inches. Packed singly. Shipping weight: 2 lbs. 3 oz.

RETAIL $6.95

456

Ca. 1937

SEAMAN Timepiece
SELF-STARTING ELECTRIC

A marine-style clock—ideal for the den, library or living room. Dark brown enameled numeral band on a solid mahogany case. Numerals, hands, sash and wheel spokes are of polished brass. Metal dial. Height, 7¾ inches. Width, 7½ inches. Depth, 3 inches. Packed singly. Shipping weight: 2 lbs. 14 oz.

RETAIL $7.50

FIGURE CLOCKS

457

1885

DON CÆSAR.

8 Day Gong Strike, Visible Escapement.

Height, 20½ inches. Dial, 4 inches. .$44 50

458

1885

DON JUAN.

Height, 21 inches. Dial, 4 inches.

8 Day Gong Strike, Visible Escapement. $44 25

459

1895

GAUL.

Height, 16 inches; Width, 20 inches; Dial, 6 inches.

8 Day, Half-Hour Strike on Cathedral Gong, with Empire Sash, Porcelain Dial, Visible Escapement.

List, $32.25

GAUL.

FIGURE CLOCKS

460

1885

GAULOIS AND AMAZON
Height, 20 inches. Dial, 4 inches.
8 Day Gong Strike, Visible Escapement....$54.50

GAULOIS AND AMAZON

461

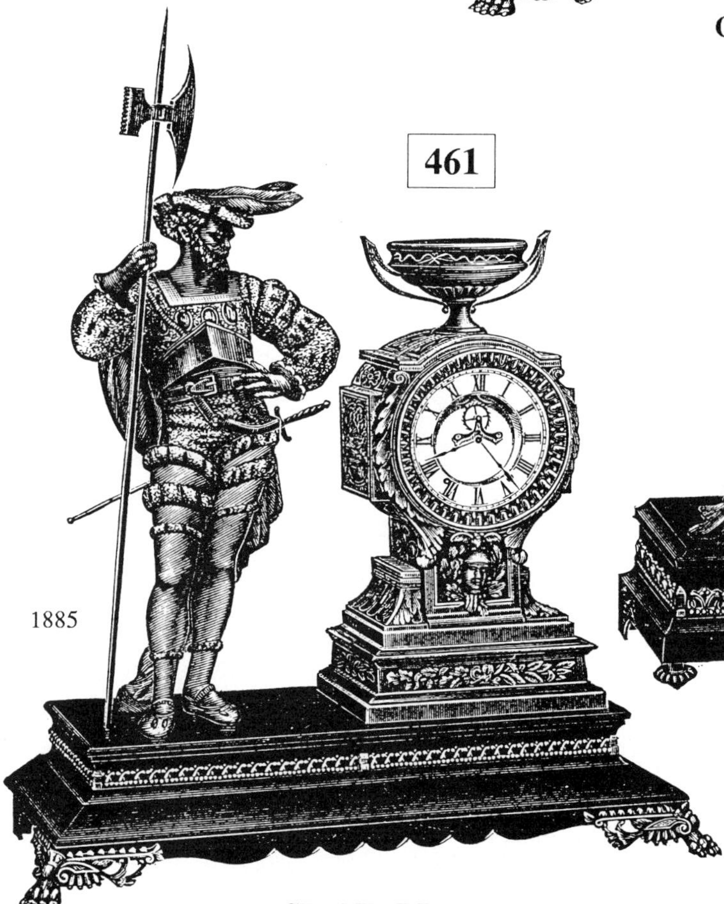

1885

SPANIARD
Height, 25 inches. Dial, 4 inches.
8 Day Gong Strike, Visible Escapement....$47.50

462

1885

PSYCHE
Height, 16 inches. Dial, 4 inches.
8 Day Gong Strike, Visible Escapement....$35.75

FIGURE CLOCKS

463

1885

RAPHAEL

Height, 15 inches. Dial, 4 inches.
8 Day Gong Strike, Visible Escapement....$39.00

464

1885

IMOGENE

Height, 17 ½ inches. Dial, 4 inches.
8 Day Gong Strike, Visible Escapement....$35.75

465

1885

VAN DYKE

Height, 15 inches. Dial, 4 inches.
8 Day Gong Strike, Visible Escapement....$39.50

466

1886

RUBENS

Height, 16½ inches. Width,4½ inches.
8-Day Gong Strike. Visible Escapement. .. $35.30

FIGURE CLOCKS

467

1900

BENVENUTO CELLINI

Height, 16 inches.
Width, 20 inches.
Dial, 6 inches.

Eight-day, Half-hour Strike, Cathedral Gong, with Empire Sash and Ivorized Porcelain Dial, Visible Escapement.

Finished in Gilt or Bronze.

List price, $23.75

468

1900

BERNARD PALISSY

Height, 16 inches.
Width, 20 inches.
Dial, 6 inches.

Eight-day, Half-hour Strike, Cathedral Gong, with Empire Sash and Ivorized Porcelain Dial, Visible Escapement.

Finished in Gilt or Bronze.

List price, $23.75

469

1900

OCTAVIUS

Height, 16 inches.
Width, 20 inches.
Dial, 6 inches.

Eight-day, Half-hour Strike, Cathedral Gong, with Empire Sash and Ivorized Porcelain Dial, Visible Escapement.

Finished in Gilt or Bronze.

List price, $22.25

470

1900

FLUTE PLAYER

Height, 16 inches.
Width, 20 inches.
Dial, 6 inches.

Eight-day, Half-hour Strike, Cathedral Gong, with Empire Sash and Ivorized Porcelain Dial, Visible Escapement.

Finished in Gilt or Bronze.

List price, $22.25

FIGURE CLOCKS

471

1889

HUNTER AND DOG.

Height, 15½ inches.

8 Day Gong Strike, Visible Escapement, Bronze....$31.50
8 Day Gong Strike, Visible Escapement, Silver.....$34.50

472

1889

CLOVIS.

Base, 22¾ inches. Height, 17 inches.

8 Day, Gong Strike, Visible Escapement, Bronze.....$40.00
8 Day, Gong Strike, Visible Escapement, Silver...... 43.00

473

1900

CARTHAGE

Height, 20 inches.
Width, 20 inches.
Dial, 6 inches.

Eight-day, Half-hour Strike, Cathedral Gong, with Empire Sash and Ivorized Porcelain Dial, Visible Escapement.

Finished in Gilt or Bronze.

List price, $27.80

CARTHAGE.

474

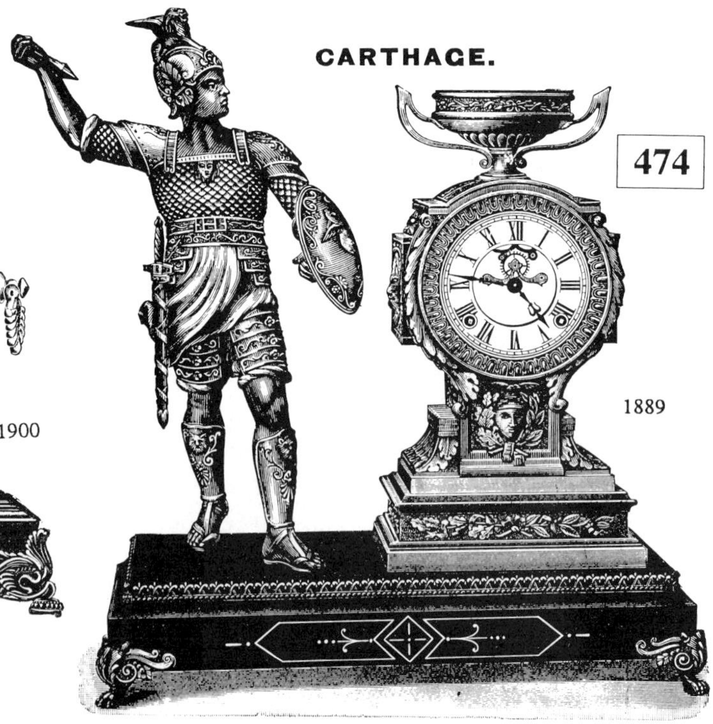

1889

Base, 17 inches. Height, 19¾ inches.

8 Day, Gong Strike, Visible Escapement., Bronze..$36.00
8 Day, Gong Strike, Visible Escapement, Silver.... 39.00

FIGURE CLOCKS

475

1889

SIR CHRISTOPHER.

Base, 16¼ inches. Height, 15⅝ inches.

8 Day Gong Strike, Visible Escapement, Bronze $38.00
8 Day Gong Strike, Visible Escapement, Silver....... 41.00

476

1900

ROMAN

Height, 20 inches.
Width, 20 inches.
Dial, 6 inches.

Eight-day, Half-hour Strike, Cathedral Gong, with Empire Sash and Ivorized Porcelain Dial, Visible Escapement.

Finished in Gilt or Bronze.

List price, $27.80

477

1900

CLOTHO

Height, 16 inches.
Width, 20 inches.
Dial, 6 inches.

Eight-day, Half-hour Strike, Cathedral Gong, with Empire Sash and Ivorized Porcelain Dial, Visible Escapement.

Finished in Gilt or Bronze.

List price, $22.25

478

1889

MIGNON.

Base, 17 inches. Height, 19¼ inches.

8 Day, Gong Strike, Visible Escapement, Bronze....$40.00
8 Day, Gong Strike, Visible Escapement, Silver..... 43 00

FIGURE CLOCKS

INFANTRY.

479

1889

Base, 17 inches. Height, 20 inches.

8 Day Gong Strike, Visible Escapement, Bronze..$40.00
8 Day Gong Strike, Visible Escapement, Silver..$43.00

480

FISHER BOY.

1889

Base, 17 inches. Height, 28 inches.

8 Day, Gong Strike, Visible Escapement, Bronze......$41.00
8 Day, Gong Strike, Visible Escapement, Silver....... 44.00

481

LULLI.

1889

Base, 17¼ inches. Height, 21 inches.

8 Day, Gong Strike, Visible Escapement, Bronze..... $41.00
8 Day, Gong Strike, Visible Escapement, Silver....... 43.00

482

PHILIP.

1889

Height, 23 inches

8 Day Gong Strike, Visible Escapement, Bronze,.$37.50
8 Day Gong Strike, Visible Escapement, Silver,...$40.50

FIGURE CLOCKS

483

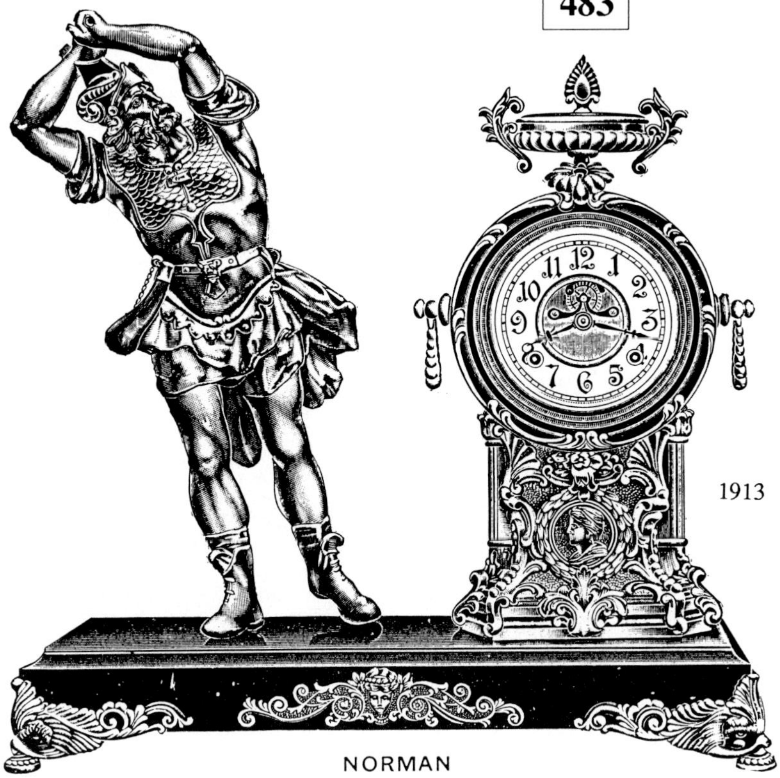

1913

NORMAN

LIST, $32.00

HEIGHT, 20½ INCHES. WIDTH, 20 INCHES.

484

NORMAN

Height, 20½ inches.
Width, 20 inches.
Dial, 6 inches.

Eight-day, Half-hour Strike, Cathedral Gong, with Empire Sash and Ivorized Porcelain Dial, Visible Escapement.

Finished in Gilt or Bronze.

List price, $27.80

1900

NORMAN

FIGURE CLOCKS

485

IVANHOE

Height, 16 inches.
Width, 20 inches.
Dial, 6 inches.

Eight-day, Half-hour Strike,
Cathedral Gong,
with Empire Sash and Ivorized
Porcelain Dial,
Visible Escapement.

Finished in Gilt or Bronze.

List price, $23.75

1900

IVANHOE

486

1906

IVANHOE EXTRA

Syrian Bronze Finish

Height, 14¾ inches. Width, 18 inches.
Six-inch Porcelain Dial, with Empire Sash.
Eight-day, Half-hour Strike, Cathedral Gong.

List Price, $29.25

487

1906

LA FONTE

Height, 16 inches. Width, 20 inches.
Six-inch V. E. Porcelain Dial.
French Sash, with Beveled Glass.
Finished in Gilt or Bronze.
Eight-day, Half-hour Strike, Cathedral Gong.

List Price, $23.75

FIGURE CLOCKS

488

1886

JEAN BART.

Height, 20 inches. Dial, 4½ inches.
8 Day Strike, Cathedral Gong, Visible Escapement .$43.00

489

1886

RUTH.

Height, 19 inches. Dial, 4½ inches.
Day Gong Strike, Visible Escapement ...$39.30

490

1906

LA FORGE

Height, 16 inches. Width, 20 inches.
Six-inch V. E. Porcelain Dial.
French Sash, with Beveled Glass.
Finished in Gilt or Bronze.
Eight-day, Half-hour Strike, Cathedral Gong.
List Price, $23.75

Also made with Gilt Sunk Center, $1.00 list extra.

491

1886

DUQUAY TRONIN.

Height, 20 inches. Dial, 4½ inches.
8 Day Strike, Cathedral Gong, Visible Escapement $43.00

FIGURE CLOCKS

BRONZE CLOCKS. 8-DAY HALF-HOUR STRIKE, CATHEDRAL GONG.
6-INCH PORCELAIN DIAL. VISIBLE ESCAPEMENT. FRENCH CAST SASH,
BEVELED GLASS. BLACK ENAMELED IRON BASE.
CLOCK, FIGURE AND TRIMMINGS FINISHED IN BRONZE OR GILT

492

BRENNUS

Height, 16 inches.
Width, 20 inches.
Dial, 6 inches.

Eight-day, Half-hour Strike,
Cathedral Gong,
with Empire Sash and Ivorized
Porcelain Dial,
Visible Escapement.

Finished in Gilt or Bronze.

List price, $22.25

1900

BRENNUS.

493

The year entered beside each clock is the year of the catalog from which the illustration was taken. Some clocks may have been produced a few years earlier.

BRENNUS

Height, 15½ inches.

	Price.
8 Day, Gong Strike, Visible Escapement, Bronze..	$28.75
8 Day, Gong Strike, Visible Escapement, Silver...	31.75

1889

BRENNUS.

FIGURE CLOCKS

494

1913

LA CARESSE

LIST, $27.60

HEIGHT, 16 INCHES. WIDTH, 20 INCHES.

495

1906

KNIGHT EXTRA

Syrian Bronze Finish

Height, 17 inches. Width, 18½ inches.
Six-inch Porcelain Dial, with Empire Sash.
Eight-day, Half-hour Strike, Cathedral Gong.

List Price, $33.35

496

1913

KNIGHT

LIST, $32.00

HEIGHT, 16 INCHES. WIDTH, 20 INCHES.

FIGURE CLOCKS

497

1889

REBECCA.

8 Day Gong Strike, Visible Escapement, Bronze. . .$36.50
8 Day Gong Strike, Visible Escapement, Silver. . . . 39.50

498

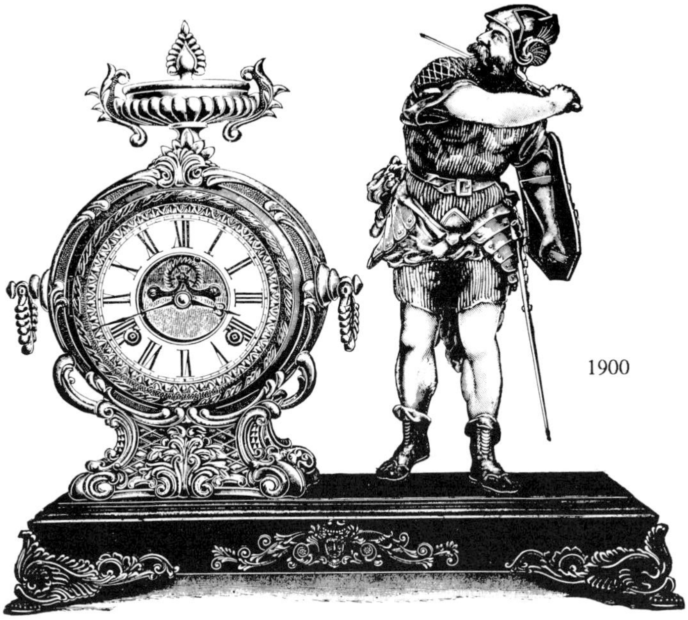

1900

SAXON

Height, 20 inches.
Width, 20 inches.
Dial, 6 inches.

Eight-day, Half-hour Strike, Cathedral Gong, with Empire Sash and Ivorized Porcelain Dial, Visible Escapement.

Finished in Gilt or Bronze.

List price, $27.80

499

1889

MARINE

8 Day, Gong Strike, Visible Escapement, Bronze. $40.00
8 Day, Gong Strike, Visible Escapement, Silver. . . 43.00

500

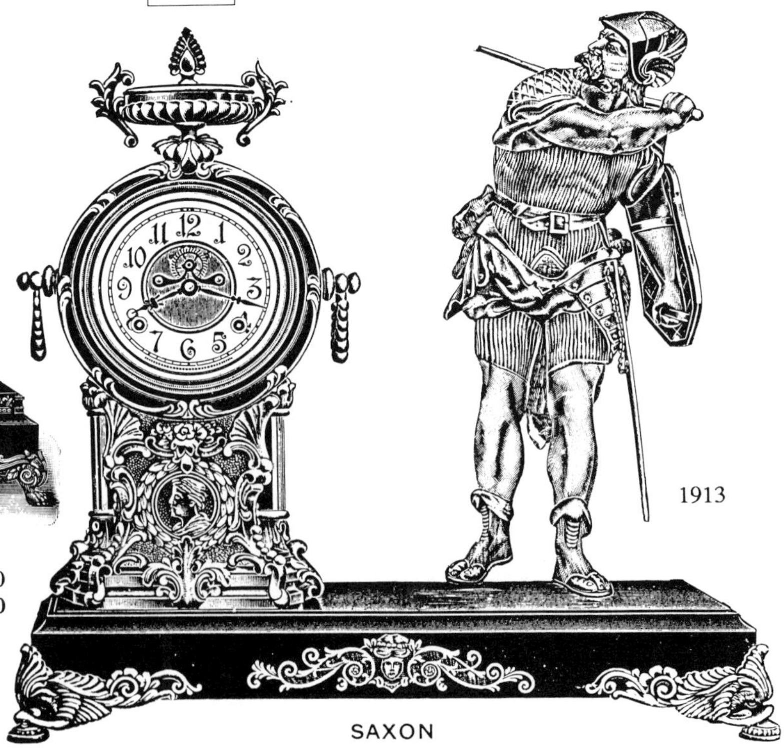

1913

SAXON

LIST, $32.00
HEIGHT, 19½ INCHES. WIDTH, 20 INCHES.

FIGURE CLOCKS

501

FAME

Height, 16 inches.
Width, 20 inches.
Dial, 6 inches.

Eight-day, Half-hour Strike,
Cathedral Gong,
with Empire Sash and Ivorized
Porcelain Dial,
Visible Escapement.

Finished in Gilt or Bronze.

List price, $22.25

1900

502

1906

FLOWER GIRL

Height, 16 inches. Width, 20 inches.
Six-inch V. E. Porcelain Dial.
French Sash, with Beveled Glass.
Finished in Gilt or Bronze.
Eight-day, Half-hour Strike, Cathedral Gong.
List Price, $23.75

503

1906

FLOWER GIRL EXTRA

Syrian Bronze Finish

Height, 15 inches. Width, 18 inches.
Six-inch Porcelain Dial, with Empire Sash.
Eight-day, Half-hour Strike, Cathedral Gong.

List Price, $29.25

FIGURE CLOCKS

These also come with a figure on top of horse.

505

1906

HORSE EXTRA
Syrian Bronze Finish

Height, 14¾ inches. Width, 18 inches.
Six-inch Porcelain Dial, with Empire Sash.
Eight-day, Half-hour Strike, Cathedral Gong.

List, Price, $27.80

504

1913

HORSE
LIST, $26.10
HEIGHT, 16 INCHES. WIDTH, 20 INCHES.

506

1900

POETRY

POETRY

Height, 16 inches.
Width, 20 inches.
Dial, 6 inches.

Eight-day, Half-hour Strike, Cathedral Gong, with Empire Sash and Ivorized Porcelain Dial, Visible Escapement.

Finished in Gilt or Bronze.

List price, $23.75

HANGING CLOCKS

507

1886

REGULATOR No. 00.

OAK OR MAHOGANY.

WOOD PENDULUM ROD, POLISHED BRASS WEIGHTS.

Code Word.		Price.
(Game)	8 Day Time, Mahogany	$90.00
(Gang)	8 " " " Sweep Seconds	97.50
(Tail)	8 " " Oak	82.50
(Tailor)	8 " " " Sweep Seconds	90.00

Weight. Cut Pinion Movement.
Height, 63 inches. Dial, 10 inches, with heavy Brass Matting.

508

1885

REGULATOR, No. 00. No. 2
Height, 77 inches. Dial, 10 inches.
8 Day Time, Sweep Seconds, $130.00

HANGING CLOCKS

509

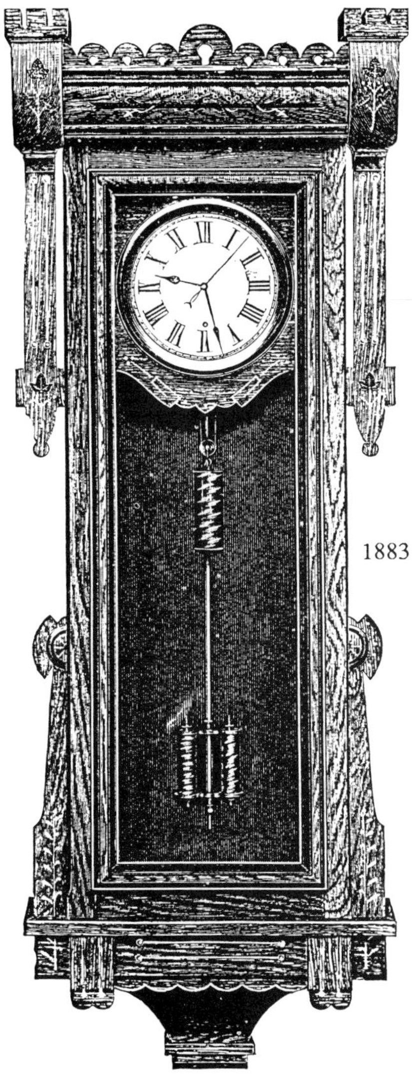

1883

REGULATOR No. 0.

BLACK WALNUT.

Length, 84 inches. Width, 31 inches. 1886

Code Word.		Price.
(Gale)	8 Day Weight Time, Walnut	$90.00
(Tag)	8 " " " Ash	90.00
(Taglock)	8 " " " Mahogany	102.00
(Tagrag)	8 " " " Strike, Walnut	105.00
(Tagsore)	8 " " " " Ash	105.00
(Tagtail)	8 " " " " Mahogany	117.00

Dial, 12 inches, with deep Brass Matting.

Sweep Second Hand. Mercurial Ball, if ordered.

510

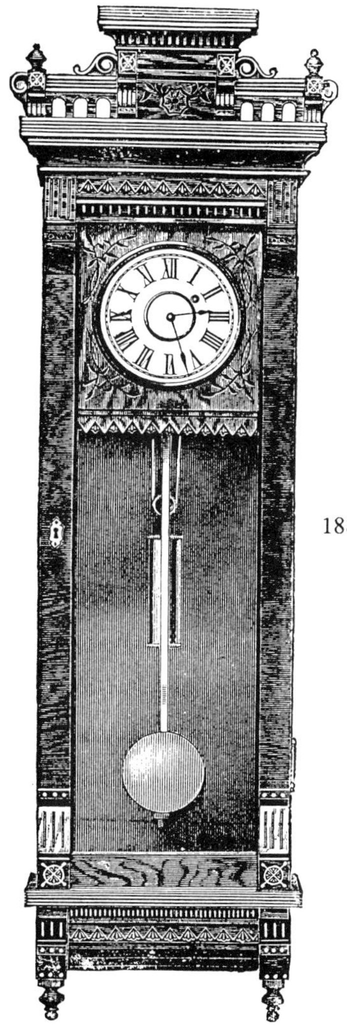

1881

REGULATOR No. 1

Weight. Walnut. Height, 58 inches. Dial, 8 inches. 8 Day Time. Finished with Brass Burnished Matting. Porcelain Dial, if ordered. This Clock is also made in Ash, Oak, or Mahogany, and finished in the prevailing styles.

HANGING CLOCKS

511

Close-up of New Haven Regulator No. 2 Movement.
{Lubertus H. Hayenga Collection}

Other movement illustrations for this clock on page 399

512

1881

REGULATOR No. 2.

WALNUT.

Code Word.	Height, 7 feet 4 inches. Width at top, 29 inches. Width at base, 24½ inches.	Price.
(Gaelic)	8 Day Time	$120.00

1886

Finished with Glass Sides, Movement of best Swiss pattern, Sweep Second Hand, Dead-beat Pin Escapement, 12-inch Porcelain Dial, Steel and Brass Pendulum Rods, adjusting to heat and cold. We will make this also in Mahogany or Ash, if so ordered.

HANGING CLOCKS

514

1883

515

1883

REGULATOR No. 6.

REGULATOR No. 3.

WALNUT.

Code Word.	Height, 7 feet 2 inches. Width at top, 27 inches. Width at base, 25 inches.	Price.	
(Gag)	8 Day Time	$120.00	1886

Finished with glass sides, movement of best Swiss pattern, sweep second hand, dead-beat pin escapement, 12-inch porcelain dial, steel and brass pendulum rods, adjusting to heat and cold.

We will make this also in mahogany or ash, if so ordered.

REGULATOR, No. 6.

EBONY FINISH, HARDWOOD POLISHED CASE. WHITE CARVINGS, FINISHED IN GOLD LEAF.

Code Word.	Height, 78 inches	Price.
(Gage)	8 Day Time	1886...$90.00

Finished with Glass Sides, Movement of best Swiss Pattern, Sweep Second Hand, Dead-beat Escapement, 12-inch Porcelain Dial, Steel and Brass Pendulum Rods, adjusting to heat and cold.

HANGING CLOCKS

The year entered beside each clock is the year of the catalog from which the illustration was taken. Some clocks may have been produced a few years earlier.

517

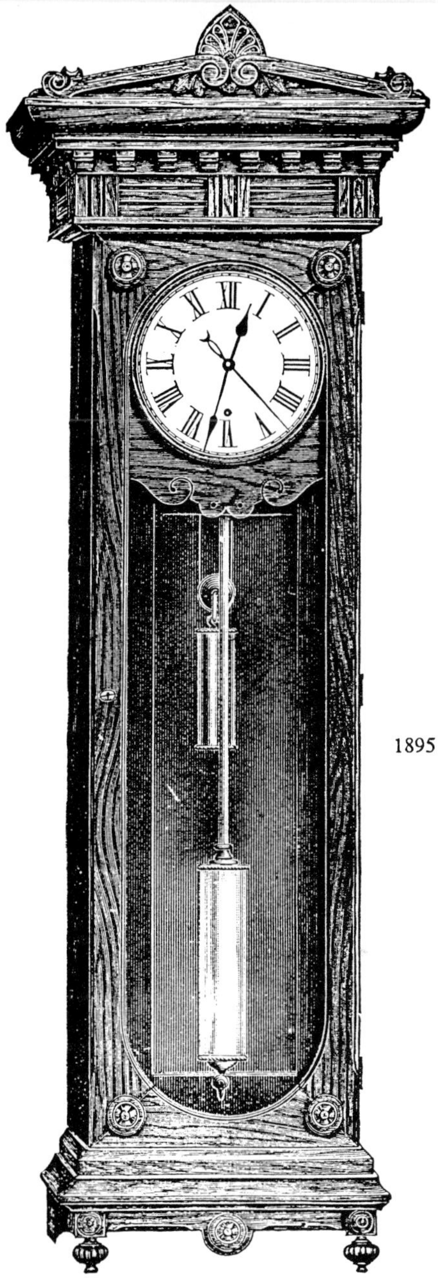

1895

REGULATOR, No. 10.

OAK, BLACK WALNUT, AND MAHOGANY.

Height, 77½ inches. Dial, 12 inches. 1886

Code Word.		Price.
(Temporist)	8 Day Time, Sweep Second	$ 70.00
(Tenable)	8 Day Time, Sweep Second, Mercurial Ball	130.00

516

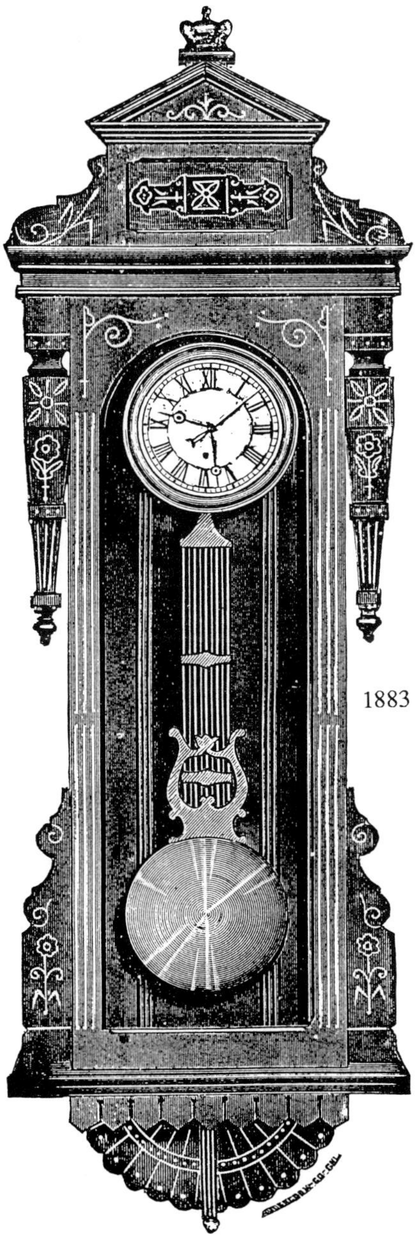

1883

REGULATOR, No. 9. SWEEP SECOND. (SWISS.)

8 Day Time. Height, 58 inches. Width, 18 inches. Dial, 9½ inches.

Finished with Glass Sides, Movement of best Swiss Pattern, Sweep Second Hand, Dead-Beat Escapement, 12 inch Porcelain Dial, Steel and Brass Pendulum Rods, adjusting to heat and cold.

HANGING CLOCKS

518

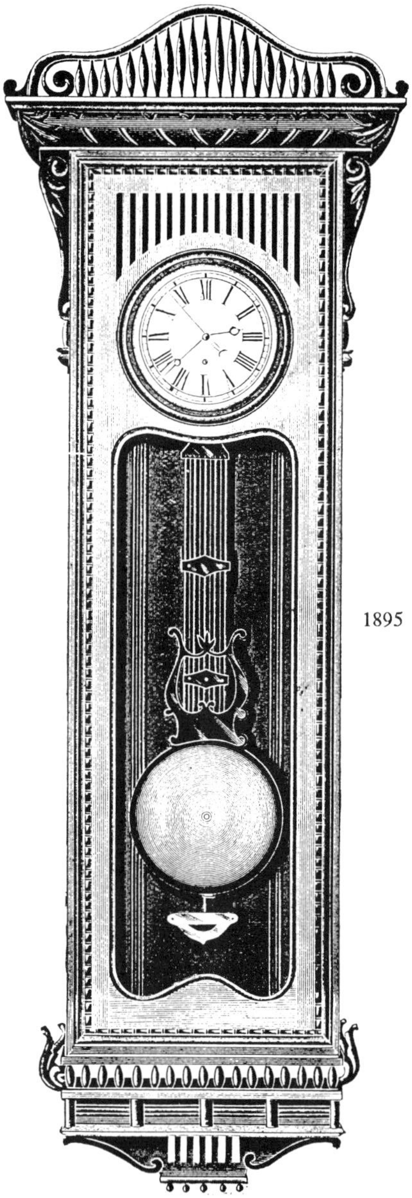

1895

519

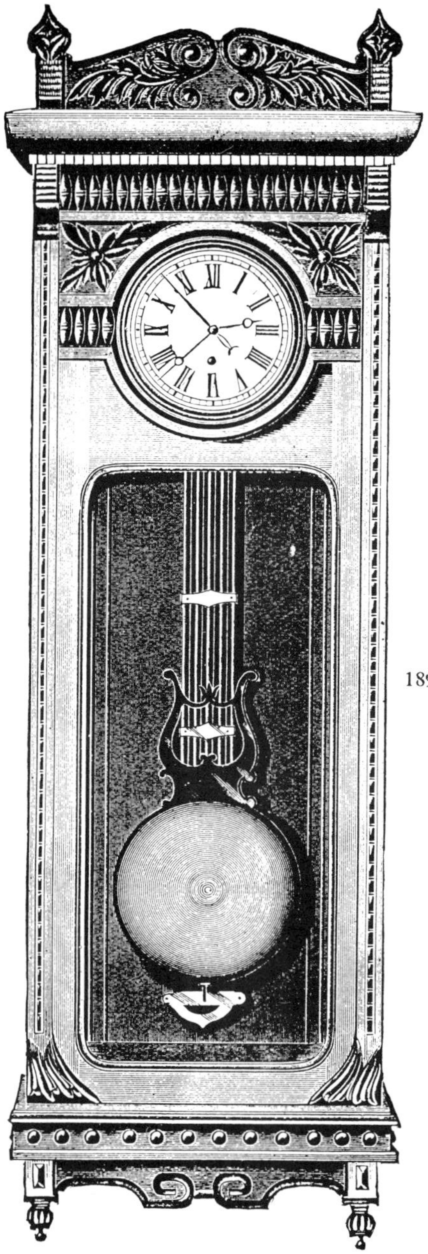

1895

520

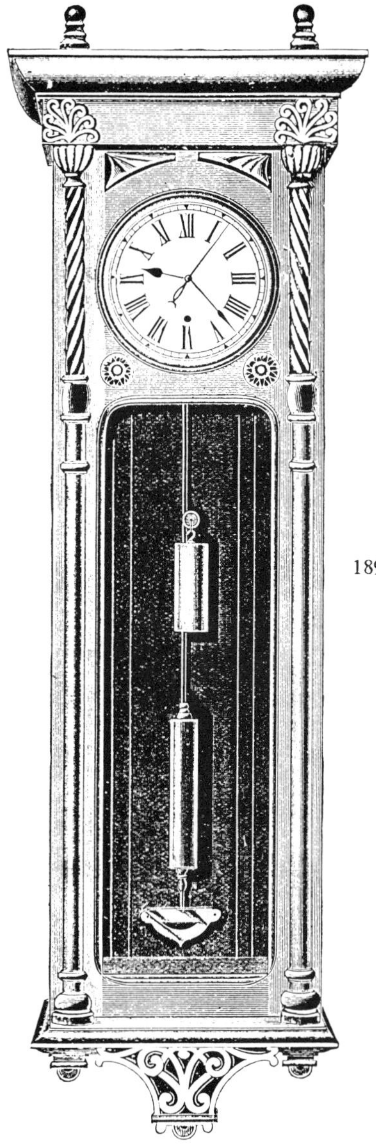

1895

THE BOSTON

SOLID OAK

EIGHT DAY TIME, SWEEP SECOND. Height, 8 feet, 1 inch. Dial, 12 inches. Fitted either with Swiss or our own Fine Regulator Movement...............................List, $110.00

FITTED WITH THREE JAR MERCURIAL PENDULUM..................$40.00 LIST EXTRA.

THE NEW YORK

SOLID OAK

EIGHT DAY TIME, SWEEP SECOND. Height, 7 feet, 2 inches. Dial, 12 inches. Fitted either with Swiss or our own Fine Regulator Movement..............List, $110.00

FITTED WITH THREE JAR MERCURIAL PENDULUM..................$40.00 LIST EXTRA.

THE ST. LOUIS

SOLID OAK, WALNUT OR CHERRY

EIGHT DAY TIME, SWEEP SECOND. Height, 6 feet, 6 inches. Dial, 12 inches. Fitted either with Swiss or our own Fine Regulator Movement................List, $90.00

FITTED WITH THREE JAR MERCURIAL PENDULUM..................$40.00 LIST EXTRA.

HANGING CLOCKS

The year entered beside each clock is the year of the catalog from which the illustration was taken. Some clocks may have been produced a few years earlier.

521

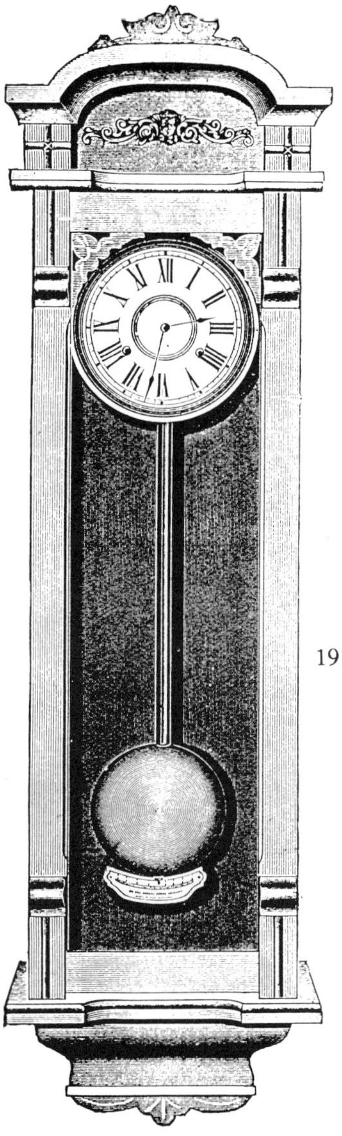

1911

COLUMBIA

HEIGHT ....48½ INCHES
WIDTH ....14¼ INCHES
DIAL ....8 INCHES
CASE, SOLID OAK OR MAHOGANY
PRICE, 8 DAY TIME ....LIST, $15.00
PRICE, 8 DAY TIME, CALENDAR
LIST, 15.45
PRICE, 8 DAY STRIKE, CATHEDRAL
GONG ....LIST, 16.50
PRICE, 8 DAY STRIKE, CATHEDRAL
GONG, CALENDAR ....LIST, 16.95
PRICE, 30 DAY TIME ....LIST, 17.20

522

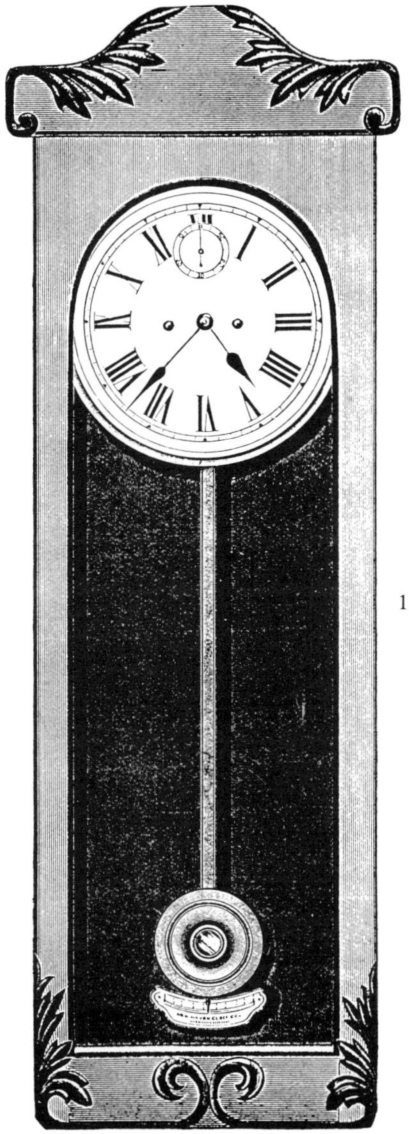

1895

GLORIANA

SOLID OAK.

Height, 42 inches. Dial, 10 inches.
30 Day Time................List, $19.70
30 Day Time, Calendar..List, $23.00

523

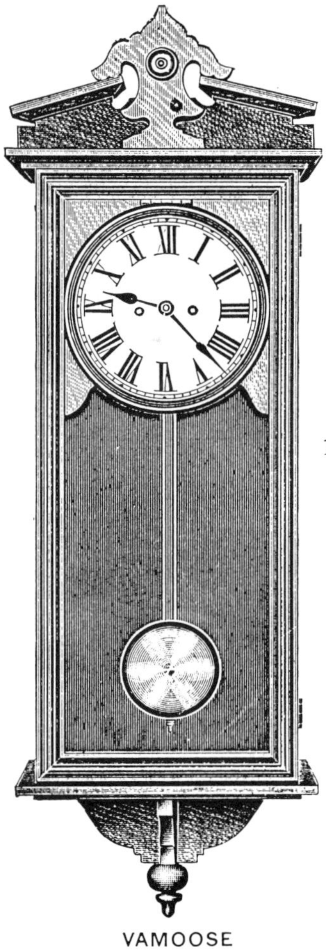

1911

VAMOOSE

HEIGHT.......................45 INCHES
WIDTH ....15¼ INCHES
DIAL ....10 INCHES
CASE, SOLID OAK OR MAHOGANY
PRICE, 8 DAY TIME ....LIST, $11.60
PRICE, 8 DAY TIME, CALENDAR
LIST, 12.05
PRICE, 8 DAY STRIKE, CATHEDRAL
GONG...........................LIST, 13.10
PRICE, 8 DAY STRIKE, CATHEDRAL
GONG, CALENDAR ....LIST, 13.55
PRICE, 30 DAY TIME ....LIST, 13.80

HANGING CLOCKS

524

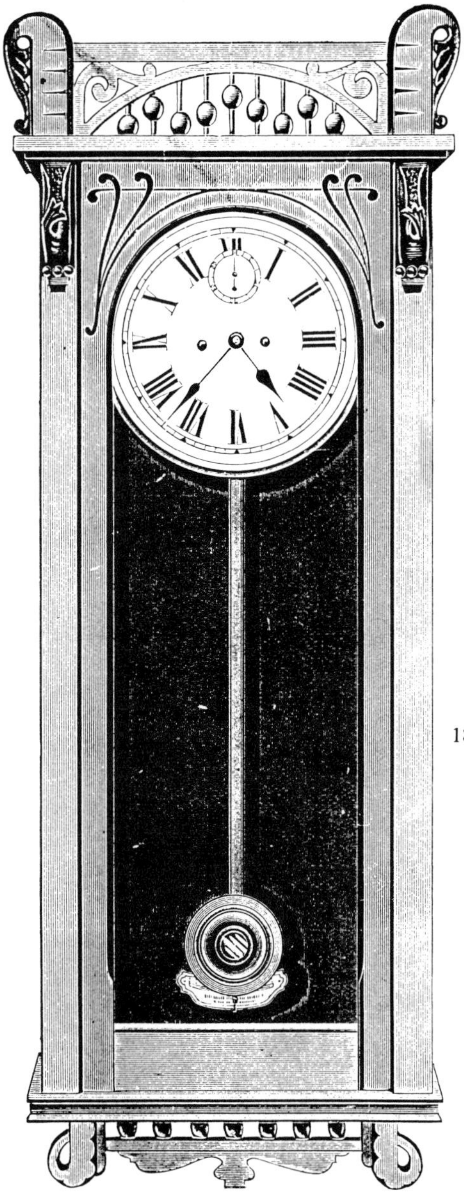

1895

BARBARA

SOLID OAK.

Height, 47 inches. Dial, 10 inches.
30 Day Time................List, $21.00
30 Day Time, Calendar..List, $24.30

525

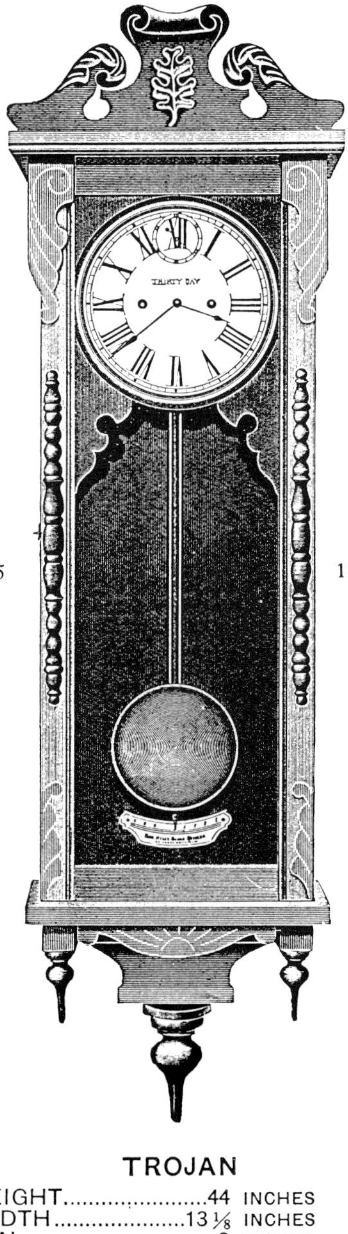

1895

TROJAN

HEIGHT.......................44 INCHES
WIDTH.....................13 1/8 INCHES
DIAL ....8 INCHES
CASE, SOLID OAK OR MAHOGANY
PRICE, 8 DAY TIME ....LIST, $13.00
PRICE, 8 DAY TIME, CALENDAR LIST, 13.45
PRICE, 8 DAY STRIKE, CATHEDRAL GONG ....LIST, 14.50
PRICE, 8 DAY STRIKE, CATHEDRAL GONG, CALENDAR........LIST, 14.95
PRICE, 30 DAY TIME ....LIST, 15.20

1911

526

1895

HEBE

SOLID OAK.

Height, 49 inches. Dial, 10 inches.
30 Day Time................List, $22.50
30 Day Time, Calendar..List, $25.80

HANGING CLOCKS

527

1878

ROUND LEVER.

1 Day or 8 Day.
Time or Strike.
4 to 8 inches.

528

1907

LAWEN KITCHEN CLOCK

Diameter, 6 inches

A small Hanging Clock especially designed for the kitchenette, bathroom or any room where a mantel clock is not practical. Also suitable for Bungalows.

The case is either White Enamel finish, fine Mahogany finish or solid Oak. Has a high grade, reliable 8-day lever time movement, a silver finished dial and Convex glass.

9819 Golden Oak Finish.........$10.90
9820 Mahogany Finish.......... 10.90
9821 White Enamel Finish...... 11.90

529

1895

INTERPID

SOLID OAK.

Height, 49 inches. Dial, 10 inches.
30 Day Time...............List, $23.80
30 Day Time, Calendar..List, $27.10

530

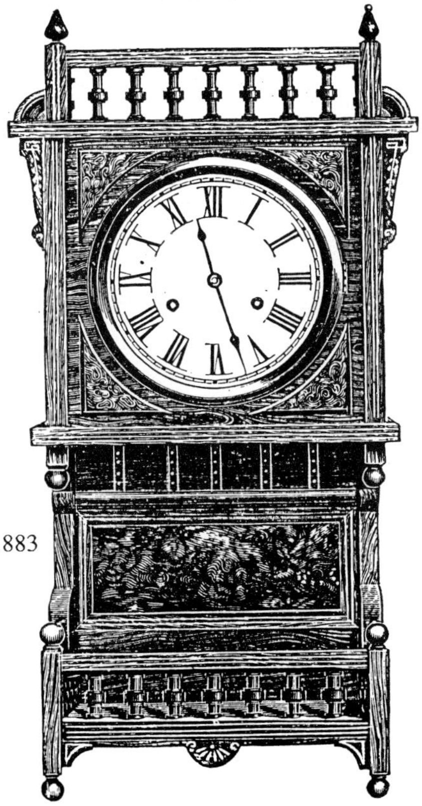

1883

AMOOR.

WALNUT.

Height, 29 inches. Dial, 8 inches.
8 Day Strike. 8 Day Strike, Alarm
8 Day Half-hour Strike on Cathedral Gong.

531

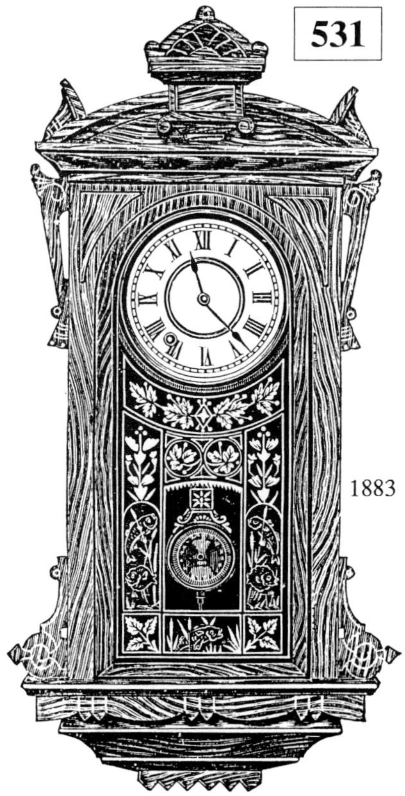

1883

MEUSE.

WALNUT HANGING.

Height, 27¼ inches. Dial, 6 inches.
8 Day Time. 8 Day Strike.
8 Day Half-hour Strike on Cathedral Gong.

HANGING CLOCKS

533

1900

1911

MAYWOOD

HEIGHT ....43½ INCHES
WIDTH ....17 INCHES
DIAL ....8 INCHES
CASE, SOLID OAK OR MAHOGANY
PRICE, 8 DAY TIME ....LIST, $10.90
PRICE, 8 DAY TIME, CALENDAR LIST, 11.35
PRICE, 8 DAY STRIKE, CATHEDRAL GONG ....LIST, 12.40
PRICE, 8 DAY STRIKE, CATHEDRAL GONG, CALENDAR........LIST, 12.85
PRICE, 30 DAY TIME ....LIST, 13.10

532

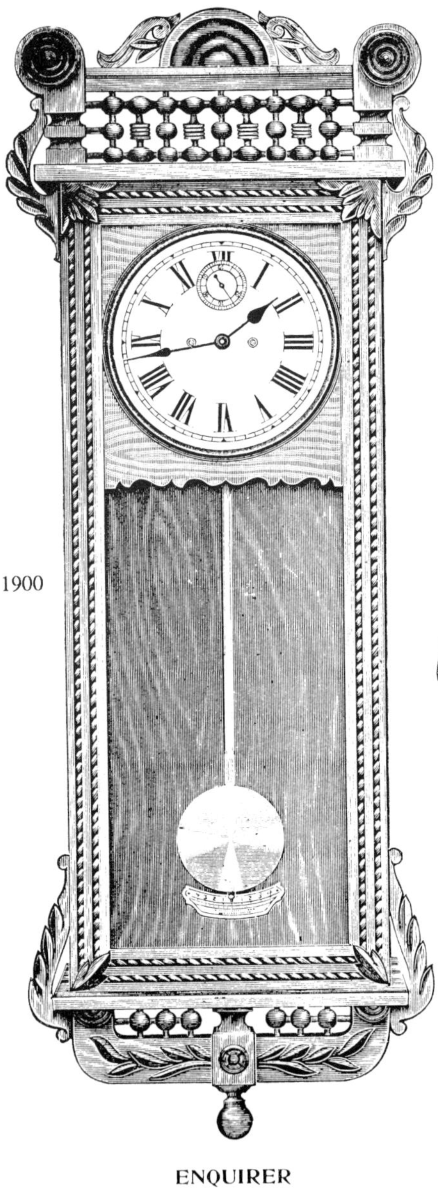

1900

ENQUIRER

Height, 50½ inches. Width, 19 inches.

Dial, 10 inches.

List price, Eight-day Time, . $18.50
" " Eight-day Strike, Gong, 20.00
" " Eight-day Time, Weight Movement, ... 22.00

534

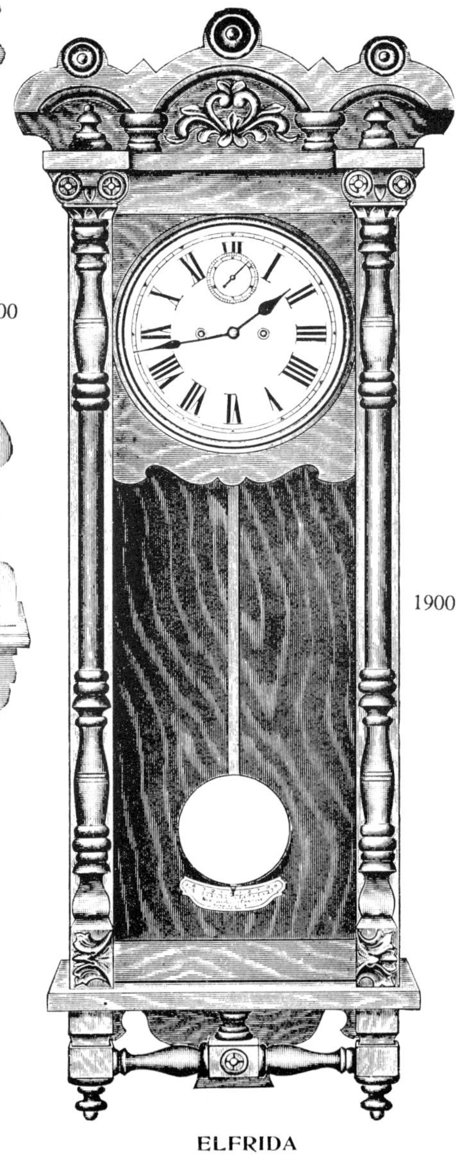

1900

ELFRIDA

Height, 49 inches. Width, 19 inches.

Dial, 10 inches.

List price, Eight-day Time, . $17.50
" " Eight-day Strike, Gong, 19.00
" " Eight-day Time, Weight Movement, ... 21.00

HANGING CLOCKS

535

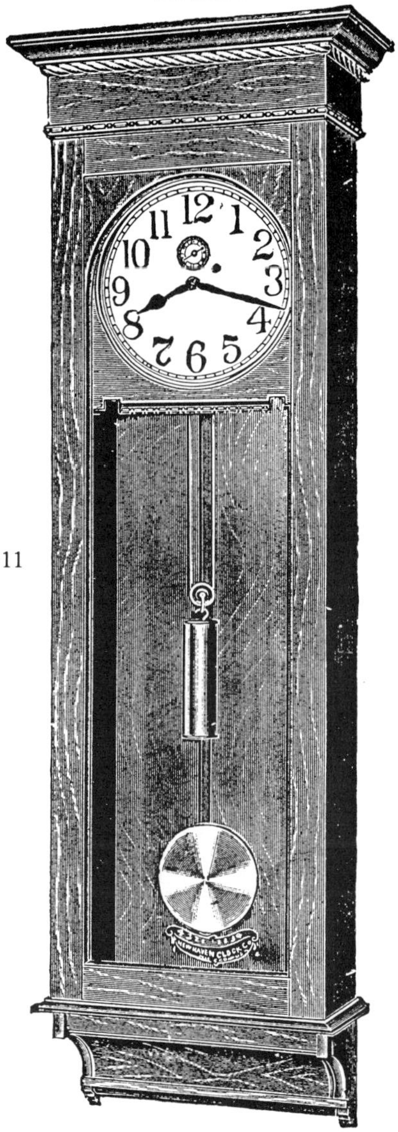

1911

STANTON

HEIGHT ....64 INCHES
WIDTH ....21 INCHES
DIAL ....12 INCHES
CASE, SOLID OAK OR MAHOGANY
MOVEMENT, 8 DAY WEIGHT TIME PIECE WITH RETAINING POWER
PRICE, ...LIST, $48.00

536

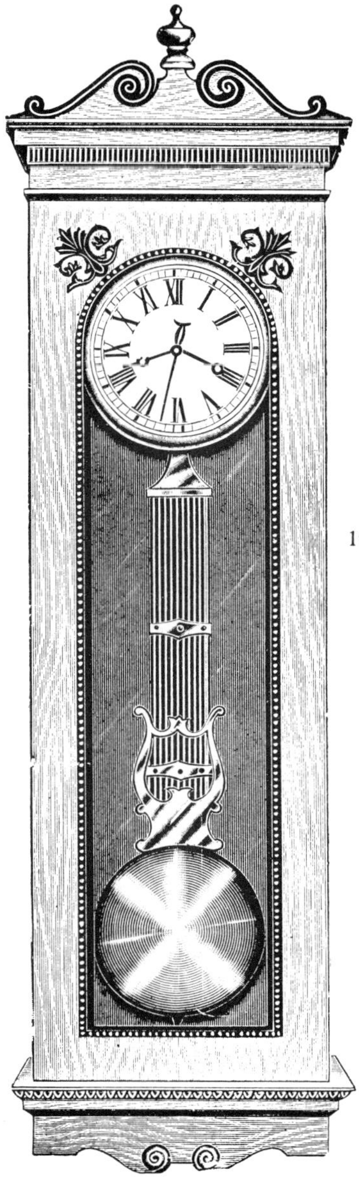

1900

THORNTON — Hanging.

Height, 74 inches. Width, 22 inches. Twelve-inch Porcelain Dial.

Oak or Cherry.

Eight-day. Fitted with Swiss Movement, Gridiron Pendulum, or Our Own Fine Regulator Movement, Wooden Pendulum and Cylindrical Ball.

Dead-beat Escapement, Sweep Second, Retaining Power.

List price, $103.00

537

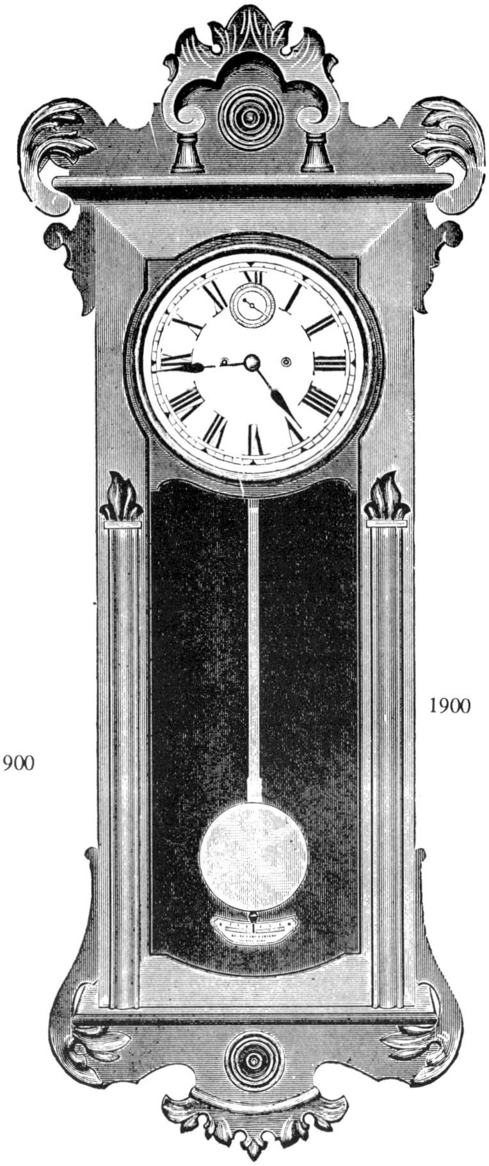

1900

GRECIAN—Oak.

Height, 51 inches. Width, 20 inches. Dial, 10 inches.

List price, Eight-day Time, . $16.75
" " Eight-day Strike, Gong, 18.25
" " Eight-day Time, Weight Movement, ... 20.25

HANGING CLOCKS

538

1883

GAMBIA. (SPRING.)

WALNUT.

Height, 49 inches. Dial, 8 inches.
8 Day Time. 8 Day Strike.
Finished with Brass Burnished Matting.
Porcelain Dial, if ordered.

This Clock is also made in Ash, Mahogany, or Oak, and finished in the prevailing styles

539

1900

ADMIRAL

Ash Burl, Polished Veneered.

Height, 62 inches. Dial, 18 inches.

List price, Eight-day Time, $50.00

540

1883

OBI. (SPRING.)

WALNUT.

Height, 49 inches. Dial, 8 inches.
8 Day Time. 8 Day Strike.
Finished with Brass Burnished Matting.
Porcelain Dial, if ordered.

This Clock is also made in Ash, Mahogany, or Oak, and finished in the prevailing styles.

HANGING CLOCKS

541

1880

BAROMETER REGULATOR.

Spring.

Height, 37 inches. 8 Day Time.
Dial, 12 inches. 8 Day Strike.

542

1907

TAMPICO—Oak.

Height 43 inches, width 15½ inches, dial 12 inches.

No. 6061 Eight-day, time........................$11 55
No. 6062 Thirty-day, time........................ 14 20

543

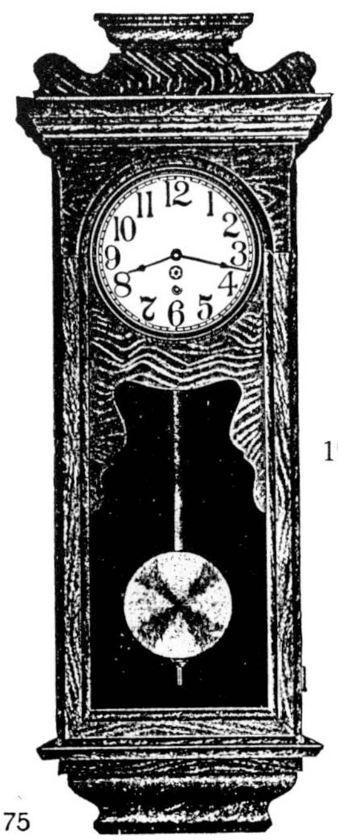

1911

SOLENT

HEIGHT ....36¾ INCHES
WIDTH ....13½ INCHES
DIAL ....8 INCHES
CASE, SOLID OAK, ANTIQUE FINISH
MOVEMENT, 8 DAY TIME PIECE
PRICE ....LIST, $13.75

HANGING CLOCKS

544

Close-up of **Saturn** Double Spring 30-Day Movement.
{Jack Wyer Collection}

545

SATURN
Oak case. 30-Day Time.
Height, 35½ inches. Dial, 12 inches.
{Jack Wyer Collection}

Saturn weight driven movement see #1545A page 395

546

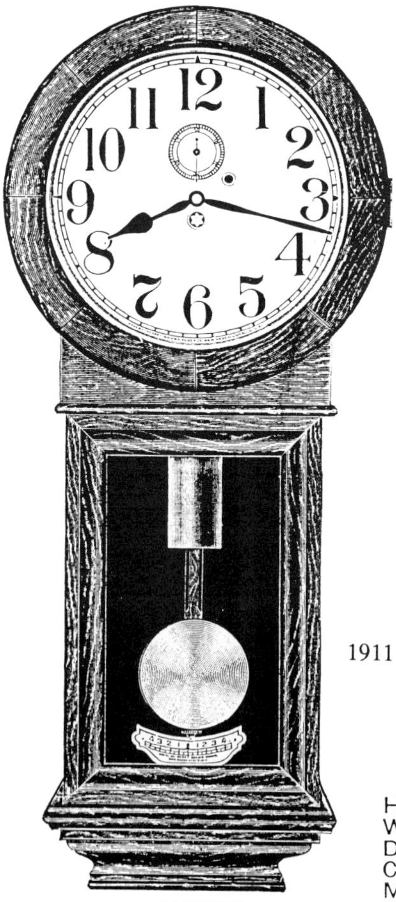

1911

SATURN

547

Saturn Beat Indicator.
{Jack Wyer Collection}

SATURN

HEIGHT ....34¾ INCHES
WIDTH ....15¼ INCHES
DIAL ....12 INCHES
CASE, SOLID OAK, ANTIQUE FINISH
MOVEMENT, 30 DAY TIME PIECE
PRICE .. ..LIST, $16.00

HANGING CLOCKS

The year entered beside each clock is the year of the catalog from which the illustration was taken. Some clocks may have been produced a few years earlier.

548

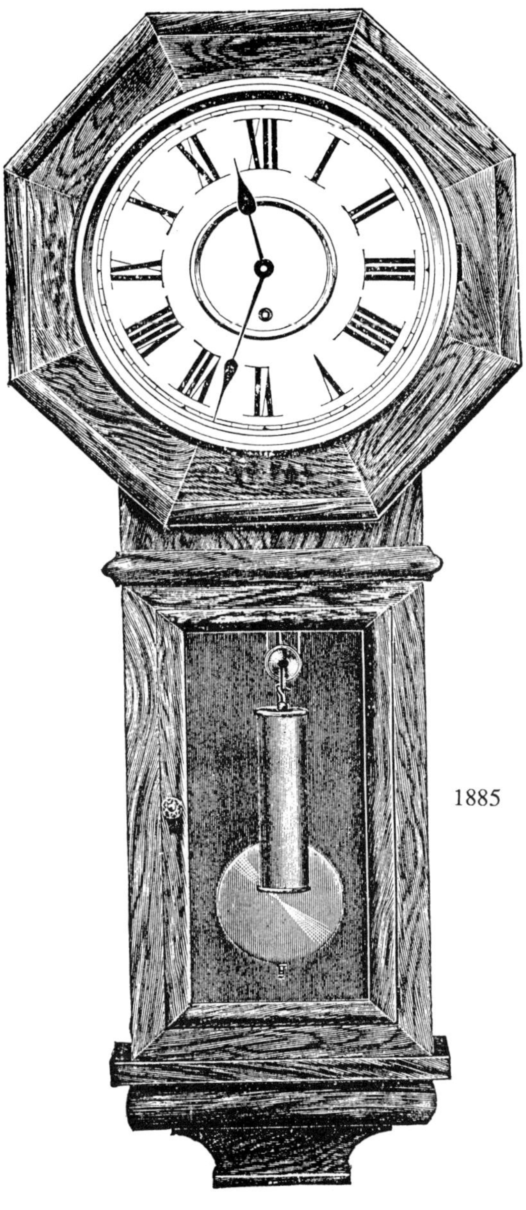

1885

OFFICE No. 1

ROSEWOOD, BLACK WALNUT, MAHOGANY OR OAK.

Height, 42 inches. Weight. Dial, 12 inches. Fitted with Standard Time Weight Movement.
8 Day Time....................................List, $21.00

549

1886

OFFICE No. 1

BLACK WALNUT OR MAHOGANY.

Height, 42 inches. Dial, 12 inches. Movement has cut and Polished Steel Pinions, Solid Frames, Screw Pillars, and Maintaining Power. Fitted with Standard Time Weight Movement.
8 Day Time.......................List, $21.00

HANGING CLOCKS

550

Close-up of **Office No. 1** front view movement.
{Jack Wyer Collection}

551

OFFICE No. 1
Walnut case. 8-Day Time.
Height, 41 inches. Dial, 12 inches.
{Jack Wyer Collection}

552

Close-up of **Office No. 1** side view movement.
{Jack Wyer Collection}

553

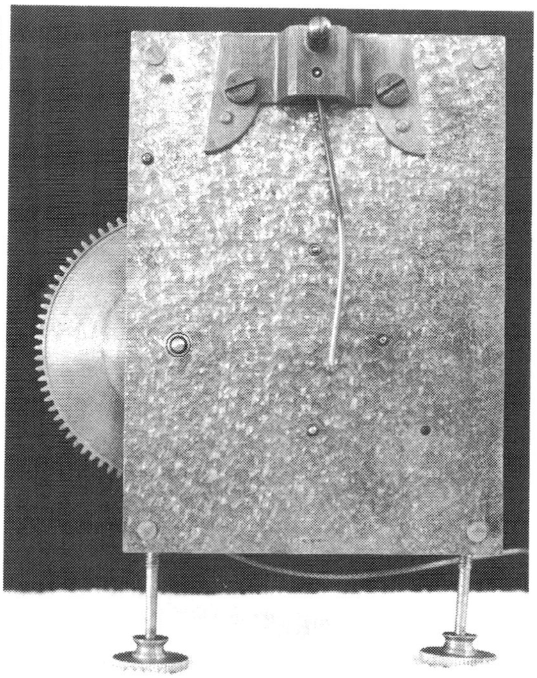

Close-up of **Office No. 1** rear view movement.
{Jack Wyer Collection}

HANGING CLOCKS

The year entered beside each clock is the year of the catalog from which the illustration was taken. Some clocks may have been produced a few years earlier.

554

1878

ACORN LEVER.

Brass or Nickel.

With French Method of Winding, Setting, and Regulating at the Back.

1 Day. Time.
1 " Alarm.
3 inch Dial.

555

1883

No. 3004

SOLID OAK OR WALNUT

Height, 47 inches. Dial, 12 inches.
30 Day Time................List, $40.00

556

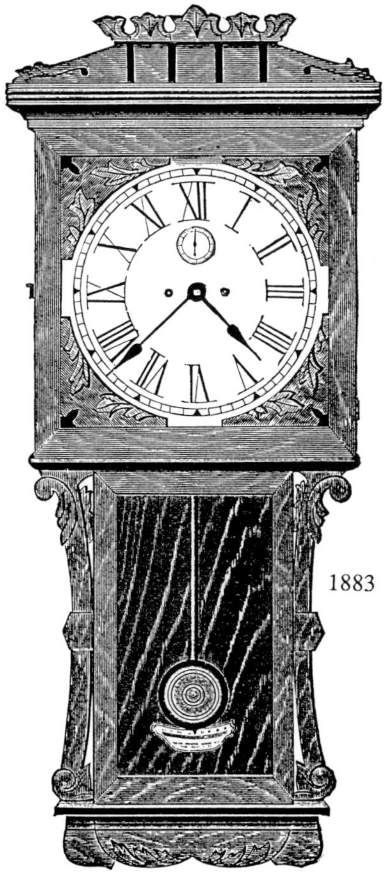

1883

COMMODORE

OAK

Height, 46 ½ inches. Dial, 14 inches.
30 Day Time................List, $22.00

HANGING CLOCKS

557

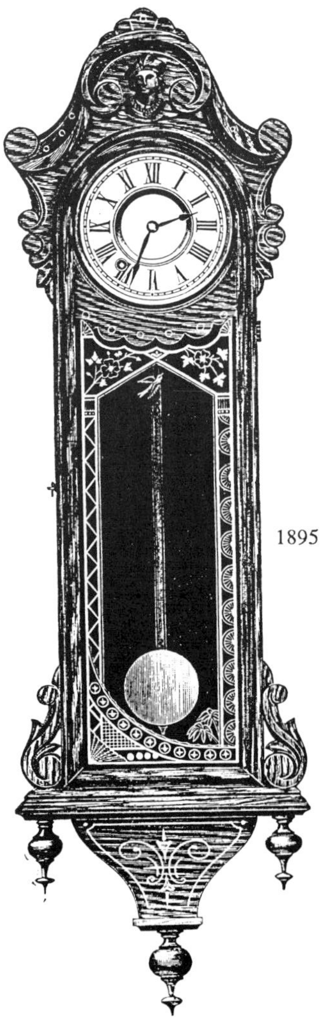

1895

DNEIPER

WALNUT.

Height, 37 inches. Dial, 6 inches.
Finished with nickel Matting.
8 Day Time. 8 Day Strike.

558

1886

OFFICE No. 2

BLACK WALNUT OR MAHOGANY.

Height, 41 inches. Dial, 12 inches.
8 Day Time..................List, $18.00

559

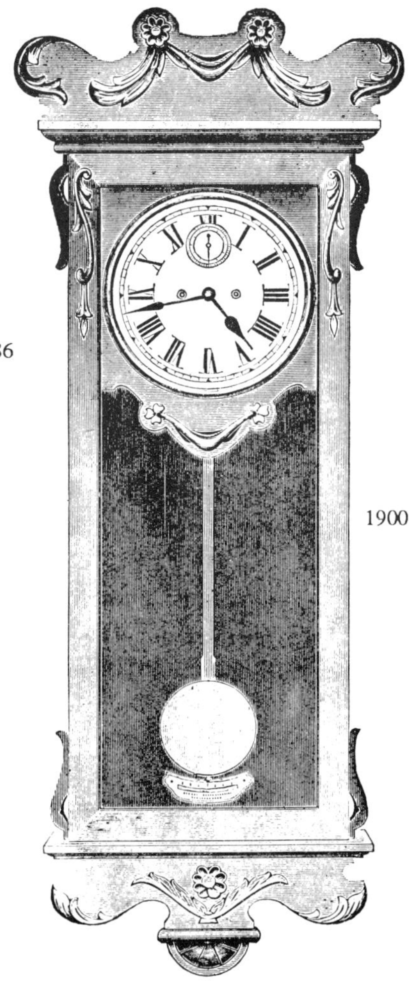

1900

AUSTRIAN — Cherry.

Height, 48¾ inches. Width, 19 inches.

Dial, 10 inches.

List price, Eight-day Time, . $16.75
" " Eight-day Strike, Gong, 18.25
" " Eight-day Time, Weight Movement, ... 20.25

HANGING CLOCKS

560

1889

VIDETTE

OAK AND BLACK WALNUT.

Height, 42 inches. Dial, 8 inches.
8 Day Time........................$15.00
8 Day Strike.......................$16.50

561

1889

NORSEMAN

OAK AND BLACK WALNUT.

Height, 47 ½ inches. Dial, 10 inches.
8 Day Time.............................$17.50
8 Day Strike............................$19.00

562

1889

SARACEN

MADE IN MAHOGANY ONLY

Height, 49 inches. Dial 10 inches.
8 Day Time...... Net $17.50
8 Day Strike .... Net $19.00

HANGING CLOCKS

The year entered beside each clock is the year of the catalog from which the illustration was taken. Some clocks may have been produced a few years earlier.

563

1889

VIKING

OAK AND BLACK WALNUT.

Height, 44 inches. Dial, 8 inches.

8 Day Time........................$16.50

8 Day Strike.......................$18.00

565

1889

HUDSON.

BLACK WALNUT AND CHERRY.

Height, 26½ inches. Dial, 5 inches.

Code Word.		Price.
(Enemy)	8 Day Strike, Cherry.......	$7.30
(Exposer)	8 Day Strike, Walnut......	8.00

564

1880

REGULATOR.

Spring.

Height, 32 inches. 8 Day Time.

Dial, 12 inches. 8 Day Strike.

Also with Alarm.

HANGING CLOCKS

566

1883

STANDARD TIME No. 1.

WALNUT, OAK, OR MAHOGANY.

Height, 42 inches. Dial, 14 inches.
8 Day T. P. Spring Movement.
8 Day T. P. Weight Movement.

567

1883

STANDARD TIME No. 2.

WALNUT, OAK, OR MAHOGANY.

Height, 45 inches. Width, 16 inches.
Dial, 12 inches.
Wooden Rod. Brass Ball.

The Standard Time system, as perfected in all its details, consists of a series of services which may be classified somewhat as follows:

1. Commercial and public service, including the dropping of time balls, the operation of flash signals, the firing of time guns, and the control of tower clocks.
2. Jewelers' service, in which the standard clock-beats are given on telegraph sounders every two seconds.
3. Private service, involving the setting and government of a series of dial time-pieces, controlled by a central standard time telegraph line.
4. The system is such that clocks of the most diverse patterns can be readily included in the same telegraphic circuit.
5. By synchronizing only at the time between two minutes before the hour, and two minutes after, and by making the signals by closing an open circuit on the telegraph lines, the danger of disturbance of the secondary clocks is avoided.

HANGING CLOCKS

568

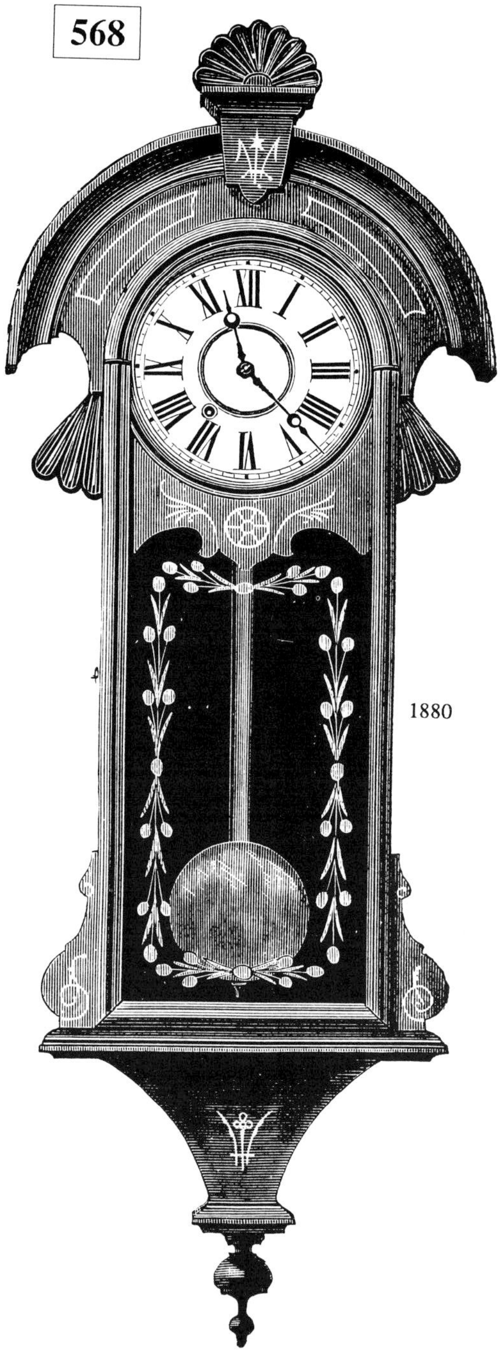

1880

569

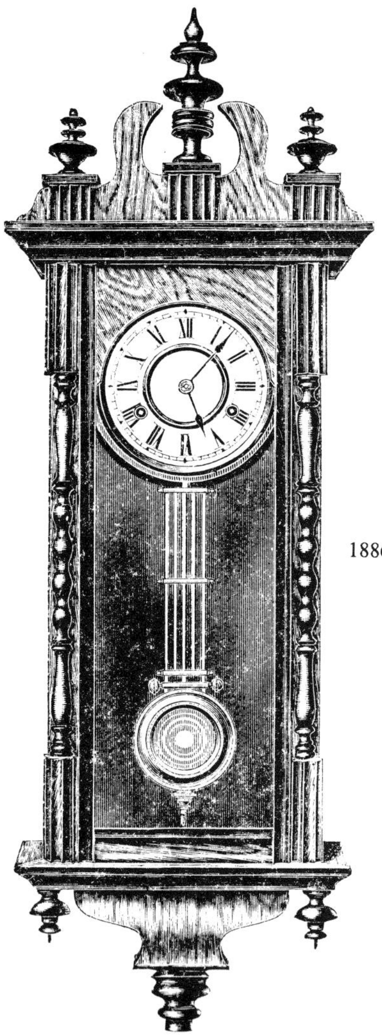

1886

570

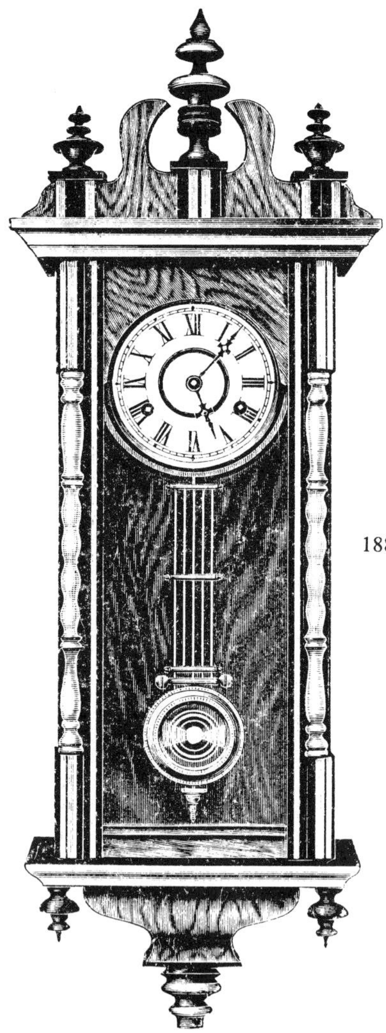

1886

CAMBRIA (SPRING).

Walnut.

Height, 42 inches. 8 Day Time.
Dial, 8 inches. 8 Day Strike.

No. 26

WALNUT, VENEERED, EBONY TRIMMINGS.

Height, 33 inches. Dial, 6 inches.
8 Day Time..............................$9.50
8 Day Time, Porcelain............$10.50
8 Day Strike............................$11.00
8 Day Strike, Porcelain............$12.00

No. 26, GILT.

WALNUT, VENEERED, EBONY AND GILT TRIMMINGS.

Height, 33 inches. Dial, 6 inches.
8 Day Time.............................$11.50
8 Day Time, Porcelain............$12.50
8 Day Strike............................$13.00
8 Day Strike, Porcelain............$14.00

HANGING CLOCKS

572

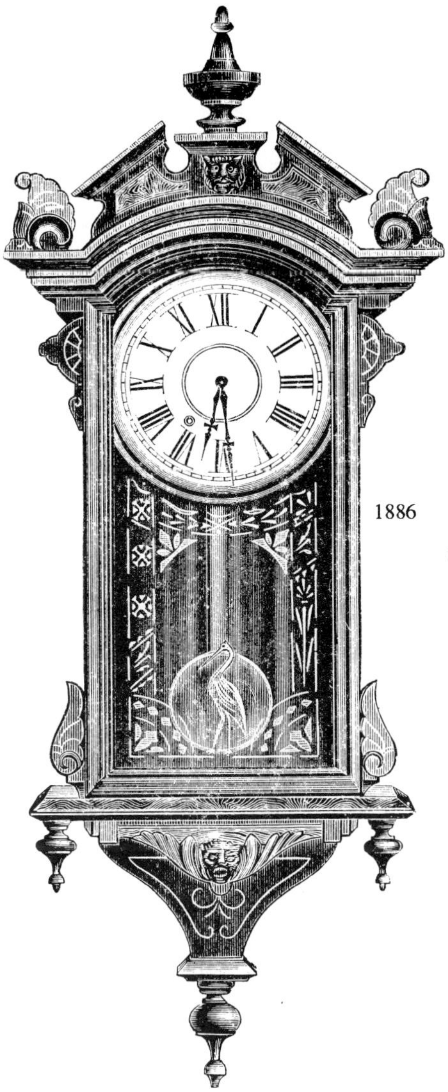

1886

REGULATOR, C. C.

WALNUT.

Height, 37 ½ inches. Dial, 8 inches.
8 Day Time..............................$10.50
8 Day Strike.............................$12.00

571

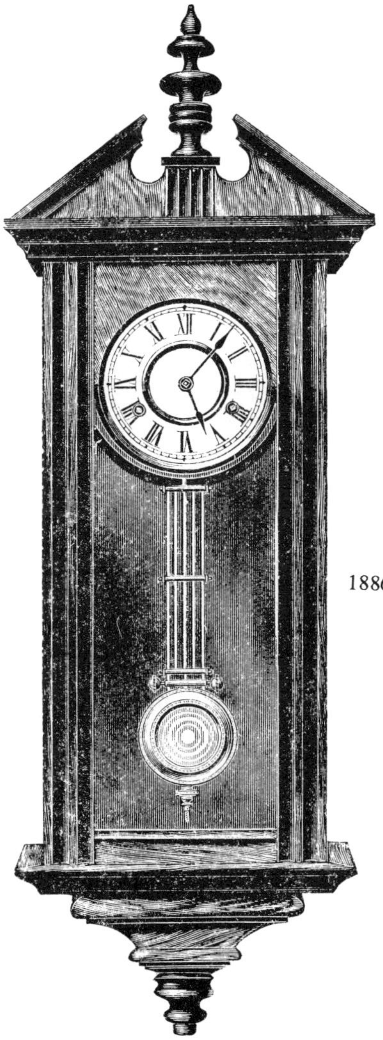

1886

No. 25

WALNUT, VENEERED, EBONY TRIMMINGS.

Height, 33 inches. Dial, 6 inches.
8 Day Time...............................$8.50
8 Day Time, Porcelain..............$9.00
8 Day Strike............................$10.00
8 Day Strike, Porcelain............$11.00

573

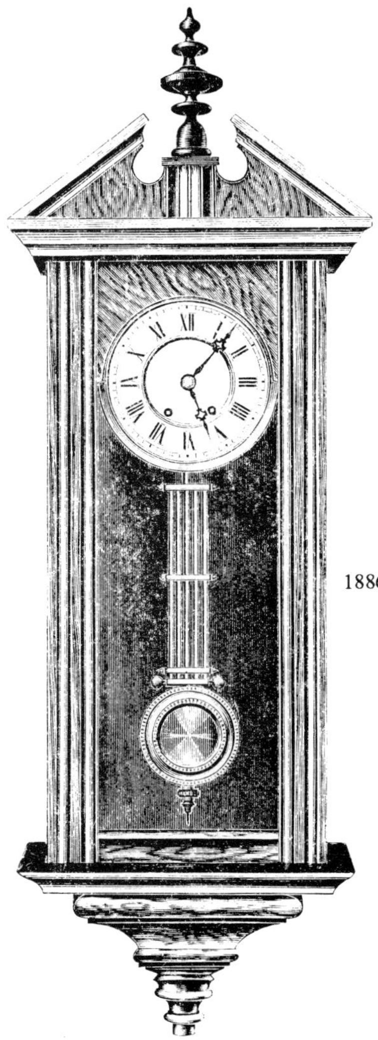

1886

No. 25, GILT.

WALNUT, VENEERED, EBONY AND GILT TRIMMINGS.

Height, 33 inches. Dial, 6 inches.
8 Day Time.............................$10.00
8 Day Time, Porcelain............$10.50
8 Day Strike............................$11.50
8 Day Strike, Porcelain............$12.50

HANGING CLOCKS

575

1920

JUNEAU
Height, 20¼ Inches, Width, 8 Inches
Dial, Silver Finish, 6 Inches
Case, Mahogany Finish
Movement, 8-Day Half-hour Strike, Cathedral Gong

574

1881

WINNIPEC

WALNUT.

Height, 35 inches. Dial, 6 inches.
8 Day Time..............................$8.50
8 Day Strike............................$10.00

577

1881

KENILWORTH

BLACK MOULDING.

Height, 24 inches. Dial, 6 inches.
8 Day Time. 8 Day Strike.
Black finished, with White Carvings.

576

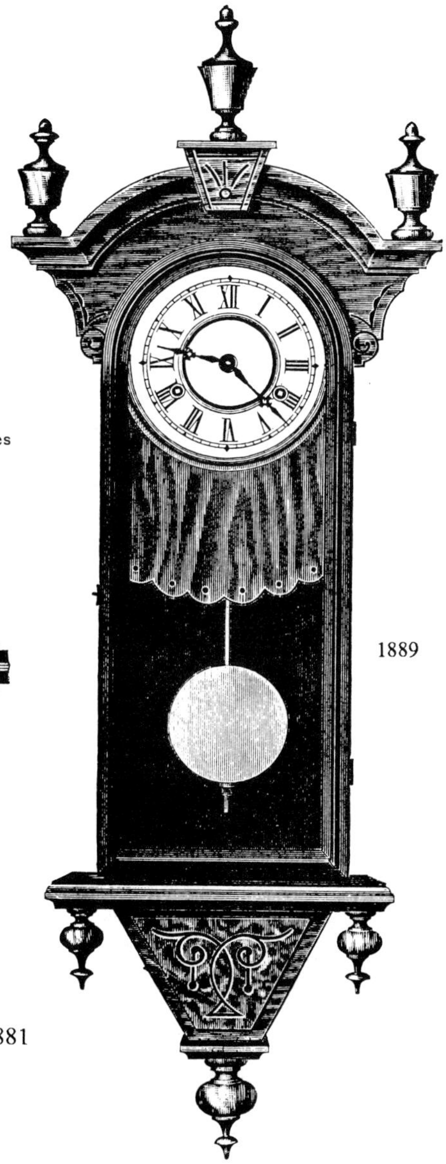

1889

WINNIPEC REGULATOR

WALNUT.

Height, 35 inches. Dial, 6 inches.
8 Day Time. 8 Day Strike.

HANGING CLOCKS

579

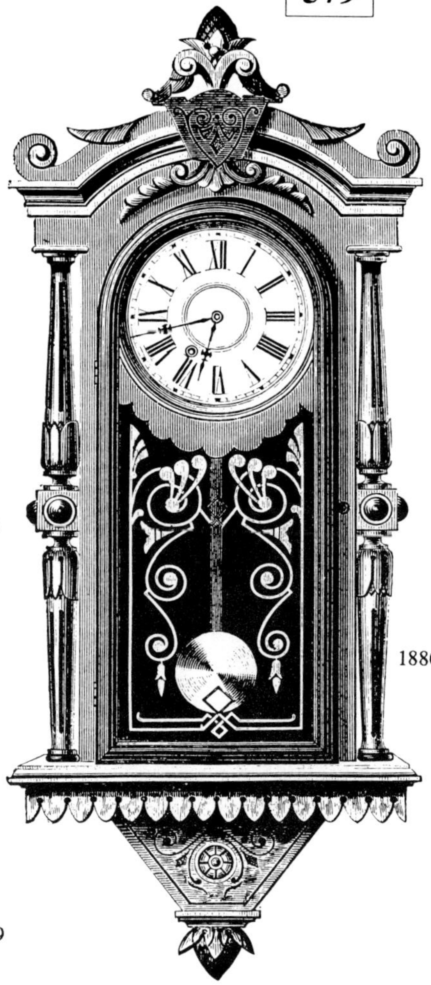

1880

REGULATOR, B. B. (SPRING.)

Walnut.

Height, 41 inches. 8 Day Time.
Dial, 8 inches. 8 Day Strike.

578

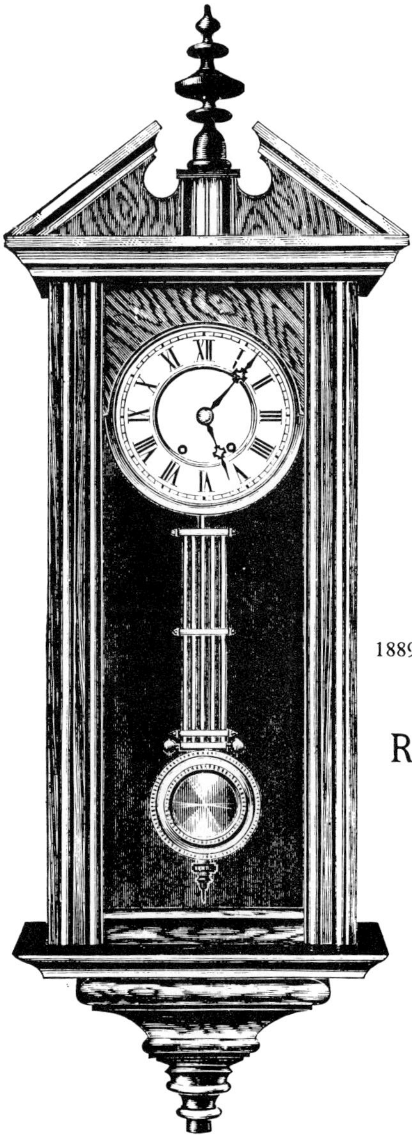

1889

FATINITZA, GILT.

WALNUT, VENEERED, EBONY AND GGILT TRIMMINGS.

Height, 33 inches. Dial, 6 inches.
8 Day Time...............................$10.00
8 Day Time, Porcelain Dial.......$11.00
8 Day Strike............................$11.50
8 Day Strike, Porcelain Dial.....$12.50

580

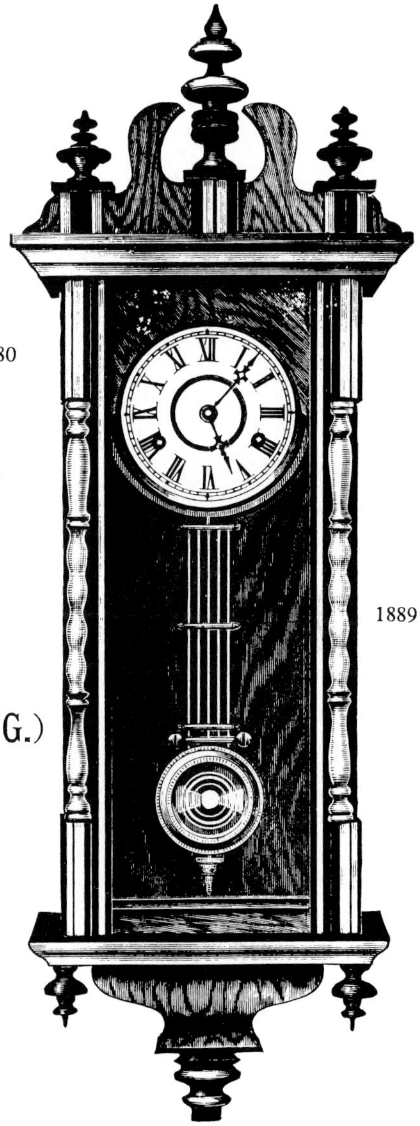

1889

FAUSTINA, GILT.

WALNUT, VENEERED, EBONY AND GILT TRIMMINGS.

Height, 33 inches. Dial, 6 inches.
8 Day Time..............................$9.50
8 Day Time, Porcelain Dial.......$10.50
8 Day Strike............................$11.00
8 Day Strike, Porcelain Dial....$12.00

HANGING CLOCKS

EIGHT-DAY SPRING TIMEPIECE. LEADED GLASS. PORCELAIN DIAL.
CASES, MAHOGANY FINISHED

581

1913

NEESE

LIST, $26.10

HEIGHT, 28¼ INCHES.
WIDTH, 8½ INCHES.
DIAL, 7 INCHES.

582

1913

NEAFUS

LIST, $21.00

HEIGHT, 27¾ INCHES.
WIDTH, 8½ INCHES.
DIAL, 7 INCHES.

583

1913

NANSEN

LIST, $22.50

HEIGHT, 34¾ INCHES.
WIDTH, 12¾ INCHES.
DIAL, 10 INCHES.

584

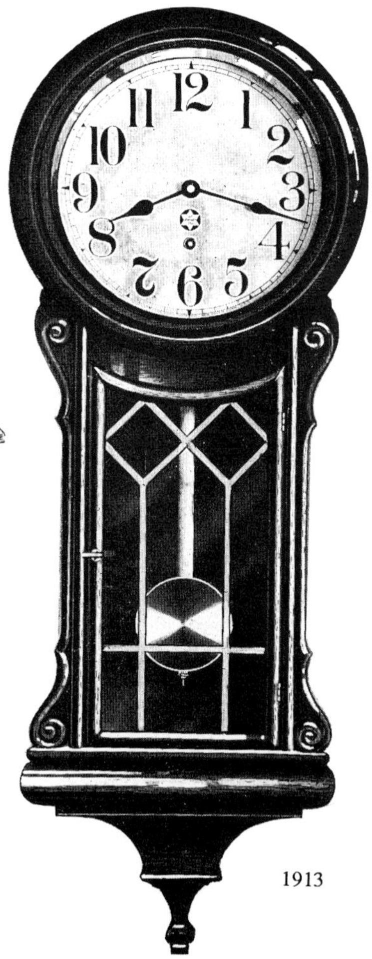

1913

NASON

LIST, $28.25

HEIGHT, 35⅜ INCHES.
WIDTH, 12¾ INCHES.
DIAL, 10 INCHES.

HANGING CLOCKS

585

1929

MARAVIA
Height, 27½ in. Width, 16 in.
12 inch Antique Dial, Ornate Hands, Polished Sash. Solid Mahogany, Hand Carved Case.
8-Day Pendulum Time Piece $53.40

586

1929

MADENA
Height 18½ inches. Width 11½ inches. 8-inch Antique Dial. Polished Sash. Solid Mahogany Hand Carved Case. Gilt Ornament. 8 day Pendulum Hour and Half-hour Rod Strike.
List $46.75

587

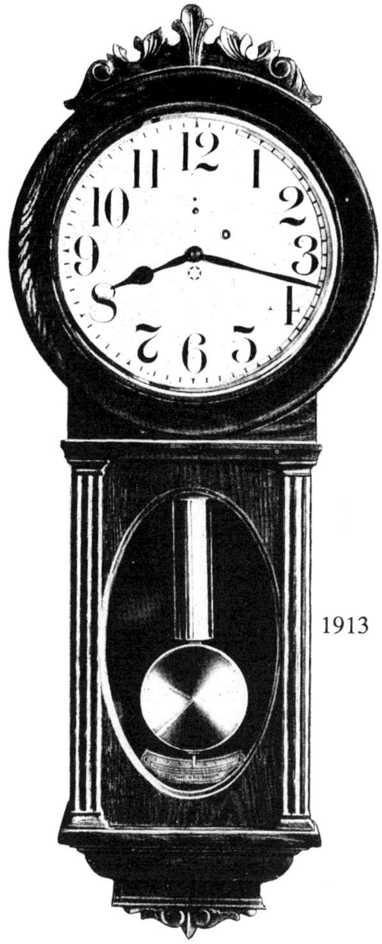

1913

GRIFFIN
(SHOWING 8-DAY WEIGHT.)
EIGHT-DAY SPRING TIMEPIECE. LIST, $25.35
HEIGHT, 38¾ INCHES. WIDTH, 15¼ INCHES.

588

1913

GRAYSON
(SHOWING 8-DAY WEIGHT.)
EIGHT-DAY SPRING TIMEPIECE. LIST, $21.00
HEIGHT, 35¼ INCHES. WIDTH, 15¼ INCHES.

589

1913

GLENOR
(SHOWING 8-DAY WEIGHT.)
EIGHT-DAY SPRING TIMEPIECE. LIST, $23.20
HEIGHT, 36½ INCHES. WIDTH, 15¼ INCHES.

	EXTRA LIST
FOR 8-DAY STRIKE, CATHEDRAL GONG . . .	$1.50
FOR 30-DAY SPRING TIMEPIECE	2.20
FOR 8-DAY CALENDAR	.45
FOR 8-DAY WEIGHT (TIMEPIECE ONLY) . . .	3.65

HANGING CLOCKS

590

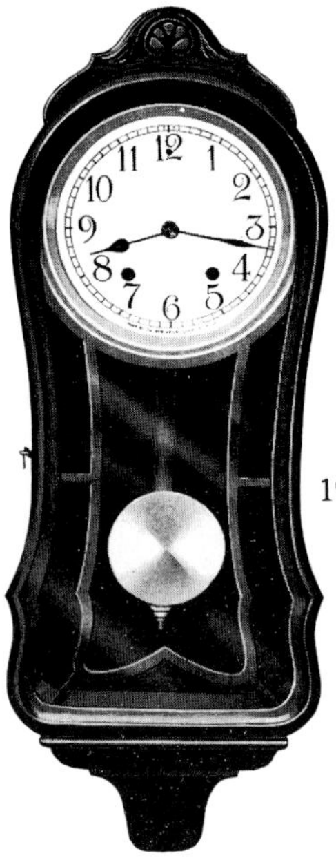

1929

JULIAN
Height, 21 Inches Width, $7\frac{1}{2}$ Inches
Dial, Silver Finish, 6 Inches
Case, Mahogany Finish
Movement, 8-Day Half-hour Strike,
Cath dral Gong

591

1929

LACONA
Height $9\frac{1}{2}$ inches. Width $6\frac{3}{4}$ inches. Convex Crystal, $5\frac{1}{2}$ inch Cream Colored Dial, 8-day Lever Timepiece Movement with Steel Cut Pinions. Modernistic Design and Panel. Pastel Colors, Rose, Yellow, Green, Blue or White.
List $3.40

592

1929

MANOR
Height $11\frac{1}{2}$ inches, width 7 inches. 6 inch Silvered Dial, 8-d y Pendulum Strike, Pastel Colors, Rose, Yellow, Blue, Green or White.
List $14.70
List, Timepiece 11.90

593

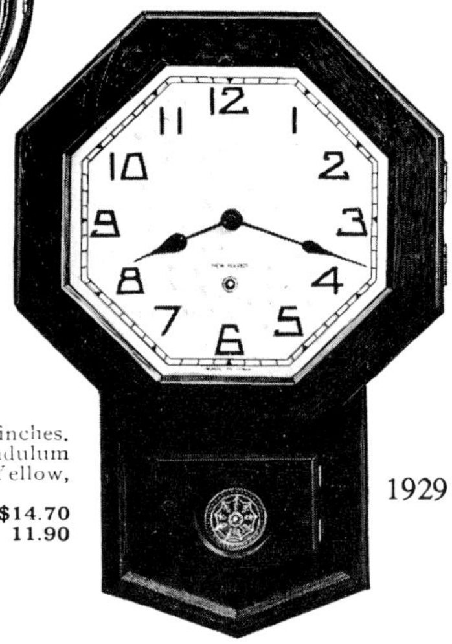

1929

12 INCH DROP OCTAGON
Height, $21\frac{1}{2}$ in. Width, 15 in.
12 inch White Dial, Solid Oak Case.
8-Day Time Movement $11.90
8-Day Strike Movement 13.95

594

1906

TAMPA—Oak

Height, 38 inches. Width, $15\frac{1}{2}$ inches.
Dial, 12 inches.

Eight-day Time	List Price,	$6.70
Eight-day, Strike	" "	7.70
Eight-day Time, Calendar	" "	7.20
Eight-day, Strike, Calendar	" "	8.20

595

1929

REFEREE
Height, $35\frac{1}{2}$ in. Width, 16 in.
12 inch White Dial, Solid Oak Case.
8-Day Time $16.00
8-Day Half-hour Strike 18.90

596

1911

REFEREE
HEIGHT ....$35\frac{1}{2}$ INCHES
WIDTH ....16 INCHES
DIAL ....12 INCHES
CASE, SOLID OAK
PRICE, 8 DAY TIME ....LIST, $7.40
PRICE, 8 DAY STRIKE, CATHEDRAL GONG LIST, 8.90
PRICE, WITH CALENDAR ....LIST EXTRA, .45

HANGING CLOCKS

597

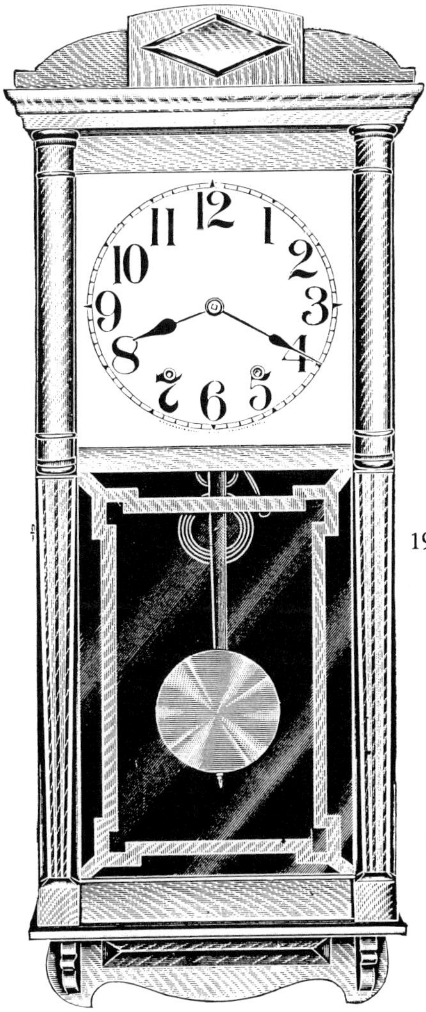

1917

MARNE
Height, 37 Inches Width, 14¾ Inches
Dial, Silver Finish, 10 Inches
Movement, 8-Day Time, 8-Day Strike Gong
30-Day Time

598

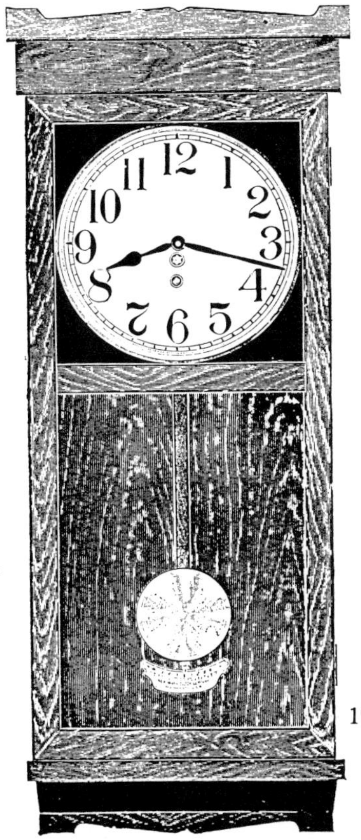

1917

NO. 3043
Height, 42 Inches Width, 16¾ Inches
Dial, 12 Inches
Case, Oak or Mahogany Finish
Movement, 8-Day Time, 8-Day Strike Gong
30-Day Time

599

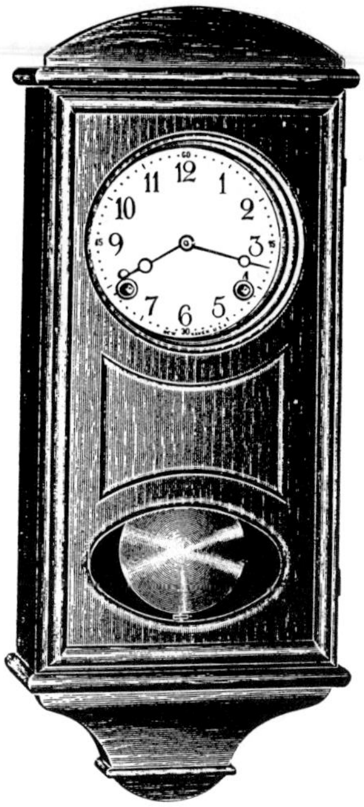

1917

JESTLA
Height, 17¾ Inches Width, 7 Inches
Dial, Porcelain, 5 Inches
Case, Mahogany Finish
Movement, 8-Day Half-hour Strike,
Cathedral Gong

600

RIVERTON

1917

RIVERTON
Height, 34½ Inches Width, 15¾ Inches
Dial, 12 Inches
Case, Solid Oak
Movement, 8-Day Time, 8-Day Strike Gong
30-Day Time
Furnished in White Enamel at Extra List

HANGING CLOCKS

601

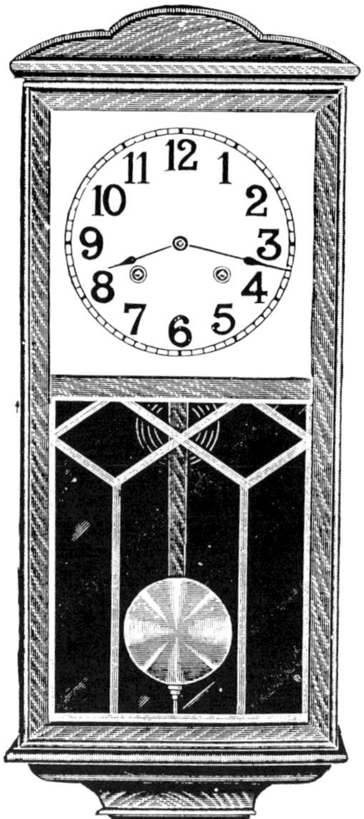

1917

MIMOSA

MIMOSA

Height, 26¾ Inches Width, 10¾ Inches
Dial, Silver Finish, 8 Inches
Case, Mahogany Finish or Oak, Leaded Glass
Movement, 8-Day Half-hour Strike,
Cathedral Gong

602

1907

MIRADA CLOCK

Height, 28¼ inches; width, 12 inches; dial, silver finish, 8 inches. Case, mahogany finish or oak, beveled glass. Movement, 8-day half-hour strike, cathedral gong.

No. 9768	Mahogany, each	$29.00
No. 9769	Golden Oak, each	29.00

603

1917

MIRADA

Height, 28¼ Inches Width, 12 Inches
Dial, Silver Finish, 8 Inches
Case, Mahogany Finish or Oak. Leaded Glass
Movement, 8-Day Half-hour Strike,
Cathedral Gong

604

1917

MADERO

Height, 28¾ Inches Width, 13 Inches
Dial, Silver Finish, 8 Inches
Case, Mahogany Finish or Oak, Leaded Glass
Movement, 8-Day Half-hour Strike,
Cathedral Gong

605

1917

MARINO

Height, 26 Inches Width, 11 Inches
Dial, Silver Finish, 8 Inches
Case, Mahogany Finish or Oak, Beveled Glass
Movement, 8-Day Half-hour Strike,
Cathedral Gong

HANGING CLOCKS

The year entered beside each clock is the year of the catalog from which the illustration was taken. Some clocks may have been produced a few years earlier.

606

1881

607

1881

REGULATOR, D. R.

Height, 31 inches. 8 Day Time.
Dial, 12 inches. 8 Day Strike.
Also with Alarm.

PROGRAMME.

EIGHT-DAY, PENDULUM, STRIKING CLOCK, FOR SCHOOLS.

The radiating wire arms on the sun wheel in the lower part of the clock, when in place between the teeth of the wheel, will strike every five minutes. To make it strike at any required interval, raise the wire arms and rest them in the small holes on the rim of the wheel around the entire circuit, counting each wire as five minutes' time.

The wheel in front behind the pendulum ball contains 144 wire springs—one for each five minutes during the twelve hours. Each spring is movable—can be left down or lifted up and the end placed in the small indentation beside it. When all are down, the clock will strike every five minutes. When all are up, it will not strike at all. The wheel is numbered the same as the dial, except in an opposite direction. (Turn the minute hand forward slowly and observe the operation of the striking part.) *Use the little wire hook to move the springs.* To set the clock to any programme, leave down the spring over the hour or minute at which you wish a signal given, and lift the others out of the way, and place their ends in the indentations to the right of them. Thus, to arrange to strike at 9.00, 9.15, 9.45, and 10.00, leave the spring over 9 down, lift the next two, and leave the third down—lift the next five and leave the sixth down—lift the next two and leave third or one over 10 down, and so on. The wheel in front containing the wire springs is movable on its axis—can be turned either to the right hand or to the left. Before attempting this, place a finger on the right hand end of the crooked lever that works the striking part, and thus lift the other end out of the way of the wire springs. If the clock should not strike at the EXACT instant, but a little too soon or too late, lift the lever out of the wires and twist the wheel a little to the left if it strikes too soon—to the right if it strikes too late.

☞ **CAUTION.**—Never turn the wheel *backward* without lifting the lever out of the way of the springs.

☞ **To regulate your Clock,** lower the pendulum ball to make the clock run slower; raise it to make it run faster.

HANGING CLOCKS

609

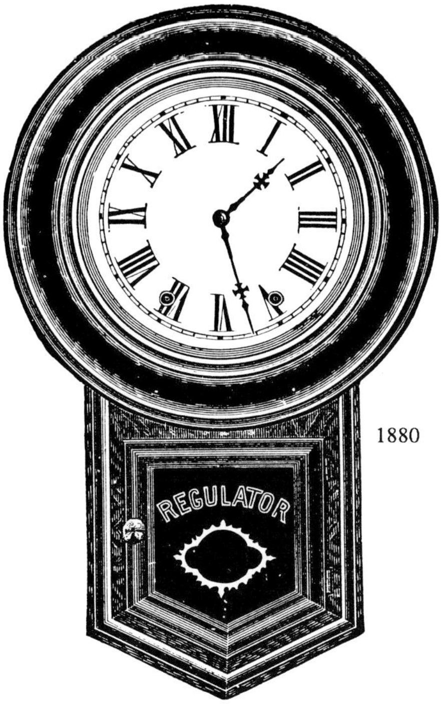

1880

SMALL REGULATOR.

Spring.

Height, 21 inches.

Dial, 8 inches.

8 Day Time. 8 Day Strike. Also with Alarm.

Also made with Gilt Edge Molding.

608

1895

VANDAL.

Oak, Hanging.

Thermometer, Barometer and Level.

Height, 28¼ inches; Dial 6 inches.

8 Day Strike, -	List, $6.50
8 Day Strike, Alarm,	List, 7.00
8 Day Strike, Gong,	List, 6.90
8 Day Strike, Gong, Alarm, List, - -	7.40

610

1883

IMPERIAL.

ROSEWOOD.

Height, 32 ½ inches. Dial, 12 inches. 8 Day Time. The Movement is extra size and weight, with solid polished frame.

HANGING CLOCKS

611

1911

BANK

HEIGHT ....33¼ INCHES
WIDTH ....18½ INCHES
DIAL ....12 INCHES
CASE, SOLID OAK
PRICE, 8 DAY TIME ....LIST, $7.25
PRICE, 8 DAY STRIKE, CATHEDRAL GONG LIST, 8.75
PRICE, WITH CALENDAR ....LIST EXTRA, .45

612

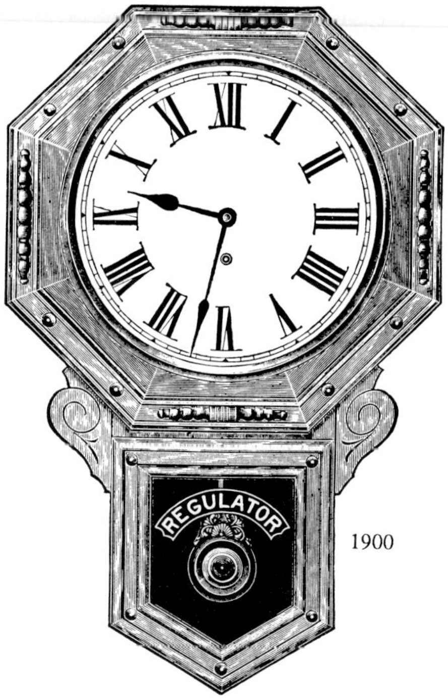

1900

BRADDOCK — Solid Oak.

Height, 27 inches. Dial, 12 inches.

List price, Eight-day Time, $6.00
" " Eight-day Strike, 7.00
" " Thirty-day Time, 8.00

613

1911

BRANFORD

HARVARD

HEIGHT ....32¼ INCHES
WIDTH ....17¼ INCHES
DIAL ....12 INCHES
CASE, SOLID OAK OR WALNUT
PRICE, 8 DAY TIME ....LIST, $10.20
PRICE, 8 DAY STRIKE, CATHEDRAL GONG LIST, 11.70
PRICE, WITH CALENDAR ....LIST EXTRA, .45

BRANFORD

	10-INCH DIAL	12-INCH DIAL
HEIGHT	22¾	27½
WIDTH	15	17½
DIAL	10	12
CASE, SOLID OAK		
PRICE, 8 DAY TIME ....LIST,	$6.10	$6.40
PRICE, 8 DAY STRIKE ....LIST,	7.10	7.40
PRICE, WITH CALENDAR LIST EXTRA,	.45	.45

614

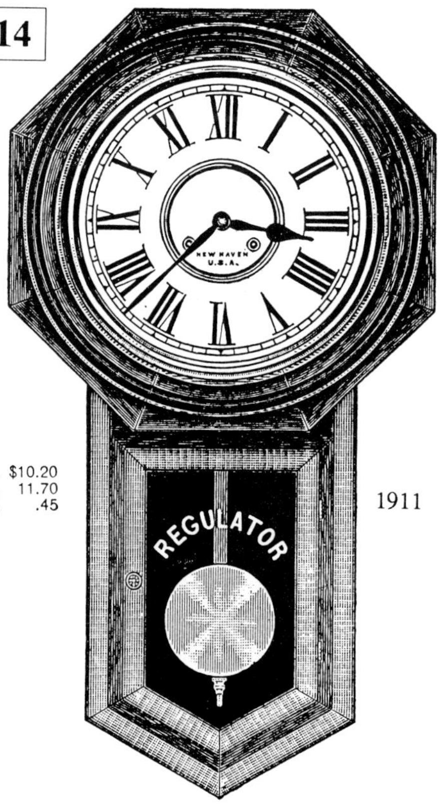

1911

HARVARD

HANGING CLOCKS

615

1911

ERIE

HEIGHT ....33½ INCHES
WIDTH ....18½ INCHES
DIAL ....12 INCHES
CASE, SOLID OAK, ANTIQUE FINISH
PRICE, 8 DAY TIME ....LIST, $7.25
PRICE, 8 DAY STRIKE, CATHEDRAL GONG LIST, 8.75
PRICE, WITH CALENDAR ....LIST EXTRA, .45

616

1900

BLAKE — Solid Oak.

Height, 27 inches. Dial, 12 inches.

List price, Eight-day Time, $5.75
" " Eight-day Strike, 6.75
" " Thirty-day Time, 7.75

617

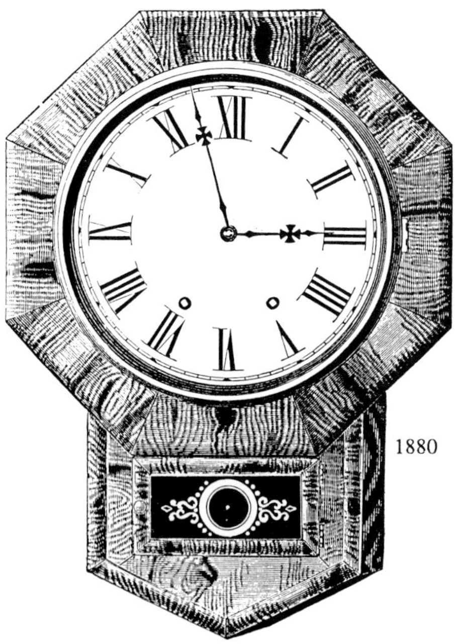

1880

DROP OCTAGON, R. C.

Height, 24 inches. Dial, 12 inches.
8 Day Time. Also with Alarm.
8 Day Strike. Also made with Gilt Edge Molding.

618

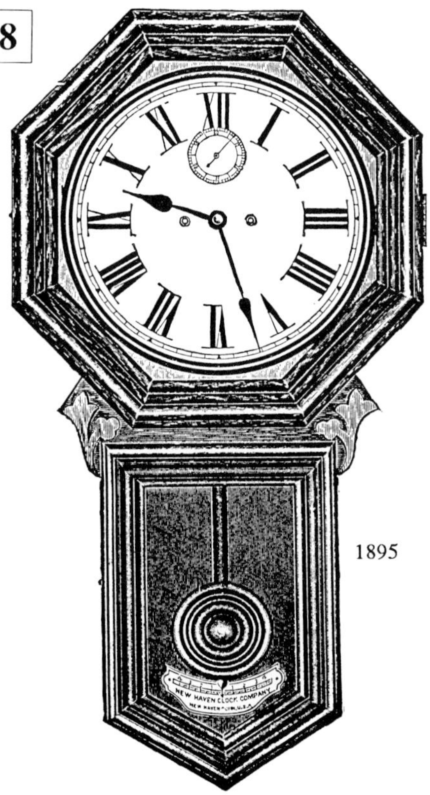

1895

BISHOP.
Light Wood or Dark Wood, with Gilt Moulding.

Height, 30½ inches; Dial, 12 inches.
8 Day Time, Maple, List, $7.00 8 Day Time, Calendar, List, $8.00
8 Day Gong, Maple, List, 9.90 30 Day " Maple " List, 9.00
Rose, with Gilt Moulding, extra, List, $1.00

HANGING CLOCKS

619

1911

DROP OCTAGON
(SOLID OAK)

	8-INCH DIAL	10-INCH DIAL	12-INCH DIAL
HEIGHT	18¼	21⅝	23⅞
WIDTH	12⅜	13⅞	16½
DIAL	8	10	12
CASE, SOLID OAK			
PRICE, 8 DAY TIME ... LIST,	$5.55	$5.90	$6.55
PRICE, 8 DAY STRIKE ... LIST,	6.55	6.90	7.55
PRICE, WITH CALENDAR ... LIST EXTRA,	.45	.45	.45
PRICE, WITH BRASS BANDS LIST EXTRA,	.60	.60	.60

DROP OCTAGON
(SOLID OAK)

620

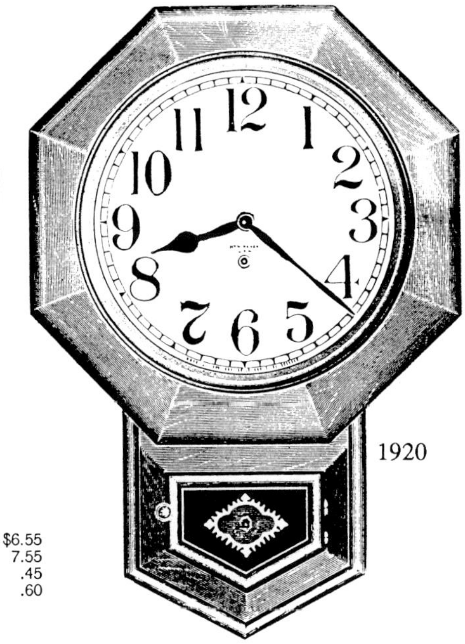

1920

DROP OCTAGON
8-Day Time, 8-Day Hour Strike
With Calendar, Extra List
Case, Solid Oak

Dial	Height	Width
8-Inch	18¼ Inches	12⅜ Inches
10-inch	22 Inches	14 Inches
12-inch	24 Inches	16½ Inches

Only 8-inch furnished in White Enamel at Extra List

621

1880

DROP OCTAGON, No. 2.

Spring.

Height, 24 inches.

Dial, 12 inches.

8 Day Time. 8 Day Strike. Also with Alarm.

622

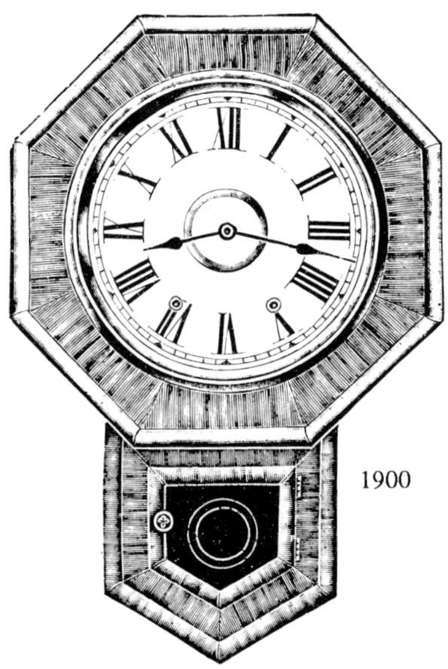

1900

10-inch DROP OCTAGON, GILT

Veneered Rosewood, or Zebra Solid Oak.
Height, 21 inches. Dial, 10 inches.

List price,	Eight-day Time,	$5.90
" "	Eight-day Strike,	6.90
" "	Eight-day Time, Calendar,	6.40
" "	Eight-day Strike, Calendar,	7.40

HANGING CLOCKS

623

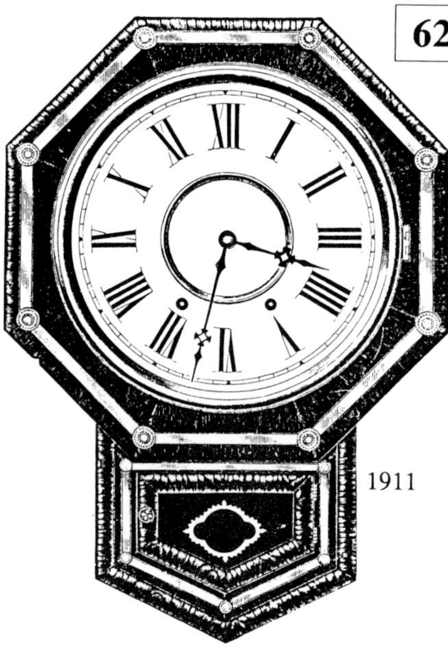

1911

624

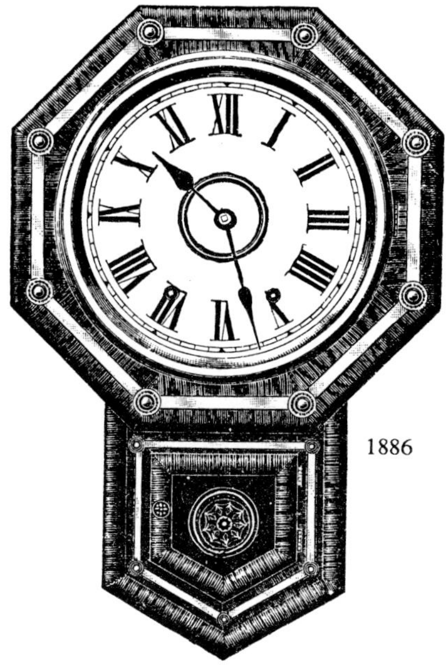

1886

12-INCH DROP OCTAGON

(VENEER, BRASS BANDS)

HEIGHT ....24 ⅛ INCHES
WIDTH ....16½ INCHES
DIAL ....12 INCHES
CASE, ROSEWOOD OR ZEBRA VENEER, WITH BRASS BANDS
PRICE, 8 DAY TIME ....LIST, $7.60
PRICE, 8 DAY STRIKE ....LIST, 8.60
PRICE, WITH CALENDAR ....LIST EXTRA, .45

SMALL DROP OCTAGON (Spring).

ROSEWOOD AND ZEBRA. BRASS BANDS.

Height, 21 inches. Dial, 10 inches.

8 Day 10 in. Time	$6.00
8 " 10 " Strike	7.50
8 " 10 " Time, Calendar	7.00
8 " 10 " Strike, "	8.50

625

1895

626

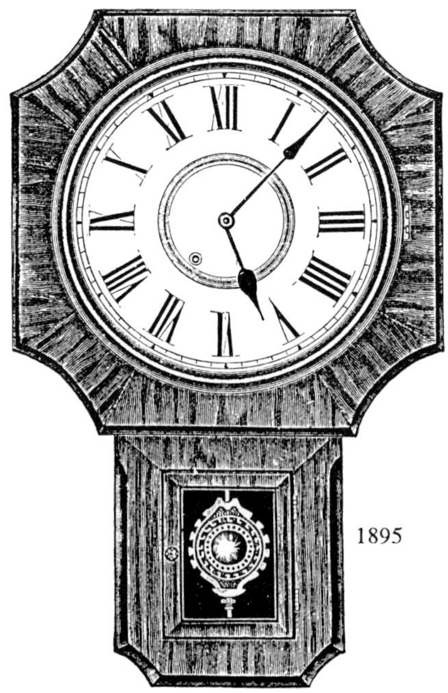

1895

EMPEROR, BRASS BANDS.

Oak.

Height, 25½ inches; Dial, 12 inches.

8 Day Time,	List, $6.60
8 Day Strike,	List, 8.90
30 Day Time,	List, 8.60

EMPEROR.

Oak.

Height, 25½ inches; Dial, 12 inches.

8 Day Time,	List, $5.80
8 Day Strike,	List, 8.10
30 Day Time,	List, 7.80

HANGING CLOCKS

627

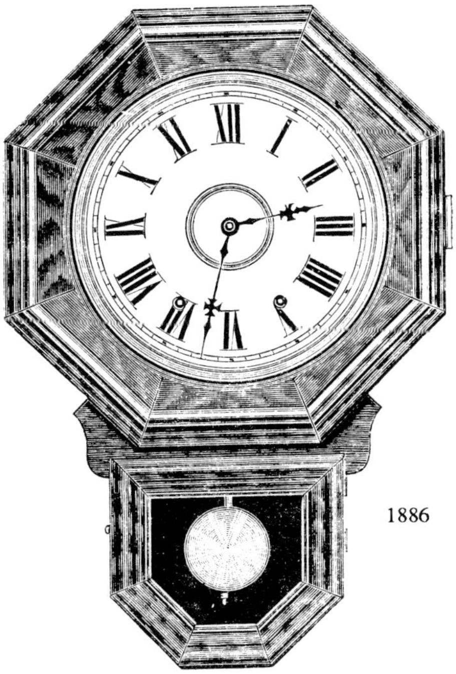

1886

SMALL DROP OCTAGON GILT (Spring).

ROSEWOOD OR ZEBRA.

8 Day 10 in. Drop Oct. Strike, Gilt	$7.80
8 " 10 " " " " Alm. .	8.40
8 " 10 " " Time, Calendar....	6.70
8 " 10 " " Strike, "	8.20
8 " 10 " " Time, Cal. Gilt . .	7 30
8 " 10 " " Strike, " . . .	8.80

628

1880

SMALL DROP OCTAGON, R. C.

Spring.

Height, 17 inches.

Dial, 8 inches.

8 Day Time. 8 Day Strike. Also with Alarm.

Also made with Gilt Edge Molding.

SMALL DROP OCTAGON

ROSEWOOD OR ZEBRA.

8 Day 10 in. Drop Octagon Time	$5.70
8 " 10 " " " " Alarm. .	6.30
8 " 10 " " " Strike........	7.20
8 " 10 " " " " Alarm . .	7.80
8 " 10 " " " Time, Gilt.....	6.30
8 " 10 " " " " " Alm.	6 90

629

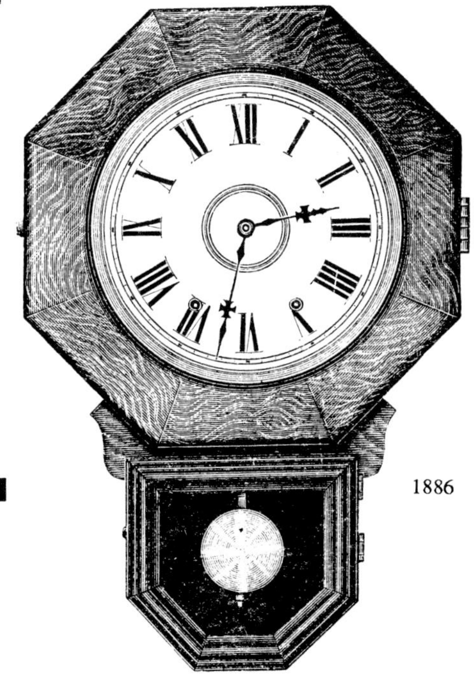

1886

SMALL DROP OCTAGON

ROSEWOOD OR ZEBRA.

HANGING CLOCKS

630

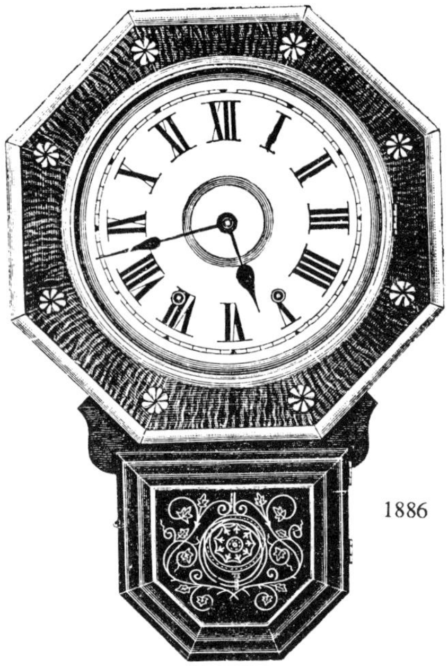

1886

SMALL DROP OCTAGON (Spring).

ROSEWOOD AND ZEBRA. GILT AND GILT BUTTONS.

Height, 21 inches. Dial, 10 inches.

8 Day 10 in. Time	$6.60
8 " 10 " " Alarm	7.20
8 " 10 " Strike	8.10
8 " 10 " " Alarm	8.70
8 " 10 " Time, Calendar	7.60
8 " 10 " Strike "	9.10

631

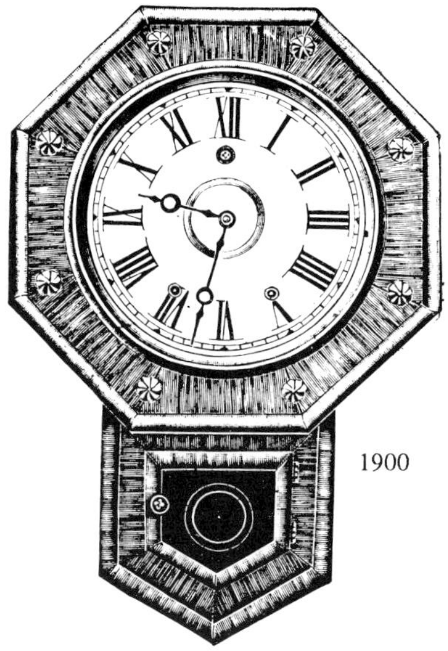

1900

10-inch DROP OCTAGON—Gilt and Gilt Buttons.

Veneered Rosewood, or Zebra Solid Oak.
Height, 21 inches. Dial, 10 inches.

List price, Eight-day Time,	$6.00
" " Eight-day Strike,	7.00
" " Eight-day Time, Calendar,	6.50
" " Eight-day Strike, Calendar,	7.50

DROP OCTAGON (Spring).

ROSEWOOD OR ZEBRA. GILT AND GILT BUTTONS.

Height, 24 inches. Dial, 12 inches.

8 Day 12 inch Time	$7.35
8 " 12 " " Alarm	7.95
8 " 12 " Strike	8.85
8 " 12 " " Alarm	9.45
8 " 12 " Time, Calendar	8.35
8 " 12 " Strike, "	9.85

632

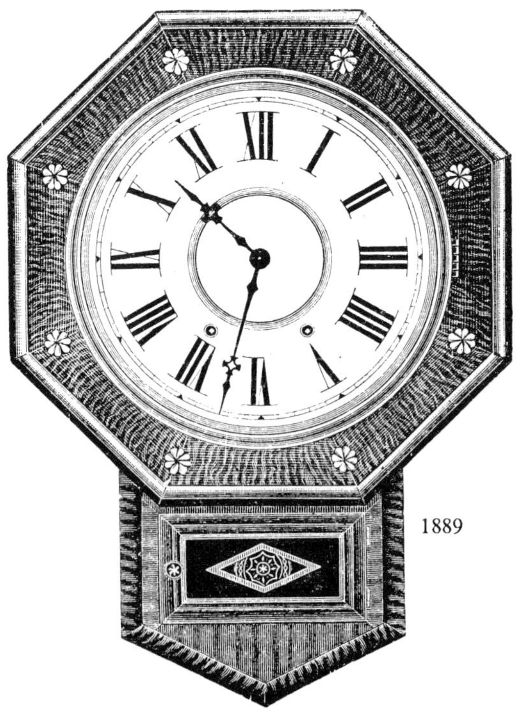

1889

DROP OCTAGON (Spring).

ROSEWOOD OR ZEBRA. GILT AND GILT BUTTONS.

HANGING CLOCKS

633

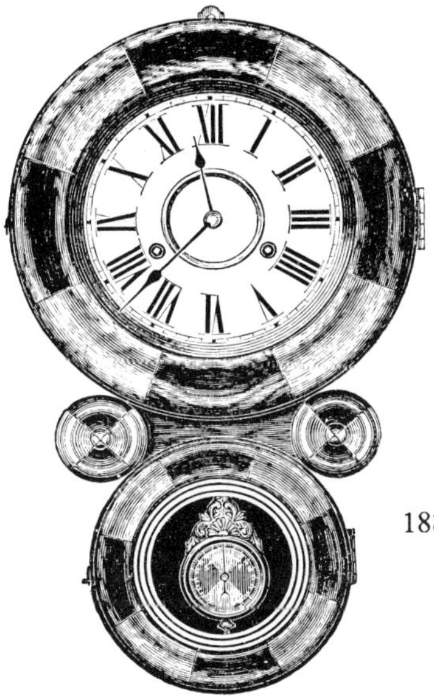

1886

8 in. SAXON MOSAIC.

Height, 20 inches. Dial, 8 inches.

Code Word.		Price.
(Gush)	8 Day 8 inch Saxon Time	$7.20
(Gut)	8 " 8 " " " Alarm	7.80
(Terras)	8 " 8 " " Strike	8.70
(Terrar)	8 " 8 " " " Alarm	9.30

634

SAXON
{The Millie & Jerry Maltz Collection}
{Photo courtesy of Jack Wyer}

635

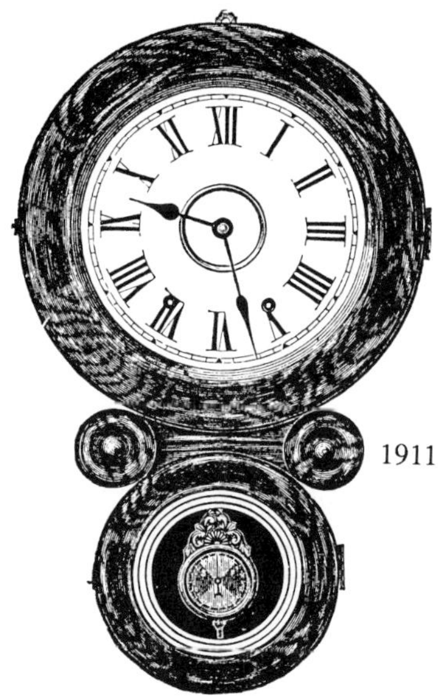

1911

SAXON

	8-INCH DIAL	10-INCH DIAL
HEIGHT	18¾	22
WIDTH	11⅞	14

CASE, ROSE OR MOSAIC VENEER

PRICE, 8 DAY TIME ....LIST,	$6.45	$6.90
PRICE, 8 DAY STRIKE ....LIST,	7.45	7.90
PRICE, WITH CALENDAR LIST EXTRA,		.45
PRICE, WITH ALARM ....LIST EXTRA,		.45

636

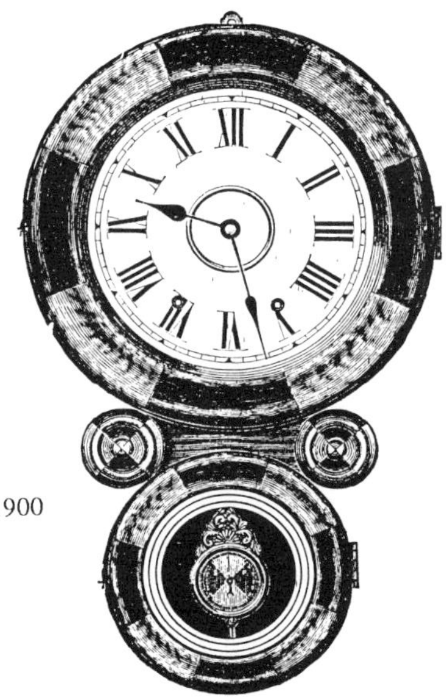

1900

10-inch SAXON MOSAIC

Height, 22 inches. Dial, 10 inches.

List price, Eight-day Time,	$6.40
" " Eight-day Strike,	7.40
" " Eight-day Time, Alarm,	6.90
" " Eight-day Strike, Alarm,	7.90
" " Eight-day Time, Calendar,	6.90
" " Eight-day Strike, Calendar,	7.90

637

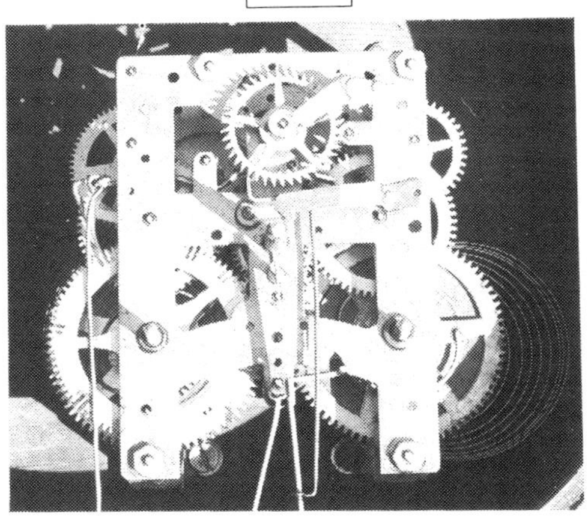

Front view of **Saxon** movement.
{The Millie & Jerry Maltz Collection}
{Photo courtesy of Jack Wyer}

638

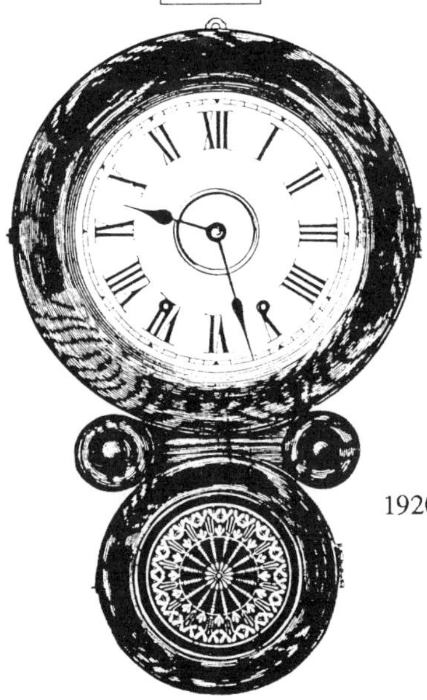

1920

SAXON
Height, 22 Inches Width, 14 Inches
Dial, 10 Inches
Case, Rose or Mosaic
8-Day Time, 8-Day Hour Strike
With Calendar, Extra List

HANGING CLOCKS

KITCHEN HANGING CLOCKS. 8-DAY, HALF-HOUR STRIKE. 6-INCH DIALS.
SOLID OAK CASES, LIGHT OR DARK FINISH.
THESE CLOCKS ARE FITTED WITH ARABIC DIALS
UNLESS ROMAN IS SPECIFIED

639

1913

MERCHANTS LINE HANGING A

LIST, $4.80

HEIGHT, 25½ INCHES. WIDTH, 15 INCHES.

640

1913

CAMDEN LINE HANGING A

LIST, $4.80

HEIGHT, 25 INCHES. WIDTH, 15 INCHES.

641

1913

MERSEY LINE HANGING A

LIST, $4.95

HEIGHT, 25 INCHES. WIDTH, 13 INCHES.

642

1913

NORWICH LINE HANGING A

LIST, $4.95

HEIGHT, 27 INCHES. WIDTH, 15½ INCHES.

HANGING CLOCKS

643

1911

CAMDEN B.

644

1911

CAMDEN C.

CAMDEN LINE HANGING

HEIGHT ....25 INCHES
WIDTH ....15 INCHES
DIAL ....6 INCHES
CASE, OAK OR DARK WOOD
PACKED, ASSORTED 6 IN A BOX
PRICE, 8 DAY HALF-HOUR STRIKE....LIST, $4.80
PRICE, 8 DAY HALF-HOUR STRIKE,
CATHEDRAL GONG ....LIST, 5.25
PRICE, WITH ALARM..............LIST EXTRA .45

645

1911

NORWICH B.

646

1911

NORWICH C.

NORWICH LINE HANGING

HEIGHT ....27 INCHES
WIDTH ......15½ INCHES
DIAL ....6 INCHES
CASE, OAK OR DARK WOOD
PACKED, ASSORTED 6 IN A BOX
PRICE, 8 DAY HALF-HOUR STRIKE...LIST, $4.95
PRICE, 8 DAY HALF-HOUR STRIKE,
CATHEDRAL GONG ....LIST, 5.40
PRICE, WITH ALARM..............LIST EXTRA .45

HANGING CLOCKS

647

1917

MATANZAS
Height, 27½ Inches Width, 14¾ Inches
Dial, 6 Inches
8-Day Half-hour Strike
With Alarm, Extra List
With Gong, Extra List
Case, Dark Wood
Packed 6 in a Box

MATANZAS

648

1917

A

649

1917

B

650

1917

C

SOLID OAK
BRAZILIAN LINE HANGING
8-Day, Half-hour Strike
8-Day, Half-hour Strike, Alarm
Height, 26 Inches Width, 13½ Inches
Dial, 6 Inches
Case, Solid Oak or Dark Wood
Packed, 4 Solid Oak and 2 Dark Wood to each Case

Fitted regularly with Roman dial, gilt glass ornamentations and trimmings. Fitted with Arabic dial, silver (frosted) glass and nickel plated mat and pendulum ball, when orders so specify.

HANGING CLOCKS

651

1917

CUBA
Height, 25⅞ Inches Width, 12¾ Inches
Dial, 6 Inches
8-Day Half-hour Strike
With Alarm, Extra List
With Gong, Extra List
Case, Dark Wood
Packed 6 in a Box

652

1900

STANDISH—Oak.

Height, 26¼ inches. Width, 13 inches.
Dial, 6 inches.

List price,	Eight-day, Strike,	$6.00
" "	Eight-day, Strike Alarm,	6.50
" "	Eight-day, Gong,	6.40
" "	Eight-day, Gong Alarm,	6.90

653

1900

SALEM—Oak.

Height, 26½ inches. Width, 12¾ inches.
Dial, 6 inches.

List price,	Eight-day, Strike,	$6.00
" "	Eight-day, Strike Alarm,	6.50
" "	Eight-day, Gong,	6.40
" "	Eight-day, Gong Alarm,	6.90

654

1900

STANTON—Oak.

Height, 28 inches. Width, 14¼ inches.
Dial, 6 inches.

List price,	Eight-day, Strike,	$6.50
" "	Eight-day, Strike Alarm,	7.00
" "	Eight-day, Gong,	6.90
" "	Eight-day, Gong Alarm,	7.40

HANGING CLOCKS

655

1895

GALLERY.

Solid Oak or Walnut.

8 Day 14 inch Time,	-	-	List, $15 00
30 Day 14 inch Time,	-	-	List, 17 00
8 Day 18 inch Time,	-	-	List, 18 00
30 Day 18 inch Time,	-	-	List, 20 00

656

1880

WALNUT & GILT GALLERY.

Spring.

8 Day, 10 inch, Walnut.
8 Day, 10 inch, Gilt.
8 Day, 12 inch, Walnut.
8 Day, 12 inch, Gilt.
8 Day, 14 inch, Walnut.
8 Day, 14 inch, Gilt.
8 Day, 18 inch, Walnut.
8 Day, 18 inch, Gilt.
8 Day, 20 inch, Walnut.
8 Day, 20 inch, Gilt.

657

1920

12-INCH GALLERY

Height, 16 Inches Width, 16 Inches

Dial, 12 Inches

Case, Solid Oak or Mahogany Finish.

Furnished in White Enamel at Extra List

658

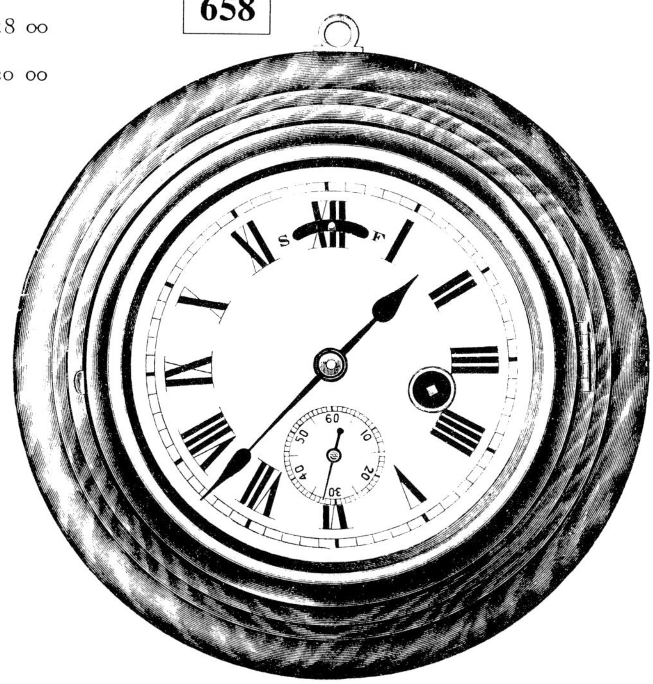

1900

YACHT LEVER

Solid Oak or Cherry.

List price,	One-day Time, Dial, 4 inches,	$2.45
" "	One-day Strike, Dial, 6 inches,	3.90
" "	Eight-day Time, Dial, 6 inches,	4.45

HANGING CLOCKS

659

1920

CANTON LEVER
Nickel or Brass
8-Inch Dial
Fitted 1-Day Time
8-Day Time
8-Day Hour Strike
Diameter 11½ Inches

660

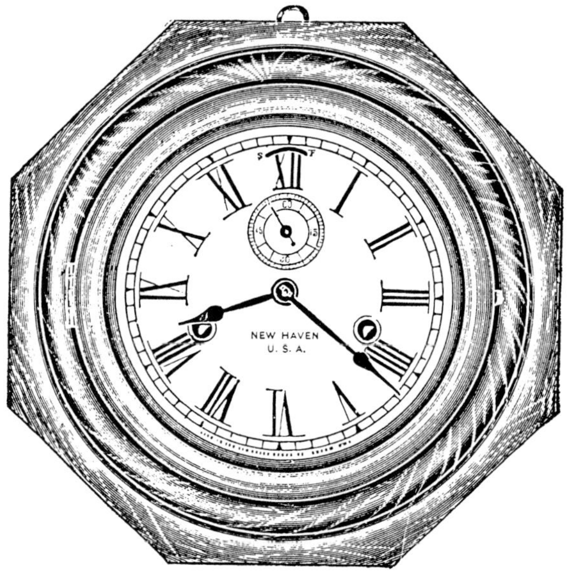

1913

WOOD LEVER
(SOLID OAK OR WALNUT)

LIST

	6 INCH DIAL	8 INCH DIAL	10 INCH DIAL	12 INCH DIAL
1 DAY TIME	$4.10	$4.60	$5.50	$6.30
1 DAY STRIKE	5.10	5.60	6.50	7.30
8 DAY TIME	5.30	5.80	6.70	7.50
8 DAY STRIKE		6.80	7.70	8.50

1 DAY TIME, WITH ALARM.........EXTRA LIST, $.45
1 DAY TIME, WITH CALENDAR...EXTRA LIST, .45

661

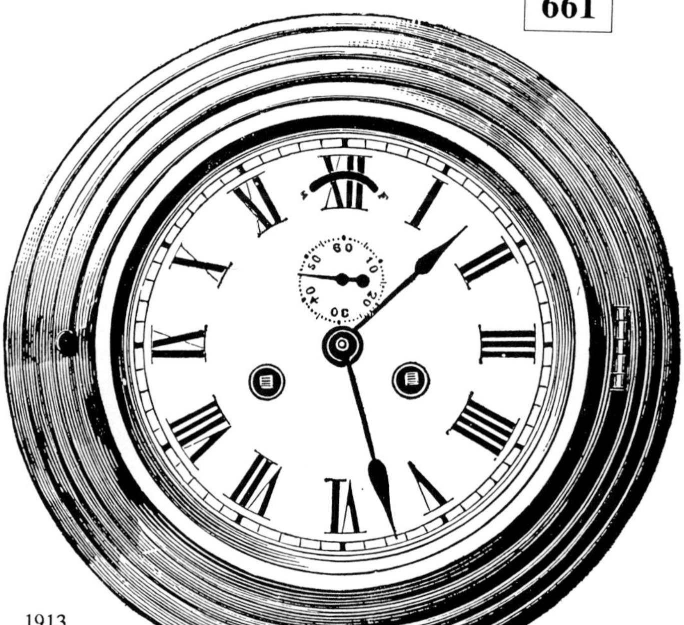

1913

CANTON LEVER
(NICKEL OR BRASS)

LIST

	6 INCH DIAL	8 INCH DIAL
1 DAY TIME	$4.10	$4.60
1 DAY STRIKE	5.10	5.60
8 DAY TIME	5.30	5.80
8 DAY STRIKE		6.80

1 DAY TIME, WITH ALARM........EXTRA LIST, $.45
1 DAY TIME, WITH CALENDAR...EXTRA LIST, .45

662

1913

ENGINE LEVER
EIGHT-DAY TIMEPIECE. LIST, $9.45
HEIGHT, 5 INCHES. WIDTH, 5 INCHES.
DIAL, PORCELAIN, 3 INCHES.
CASE, HEAVY BRASS, DUST- AND WATER-PROOF.
FINISH, POLISHED BRASS, NICKEL OR BLACK ENAMEL.

HANGING CLOCKS

663

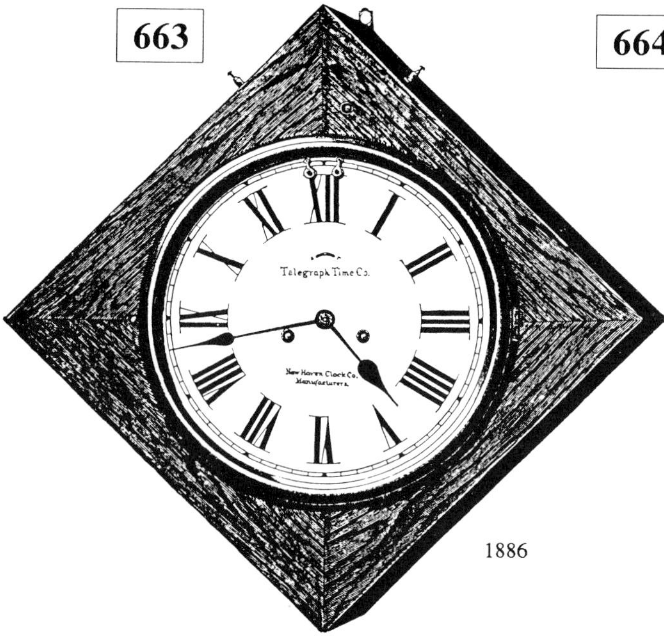

1886

SYNCHRONIZED LEVER.

BLACK WALNUT, ASH, AND MAHOGANY.

Height, 22½ inches. Dial, 12 inches.

(Taurus) 8 Day Time .... Net. $4.00

This Clock is fitted with our new Duplex Movement and the Wilson Synchronizer; manufactured under the patent of the Standard Time Co.

Magnet and Batteries extra.

664

1885

GLASS DIAL.

NUMERALS ETCHED IN BLUE OR RED ON TRANSLUCENT GLASS, OR IN PLAIN GOLD ON CLEAR GLASS.

TWELVE-INCH DIAL ON FOURTEEN-INCH PLATE, GOLD HANDS.

1 Day Time. $9.50

Swung in brass standing frame, $2.00 extra, net.

This movement has cut and polished steel pinions, double roller club tooth escapement, straight line lever, screw pillars.

665

1895

666

1889

EAGLE.

CARVED OAK OR CHERRY.

Height, 40 inches.

Dial, 17 inches.

8 Day Time, Oak......$49.50
8 Day Time, Cherry... 49.50

HANGING CLOCKS

667

1895

TRANSPARENT AND TRANSLUCENT DIAL CLOCKS.

GLASS DIAL

12 inches Diameter. Beveled Plate Glass.
30 Day Time...List, $10.00

668

1929

LUCERNE
Height 10¼ inches. Width 10 inches. Convex Crystal. 5½ inch Cream Colored Dial. 8-day Lever Timepiece Movement with Steel Cut Pinions. Futuristic Design. Pastel Colors, Rose, Yellow, Blue, Green or White.
List **$8.40**

669

1929

12 INCH BALCONY
Width, 14½ inches. 12 inch Dial.
Oak or Mahogany Finish.
8-Day Pendulum Movement **$13.95**

670

1920

14-INCH BALCONY
Height, 19 Inches Width, 19 Inches
Dial, 14 Inches
Case, Solid Oak or Mahogany Finish

671

1920

12-INCH BALCONY
Height, 15 Inches Width, 15 Inches
Dial, 12 Inches
Case, Solid Oak or Mahogany Finish.
Furnished in White Enamel at Extra List

BALCONY CLOCKS
Fitted with Reliable 8-Day Pendulum Time Movements

HANGING CLOCKS

672

1900

WOOD LEVER
Solid Oak or Mahogany Finish
6, 8-, 10- or 12-Inch Dial
Fitted 1-Day Time, 1-Day Hour Strike
8-Day Time, 8-Day Hour Strike
Furnished in White Enamel at Extra List
6 inch—Height 9 in. Width 9 in.
8 inch—Height 11½ in. Width 11½ in.
10 inch—Height 13½ in. Width 13½ in.
12 inch—Height 15 in. Width 15 in.
6-in. packed 12 Clocks in a case.
8-in., 10-in. and 12-in. packed 6 Clocks in a case.

1920

673

1880

ROUND CORNER OCTAGON LEVER.

Size: 4, 6, 8, 10 and 12 inch Dial.

1 Day or 8 Day Time, Time Alarm or Strike.

674

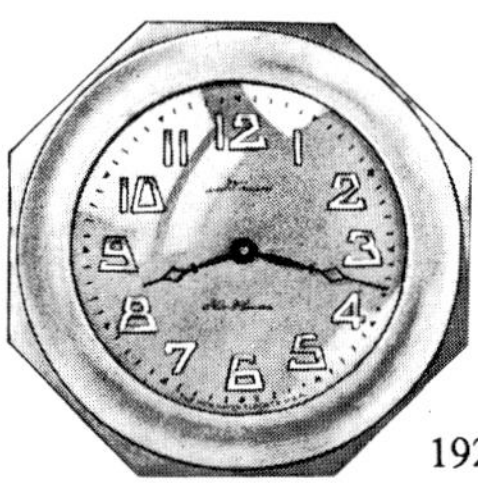

1929

KITCHENETTE
Width 4¾ inches, 3½ inch Silvered Dial with Raised Numerals. 8-day Jeweled Lever Movement, Convex Crystal, Pastel Colors, Rose, Blue, Green, Yellow or White.
List $11.50

675

1929

LARINA
Width 8½ inches. 5½ inch Cream Colored Dial. Convex Crystal. 8 day Lever Timepiece Movement with Steel Cut Pinions. Pastel Colors, Rose, Yellow, Blue, Green or White.
List $7.75

676

1886

WOOD LEVER.

POLISHED VENEERED.

1 Day 4 in. Oct. R'nd Cor. Lev. Time..........	$2.70
1 " 6 " " "	3.55
1 " 6 " " " Alarm...	4.15
1 " 6 " " Strike..	4.15
1 " 8 " " Time..........	4.15
1 " 8 " " " Alarm...	4.75
1 " 8 " " Strike......	4.75
1 " 10 " " Time..........	4.85
1 " 10 " " Strike.........	5.45
8 " 6 " " Time..........	4.55
8 " 8 " " "	5.25
8 " 8 " " Strike...... ..	7.25
8 " 10 " " Time..........	5.95
8 " 10 " " Strike.........	7.95
8 " 12 " " Time..........	6.65
8 " 12 " " Strike.........	8.65
1 " 10 " Strike, Calendar...	5.95
8 " 10 " " Time, " ...	5.75

1889

677

1929

LAWEN
Width 7¼ inches, 5¼ inch Silvered Dial, 8 day Jeweled Lever Movement, Convex Crystal. Pastel colors, Rose, Blue, Green, Yellow, White or Mahogany Finish.
List $11.50

678

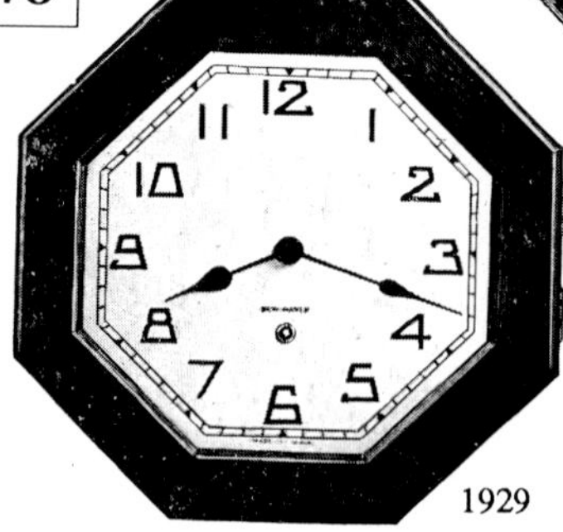

1929

12 INCH GALLERY
Width, 15 inches. 12 inch White Dial. Solid Oak Case of Octagon Design. 8-Day Pendulum Movement. $13.30

HANGING CLOCKS

679

1886

No. 1012

ANTIQUE BRASS, WITH ROYAL COPPER CENTER.
Height, 17 inches. Dial, 6 inches.
Width, 18 inches.
8 Day Time $33.50

680

PAN
LENGTH..............11¾ INCHES
WIDTH..................6¼ INCHES
DIAL......................4½ INCHES, GILT FIGURES, FANCY GILT HANDS
FINISH, DULL BLACK
MOVEMENT, 1 DAY TIME PIECE
PRICE......................................LIST, $2.35

1911

PAN

681

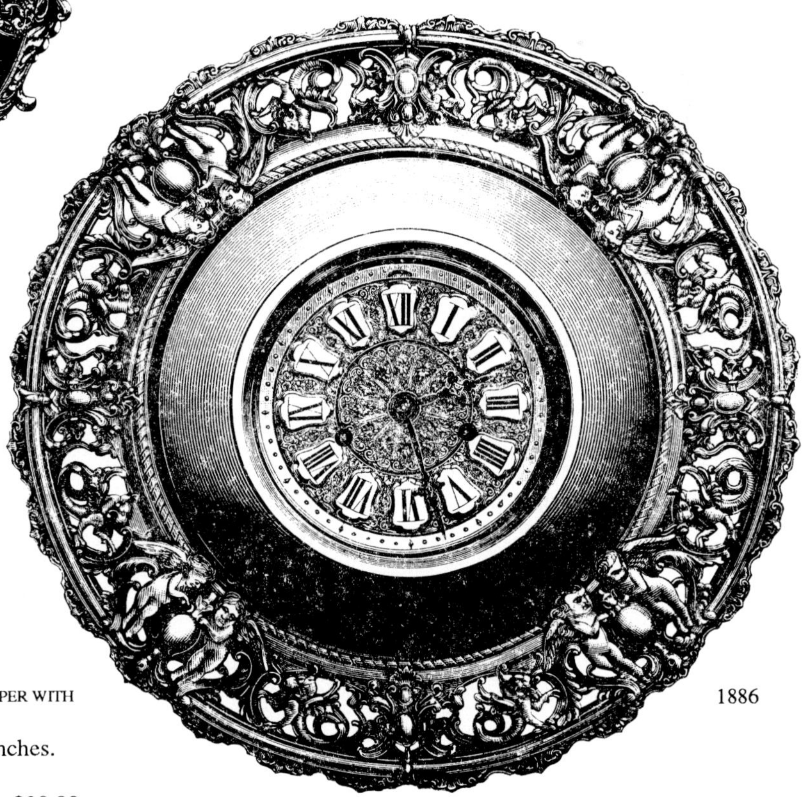

1886

No. 1013 - 1014

No. 1013 - 1014

FINISHED IN ANTIQUE BRASS, ALSO IN ROYAL COPPER WITH SILVER ORNAMENTS.
Height of Clock, 17 inches. Dial, 6 inches.
Length over all, 26 inches
Antique Brass finished, 15 Day Time....... $20.00
Copper and Silver finished, 15 Day Time. $26.25

HANGING CLOCKS

683

1889

684

AIDA.

ANTIQUE BRASS.

Height, 12¾ inches.

Dial, 6 inches.

15 Day Time.....$12.75

682

1885

1889

MANDARIN.

POLISHED BRASS.

1	Day Strike,	Open Dial	....	$10 25
1	" "	Closed "	....	10 25
8	" "	" "	....	12 75
8	" "	Open "	....	12 75

CHATELAINE.

ANTIQUE BRASS FINISH.

Height, 14 inches.   Dial, 6 inches.

Length over all, 21 inches.

15 Day Time............$14.75

HANGING CLOCKS

685

1885

STAR LEVER.

BLACK WALNUT, MAHOGANY OR CHESTNUT.

Dial, 6 inches.

1	Day 6 in.	Star Lever, Time.	$4 75
1	" 6 "	" " Alarm.	5 35
1	" 6 "	" " Strike.	5 25
8	" 6 "	" " Time.	5 25
1	" 6 "	Rosewood Lever. Time.	5 25
1	" 6 "	" " " Alarm.	5 85

686

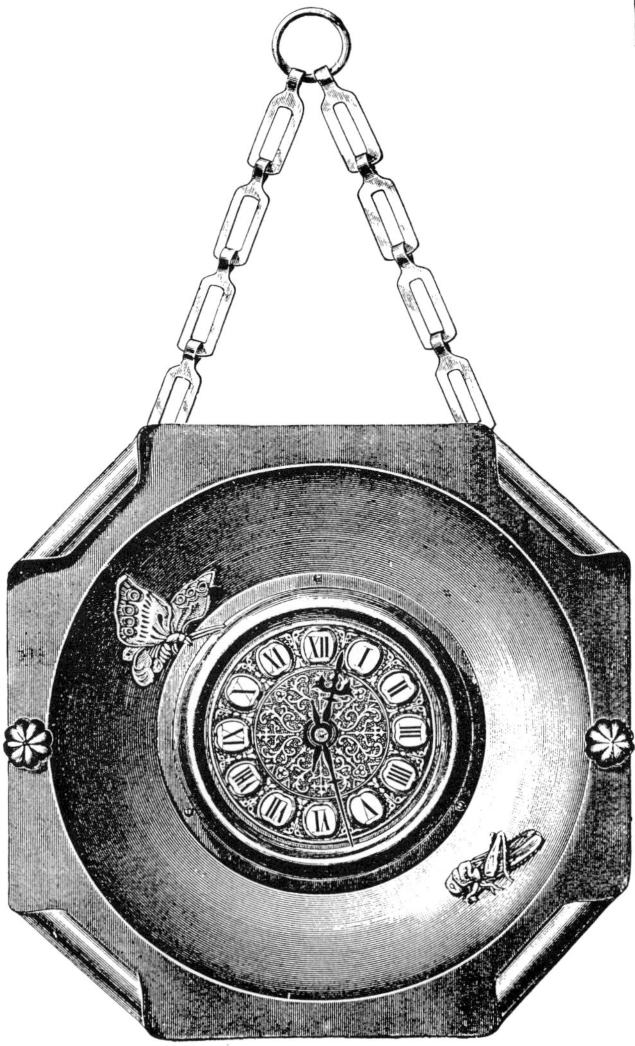

1885

No. 1022.

ROYAL COPPER WITH SOLID SILVER ORNAMENTS.

Height over all, 19 inches. Dial, 4½ inches.

8 Day Time $19 50

687

1895

VENTURA.

Light or Dark Wood.

8 Day Time, List, $5.60
30 Day Time, List, 7.60

INKSTAND CLOCKS

The year entered beside each clock is the year of the catalog from which the illustration was taken. Some clocks may have been produced a few years earlier.

689

1886

SECRETARY INK.

FINISHED IN BRASS WITH CUT-GLASS INKS.

Code Word.	Price.
(Baal) 1 Day Time	$10.50

688

1880

MAINTENON.

1 Day Time. 1 Day Time, Calendar.

See page 454 number 1921 for novelty **Ink Stand** clock.

690

1880

LOUIS.

1 Day Time. 1 Day Time, Calendar.

These Clocks are made of solid brass, polished, designed and finished in the fashion of Louis Quatorze. They have no plating or lacquering of any kind to tarnish, and are, in every way, solid and artistic.

INKSTAND CLOCKS

691

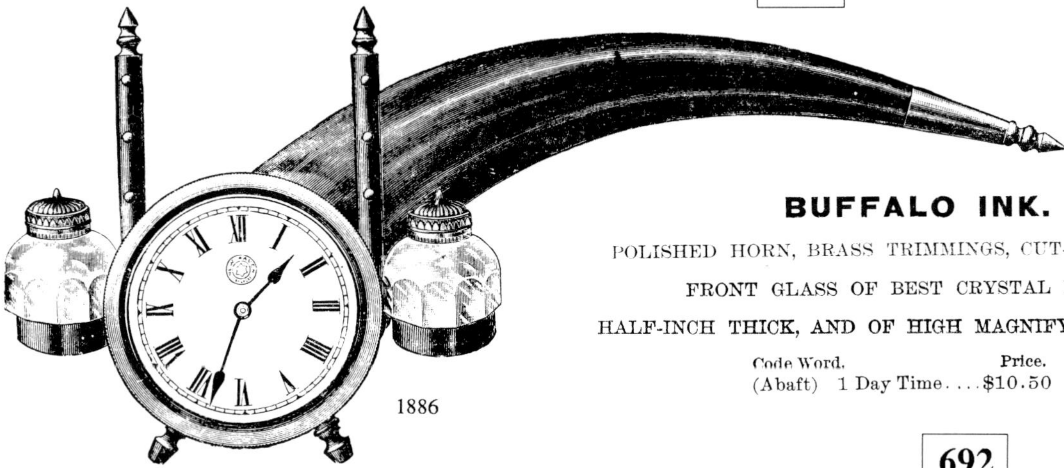

1886

BUFFALO INK.

POLISHED HORN, BRASS TRIMMINGS, CUT-GLASS INKS.

FRONT GLASS OF BEST CRYSTAL LENS,

HALF-INCH THICK, AND OF HIGH MAGNIFYING POWER.

Code Word.	Price.
(Abaft) 1 Day Time....	$10.50

692

SCRIBE.

Height, 9 inches. Dial. 4 inches.

Code Word.	Price.
(Saloon) 1 Day Time, Calendar	$6.00

Finished in black and gold, bronze and gold, and verde and black.

The Clock is arranged to be removed for winding without disturbing the inks.

1886

SCRIBE.

693

1889

THE SCRIBE (Ink Stand).

BRASS FINISH. REVERSIBLE INKS.

Code Word.	Price.
(Dilate) 1 Day Time........	$6.25

THE SCRIBE (Ink Stand).

LEATHER CASES

694

1911

PANZA
(In Leather Case)
HEIGHT ....3 1/8 inches
WIDTH ....3 1/8 inches
DIAL, Porcelain ......2 inches
MOVEMENT, 1 day Time Piece
PRICE, with Leather Case ....List, $6.85

696

1911

JASPER CLOCK
(In Leather Case)
HEIGHT ....3 1/4 inches
WIDTH ....3 inches
DIAL, Porcelain ..1 1/2 inches, with Magnifying Lenses
MOVEMENT, 1 day Time Piece
PRICE, with Leather Case ....List, $13.75

695

1911

MAGNUS
(In Leather Case)
HEIGHT ....4 inches
WIDTH ....3 3/8 inches
DIAL, Porcelain ......2 inches
MOVEMENT, 1 day Time Piece
PRICE, with Leather Case ....List, $5.85

697

COURIER
HEIGHT ....3 inches 1911
WIDTH ....3 inches
DIAL, Porcelain ....2 inches
MOVEMENT, 1 day Intermittent Alarm
PRICE, with Leather Case List, $4.35

698

STING TIME
(In Leather Case) 1911
HEIGHT ....3 1/8 inches
WIDTH ....3 1/8 inches
DIAL, White ....2 inches
MOVEMENT, 1 day Time Piece
PRICE, with Leather Case List, $3.95

699

SHOWING KENNETH OR KENWICK
(In Leather Case)
HEIGHT ....5 inches
WIDTH ....3 3/8 inches 1911
DIAL, Porcelain ..1 3/4 inches
PRICE, with Leather Case, 1 day Time Piece ....List, $8.15
PRICE, with Leather Case, 1 day Strike ....List, 9.65
PRICE, with 1 Day Intermittent Alarm..................................List extra, 1.50

700

JUNIOR TATTOO
(In Leather Case) 1911
HEIGHT ....3 1/4 inches
WIDTH ....3 inches
DIAL, White ....2 inches
MOVEMENT, 1 day Intermittent Alarm
PRICE, with Leather Case List, $5.25

701

LEATHERN 1911
HEIGHT ....3 inches
WIDTH ....3 inches
DIAL, Porcelain ....2 inches
MOVEMENT, 1 day Time Piece
PRICE, with Leather Case List, $3.50

LONGCASE CLOCKS & STANDING REGULATORS

702

1883

703

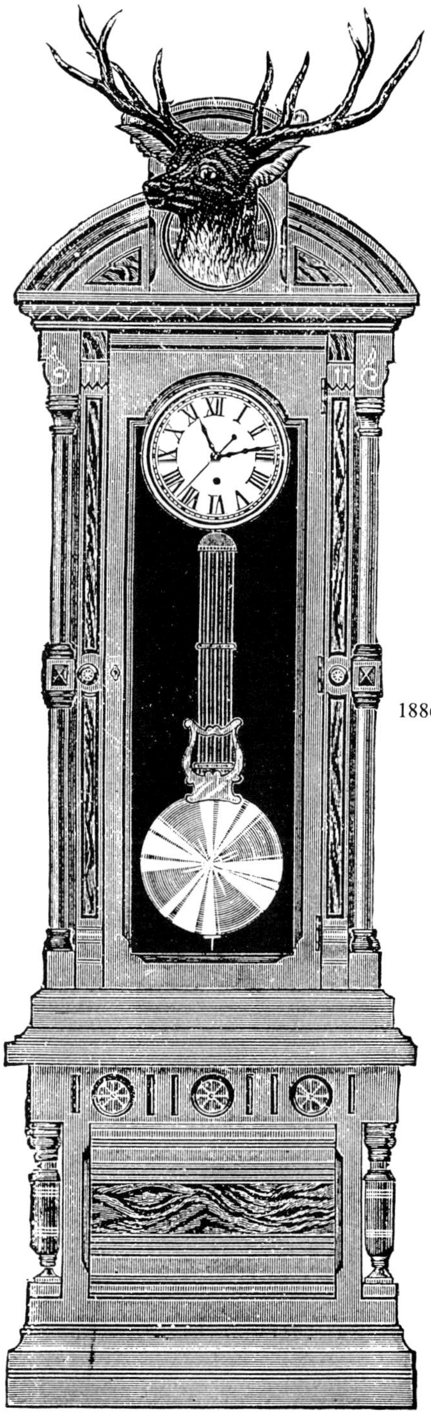

1886

REGULATOR No. 8

WALNUT

Height, 10 feet 6 inches.

Finished with Glass Sides, Movement of best Swiss pattern, Sweep Second Hand, Dead-Beat Escapement, 12 inch Porcelain Dial, Steel and Brass Pendulum Rods, adjusting to heat and cold.

LONGCASE CLOCKS & STANDING REGULATORS

704

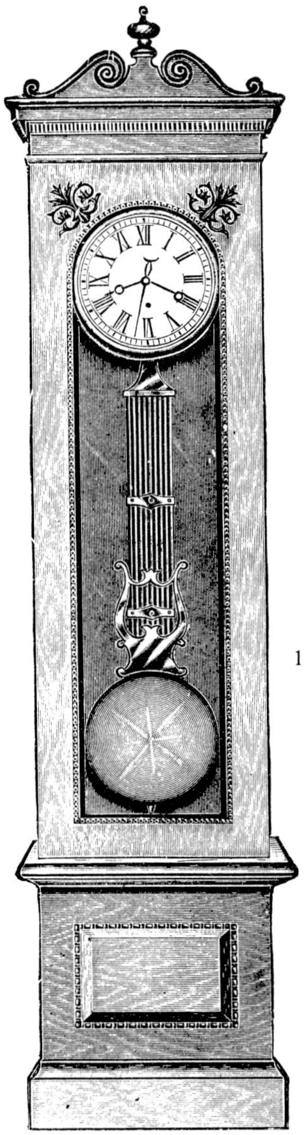

1900

THORNTON.
Solid Oak.

EIGHT DAY TIME, SWEEP SECOND.

Height, 7 feet, 10 inches; Dial, 12 inches.

Fitted either with Swiss or our own Fine Regulator Movement, - - List, $80.00

1895

705

1900

CHIPPENDALE

Antique Oak, Mahogany or Walnut. Silver Dial. Height, 8 feet. Dial, 12 inches. Fitted with finest Weight Movement. 8 Day Strike. Cathedral Gong. List price, ... $190.00

Earlier examples of this model had rebuilt *ca.* 1850 English bell-strike movements. Later, New Haven movements were used.

706

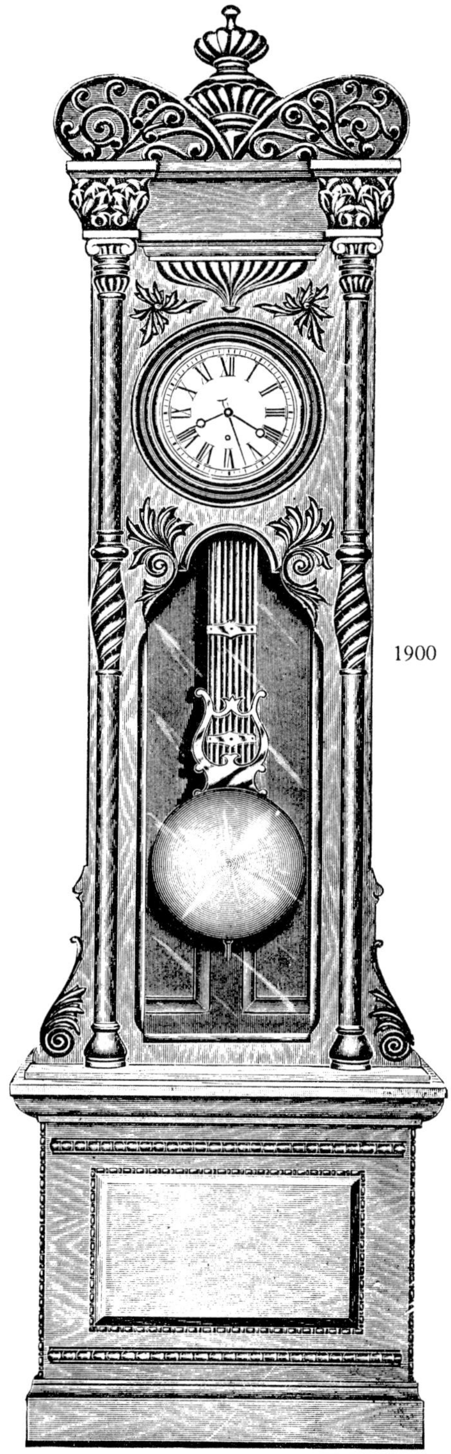

1900

GIANT — Solid Oak.

Height, 10 feet 3 inches. Dial, 12 inches.
Width of Base, 3 feet.
Dead Beat Pin Escapement and Retaining Power with Sweep Second.
List price, with our own Movement, Round Pendulum Ball and Wooden Rod, $160.00
List price, with Swiss Movement, Gridiron Pendulum, as illustrated, $190.00
Fitted with Three-Jar Mercurial Pendulum, Extra list, $40.00

LONGCASE CLOCKS & STANDING REGULATORS

707

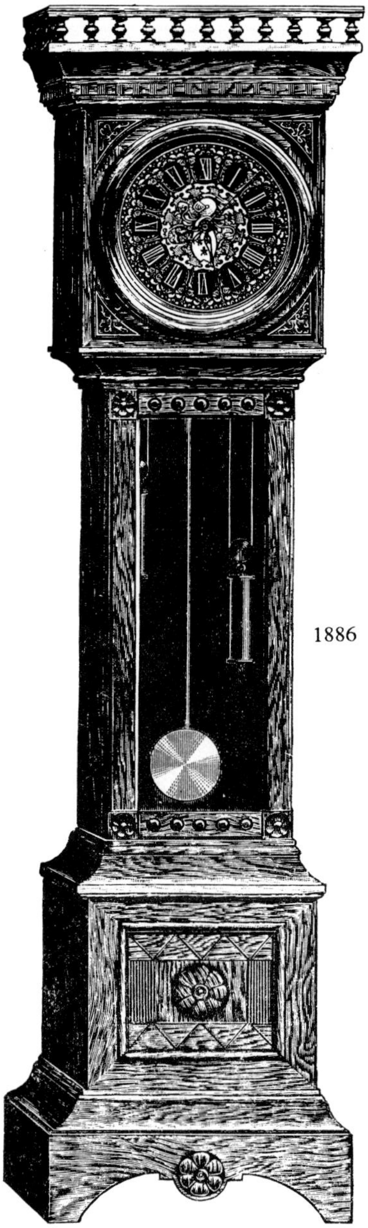

1886

HALL CLOCK No. 1.

OAK, MAHOGANY, OR WALNUT,

HAND CARVED, CABINET FINISHED.

Code Word.	Height, 84 inches. Alabaster Dial, 12 inches, Hand Painted.		Price.
(Slave)	8 Day Strike on Gong, Oak	Net.	$65.00
(Sleigh)	8 " " " " Ebony	Net.	65.00
(Slice)	8 " " " " Walnut	Net.	65.00

Seconds Pendulum, with cut and polished steel pinions, and highly finished movement.

708

1920

WELLINGTON
Height, 88 Inches Width, 21 Inches
Dial, Etched Silver, 12 Inches
Case, Solid Mahogany
Highly Polished
Movement, 8-Day Weight, Half-hour Strike,
Cathedral Gong

709

1886

REGULATOR No. 4

WALNUT

Height, 10 feet 2½ inches. Width at top, 35 inches. Width at base, 35 inches.

Code Word.		Price.
(Gallic)	8 Day Time	$135.00

Finished with Glass Sides, Movement of best Swiss pattern, Sweep Second Hand, Dead-beat Escapement, 12-inch Porcelain Dial, Steel and Brass Pendulum Rods, adjusting to heat and cold.

LONGCASE CLOCKS & STANDING REGULATORS

711

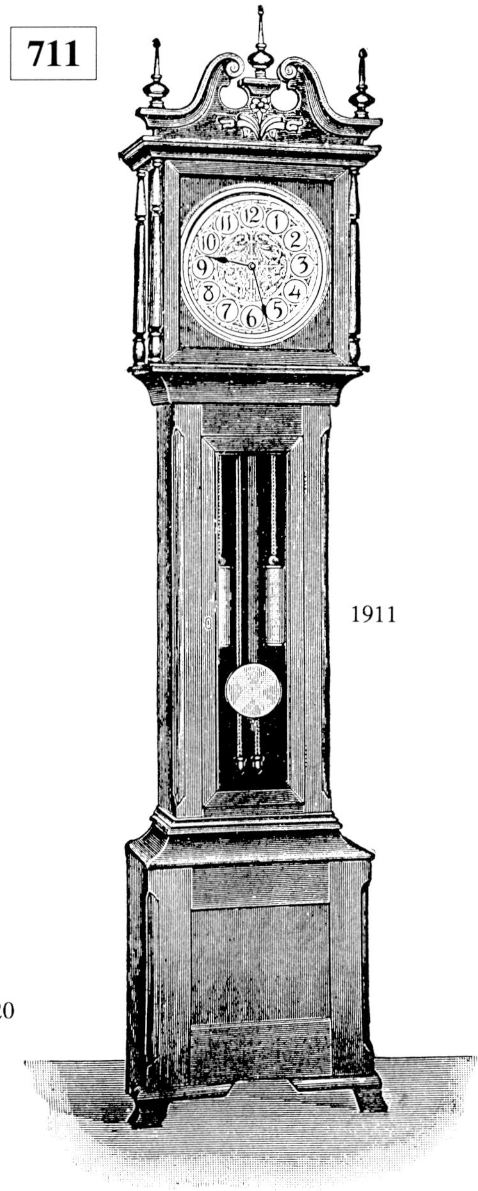

1911

712

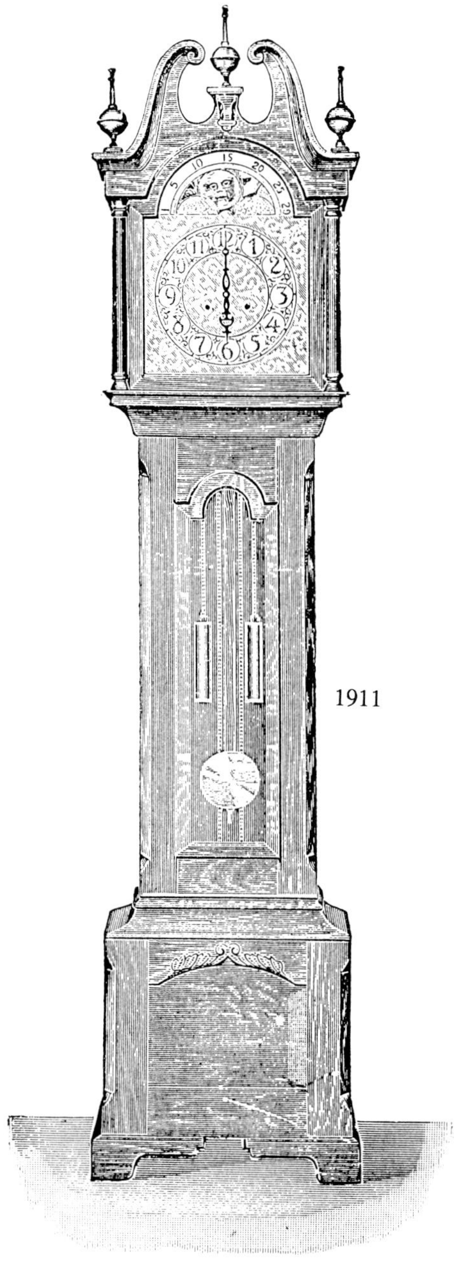

1911

710

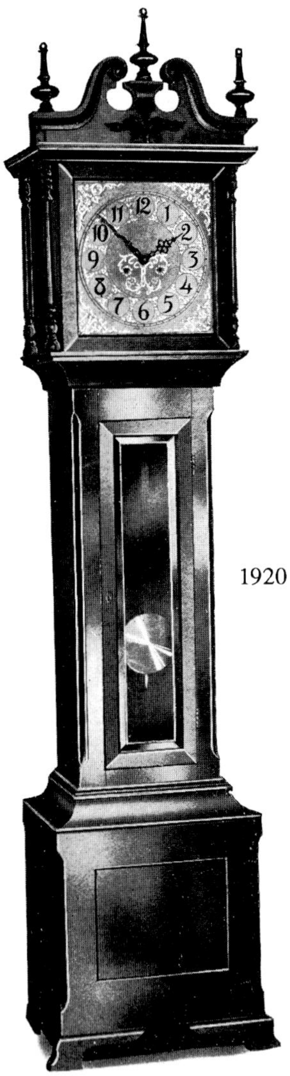

1920

WILMINGTON

HEIGHT ....91 INCHES
WIDTH ....21½ INCHES
DIAL, ETCHED METAL ....12 INCHES
CASE, ANTIQUE OAK OR MAHOGANY WITH GILT TOP ORNAMENTS
MOVEMENT, 8 DAY WEIGHT, HALF-HOUR STRIKE, CATHEDRAL GONG
PRICE ....LIST, $75.50
PRICE, FITTED WITH WESTMINSTER CHIME MOVEMENT, CHIMING ON THE QUARTER-HOURS ON CATHEDRAL GONGS
LIST EXTRA, $72.50

SHERBROOKE

HEIGHT ....95 INCHES
WIDTH ....22 INCHES
DIAL, ETCHED METAL ....12 INCHES, WITH MOON PHASES
CASE, ANTIQUE OAK OR MAHOGANY WITH GILT TOP ORNAMENTS
MOVEMENT, 8 DAY WEIGHT, HALF-HOUR STRIKE, CATHEDRAL GONG
PRICE ....LIST, $159.50
PRICE, FITTED WITH WESTMINSTER CHIME MOVEMENT, CHIMING ON THE QUARTER HOURS ON CATHEDRAL GONGS .... LIST EXTRA, $72.50

WILMINGTON
Height, 91 Inches Width, 21½ Inches
Wood Top Ornaments
Dial, Etched Silver, 12 Inches
Case, Solid Mahogany
Highly Polished
Movement, 8-Day Weight, Half-hour Strike, Cathedral Gong

LONGCASE CLOCKS & STANDING REGULATORS

714

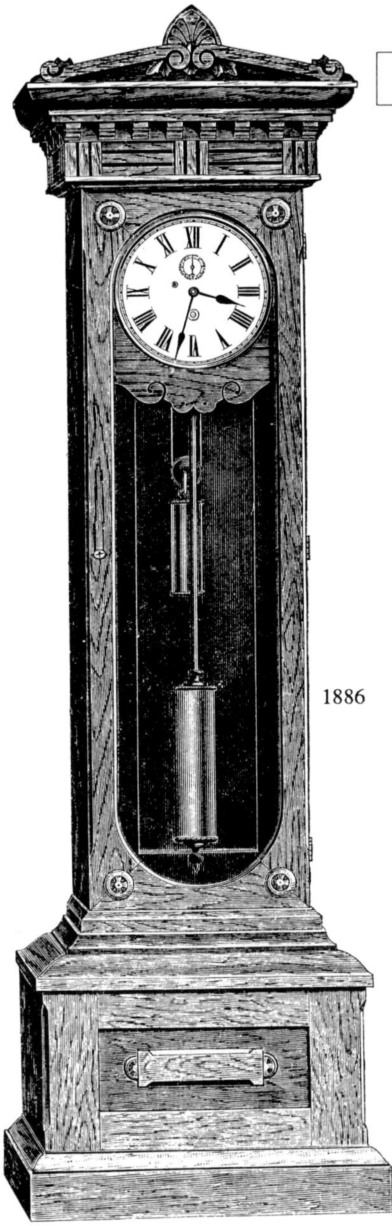

1886

REGULATOR No. 10
WITH BASE. OAK, BLACK WALNUT, AND MAHOGANY.
Height, 90 inches. Dial, 12 inches.
8 Day Time $82.00

713

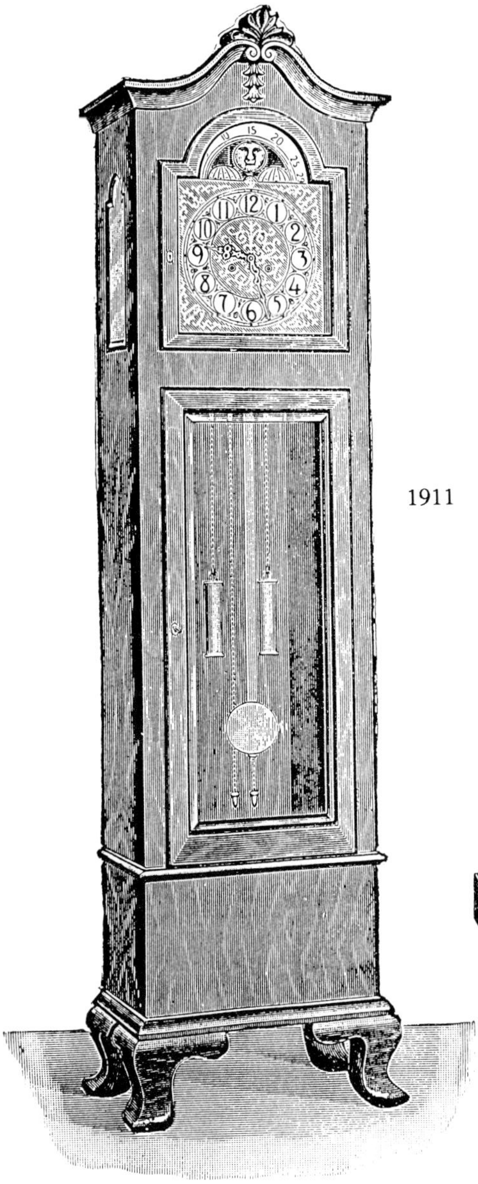

1911

ANNAPOLIS

HEIGHT ....91 INCHES
WIDTH ....23 INCHES
DIAL, ETCHED METAL.........12 INCHES, WITH MOON PHASES
CASE, ANTIQUE OAK OR MAHOGANY
MOVEMENT, 8 DAY WEIGHT, HALF-HOUR STRIKE, CATHEDRAL GONG
PRICE ....LIST, $159.50
PRICE, FITTED WITH WESTMINSTER CHIME MOVEMENT, CHIMING ON THE QUARTER-HOURS ON CATHEDRAL GONGS.............LIST EXTRA, $72.50

715

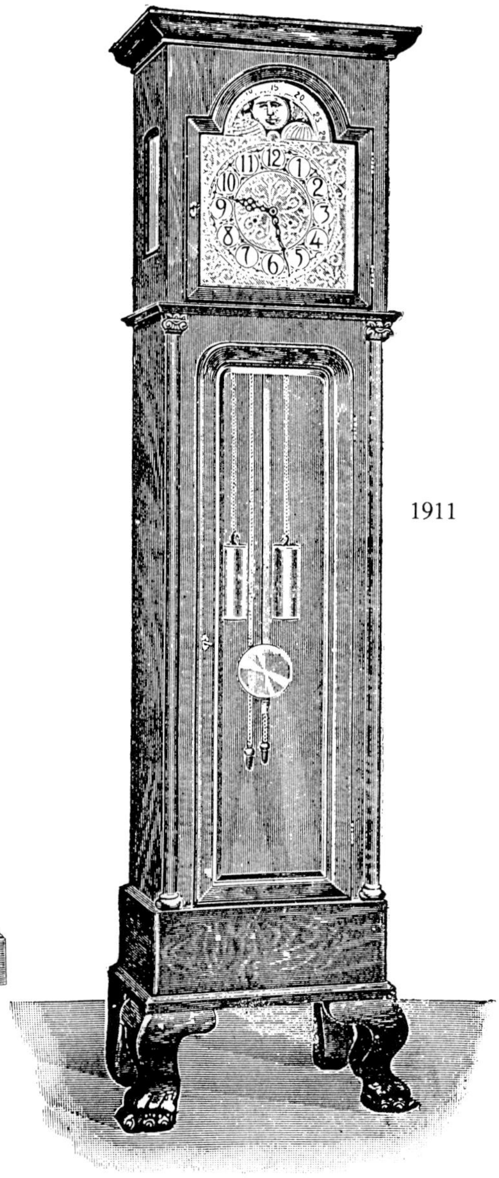

1911

MINNEAPOLIS

HEIGHT ....91 INCHES
WIDTH ....25 INCHES
DIAL, ETCHED METAL ....12 INCHES, WITH MOON PHASES
CASE, ANTIQUE OAK OR MAHOGANY
MOVEMENT, 8 DAY WEIGHT, HALF-HOUR STRIKE, CATHEDRAL GONG
PRICE ....LIST, $191.50
PRICE, FITTED WITH WESTMINSTER CHIME MOVEMENT, CHIMING ON THE QUARTER-HOURS ON CATHEDRAL GONGS ....LIST EXTRA $72.50

LONGCASE CLOCKS & STANDING REGULATORS

The year entered beside each clock is the year of the catalog from which the illustration was taken. Some clocks may have been produced a few years earlier.

716

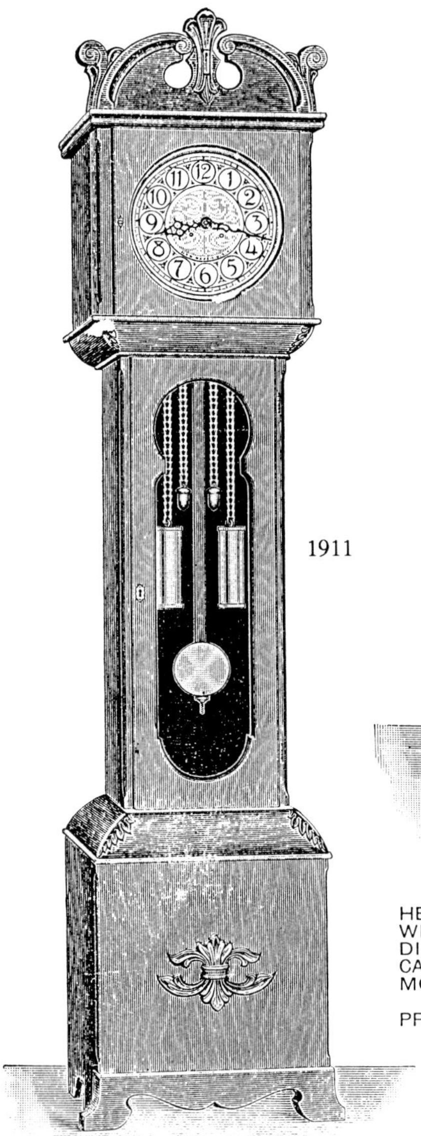

1911

717

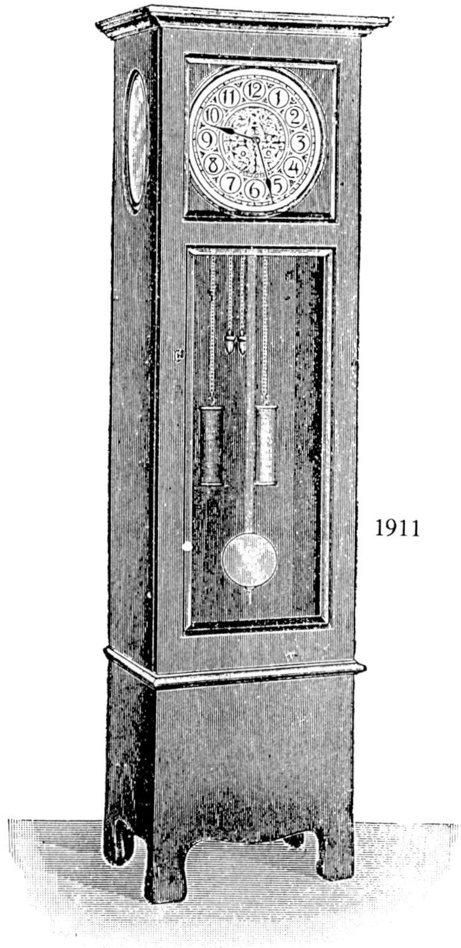

1911

BROOKSDALE

HEIGHT ....72½ INCHES
WIDTH ....20½ INCHES
DIAL, ETCHED METAL ....12 INCHES
CASE, SOLID ANTIQUE OAK OR MAHOGANY
MOVEMENT, 8 DAY WEIGHT, HALF-HOUR STRIKE, CATHEDRAL GONG
PRICE .... LIST, $79.75

718

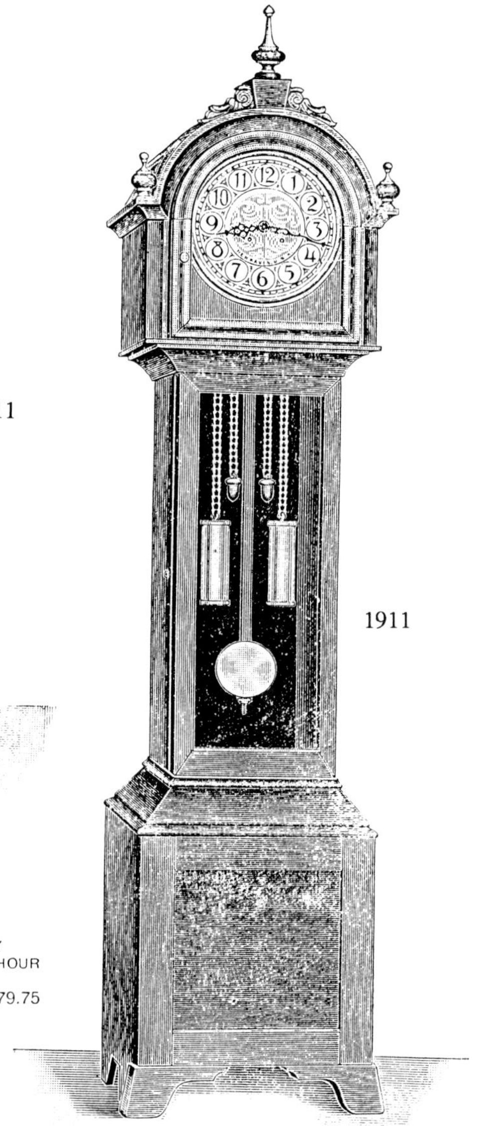

1911

ARLINGTON

HEIGHT ....90½ INCHES
WIDTH ....21 INCHES
DIAL, ETCHED METAL ....12 INCHES
CASE, SOLID MAHOGANY
MOVEMENT, 8 DAY WEIGHT, HALF-HOUR STRIKE, CATHEDRAL GONG
PRICE .... LIST $75.50

SHEFFIELD

HEIGHT ....90½ INCHES
WIDTH ....21¾ INCHES
DIAL, ETCHED METAL ....12 INCHES
CASE, SOLID MAHOGANY
MOVEMENT, 8 DAY WEIGHT, HALF-HOUR STRIKE, CATHEDRAL GONG
PRICE .... LIST, $75.50

LONGCASE CLOCKS & STANDING REGULATORS

719

1895

BLENHEIM
Solid oak. Height, 7 feet. Dial, 12 inches.
8 Day Gong. Half-Hour Strike.
List, $30.00

720

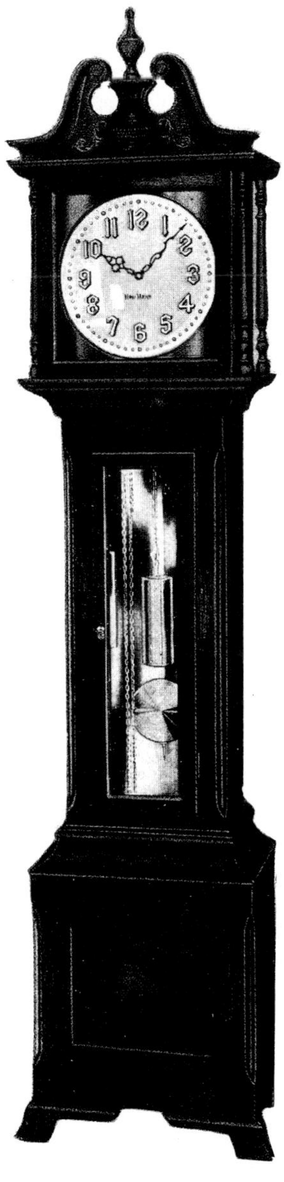

CHAPEL CHIME NO. 1 HALL CLOCK
Height, 74 inches Width, 18 inches
Dial—12 inch Silver with Raised Gold Numerals
Case—Solid Mahogany, Hand Rubbed, Antique Finish
Movement—8 Day Weight, Chain Wind, Full Westminster Chime, Striking the Quarters on Rods, the Hour Striking on Four Blended Notes.

721

Front view of **Hall Clock No. 1** movement *ca.* 1915 with mounting boards shown. Attachment of side arm at upper left of movement is for strike hammer to strike overhead gong.
{Photo courtesy of Tom Spittler}

MANTEL CLOCKS - CABINET CLOCKS

722

Side view of **Cabinet No. 1.**
{Terry & Carole Looft Collection}

723

CABINET No. 1
Height, 17¼ inches. Width, 10 inches. Depth, 5¾ inches. Dial, 4½ inches. Moderately carved mahogany clock with porcelain dial. Heavy brass bezel. 8-day time and strike. Movement mounted to back panel of clock.
{Terry & Carole Looft Collection}

724

Front view of **Cabinet No. 1** movement.
{Terry & Carole Looft Collection}

725

Close-up of **Cabinet No. 1** label.
{Terry & Carole Looft Collection}

MANTEL CLOCKS - CABINET CLOCKS

726

1895

CATO
POLISHED OAK.
Height, 15 inches. Width, 12 inches.
6-inch Gilt Dial. List, ... $7.00

727

1895

SCIPIO
POLISHED OAK.
Height, 16 inches. Width, 12 inches.
6-inch Gilt Dial. List, ... $7.00

728

1895

NERO
POLISHED OAK.
Height, 16 inches. Width, 12 inches.
6-inch Gilt Dial. List, ... $7.00

729

1883

CABINET No. 1
WALNUT, EBONY, OR MAHOGANY.
Height, 17 ¾ inches. Dial 5 inches. 8 Day,
Striking Hour and Half-Hour on Cathedral Gong.
French Sash and Porcelain Dial.

730

1883

CABINET No. 2
WALNUT, EBONY, OR MAHOGANY.
Height, 17 ½ inches. Dial 5 inches. 8 Day,
Striking Hour and Half-Hour on Cathedral Gong.
French Sash and Porcelain Dial.

MANTEL CLOCKS - CABINET CLOCKS

731

1895

PYRRHUS
POLISHED OAK.
Height, 15 ½ inches. Width, 12 inches.
6-inch Gilt Dial. List, ... $6.75

732

1895

VIRGIL
POLISHED OAK.
Height, 15 ¼ inches. Width, 12 inches.
6-inch Gilt Dial. List, ... $6.75

733

1895

HORACE
POLISHED OAK.
Height, 14 ¾ inches. Width, 12 inches.
6-inch Gilt Dial. List, ... $6.75

734

1883

CABINET No. 3
WALNUT, EBONY, OR MAHOGANY.
Height, 16 ¼ inches. Dial 5 inches. 8 Day,
Striking Hour and Half-Hour on Cathedral Gong.
French Sash and Porcelain Dial.

735

1883

PETREL
LIGHT BAMBOO
Height, 17 inches. Dial 5 inches.
1 Day, Strike. 8 Day, Strike.
Made also with Alarm.

MANTEL CLOCKS - CABINET CLOCKS

736

1884

CABINET No. 6

Walnut. Height, 16 inches. Dial, 5 inches.

8-Day, Strike.................................... $5.29

8-Day, Strike, Alarm......................... $5.74

736-A

FUNSTON—Oak Cabinet.

Height, 15½ inches. Width, 11¾ inches.
Fitted with six-inch White, Gilt
or Ivorine Dial.
Barometer, Thermometer and Spirit Level.
Eight-day, Half-hour Strike,
Cathedral Gong.

List price, $7.05

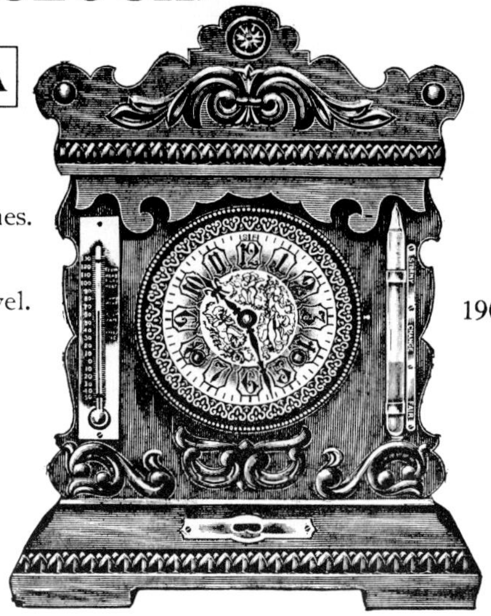

1900

FUNSTON

737

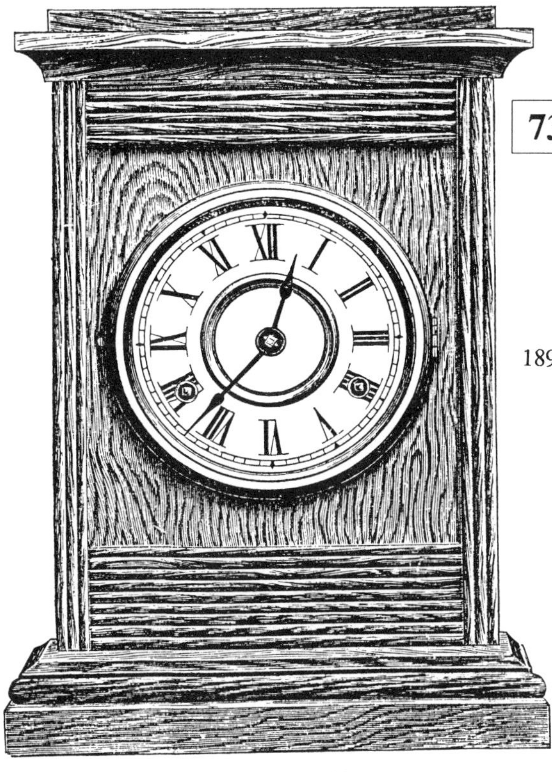

1895

CABINET No. 4.

ASH OR MAHOGANY, POLISHED FINISH.

Height, 14½ inches. Dial, 5 inches.

White, Black or Red Dials.

8 Day Strike, Ash...$6 75
8 " " Alarm, Ash..7 35
8 " " Cathedral Gong, Ash....................................7 25
8 " " Cathedral Gong, Alarm, Ash.........................7 85
8 " " Mahogany...7 15
8 " " Alarm, Mahogany.......................................7 75
8 " " Cathedral Gong, Mahogany..........................7 65
8 " " " " Alarm, Mahogany..................................8 25

738

1886

CABINET No. 5.

OAK OR MAHOGANY, POLISHED FINISH.

Height, 13¾ inches. Dial, 5 inches.

8 Day Strike, Oak...$7.50
8 Day Strike, Alarm, Oak.................................8.10
8 Day Strike, Cathedral Gong, Oak....................8.00
8 Day Strike, Cathedral Gong, Alarm, Oak..........8.60
8 Day Strike, Mahogany..................................7.90
8 Day Strike, Alarm, Mahogany.........................8.50
8 Day Strike, Cathedral Gong, Mahogany............8.40
8 Day Strike, Cathedral Gong, Alarm, Mahogany....9.00

MANTEL CLOCKS - CABINET CLOCKS

739

1900

CLIFTON—Oak Cabinet.

Height, 11½ inches. Width, 13¼ inches.
Fitted with six-inch White, Gilt
or Ivorine Dial.
Eight-day, Half-hour Strike,
Cathedral Gong.

List price, $8.35

740

1900

CALIBAN—Oak Cabinet

Height, 14¾ inches. Width, 11 inches.
Fitted with six-inch White, Gilt
or Ivorine Dial.
Eight-day, Half-hour Strike,
Cathedral Gong.

List price, $7.65

741

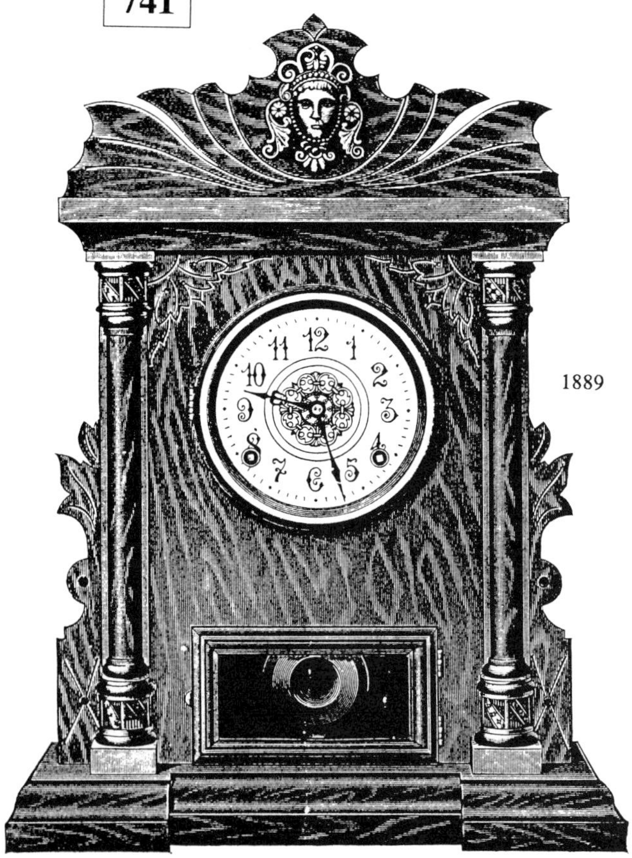

1889

AURANIA

ANTIQUE OAK AND BLACK WALNUT.
HEAVILY GILT ORNAMENTS.
Height, 19 ¼ inches. Dial 5 inches.
8 Day, Cathedral Gong. List, ... $9.75
8 Day, Cathedral Gong, Alarm. List, $10.35

742

1900

CORBIN—Cabinet Oak.

Height, 15¼ inches. Width, 11¾ inches.
Six-inch White, Gilt or Ivorine Dial.
Eight-day, Half-hour Strike, Gong.

List price, $7.65

MANTEL CLOCKS - CABINET CLOCKS

744

1895

ANTONY

SOLID OAK WITH CAST ORNAMENTS.

Height, 10 inches. Width, 15 inches.
Dial 5 inches. 8 Day, Half-Hour Strike
on Cathedral Gong. List, ... $19.00

745

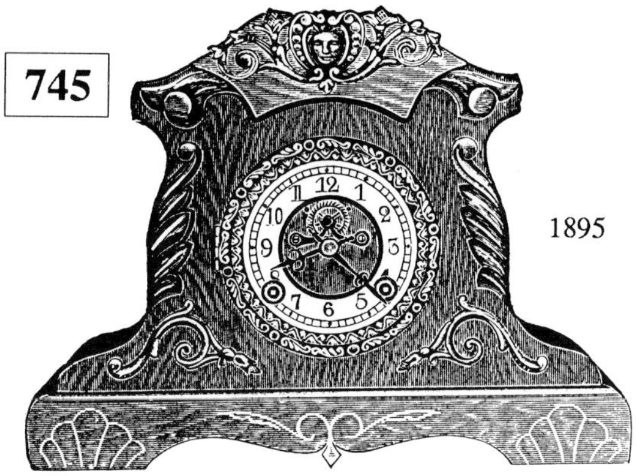

1895

SULLA

SOLID OAK WITH CAST ORNAMENTS.

Height, 11 inches. Width, 14 inches.
Dial 5 inches. 8 Day, Half-Hour Strike
on Cathedral Gong. List, ... $17.00

746

1885

CABINET No. 7.

ASH.

Black or White Dial.

Height, 16 inches. Dial, 5inches.

8 Day Strike............................	$7 20
8 " " Alarm.......................	7 80
8 " " Cathedral Gong.............	7 70
8 " " " " Alarm.........	8 30

747

1889

ARAB

ANTIQUE OAK AND MAHOGANY. IVORIZED DIAL.

Height, 18 ½ inches. Dial 4 inches.
15 Day Striking Hours and Half Hours on
Cathedral Gong, New Cut Pinion Movement.
List, ... $35.00

MANTEL CLOCKS - CABINET CLOCKS

VREDA.

POLISHED ANTIQUE OAK.

Height, 20½ inches. Dial, 8 inches.

8 Day, Striking Hour and Half-hour on Cathedral Gong . . . $12.50

749

1889

VREDA

748

1885

CABINET No. 9.

OAK.

Richly Ornamented with Cast Brass Trimmings, in Antique Finish.

Height, 21¾ inches. Dial, 4½ inches

8 Day, Striking Hour and Half-hour on Cathedral Gong.. $31 25

8 " " " " " Visible Escapement... 33 75

750

1889

VALKYRIE.

POLISHED ANTIQUE OAK.

Height, 21¼ inches. Dial, 8 inches.

8 Day, Striking Hour and Half-hour on Cathedral Gong. . $11.50

MANTEL CLOCKS - CABINET CLOCKS

751

1890

SERVIA.

Antique Oak or Black Walnut. Heavily Gilt Ornaments.

Height, 19 inches. Dial, 5 inches. Fitted with Cathedral Gong.

List $9.00. Alarm, $9.60.

Silver Dial, $9.50. Alarm, $10.10.

752

1890

GALLIA.

Antique Oak or Black Walnut. Heavily Gilt Ornaments.

Height, 19 inches. Dial, 5 inches. Fitted with Cathedral Gong.

List, $8.50. Alarm, $9.10.

Silver Dial, $9.00. Alarm, $9.60.

753

UMBRIA.

ANTIQUE OAK AND BLACK WALNUT.

Height, 18 inches.

Base, 22½ inches. Dial, 5 inches.

8 Day Cathedral Gong. $12.00

Silver Dial, fifty cents list extra.

1890

UMBRIA.

MANTEL CLOCKS - CABINET CLOCKS

The year entered beside each clock is the year of the catalog from which the illustration was taken. Some clocks may have been produced a few years earlier.

754

1881

HECLA.

WALNUT VENEER.

Height, 15 inches.

Dial, 8 inches.

8 Day Strike.

Plain Rich Library Clock, of extra width, Highly Polished.

755

1917

AUSPREY
8-Day Half Hour Strike
Cathedral Gong
Height, $8\frac{7}{8}$ Inches Width, $15\frac{3}{4}$ Inches
6 Inch Metal Dial, Bowed Glass
Mahogany Finish
Fitted with Porcelain Dial at Extra List

756

1900

MENDON

757

1900

MELITA

758

1900

MEDEA

Height, $15\frac{1}{4}$ inches. Width, $11\frac{3}{4}$ inches. Dial, 6 inches.
Eight-day, Half-hour Strike, Cathedral Gong. White, Fancy Gilt or Ivorine Dial.
List price each, $5.55

MANTEL CLOCKS - CABINET CLOCKS

759

1895

POMPEY
SOLID OAK WITH CAST ORNAMENTS.
Height, 9 ½ inches. Width, 14 inches.
Dial 5 inches. 8 Day, Half-Hour Strike
on Cathedral Gong. List, ... $20.00

760

1885

BLACK BIRD.
OAK, WALNUT OR MAHOGANY.
Dial, 5 inches. Height, 18 inches.
Code Word. Price.
(Mastodon) 8 Day Strike, Cathedral
Gong $11 15

761

1885

CABINET No. 10.
MAHOGANY.
Height, 16 inches. Dial, 4 inches.
Richly Ornamented with Cast Brass Trimmings, in Antique Finish.
8 Day Striking Hour and Half-hour on Cathedral Gong $14 30

762

1889

ETRURIA.
ANTIQUE OAK AND BLACK WALNUT.
Height, 18 inches. Base, 21 inches. Dial, 5 inches.
8 Day, Cathedral Gong.............$11.00
Silver Dial, fifty cents list extra.

MANTEL CLOCKS - CABINET CLOCKS

763

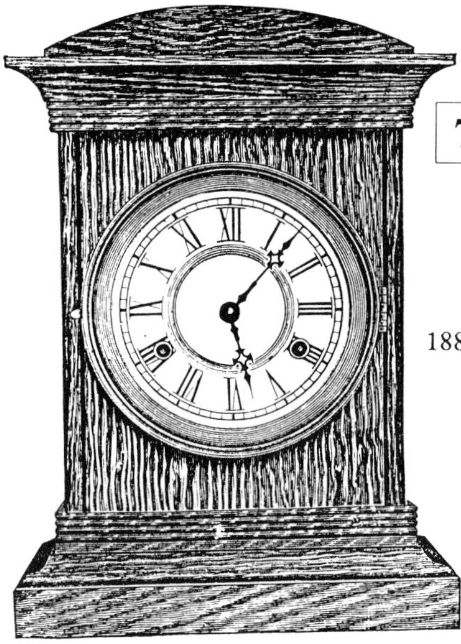

1885

THRUSH.

OAK, WALNUT OR MAHOGANY.

Dial, 5 inches. Height, 17½ inches.

	Price.
8 Day Strike, Cathedral Gong..	$13 00

764

1885

MARTIN.

OAK OR CHERRY.

Height, 9¾ inches. Dial, 5 inches.

	Price.
1 Day Half-hour Strike	$7 15
8 " Bell Strike	9 60
8 " Time	7 25

765

1889

RUSSIA

ANTIQUE OAK AND BLACK WALNUT. HEAVILY GILT ORNAMENTS.

Height, 18 ½ inches. Dial 5 inches. 8 Day, Cathedral Gong. List, ... $8.00. 8 Day, Cathedral Gong, Alarm... List, $8.60.

766

1889

ARROW

WALNUT, BRASS TRIMMINGS.

Height, 16 ¼ inches. Dial 5 inches. 8 Day Strike. List, $7.50. 8 Day Strike, Alarm. List, $8.10. 8 Day Strike Cathedral Gong. List, $8.00. 8 Day Strike, Cathedral Gong, Alarm. List, $8.60

MANTEL CLOCKS - CABINET CLOCKS

767

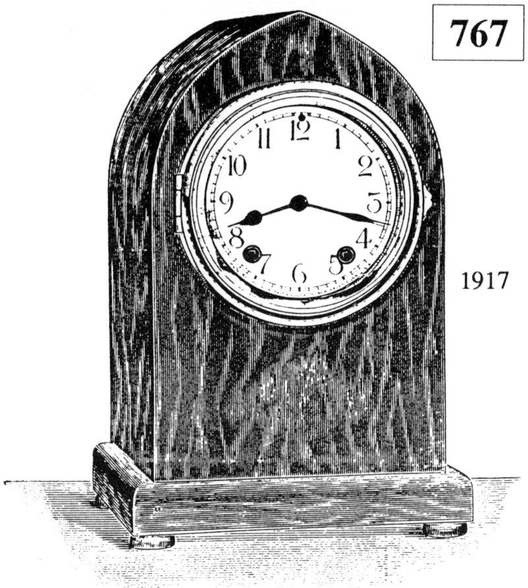

1917

768

1917

MORREA

Height, 8½ Inches Width, 7⅛ Inches
Dial, Porcelain, 5 Inches
Case, Solid Mahogany
Cast Sash, Beveled Bowed Glass
Also fitted with Silver Dials at Extra List

MYRA

Height, 9¾ Inches Width, 6¼ Inches
Dial, Porcelain, 5 Inches
Case, Solid Mahogany
Cast Sash, Beveled Bowed Glass
Also fitted with Silver Dials at Extra List
Furnished in Mahogany, Brown Mahogany, Circassian Walnut Finish, Golden and Fumed Oak, at same list.

769

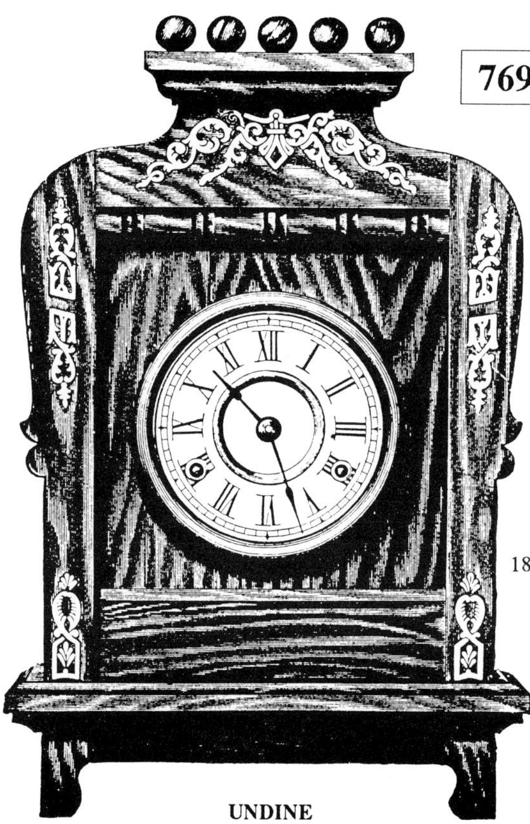

1889

UNDINE

WALNUT, BRASS TRIMMINGS.

Height, 17 inches. Dial 5 inches. 8 Day Strike. List, $7.75
8 Day Strike, Alarm. List, $8.35. 8 Day Strike Cathedral Gong. List, $8.25. 8 Day Strike, Cathedral Gong, Alarm. List, $8.85

770

1889

LURINE

ANTIQUE OAK WITH HAND CHASED BRASS ORNAMENTATIONS.

Height, 15 ½ inches. Dial 5 inches. 15 Day, Striking Hours and Half-Hours on Cathedral Gong........... List, $30.00

MANTEL CLOCKS - CABINET CLOCKS

771

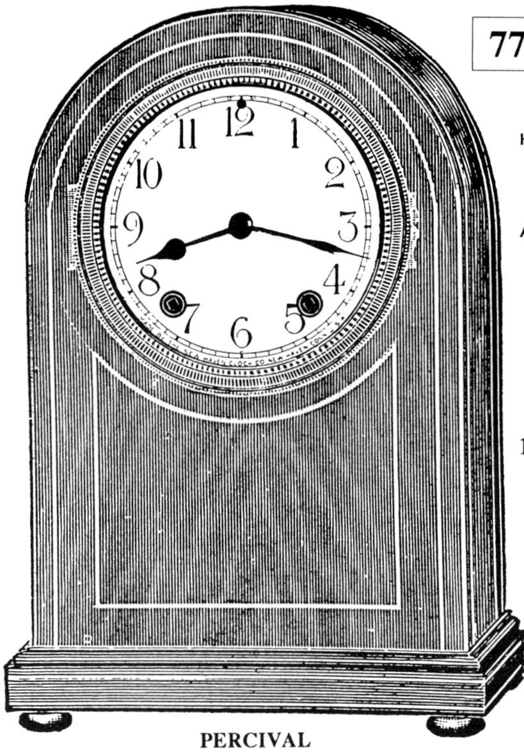

1917

PERCIVAL

PERCIVAL
Height, 11¼ Inches Width, 7½ Inches
Dial, Porcelain, 5 Inches
Case, Solid Oak or Mahogany, Marquetry Inlaid
Cast Sash, Beveled Bowed Glass
Also fitted with Silver Dials at Extra List

772

1911

EDINA

HEIGHT ....6¼ INCHES
WIDTH ....4½ INCHES
DIAL, PORCELAIN........2 INCHES
CASE, SOLID MAHOGANY WITH INLAID MARQUETRY DECORATIONS
PRICE, 1 DAY TIME PIECE ....LIST, $4.35
PRICE, 8 DAY TIME PIECE ....LIST EXTRA, 1.50
PRICE, WITH 1 DAY INTERMITTENT ALARM MOVEMENT ....LIST EXTRA, 1.50

773

1885

CABINET No. 516

ASH

Height, 16 ½ inches. Dial 5 inches.
8 Day Strike................ List, $5.25
8 Day Strike, Alarm...... List, $5.85

774

1889

AMPHION (Also called **CABINET No. 8**)

OAK OR MAHOGANY.

Height, 18 ½ inches. Dial 5 inches.
8 Day Striking Hours and Half-Hours on Cathedral Gong. List, $15.00. 8 Day Striking Hours and Half-Hours on Cathedral Gong with French Sash and Porcelain Dial. List, $16.50

MANTEL CLOCKS - CABINET CLOCKS

The year entered beside each clock is the year of the catalog from which the illustration was taken. Some clocks may have been produced a few years earlier.

775

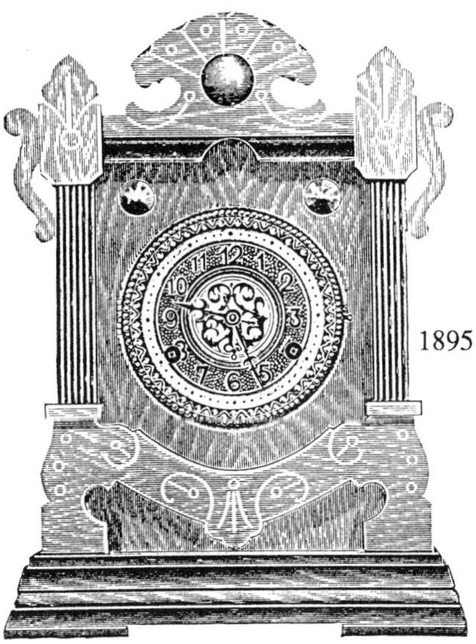

1895

ROMULUS
OAK.
Height, 15 ¼ inches. Width, 12 inches.
6-inch Gilt Dial. List, ... $6.00

776

1895

REMUS
OAK.
Height, 15 ¼ inches. Width, 12 inches.
6-inch Gilt Dial. List, ... $6.00

777

1895

CASSIUS
OAK.
Height, 15 ½ inches. Width, 11 ¼ inches.
6-inch Gilt Dial. List, ... $6.00

778

1895

BRUTUS
OAK.
Height, 15 ½ inches. Width, 11 ¼ inches.
6-inch Gilt Dial. List, ... $6.00

779

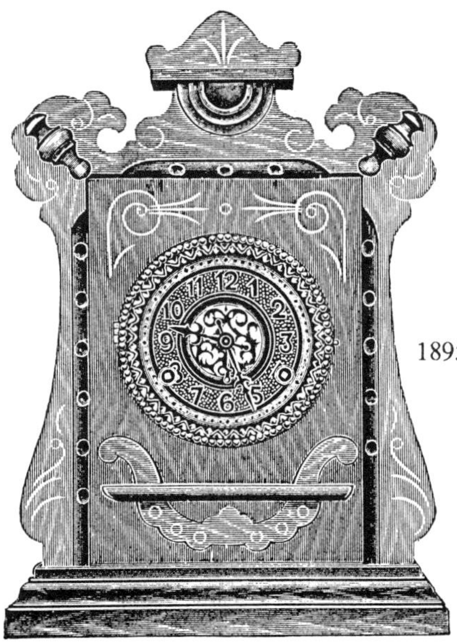

1895

MARIUS
OAK.
Height, 15 ¾ inches. Width, 11 ½ inches.
6-inch Gilt Dial. List, ... $6.00

780

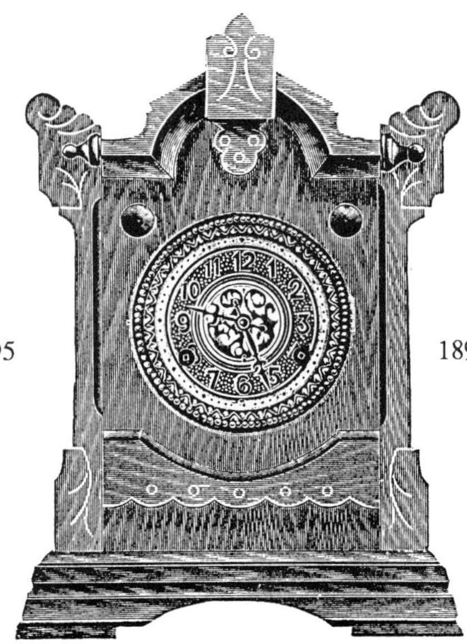

1895

CADMUS
OAK.
Height, 15 ¼ inches. Width, 11 ¼ inches.
6-inch Gilt Dial. List, ... $6.00

MANTEL CLOCKS - CABINET CLOCKS

781

1917

STUART
(William and Mary Design)
Height, 10¼ Inches Width, 9½ Inches
Dial, 7 Inches, Metal, Silver Finish
Case, Solid Mahogany and Fumed Oak

782

1917

JACOBEAN No. 1
Height, 10½ Inches Width, 9½ Inches
Dial, 7 Inches, Metal, Silver Finish
Case, Fumed Oak

783

1917

COLONIAL
Height, 9¾ Inches Width, 8¾ Inches
Dial, 7 Inches, Metal, Silver Finish
Case, Solid Mahogany

784

1917

ADAM No. 2
Height, 8¾ Inches Width, 9¼ Inches
Dial, 7 Inches, Metal, Silver Finish
Case, Solid Mahogany

Movements are Rack and Snail, 8-Day Half Hour Strike Cathedral Gong. Plates and Wheels Polished
Clocks on this page are exact reproductions of the designers art prevailing in the Seventeenth, Eighteenth and Nineteenth Centuries.

MANTEL CLOCKS - CABINET CLOCKS

785

1911

PIEDMONT

HEIGHT ....13 INCHES
WIDTH ....9¼ INCHES
DIAL, PORCELAIN.......................7 INCHES, FRENCH CAST SASH, BEVELED BOWED GLASS
CASE, SOLID MAHOGANY, MARQUETRY INLAID
MOVEMENT, 8 DAY HALF-HOUR STRIKE, CATHEDRAL GONG
PRICE ....LIST, $21.75

786

1911

OLIVIA

HEIGHT ....12½ INCHES
WIDTH ....9 INCHES
DIAL, PORCELAIN ......6 INCHES, FRENCH CAST SASH, BEVELED GLASS
CASE, SOLID MAHOGANY
FINISH, HIGHLY POLISHED
MOVEMENT, 8 DAY HALF-HOUR STRIKE, CATHEDRAL GONG
PRICE ....LIST, $16.00

787

1889

BANSHEE

CHERRY, WITH BRASS ORNAMENTS.

Height, 13 ½ inches. Dial 5 inches.
8 Day Strike................ List, $4.00
8 Day Strike, Alarm...... List, $4.60

788

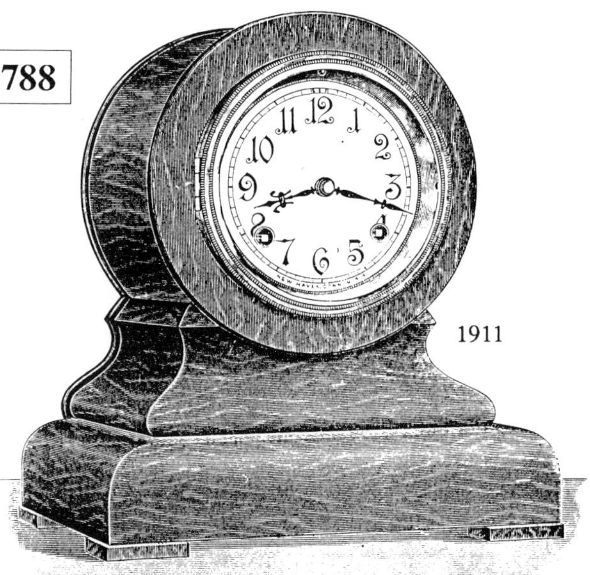

1911

OLEAN

HEIGHT ....11¼ INCHES
WIDTH ....10¼ INCHES
DIAL, PORCELAIN ......6 INCHES, FRENCH CAST SASH, BEVELED GLASS
CASE, SOLID MAHOGANY
FINISH, HIGHLY POLISHED
MOVEMENT, 8 DAY HALF-HOUR STRIKE, CATHEDRAL GONG
PRICE ....LIST, $16.00

MANTEL CLOCKS - CABINET CLOCKS

789

790

AUBREY 1911

HEIGHT ....12 1/8 INCHES
WIDTH ....8 1/4 INCHES
DIAL, PORCELAIN ......6 INCHES, FRENCH CAST SASH, BEVELED GLASS
CASE, MAHOGANY FINISH
MOVEMENT, 8 DAY HALF-HOUR STRIKE, CATHEDRAL GONG
PRICE ....LIST, $10.15

AUBURN 1911

HEIGHT ....12 1/4 INCHES
WIDTH ....8 3/8 INCHES
DIAL, PORCELAIN ......6 INCHES, FRENCH CAST SASH, BEVELED GLASS
CASE, MAHOGANY FINISH
MOVEMENT, 8 DAY HALF-HOUR STRIKE, CATHEDRAL GONG
PRICE ....LIST, $10.15

792

791

1911

793

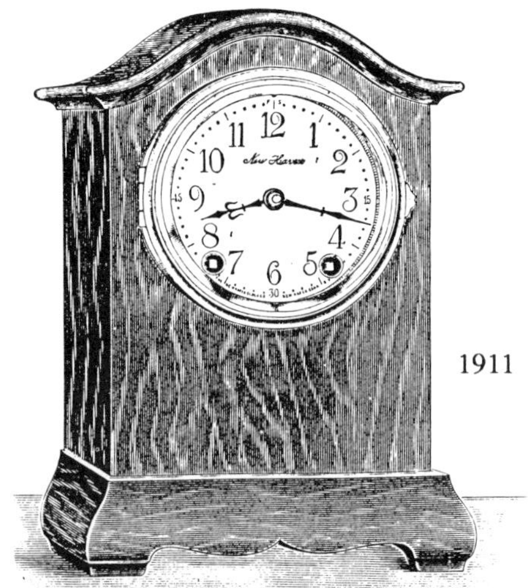

1911

MALO

HEIGHT ....9 1/4 INCHES
WIDTH ....6 1/2 INCHES
DIAL, PORCELAIN........4 INCHES, FRENCH CAST SASH, BEVELED GLASS
CASE, SOLID MAHOGANY
MOVEMENT, 8 DAY HALF-HOUR STRIKE, CATHEDRAL GONG
PRICE ....LIST, $17.50

AURORA 1911

HEIGHT ....12 1/2 INCHES
WIDTH ....8 1/4 INCHES
DIAL, PORCELAIN ......6 INCHES, FRENCH CAST SASH, BEVELED GLASS
CASE, MAHOGANY FINISH
MOVEMENT, 8 DAY HALF-HOUR STRIKE, CATHEDRAL GONG
PRICE ....LIST, $10.15

MERTON

HEIGHT ....9 1/2 INCHES
WIDTH ....6 3/4 INCHES
DIAL, PORCELAIN........4 INCHES, FRENCH CAST SASH, BEVELED GLASS
CASE, SOLID MAHOGANY
MOVEMENT, 8 DAY HALF-HOUR STRIKE, CATHEDRAL GONG
PRICE ....LIST, $17.50

MANTEL CLOCKS - CABINET CLOCKS

794

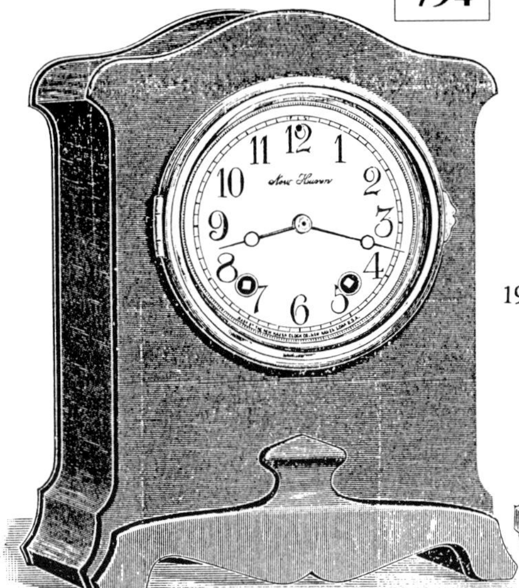

1911

795

1911

EFFINGTON

HEIGHT ....11 INCHES
WIDTH ....9 INCHES
DIAL, PORCELAIN ......6 INCHES, FRENCH CAST SASH, BEVELED GLASS
CASE, SOLID MAHOGANY
MOVEMENT, 8 DAY HALF-HOUR STRIKE, CATHEDRAL GONG
PRICE ....LIST, $20.30

ORVILLE

HEIGHT ....12¾ INCHES
WIDTH ....16 INCHES
DIAL, PORCELAIN ......6 INCHES, FRENCH CAST SASH, BEVELED GLASS
CASE, SOLID MAHOGANY
MOVEMENT, 8 DAY HALF-HOUR STRIKE, CATHEDRAL GONG
PRICE..LIST, $20.30

796

1911

797

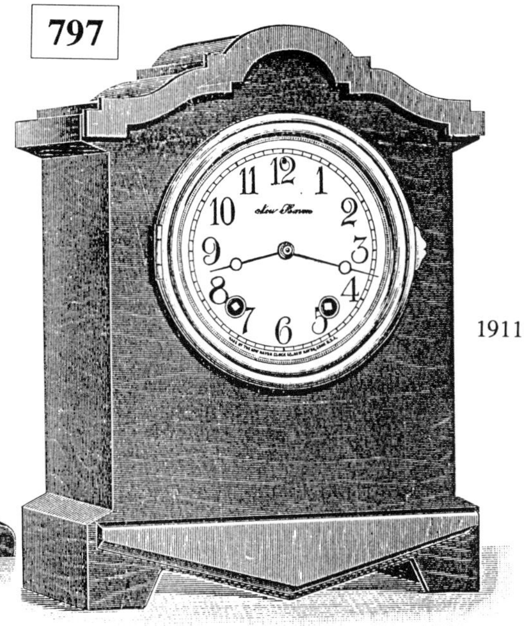

1911

OSBORNE

HEIGHT ....12 INCHES
WIDTH ....15½ INCHES
DIAL, PORCELAIN ......6 INCHES, FRENCH CAST SASH BEVELED GLASS
CASE, SOLID MAHOGANY
MOVEMENT, 8 DAY HALF-HOUR STRIKE, CATHEDRAL GONG
PRICE..LIST, $20.30

EGGLESTON

HEIGHT ....11¾ INCHES
WIDTH ....8⅝ INCHES
DIAL, PORCELAIN ......6 INCHES, FRENCH CAST SASH, BEVELED GLASS
CASE, SOLID MAHOGANY
MOVEMENT, 8 DAY HALF-HOUR STRIKE, CATHEDRAL GONG
PRICE..LIST, $20.30

MANTEL CLOCKS - CABINET CLOCKS

798

1917

AURINE
8-Day Half Hour Strike
Cathedral Gong
Height, 13 Inches Width, 8½ Inches
6 Inch Metal Dial, Bowed Glass
Mahogany Finish

799

1911

ORISON
HEIGHT ....10 INCHES
WIDTH ....14¾ INCHES
DIAL, PORCELAIN ......6 INCHES, FRENCH CAST SASH, BEVELED GLASS
CASE, SOLID MAHOGANY
MOVEMENT, 8 DAY HALF-HOUR STRIKE, CATHEDRAL GONG
PRICE ....LIST, $20.30

800

1911

OBERLIN
HEIGHT ....11¼ INCHES
WIDTH ....14½ INCHES
DIAL, PORCELAIN ......6 INCHES, FRENCH CAST SASH, BEVELED GLASS
CASE, SOLID MAHOGANY
MOVEMENT, 8 DAY HALF-HOUR STRIKE, CATHEDRAL GONG
PRICE...LIST, $20.30

801

1917

AUBERT
8-Day Half Hour Strike
Cathedral Gong
Height, 12¾ Inches Width, 8½ Inches
6 Inch Metal Dial, Bowed Glass
Mahogany Finish

MANTEL CLOCKS - CABINET CLOCKS

802

1917

AUSTELL
8-Day Half Hour Strike
Cathedral Gong
Height, 10 Inches Width, 8 Inches
6 Inch Metal Dial, Bowed Glass
Mahogany Finish, or Fumed or Golden Oak
Fitted with Porcelain Dial at Extra List

803

1911

ORIOLE
HEIGHT ....11¼ INCHES
WIDTH ....18½ INCHES
DIAL, PORCELAIN ......6 INCHES, FRENCH CAST SASH, BEVELED GLASS
CASE, SOLID MAHOGANY
MOVEMENT, 8 DAY HALF-HOUR STRIKE, CATHEDRAL GONG
PRICE ....LIST, $20.30

804

1911

OREGON
HEIGHT ....10¼ INCHES
WIDTH ....15 INCHES
DIAL, PORCELAIN ......6 INCHES, FRENCH CAST SASH, BEVELED GLASS
CASE, SOLID MAHOGANY
MOVEMENT, 8 DAY HALF-HOUR STRIKE, CATHEDRAL GONG
PRICE ....LIST, $20.30

805

1917

AUSTRO
8-Day Half Hour Strike
Cathedral Gong
Height, 9¼ Inches Width, 8¼ Inches
6 Inch Metal Dial, Bowed Glass
Mahogany Finish
Fitted with Porcelain Dial at Extra List

MANTEL CLOCKS - CABINET CLOCKS

806

1917

AURIS
8-Day Half Hour Strike
Cathedral Gong
Height, 10⅛ Inches Width, 8⅞ Inches
6 Inch Metal Dial, Bowed Glass
Mahogany Finish
Fitted with Porcelain Dial at Extra List

807

1911

OSTEND
HEIGHT ....11¾ INCHES
WIDTH ....18 INCHES
DIAL, PORCELAIN ......6 INCHES, FRENCH CAST SASH, BEVELED GLASS
CASE, SOLID MAHOGANY
MOVEMENT, 8 DAY HALF-HOUR STRIKE, CATHEDRAL GONG
PRICE ....LIST, $20.30

808

1911

OPORTO
HEIGHT ....11 INCHES
WIDTH ....18 INCHES
DIAL, PORCELAIN ......6 INCHES, FRENCH CAST SASH, BEVELED GLASS
CASE, SOLID MAHOGANY
MOVEMENT, 8 DAY HALF-HOUR STRIKE, CATHEDRAL GONG
PRICE ....LIST, $20.30

809

1917

AUSTON
8-Day Half Hour Strike
Cathedral Gong
Height, 10¼ Inches Width, 8¾ Inches
6 Inch Metal Dial, Bowed Glass
Mahogany Finish, or Fumed or Golden Oak
Fitted with Porcelain Dial at Extra List

MANTEL CLOCKS - CABINET CLOCKS

EIGHT-DAY CABINET CLOCKS. HALF-HOUR STRIKE, CATHEDRAL GONG

810

1913

811

1913

DELMAR

LIST, $13.75

HEIGHT, 12¼ INCHES. WIDTH, 14½ INCHES.
DIAL, PORCELAIN, 7 INCHES.
FRENCH CAST SASH, BEVELED, BOWED GLASS.
CASE, SOLID OAK, FLEMISH FINISH.

LORNE

LIST, $8.75

HEIGHT, 11⅜ INCHES. WIDTH, 10 INCHES.
DIAL, 6 INCHES.
CASE, MAHOGANY FINISH.

812

1913

813

1913

LOPEZ

LIST, $7.25

HEIGHT, 11⅜ INCHES. WIDTH, 10 INCHES.
DIAL, 6 INCHES.
CASE, SOLID OAK, WAX FLEMISH FINISH.

DELHI

LIST, $14.50

HEIGHT, 12 INCHES. WIDTH, 15¾ INCHES.
DIAL, PORCELAIN, 7 INCHES.
FRENCH CAST SASH, BEVELED, BOWED GLASS.
CASE, SOLID OAK, FLEMISH FINISH.

MANTEL CLOCKS - CABINET CLOCKS

815

1920

PALLAS
Height, 9 Inches Width, 8¾ Inches
Dial, Silver Finish, 6 Inches
Cast Sash, Beveled Convex Glass

816

1920

PORTIA
Height, 9¼ Inches Width, 8 Inches
Dial, Silver Finish, 6 Inches
Cast Sash, Beveled Convex Glass

CABINETS

Solid Mahogany Cases. Movements are Rack and Snail, 8-Day Half Hour Strike, Cathedral Gong Plates and Wheels Polished

Above Clocks also furnished in Brown Mahogany

814

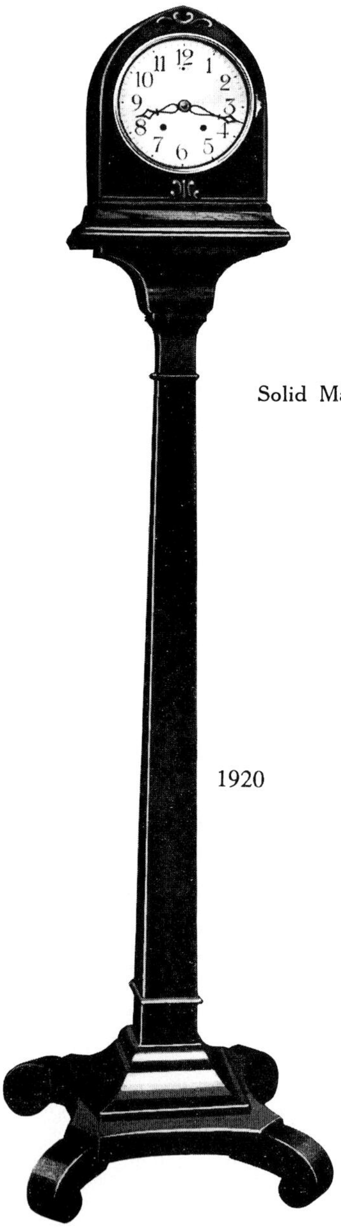

1920

PEDESTAL CLOCK No. 1
(Montross Clock attached to a Solid Mahogany Pedestal)
Height, 64½ Inches Width, 15 Inches

817

1920

MONTROSS
Height, 11¾ Inches Width, 9½ Inches
Dial, Silver Finish, 7½ Inches
Cast Sash, Beveled Glass.
Also furnished in Brown Mahogany.

818

1917

WESLEY
Height, - - - - - - 6¾ Inches
Width, - - - - - - - 5 Inches

8-Day Lever Time Movements.

3½-inch Porcelain Dials. Beveled Glass

JEWELED LEVER STRIKE CABINETS

Solid Mahogany Cases. Fine 8-Day Lever Gong, Rack and Snail Strike Movements with Jeweled Balance and Heavy Nickeled Damaskeened Plates

MANTEL CLOCKS - CABINET CLOCKS

The year entered beside each clock is the year of the catalog from which the illustration was taken. Some clocks may have been produced a few years earlier.

819

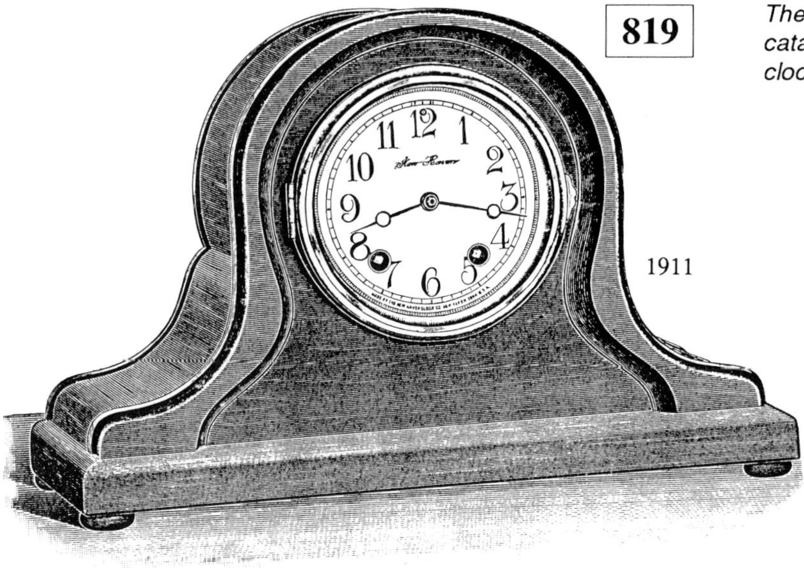

1911

ORIN

HEIGHT ....10 INCHES
WIDTH ....14¾ INCHES
DIAL, PORCELAIN ......6 INCHES, FRENCH CAST SASH, BEVELED GLASS
CASE, SOLID MAHOGANY
MOVEMENT, 8 DAY HALF-HOUR STRIKE, CATHEDRAL GONG
PRICE ....LIST, $20.30

820

JACOBEAN No. 53

Height, - - - - - - 7¼ Inches
Width, - - - - - - 5½ Inches

822

1911

821

ADAM No. 52

Height, - - - - - - 6½ Inches
Width, - - - - - - 4¾ Inches
Case, - - - - - Marquetry Inlaid

OLYMPUS

HEIGHT ....11 INCHES
WIDTH ....18¾ INCHES
DIAL, PORCELAIN ......6 INCHES, FRENCH CAST SASH, BEVELED GLASS
CASE, SOLID MAHOGANY
MOVEMENT, 8 DAY HALF-HOUR STRIKE, CATHEDRAL GONG
PRICE..LIST, $20.30

MANTEL CLOCKS - CABINET CLOCKS

823

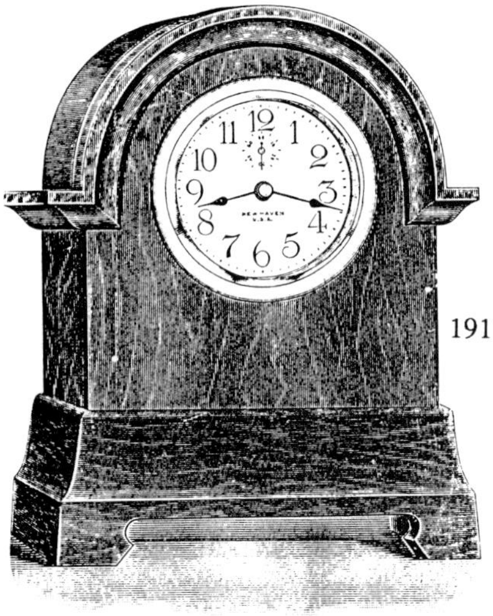

1917

LOMBARD
HEIGHT ....7⅜ INCHES
WIDTH ....6 INCHES
DIAL, PORCELAIN ....3 INCHES
CASE, SOLID MAHOGANY
MOVEMENT, 8 DAY TIME PIECE
PRICE ....LIST, $7.25

824

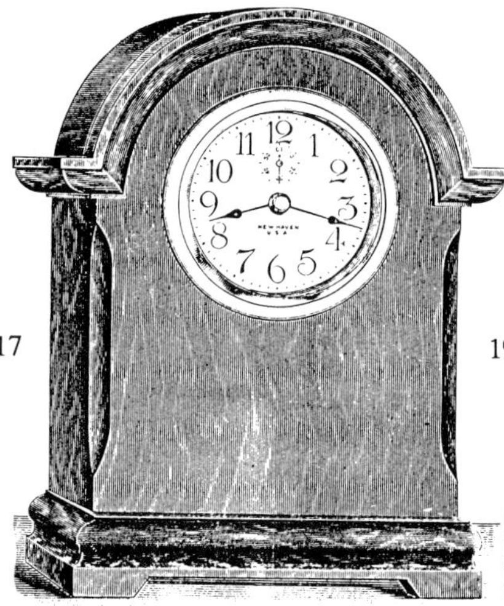

1911

LOANDA
HEIGHT ....8⅛ INCHES
WIDTH.......................6⅛ INCHES
DIAL, PORCELAIN ....3 INCHES
CASE, SOLID MAHOGANY
MOVEMENT, 8 DAY TIME PIECE
PRICE ....LIST, $7.25

825

1911

LORRAINE
HEIGHT ....8¼ INCHES
WIDTH ....8⅜ INCHES
DIAL, PORCELAIN ....3 INCHES
CASE, SOLID MAHOGANY
MOVEMENT, 8 DAY TIME PIECE
PRICE ....LIST, $8.00

826

1917

IRENE
8-Day Half Hour Strike
Height, 13¾ Inches Width, 8¼ Inches
Dial, 5 Inches, Metal, Silver Finish.
Case, Silvered Oak

827

1917

IRWIN
8-Day Half Hour Strike
Height, 13¾ Inches Width, 8¾ Inches
Dial, 5 Inches, Metal, Silver Finish.
Case, Silvered Oak

828

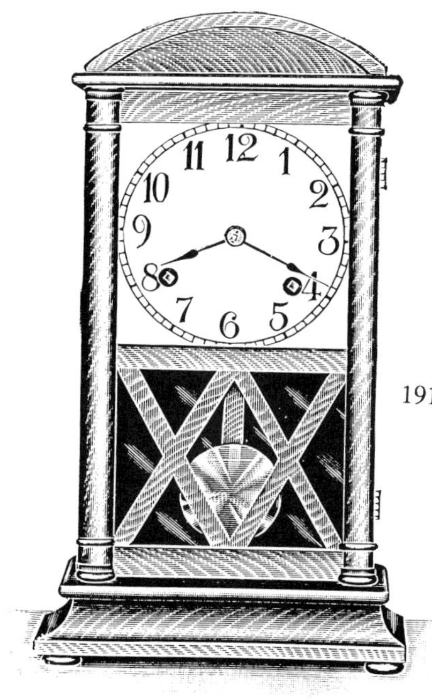

1917

INWOOD
8-Day Half Hour Strike
Height, 13¾ Inches Width, 7½ Inches
Dial, 5 Inches, Metal, Silver Finish.
Case, Silvered Oak

MANTEL CLOCKS - ENAMELED IMITATION CASES

Enameled Imitation Marble and Bronze Clocks.

THESE Clocks are all of Superior Finish, fitted with Half-hour Slow Striking Movement with Cathedral Gong on Sounding-Board. This movement is guaranteed, and is equal to any in the market. Our new Triple Plate Finish on the Bronze Figure Clocks is unsurpassed in bringing out the artistic detail. Our Fittings of the Enameled Imitation Marbles are as follows:—

- A Plain White Paper Dial.
- C V. E. with White Paper Dial and Gilt Center.
- E Black Dial and Gilt Center.
- G V. E. Black Dial and Gilt Center.
- H Porcelain Dial and Porcelain Center.
- L V. E. with Porcelain Center.
- M V. E. Gilt Dial and Gilt Center.
- O Plain Gilt Dial.

829

1890

HENRIETTA

BRONZE OR SILVER TRIMMINGS.

Height, 12 inches. Dial 5½ inches.

8 Day Gong Strike, Visible Escapement, French Sash, Porcelain Dial.. $29.00

MANTEL CLOCKS - ENAMELED IMITATION CASES

Enameled Imitation Marble and Bronze Clocks.

830

1890

ORIVA

Height, 10 inches. Dial 4 inches.
8 Day Gong Strike, Porcelain Dial $17.00
8 Day Gong Strike, Visible Escapement. $19.50

832

1890

GALATEA

FITTED WITH STRUCK SASH AND PAPER DIAL.
Height, 10 inches. Dial 5 inches.
8 Day Strike, Cathedral Gong, Visible Escapement.. $16.00
8 Day Strike, Cathedral Gong, Plain Dial.. $14.50

831

1890

MOHICAN

Height, 11 inches. Dial 4 inches.
8 Day Gong Strike, Visible Escapement,
French Sash and Porcelain Dial.. $24.00

833

1890

MONA

BRONZE OR SILVER TRIMMINGS.
Height, 11½ inches. Dial 5 inches.
8 Day Strike, Cathedral Gong, Visible Escapement.. $17.00
8 Day Strike, Cathedral Gong.. $15.50

MANTEL CLOCKS - ENAMELED IMITATION CASES

Enameled Imitation Marble and Bronze Clocks.

834

1890

HUNTRESS

Height, 11 inches. Dial 5½ inches.
8 Day Strike, Cathedral Gong, Visible Escapement,
French Sash, Porcelain Dial.. $33.00

835

1890

HILDEGARDE

BRONZE OR SILVER TRIMMINGS.
Height, 10 inches. Dial 5 inches.
8 Day Strike, Cathedral Gong, Visible Escapement.. $15.00
8 Day Strike, Cathedral Gong.. $13.50

836

1890

IANTHE

Height, 10 inches. Dial 5½ inches.
8 Day Strike, Cathedral Gong, Visible Escapement, French Sash and
Porcelain Dial.. $24.00

MANTEL CLOCKS - ENAMELED IMITATION CASES

Enameled Imitation Marble and Bronze Clocks.

837

1890

GENESTA

FITTED WITH STRUCK SASH AND PAPER DIAL.
Height, 10 inches. Dial 5 inches.
8 Day Strike, Cathedral Gong, Visible Escapement.. $16.75
8 Day Strike, Cathedral Gong, Plain Dial.. $15.25

838

1890

GLADYS

FITTED WITH STRUCK SASH AND PAPER DIAL.
Height, 10 inches. Dial 5 inches.
8 Day Strike, Cathedral Gong, Visible Escapement.. $15.00
8 Day Strike, Cathedral Gong, Plain Dial.. $13.50

839

1890

FLORENCE

ENAMELED IMITATION MARBLE.
STRUCK SASH AND METAL DIAL.
Height, 9¾ inches. Dial 5 inches.
8 Day Strike, Cathedral Gong.. $11.10
8 Day Strike, Visible Escapement, Cathedral Gong.. $12.60

840

1890

SIREN

BRONZE OR SILVER TRIMMINGS.
Height, 10 inches. Dial 5 inches.
8 Day Strike, Cathedral Gong, Visible Escapement.. $18.00
8 Day Strike, Cathedral Gong.. $16.50

MANTEL CLOCKS - ENAMELED IMITATION CASES

Enameled Imitation Marble and Bronze Clocks.

841

1890

HERMIT

Height, 10 inches. Dial 5½ inches.
8 Day Strike, Cathedral Gong, Visible Escapement,
French Sash, Porcelain Dial.. $23.00

842

1890

HALCYON

BRONZE OR SILVER TRIMMINGS.
Height, 10½ inches. Dial 4½ inches.
8 Day Gong Strike, Visible Escapement,
French Sash, Porcelain Dial.. $19.50

843

1890

FAVORITA

Height, 10 inches. Dial 4½ inches.
8 Day Strike, Cathedral Gong, Visible Escapement,
French Sash, Porcelain Dial.. $16.50

MANTEL CLOCKS - ENAMELED IMITATION CASES

Enameled Imitation Marble and Bronze Clocks.

844

1890

REGINA

BRONZE OR SILVER TRIMMINGS.
Height, 11½ inches. Dial 4½ inches.
8 Day Gong Strike, Visible Escapement, French Sash, Porcelain Dial.. $35.00

845

1890

GRETCHEN

Height, 12 inches. Dial 4½ inches.
8 Day Gong Strike, Visible Escapement,
French Sash, Porcelain Dial.. $26.00

846

1890

FORTUNA

FITTED WITH STRUCK SASH AND PAPER DIAL.
Height, 10 inches. Dial 5 inches.
8 Day Strike, Cathedral Gong, Visible Escapement, $14.50
8 Day Strike, Cathedral Gong, Plain Dial.. $13.00

MANTEL CLOCKS - ENAMELED IRON CASES

These Clocks are all of superior finish, made all Iron, or Iron and Imitation Marble, making a clock very similar to the French Marble. Fitted with Porcelain Dial and French Sash, half-hour Slow Strike Movement on a Cathedral Gong, and the Patent Regulator Attachment, by means of which the Clock is accurately regulated from the front, without disturbing the position of the Clock.

847

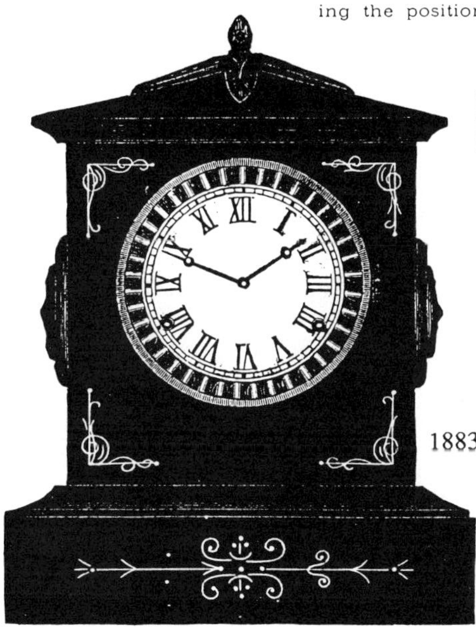

1883

WINDEMERE.

IRON.

Bronze Ornaments.

848

1883

ABBOTTSFORD.

IRON.

Inlaid with Imitation Marble.

Height, 12 inches. Dial, 4 inches.
8 Day Strike. 8 Day Half-hour Strike on Cathedral Gong.

849

1883

COUPE H.

Height, 9½ inches.

850

1883

MELROSE.

IRON.

Height, 10½ inches. Dial, 4 inches.
8 Day Strike.
8 Day Half-Hour Strike on Cathedral Gong.
Gilt Ornaments.

851

1883

COUPE H.

Height, 9½ inches.

MANTEL CLOCKS - ENAMELED IRON CASES

852

1883

IVANHOE.

IRON.

Height, 17½ inches.

Width at base, 12½ inches. Dial, 5½ inches.

8 Day Half-hour Strike on Cathedral Gong.

853

1883

WAVERLY.

IRON.

Height, 10¼ inches.

Width at base, 15 inches. Dial, 4½ inches.

8 Day Strike.

8 Day Half-hour Strike on Cathedral Gong.

854

1883

ABERDEEN.

IRON.

8 Day Strike.

8 Day Half-hour Strike on Cathedral Gong.

Height, 11 inches. Dial, 5 inches.

Gilt Side Ornaments and Imitation Marble Pillars.

855

1883

CUMBERLAND.

IRON.

Height, 14 inches. Dial, 5 inches.

8 Day Strike.

8 Day Half-hour Strike on Cathedral Gong.

Bronze Ornaments inlaid with Imitation Marble.

MANTEL CLOCKS - ENAMELED IRON CASES

856

1883

ABERFOYLE.

IRON.

Height, 14½ inches. Width at base, 9 inches.

8 Day Strike. Dial, 4½ inches.

8 Day Half-hour Strike on Cathedral Gong.

857

1883

BELFAST.

IRON.

Height, 12 inches. Dial, 6 inches.

8 Day Strike.

8 Day Half-hour Strike on Cathedral Gong.

Bronze Ornaments.

858

1883

MUNICH.

IRON.

Height, 12 inches. Dial, 5 inches.

8 Day Strike.

8 Day Half-hour Strike on Cathedral Gong.

Bronze Ornaments.

859

1883

TRAFALGAR.

IRON.

Height, 13½ inches. Dial, 5 inches.

8 Day Strike.

8 Day Half-hour Strike on Cathedral Gong.

Bronze Ornaments. Imitation Marble Side Pillars.

MANTEL CLOCKS - ENAMELED IRON CASES

860

1895

DE SOTO.

Height, 9½ inches; Width, 11¼ inches; Dial, 4 inches.
8 Day, French Cup Bell, Hour Strike. List, $9.00

861

1911

KENMORE
(SHOWING PEARL DIAL)

HEIGHT ....11½ INCHES
WIDTH ....11 INCHES
DIAL, IVORINE OR PEARL...........6 INCHES
CASE, BLACK ENAMELED IRON
FINISH, ON TRIMMINGS, GILT OR BRONZE
MOVEMENT, 8 DAY HALF-HOUR STRIKE, CATHEDRAL GONG
PRICE ....LIST, $11.60
PRICE, WITH VISIBLE ESCAPEMENT ....LIST EXTRA, .60
PRICE, WITH FRENCH CAST SASH, BEVELED GLASS AND PORCELAIN DIAL ....LIST EXTRA, 1.50

862

1895

BOLIVIA.

Height, 8¾ inches. Width, 11½ inches; Dial 4 inches.
8 Day, French Cup Bell, Hour Strike, - - List, $9.25
Empire Sash, Extra, List, 1.50

863

HONDURAS. 1895

Height, 8½ inches; Width, 10 inches; Dial, 4 inches.
8 Day, French Cup Bell, Hour Strike, List, $7.50

864

1895

DOMINICA.

Height, 9¼ inches; Width, 10½ inches; Dial, 4 inches.
8 Day, French Cup Bell, Hour Strike. List, $8.50

865

1911

ARAGON

HEIGHT ....10¾ INCHES
WIDTH ....9¾ INCHES
DIAL, IVORINE OR PEARL ....6 INCHES
CASE, BLACK ENAMELED IRON
FINISH, ON TRIMMINGS, GILT OR BRONZE
MOVEMENT, 8 DAY HALF-HOUR STRIKE, CATHEDRAL GONG
PRICE ....LIST, $10.85
PRICE, WITH VISIBLE ESCAPEMENT ....LIST EXTRA, .60
PRICE, WITH FRENCH CAST SASH, BEVELED GLASS AND PORCELAIN DIAL ....LIST EXTRA, 1.50

866

1895

COLONNA.

Gilt Ornaments. Marbloid Columns.

Height, 10 inches; Width 15 inches; Fancy Gilt or Plain White, 6 inch Dial.

8 Day, Half-Hour Strike, Cathedral Gong, - List, $12.60
Visible Escapement, - - - - - Extra, List, .50
Porcelain Dial, - - - - - Extra, List, 1.50
Empire Sash, Convex Glass, Silver Dial with Gilt Overlay, Extra, List, 2.00

MANTEL CLOCKS - ENAMELED IRON CASES

867

1895

WASHINGTON.

Height, 10½ inches; Width, 16 inches; Fancy Gilt or Plain White, 6 inch Dial.

8 Day Half-Hour Strike, Cathedral Gong,	List,	$12.70
Visible Escapement,	Extra, List,	.50
Porcelain Dial,	Extra, List,	1.50
Empire Sash, Convex Glass, Silver Dial with Gilt Overlay,	Extra, List,	2.00

868

1900

CONSTANCE

Black or Malachite Finish with Gilt Trimmings.
Height, 12¾ inches. Width, 9¾ inches.
Eight-day, Half-hour Strike, Cathedral Gong. Four-inch Ivorine Dial, with Gilt Center.

List price, $15.30

Four-inch Ivorized Porcelain Dial, Gilt Center, Empire Sash and Heavy Beveled Glass, extra list, $1.50

Fitted, to order only, with Jeweled Sash in Brilliant, Ruby, Emerald, Amethyst, Sapphire, or alternate stones, at extra list of $2.80

869

1900

CORYDON

Finished in Black, Smoke or Malachite, with Gilt or Bronze Trimmings.
Height, 15 inches.
Width, 9 inches.
Six-inch fancy Gilt, White or Ivorine Dial.
Eight-day, Half-hour Strike, Cathedral Gong.

List price, $10.50

Extra List for Ivorized Porcelain Dial, with Empire Sash and Heavy Beveled Glass, $1.50
Extra list for Visible Escapement, 50 cts.
Extra list for Pearl Dial, 30 cts.

870

1895

MONTROSE.

Gilt Ornaments. Marbloid Columns.

Height, 10 inches; Width 14¾ inches; Fancy Gilt or Plain White, 6 inch Dial.

8 Day, Half-Hour Strike, Cathedral Gong,	List,	$11.90
Visible Escapement,	Extra, List,	.50
Porcelain Dial,	Extra, List,	1.50
Empire Sash, Convex Glass, Silver Dial with Gilt Overlay,	Extra, List,	2.00

MANTEL CLOCKS - ENAMELED IRON CASES

The year entered beside each clock is the year of the catalog from which the illustration was taken. Some clocks may have been produced a few years earlier.

871

1895

NEW MONACO

Height, 11 ¼ inches. Width, 12 inches.
Fancy Gilt or Plain White, Dial, 6 inches.
8 Day, Half-Hour Strike, Cathedral Gong, $13.00
Visible Escapement, ...Extra, List .50
Porcelain Dial, ...Extra List, 1.50
Empire Sash, Convex Glass, Silver Dial
with Gilt Overlay, ...Extra, List $2.00

872

1895

CZARINA

Height, 10 inches. Width, 14 ¾ inches.
Fancy Gilt or Plain White, Dial, 6 inches.
8 Day, Half-Hour Strike, Cathedral Gong, $14.70
Visible Escapement, ...Extra, List .50
Porcelain Dial, ...Extra List, 1.50
Empire Sash, Convex Glass, Silver Dial
with Gilt Overlay, ...Extra, List $2.00

873

1895

MARGARITA

Dial, 4 inches. Height, 8 ½ inches.
Width, 11 ½ inches.
8 Day, French Cup Bell,
Hour Strike .. List, $8.50

874

1895

CUBA

Height, 8 inches. Width, 10 inches.
Dial, 4 inches.
8 Day, French Cup Bell, .. List, $8.10
Empire Sash, ...Extra, List $2.00

875

1900

ROSALIND

Finished in Black, Smoke or Malachite with Gilt or Bronze Trimmings.
Height, 10⅝ inches. Width, 16⅜ inches.
Eight-day, Half-hour Strike, Cathedral Gong. Six-inch fancy Gilt, White or Ivorine Dial.

List price, $12.50

Extra list for Ivorized Porcelain Dial with Empire Sash, Heavy Beveled Glass, $1.50
Extra list for Visible Escapement, 50 cents.
Extra list for Pearl Dial, 30 cents.

MANTEL CLOCKS - ENAMELED IRON CASES

877

1895

SINBAD

Height, 11 ½ inches. Width, 14 ½ inches.
Fancy Gilt or Plain White, Dial, 6 inches.
8 Day, Half-Hour Strike, Cathedral Gong, $13.30
Visible Escapement, ...Extra, List .50
Porcelain Dial, ...Extra List, 1.50
Empire Sash, Convex Glass, Silver Dial
with Gilt Overlay, ...Extra, List $2.00

876

1895

THETIS

Height, 11 ½ inches. Width, 12 ½ inches.
Fancy Gilt or Plain White, Dial, 6 inches.
8 Day, Half-Hour Strike, Cathedral Gong, $13.30
Visible Escapement, ...Extra, List .50
Porcelain Dial, ...Extra List , 1.50
Empire Sash, Convex Glass, Silver Dial
with Gilt Overlay, ...Extra, List $2.00

879

1895

POCAHONTAS

Height, 11 inches. Width, 18 ¾ inches.
Fancy Gilt or Plain White, Dial, 6 inches.
8 Day, Half-Hour Strike, Cathedral Gong, $16.80
Visible Escapement, ...Extra, List .50
Porcelain Dial, ...Extra List, 1.50
Empire Sash, Convex Glass, Silver Dial
with Gilt Overlay, ...Extra, List $2.00

878

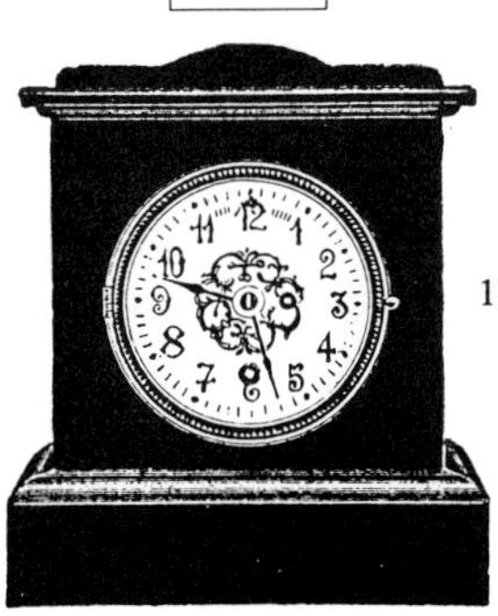

1895

HAYTI

Dial, 4 inches. Height, 9 inches.
Width, 7 ½ inches.
8 Day, French Cup Bell,
Hour Strike .. List, $6.50

MANTEL CLOCKS - ENAMELED IRON CASES

The year entered beside each clock is the year of the catalog from which the illustration was taken. Some clocks may have been produced a few years earlier.

880

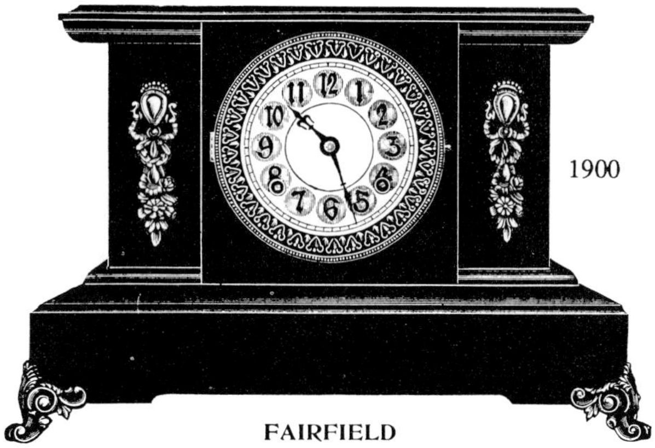

1900

FAIRFIELD

Finished in Black, Smoke or Malachite, with Gilt or Bronze Trimmings.
Height, 10½ inches. Width, 15½ inches.
Eight-day, Half-hour Strike, Cathedral Gong.
Six-inch Fancy Gilt, White or Ivorine Dial.

List price, $10.75

Extra list for Ivorized Porcelain Dial with Empire Sash and Heavy Beveled Glass, $1.50
Extra list for Visible Escapement, 50 cents.
Extra list for Pearl Dial, 30 cents.

881

1900

FAIRMONT

Finished in Black, Smoke or Malachite, with Gilt or Bronze Trimmings.
Height, 10½ inches. Width, 15½ inches.
Eight-day, Half-hour Strike, Cathedral Gong.
Six-inch Fancy Gilt, White or Ivorine Dial.

List price, $10.75

Extra list for Ivorized Porcelain Dial with Empire Sash and Heavy Beveled Glass, $1.50
Extra list for Visible Escapement, 50 cents.
Extra list for Pearl Dial, 30 cents.

882

1911

TYNDALL

HEIGHT ....13¼ INCHES
WIDTH ....9 INCHES
DIAL, IVORINE OR PEARL ....6 INCHES
CASE, BLACK ENAMELED IRON
FINISH, ON TRIMMINGS, GILT OR BRONZE
MOVEMENT, 8 DAY HALF-HOUR STRIKE, CATHEDRAL GONG
PRICE ....LIST, $11.60
PRICE, WITH VISIBLE ESCAPEMENT ....LIST EXTRA, .60
PRICE, WITH FRENCH CAST SASH, BEVELED GLASS AND PORCELAIN DIAL ....LIST EXTRA, 1.50

883

1895

BACCARAT

Height, 11 ¼ inches. Width, 16 inches.
Fancy Gilt or Plain White, Dial, 6 inches.
8 Day, Half-Hour Strike, Cathedral Gong, $16.00
Visible Escapement,.....................Extra, List .50
Porcelain Dial, ...Extra List, 1.50
Empire Sash, Convex Glass, Silver Dial
with Gilt Overlay, ...Extra, List $2.00

MANTEL CLOCKS - ENAMELED IRON CASES

884

1911

FREMONT

FREMONT
(Showing Pearl Dial)
HEIGHT ....10½ inches
WIDTH ....15½ inches
DIAL, Ivorine or Pearl ....6 inches
CASE, Black Enameled Iron
FINISH, on Trimmings, Gilt or Bronze
MOVEMENT, 8 day Half-hour Strike, Cathedral Gong
PRICE ....list, $11.20
PRICE, with Visible Escapement ....list extra, .60
PRICE, with French Cast Sash, Beveled Glass
and Porcelain Dial ....list extra, 1.50

885

1914

TALMAGE
LIST, $11.60
HEIGHT, 14 inches. WIDTH, 9½ inches.

886

1911

NEWARK
HEIGHT...............................11½ inches
WIDTH..................................10½ inches
DIAL, Porcelain ....6 inches, French Cast
Sash, Beveled Glass
CASE, Black Enameled Iron
FINISH, on Trimmings, Gilt or Bronze
MOVEMENT, 8 day Half-hour Strike, Cathedral Gong
PRICE ....list, $13.00
PRICE, with Visible Escapement ....list extra, .60

887

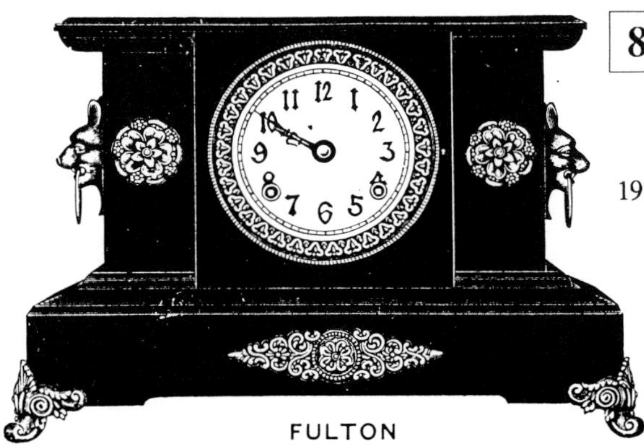

1911

FULTON
HEIGHT ....10½ inches
WIDTH ....15½ inches
DIAL, Ivorine or Pearl ....6 inches
CASE, Black Enameled Iron
FINISH, on Trimmings, Gilt or Bronze
MOVEMENT, 8 day Half-hour Strike, Cathedral Gong
PRICE ....list, $11.20
PRICE, with Visible Escapement ....list extra, .60
PRICE, with French Cast Sash, Beveled Glass
and Porcelain Dial ....list extra, 1.50

888

1914

WARWICK
LIST, $11.60
HEIGHT, 11¼ inches. WIDTH, 12¼ inches.

MANTEL CLOCKS - ENAMELED IRON CASES

889

1911

WALLACE

HEIGHT ....13¾ INCHES
WIDTH ....11½ INCHES
DIAL, PORCELAIN .....6 INCHES
FRENCH CAST SASH, BEVELED GLASS
CASE, BLACK ENAMELED IRON
FINISH, ON TRIMMINGS, GILT OR BRONZE
MOVEMENT, 8 DAY HALF-HOUR STRIKE, CATHEDRAL GONG
PRICE ....LIST, $19.60
PRICE, WITH VISIBLE ESCAPEMENT...LIST EXTRA, $ .60

890

1911

WEXFORD

HEIGHT ....16¼ INCHES
WIDTH...................14 INCHES
DIAL, PORCELAIN .....6 INCHES
FRENCH CAST SASH, BEVELED GLASS
CASE, BLACK ENAMELED IRON
FINISH, ON TRIMMINGS, GILT OR BRONZE
MOVEMENT, 8 DAY HALF-HOUR STRIKE, CATHEDRAL GONG
PRICE ....LIST, $21.75
PRICE, WITH VISIBLE ESCAPEMENT ...LIST EXTRA, $ .60

891

1913

WALTHAM

LIST, $18.90
HEIGHT, 14 INCHES. WIDTH, 11½ INCHES.

892

1911

CZARIN

HEIGHT ....10½ INCHES
WIDTH ....16 INCHES
DIAL, IVORINE OR PEARL ....6 INCHES
CASE, BLACK ENAMELED IRON
FINISH, ON TRIMMINGS, GILT OR BRONZE
MOVEMENT, 8 DAY HALF-HOUR STRIKE, CATHEDRAL GONG
PRICE ....LIST, $13.00
PRICE, WITH VISIBLE ESCAPEMENT ....LIST EXTRA, .60
PRICE, WITH FRENCH CAST SASH, BEVELED GLASS AND PORCELAIN DIAL ....LIST EXTRA, 1.50

MANTEL CLOCKS - ENAMELED IRON CASES

The year entered beside each clock is the year of the catalog from which the illustration was taken. Some clocks may have been produced a few years earlier.

893

1911

WABASH
(Showing Visible Escapement)
HEIGHT................................12¼ inches
WIDTH ....11 inches
DIAL, Porcelain ....6 inches, French Cast Sash, Beveled Glass
CASE, Black Enameled Iron
FINISH, on Trimmings, Gilt or Bronze
MOVEMENT, 8 day Half-hour Strike, Cathedral Gong
PRICE ....list, $13.00
PRICE, with Visible Escapement ......list extra, .60

894

1911

NEWBERNE
HEIGHT................................12½ inches
WIDTH ....10½ inches
DIAL, Porcelain ....6 inches, French Cast Sash, Beveled Glass
CASE, Black Enameled Iron
FINISH, on Trimmings, Gilt or Bronze
MOVEMENT, 8 day Half-hour Strike, Cathedral Gong
PRICE ....list, $13.00
PRICE, with Visible Escapement ....list extra, .60

895

1911

FAYETTE
(Showing Pearl Dial)
HEIGHT ....10½ inches
WIDTH ....16 inches
DIAL, Ivorine or Pearl ....6 inches
CASE, Black Enameled Iron
FINISH, on Trimmings, Gilt or Bronze
MOVEMENT, 8 day Half-hour Strike, Cathedral Gong
PRICE ....list, $11.20
PRICE, with Visible Escapement ....list extra, .60
PRICE, with French Cast Sash, Beveled Glass and Porcelain Dial ....list extra, 1.50

896

1911

FERNCLIFF
HEIGHT ....10½ inches
WIDTH ....16 inches
DIAL, Ivorine or Pearl ....6 inches
CASE, Black Enameled Iron
FINISH, on Trimmings, Gilt or Bronze
MOVEMENT, 8 day Half-hour Strike, Cathedral Gong
PRICE ....list, $11.20
PRICE, with Visible Escapement ....list extra, .60
PRICE, with French Cast Sash, Beveled Glass and Porcelain Dial ....list extra, 1.50

MANTEL CLOCKS - ENAMELED IRON CASES

897

1900

898

1900

CHISELHURST

Finished in Black, Smoke or Malachite, with Gilt or Bronze Trimmings.
Height, 10¾ in. Width, 10⅞ in.
Eight-day, Half-hour Strike, Cathedral Gong.
Four-inch Ivorine Dial, Gilt Center.

List price, $11.85

Fitted, to order, with Jeweled Sash in Brilliant, Ruby, Emerald, Amethyst, Sapphire, or alternate stones, at extra list $2.80
Extra list for Ivorized Porcelain Dial, Empire Sash, Beveled Glass, $1.50
Extra list for Visible Escapement, 50 cents.
Extra list for Pearl Dial, 30 cents.

CHATEAU

Height, 9⅞ inches. Width, 11½ inches.
Finished in Black, Smoke or Malachite with Gilt Trimmings.
Eight-day, Half-hour Strike, Cathedral Gong. Four-inch Ivorine Dial with Gilt Center.

List price, $12.50

Ivorized Porcelain Dial with Empire Sash and Heavy Beveled Glass, extra list, $1.50
Fitted, to order only, with Jeweled Sash in Brilliant, Ruby, Emerald, Amethyst, Sapphire, or alternate stones, at extra list of $2.80.

899

1900

900

1900

ARNO

Finished in Black or Green Marble with Cast Brass Ornaments Finished in French Satin Gilt.
Height, 13 inches. Width, 11 inches.
Eight-day, Half-hour Strike, Cathedral Gong. Four-inch Gilt or Enameled Porcelain Dial with Gilt Center, Empire Sash and Heavy Beveled Glass.

List price, $16.75

Fitted, to order only, with Jeweled Sash in Brilliant, Ruby, Emerald, Amethyst, Sapphire, or alternate stones, at extra list of $2.80

CHARLTON

Finished in Black, Smoke or Malachite with Gilt Ornaments.
Height, 14⅝ inches. Width, 9 inches.
Eight-day, Half-hour Strike, Cathedral Gong.
Four-inch Ivorine Dial with Gilt Center.

List price, $11.25

Ivorized Porcelain Dial with Empire Sash and Heavy Beveled Glass, extra list, $1.50
Fitted, to order only, with Jeweled Sash in Brilliant, Ruby, Emerald, Amethyst, Sapphire, or alternate stones at extra list of $2.80

MANTEL CLOCKS - ENAMELED IRON CASES

902

1900

OLINDA

Finished in Black, Smoke or Malachite, with Gilt or Bronze Trimmings.
Height, 12¼ inches.
Width, 11½ inches.
Six-inch fancy Gilt, White or Ivorine Dial.
Eight-day, Half-hour Strike,
Cathedral Gong.

List price, $11.25

Extra list for Ivorized Porcelain Dial, with Empire Sash and Heavy Beveled Glass, $1.50
Extra list for Visible Escapement, 50 cts.
Extra list for Pearl Dial, 30 cts.

901

1900

TEMPEST

Height, 19¼ inches.
Width, 11¼ inches.

French Satin Gold Finish, with Black or Green Enameled Iron Base.

Eight-day, Half-hour Strike, Cathedral Gong, Empire Sash and Heavy Beveled Glass, Three-inch Ivorized Porcelain Dial with Gilt Center, or Decorated Porcelain Dial.

List price, Satin Gold, $42.00
" " Bronze, 33.50

903

1900

904

1900

905

1900

ÆGEAN SET

Black Enamel or Malachite Finish, French Gilt or Bronze Trimmings.

Height, 20 inches.
Width, 12¼ inches.
Height of Candelabra, 22¼ inches.

Eight-day, Half-hour Strike, Cathedral Gong. Four-inch Ivorized Porcelain Dial with Gilt Center, or Decorated Porcelain Dial, Empire Sash and Heavy Beveled Glass.

List price, Clock, $26.50
" " Candelabra, 19.50
" " Set complete, 46.00

Fitted with Jeweled Sash in Brilliant, Ruby, Emerald, Amethyst, Sapphire, or alternate stones, at extra list of $2.80.

MANTEL CLOCKS - ENAMELED IRON CASES

906

1900

LELAND

Finished in Black, Smoke or Malachite, with Gilt or Bronze Trimmings.
Height, 12 1/8 in. Width, 9 3/4 in.
Eight-day, Half-hour Strike, Cathedral Gong.
Six-inch fancy Gilt, White or Ivorine Dial.

List price, $11.85

Extra list for Ivorized Porcelain Dial, Empire Sash, Beveled Glass, $1.50
Extra list for Visible Escapement, 50 cents.
Extra list for Pearl Dial, 30 cents.

907

1900

MONACO

Height, 11 3/8 inches. Width, 12 1/4 inches. Dial, 6 inches.
Black Enamel, Smoke or Malachite Finish with Gilt Trimmings.
Eight-day, Half-hour Strike, Cathedral Gong. Six-inch fancy Gilt, plain White or Pearl Dial.

List price, $11.85

Extra list for Ivorized Porcelain Dial with Empire Sash and Heavy Beveled Glass, $1.50

Extra list for Visible Escapement, 50 cents.
Extra list for Pearl Dial, 30 cents.

908

1900

HARCOURT

Finished in Black, Smoke or Malachite with Gilt or Bronze Trimmings.
Height, 10 5/8 inches. Width, 16 3/8 inches.
Eight-day, Half-hour Strike, Cathedral Gong. Six-inch fancy Gilt, White or Ivorine Dial.

List price, $12.50

Extra list for Ivorized Porcelain Dial with Empire Sash, Heavy Beveled Glass, $1.50
Extra list for Visible Escapement, 50 cents.
Extra list for Pearl Dial, 30 cents.

909

1900

TITAN

Finished in Black, Smoke or Malachite, with Gilt or Bronze Trimmings.
Height, 13 1/2 inches.
Width, 9 inches.
Six-inch fancy Gilt, White or Ivorine Dial.
Eight-day, Half-hour Strike, Cathedral Gong.

List price, $10.50

Extra list for Ivorized Porcelain Dial, with Empire Sash and Heavy Beveled Glass, $1.50
Extra list for Visible Escapement, 50 cts.
Extra list for Pearl Dial, 30 cts.

MANTEL CLOCKS - ENAMELED WOOD CASES

The year entered beside each clock is the year of the catalog from which the illustration was taken. Some clocks may have been produced a few years earlier.

910

1911

ELBRIDGE

HEIGHT................................12 1/8 INCHES
WIDTH................................14 3/4 INCHES
DIAL, IVORINE OR PEARL..........6 INCHES
CASE, BLACK ENAMELED WOOD, WITH BRONZE TRIMMINGS OR DARK STEEL, WITH GILT TRIMMINGS
MOVEMENT, 8 DAY HALF-HOUR STRIKE, CATHEDRAL GONG
PRICE ....LIST, $8.75

911

1911

ELSMERE

HEIGHT................................12 1/8 INCHES
WIDTH................................14 3/4 INCHES
DIAL, IVORINE OR PEARL..........6 INCHES
CASE, BLACK ENAMELED WOOD, WITH BRONZE TRIMMINGS OR DARK STEEL, WITH GILT TRIMMINGS
MOVEMENT, 8 DAY HALF-HOUR STRIKE, CATHEDRAL GONG
PRICE ....LIST, $8.75

912

1911

EDGEBROOK

HEIGHT................................10 1/4 INCHES
WIDTH................................15 1/2 INCHES
DIAL, IVORINE OR PEARL..........6 INCHES
CASE, BLACK ENAMELED WOOD, WITH BRONZE TRIMMINGS OR DARK STEEL, WITH GILT TRIMMINGS
MOVEMENT, 8 DAY HALF-HOUR STRIKE, CATHEDRAL GONG
PRICE ....LIST, $8.25

913

1911

ELSWORTH

HEIGHT................................11 1/2 INCHES
WIDTH................................15 1/4 INCHES
DIAL, IVORINE OR PEARL..........6 INCHES
CASE, BLACK ENAMELED WOOD, WITH BRONZE TRIMMINGS OR DARK STEEL, WITH GILT TRIMMINGS
MOVEMENT, 8 DAY HALF-HOUR STRIKE, CATHEDRAL GONG
PRICE ....LIST, $8.75

MANTEL CLOCKS - ENAMELED WOOD CASES

914

1911

WOODBINE

HEIGHT ....12¼ INCHES
WIDTH ....18½ INCHES
DIAL, IVORINE OR PEARL ....6 INCHES
CASE, BLACK ENAMELED WOOD WITH MARBLEIZED MOULDINGS AND PILLARS
FINISH, ON TRIMMINGS, GILT OR BRONZE
MOVEMENT, 8 DAY HALF-HOUR STRIKE, CATHEDRAL GONG
PRICE ....LIST, $10.15

915

1911

LORING

HEIGHT ....11½ INCHES
WIDTH ....15½ INCHES
DIAL, IVORINE OR PEARL ....6 INCHES
CASE, BLACK ENAMELED WOOD, WITH MARBLEIZED MOULDINGS AND PILLARS
FINISH, ON TRIMMINGS, GILT OR BRONZE
MOVEMENT, 8 DAY HALF-HOUR STRIKE, CATHEDRAL GONG
PRICE...LIST, $8.00

916

1911

CONCORD

HEIGHT ....11 INCHES
WIDTH ....15¾ INCHES
DIAL, IVORINE OR PEARL ....6 INCHES
CASE, BLACK ENAMELED WOOD, WITH MARBLEIZED PILLARS
FINISH, ON TRIMMINGS, GILT OR BRONZE
MOVEMENT, 8 DAY HALF-HOUR STRIKE, CATHEDRAL GONG
PRICE ....LIST, $8.00

917

1911

WOODSIDE

HEIGHT.............................. 12¼ INCHES
WIDTH ....18¼ INCHES
DIAL, IVORINE OR PEARL ....6 INCHES
CASE, BLACK ENAMELED WOOD WITH MARBLEIZED MOULDINGS AND PILLARS
FINISH, ON TRIMMINGS, GILT OR BRONZE
MOVEMENT, 8 DAY HALF-HOUR STRIKE, CATHEDRAL GONG
PRICE ....LIST, $10.15

MANTEL CLOCKS - ENAMELED WOOD CASES

918

1911

BINGEN

HEIGHT ....10 INCHES
WIDTH ....12½ INCHES
DIAL, IVORINE OR PEARL ....6 INCHES
CASE, BLACK ENAMELED WOOD
MOVEMENT, 8 DAY HALF-HOUR STRIKE, CATHEDRAL GONG
PRICE ....LIST, $6.10

919

1911

WOODLAKE

HEIGHT.............................. 12¼ INCHES
WIDTH..................................18¼ INCHES
DIAL, IVORINE OR PEARL ....6 INCHES
CASE, BLACK ENAMELED WOOD WITH MARBLEIZED MOULDINGS AND PILLARS
FINISH, ON TRIMMINGS, GILT OR BRONZE
MOVEMENT, 8 DAY HALF-HOUR STRIKE, CATHEDRAL GONG
PRICE ....LIST, $10.15

920

1900

PRESCOTT

Height, 11 inches. Width, 16 inches.
This is the only Blackwood that is fitted with Alarm attachment.

List price, Eight-day Strike, $9.05
" " Eight-day Strike, Alarm, 9.55

921

1900

JOLIET

Height, 11¾ inches. Width, 10 inches.

List price, $8.05

Enameled Wood, finished in Black, Smoke or Malachite, Bronze or Gilt Trimmings.
Eight-day, Half-hour Strike, Cathedral Gong, plain White, fancy Gilt or Ivorine Dial.
Ivorized Porcelain Dial with Empire Sash and Heavy Beveled Glass, extra list, $1.50.

Visible Escapement, extra list, 50 cents.

MANTEL CLOCKS - ENAMELED WOOD CASES

922

1911

WOODVALE

HEIGHT ....12¼ INCHES
WIDTH ....19 INCHES
DIAL, IVORINE OR PEARL ....6 INCHES
CASE, BLACK ENAMELED WOOD WITH MARBLEIZED MOULDINGS AND PILLARS
FINISH, ON TRIMMINGS, GILT OR BRONZE
MOVEMENT, 8 DAY HALF-HOUR STRIKE, CATHEDRAL GONG
PRICE ....LIST, $10.15

WOODVALE

923

WOODMONT

HEIGHT ....12¼ INCHES
WIDTH ....19 INCHES
DIAL, IVORINE OR PEARL ....6 INCHES
CASE, BLACK ENAMELED WOOD WITH MARBLEIZED MOULDINGS AND PILLARS
FINISH, ON TRIMMINGS, GILT OR BRONZE
MOVEMENT, 8 DAY HALF-HOUR STRIKE, CATHEDRAL GONG
PRICE ....LIST, $10.15

1911

WOODMONT

924

1900

CANFIELD

Height, 10½ inches. Width, 13¼ inches. Dial, 6 inches.
List price, $7.90

925

1900

CRANSTON

Height, 10½ inches. Width, 13¼ inches.
List price, $7.65

Enameled Wood, finished in Black, Smoke or Malachite, Bronze or Gilt Trimmings.
Eight-day, Half-hour Strike, Cathedral Gong, plain White, fancy Gilt or Ivorine Dial.
Ivorized Porcelain Dial with Empire Sash and Heavy Beveled Glass, extra list, $1.50.
Visible Escapement, extra list, 50 cents.

MANTEL CLOCKS - ENAMELED WOOD CASES

927

1911

CARMONA

HEIGHT ....13 INCHES
WIDTH ....18 INCHES
DIAL, IVORINE OR PEARL ....6 INCHES
CASE, BLACK ENAMELED WOOD WITH MARBLEIZED MOULDINGS AND PILLARS
FINISH, ON TRIMMINGS, GILT OR BRONZE
MOVEMENT, 8 DAY HALF-HOUR STRIKE, CATHEDRAL GONG
PRICE ....LIST, $10.15

926

1900

ALEXANDER

Finished in Black, Smoke or Malachite, with Marbleized Moldings, Gilt or Bronze Trimmings.

Height, 11¾ inches. Width, 16 inches.

Eight-day Half-hour Strike, Cathedral Gong, with Six-inch White, Ivorine or Fancy Gilt Dial.

List price, $9.75

Extra list for Ivorized Porcelain Dial, Empire Sash and Heavy Beveled Glass, $1.50
Extra list for Visible Escapement, 50 cents.
Extra list for Pearl Dial, 30 cents.

928

1911

OTTAWA

HEIGHT ....11¾ INCHES
WIDTH ....16¾ INCHES
DIAL, IVORINE OR PEARL ....6 INCHES
CASE, BLACK ENAMELED WOOD WITH MARBLEIZED MOULDINGS AND PILLARS
FINISH, ON TRIMMINGS, GILT OR BRONZE
MOVEMENT, 8 DAY HALF-HOUR STRIKE, CATHEDRAL GONG
PRICE ....LIST, $9.75

929

1900

NUMA

Finished in Black, Smoke or Malachite, with Marbleized Moldings, Gilt or Bronze Trimmings.
Height, 11 inches. Width, 16 inches.
Eight-day, Half-hour Strike, Cathedral Gong, with Six-inch White, Ivorine or Fancy Gilt Dial.

List price, $9.75

Extra list for Ivorized Porcelain Dial with Empire Sash and Heavy Beveled Glass, $1.50
Extra list for Visible Escapement, 50 cents.
Extra list for Pearl Dial, 30 cents.

MANTEL CLOCKS - ENAMELED WOOD CASES

930

1900

BANCROFT

Finished in Black, Smoke or Malachite with Gilt or Bronze Trimmings.

Height, 11 inches. Width, 16 inches.

Eight-day, Half-hour Strike, Cathedral Gong, with Six-inch White, Ivorine or Fancy Gilt Dial.

List price, $9.75

Extra list for Ivorized Porcelain Dial with Empire Sash and Heavy Beveled Glass, $1.50
Extra list for Visible Escapement, 50 cents.
Extra list for Pearl Dial, 30 cents.

931

1911

EDGEWOOD

HEIGHT ....10¼ INCHES
WIDTH ....14½ INCHES
DIAL, IVORINE OR PEARL ....6 INCHES
CASE, BLACK ENAMELED WOOD, WITH BRONZE TRIMMINGS OR DARK STEEL, WITH GILT TRIMMINGS
MOVEMENT, 8 DAY HALF-HOUR STRIKE, CATHEDRAL GONG
PRICE ....LIST, $8.00.

932

1911

WOODBRIDGE

HEIGHT ....11¾ INCHES
WIDTH .... 18 INCHES
DIAL, IVORINE OR PEARL ....6 INCHES
CASE, BLACK ENAMELED WOOD WITH MARBLEIZED MOULDINGS AND PILLARS
FINISH, ON TRIMMINGS, GILT OR BRONZE
MOVEMENT, 8 DAY HALF-HOUR STRIKE, CATHEDRAL GONG
PRICE ....LIST, $10.15

933

1911

EDGEMONT

HEIGHT ....10¼ INCHES
WIDTH ....15⅛ INCHES
DIAL, IVORINE OR PEARL ....6 INCHES
CASE, BLACK ENAMELED WOOD, WITH BRONZE TRIMMINGS OR DARK STEEL, WITH GILT TRIMMINGS
MOVEMENT, 8 DAY HALF-HOUR STRIKE, CATHEDRAL GONG
PRICE ....LIST, $8.00

MANTEL CLOCKS - ENAMELED WOOD CASES

The year entered beside each clock is the year of the catalog from which the illustration was taken. Some clocks may have been produced a few years earlier.

934

1911

SENECA

(SHOWING PEARL DIAL)

HEIGHT ....11 INCHES
WIDTH ....16 INCHES
DIAL, IVORINE OR PEARL ....6 INCHES
CASE, BLACK ENAMELED WOOD WITH MARBLEIZED MOULDINGS AND PILLARS
FINISH, ON TRIMMINGS, GILT OR BRONZE
MOVEMENT, 8 DAY HALF-HOUR STRIKE, CATHEDRAL GONG
PRICE ....LIST, $9.45

936

1911

PEMBROKE

HEIGHT ....11 INCHES
WIDTH ....16 INCHES
DIAL, IVORINE OR PEARL ....6 INCHES
CASE, BLACK ENAMELED WOOD WITH MARBLEIZED MOULDINGS
FINISH, ON TRIMMINGS, GILT OR BRONZE
MOVEMENT, 8 DAY HALF-HOUR STRIKE, CATHEDRAL GONG
PRICE ....LIST, $9.15

935

1911

CLAYPOOL

HEIGHT ....11½ INCHES
WIDTH ....18½ INCHES
DIAL, IVORINE OR PEARL ....6 INCHES
CASE, BLACK ENAMELED WOOD WITH MARBLEIZED MOULDINGS AND PILLARS
FINISH, ON TRIMMINGS, GILT OR BRONZE
MOVEMENT, 8 DAY HALF-HOUR STRIKE, CATHEDRAL GONG
PRICE ....LIST, $9.45

937

1900

BURLINGTON

Finished in Black, Smoke or Malachite,
Gilt or Bronze Trimmings.
Height, 11 inches. Width, 16 inches.
Eight-day, Half-Hour Strike, Cathedral Gong, with Six-inch White, Ivorine or Fancy Gilt Dial.

List price, $9.75

Extra list for Ivorized Porcelain Dial with Empire Sash and Heavy Beveled Glass, $1.50
Extra list for Visible Escapement, 50 cents.
Extra list for Pearl Dial, 30 cents.

MANTEL CLOCKS - ENAMELED WOOD CASES

938

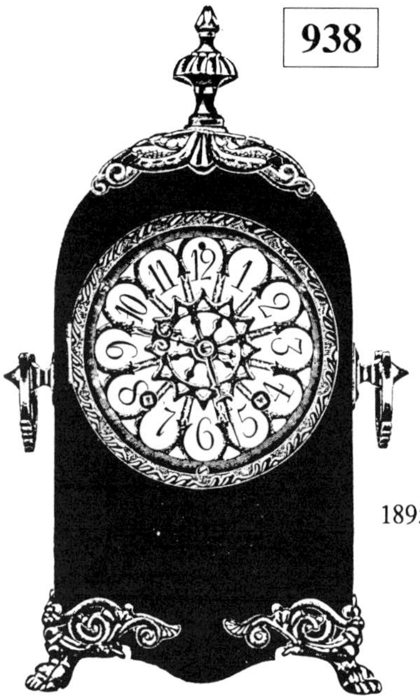

1895

RIVIERA.

Gilt Trimmings. Black, Mahogany, Oak or Malachite Finish.

Height, 11½ inches; Width, 6¾ inches.
4 inch Silver Dial with Gilt Overlay.

8 Day, French Cup Bell, Hour Strike, Black, - - - List, $10.00
8 Day, French Cup Bell, Hour Strike, Malachite, Oak or Mahogany, List, 10.75

939

1895

Fitted with our New "Crown" Movements.

MALVERN.

Height, 11½ inches; Width, 15½ inches; Fancy Gilt or Plain White, 6 inch Dial.

8 Day, Half Hour Strike, Cathedral Gong, $14.70

940

1914

BRIGHTON

LIST, $6.50

HEIGHT, 10 INCHES. WIDTH, 12½ INCHES.

941

1900

MANTUA

Finished in Black, Smoke or Malachite, with Marbleized Moldings, Gilt or Bronze Trimmings.
Height, 11 inches. Width, 16 inches.
Eight-day, Half-hour Strike, Cathedral Gong, with Six-inch White, Ivorine or Fancy Gilt Dial.

List price, $9.75

Extra list for Ivorized Porcelain Dial with Empire Sash and Heavy Beveled Glass, $1.50
Extra list for Visible Escapement, 50 cents.
Extra list for Pearl Dial, 30 cents.

MANTEL CLOCKS - ENAMELED WOOD CASES

942

1895

SEVILLE.

Black, Mahogany, Oak or Malachite Finish.

Height, 8½ inches; Width, 7¾ inches; Dial, 5 inches.

8 Day, Striking Hours and Half-Hours on Cathedral Gong, Black. - - - List, $10.00
8 Day, Striking Hours and Half-Hours on Cathedral Gong, Oak, Mahogany or Malachite. List, 10.75
Visible Escapement. - - Extra, List, .50

943

1895

MILAN.

Black, Mahogany, Oak or Malachite Finish.

Height, 11 inches; Width, 11 inches; Fancy Gilt, Plain White or Mother of Pearl, 6 inch Dial.

8 Day, Half-Hour Strike, Cathedral Gong, Black, - - - List, $13.00
8 Day, Half-Hour Strike, Cathedral Gong, Oak, Mahogany or Malachite, - List, 13.75
Visible Escapement, - - - - - - - Extra, List, .50
Porcelain Dial, - - - - - - Extra, List, 1.50
Empire Sash, Convex Glass, Silver Dial with Gilt Overlay. - Extra, List, 2.00

944

1895

ARMOREL.

Black, Mahogany, Oak or Malachite.

Height, 12 inches; Width, 10 inches; Fancy Gilt, Plain White or Mother of Pearl, 6 inch Dial.

8 Day, Half-Hour Strike, Cathedral Gong, Black, List, $15.60
8 Day, Half-Hour Strike, Cathedral Gong, Mahogany, Oak or Malachite, - - List, 16.35
Visible Escapement, - - Extra, List, .50
Porcelain Dial, - - - Extra, List, 1.50
Empire Sash, Convex Glass, Silver Dial with Gilt Center, - - - Extra, List, 2.00

945

1895

TYNDALL.

Black, Mahogany, Oak or Malachite.

Height, 13 inches; Width, 9 inches; Fancy Gilt, Plain White or Mother of Pearl, 6 inch Dial.

8 Day, Half-Hour Strike, Cathedral Gong, Black, List, $13.00
8 Day, Half-Hour Strike, Cathedral Gong, Mahogany, Oak or Malachite, - - List, 13.75
Visible Escapement, - - Extra, List, .50
Porcelain Dial, - - - Extra, List, 1.50
Empire Sash, Convex Glass, Silver Dial with Gilt Overlay, - - Extra, List, 2.00

MANTEL CLOCKS - ENAMELED WOOD CASES

946

1895

CASTILE.

Gilt Trimmings. Black, Mahogany, Oak or Malachite Finish.

Height, 10½ inches; Width, 6½ inches.
3½ inch Porcelain Dial.

8 Day, French Cup Bell, Hour Strike, Black, - - - List, $8.40
8 Day, French Cup Bell, Hour Strike, Malachite, Oak or Mahogany, List, 9.15

947

1895

ROSETTI.

Height, 11¾ inches; Width, 16½ inches; Fancy Gilt or Plain White, 6 inch Dial.

8 Day, Half-Hour Strike, Cathedral Gong, List, $10.50

948

1895

DANTE.

Height, 11 inches; Width, 15½ inches; Fancy Gilt or Plain White, 6 inch Dial.

8 Day, Half-Hour Strike, Cathedral Gong, List, $11.00

949

1895

CORDOVA.

Black, Mahogany, Oak or Malachite Finish.

Height, 9 inches; Width, 7¾ inches; Dial, 5 inches.

8 Day, Striking Hours and Half-hours on Cathedral Gong, Black. - - - List, $10.00
8 Day, Striking Hours and Half-hours on Cathedral Gong, Oak, Mahogany or Malachite. - List, 10.75
Visible Escapement. - - Extra, List, .50

MANTEL CLOCKS - MARBLE CASES

The marble clocks on pages 247 to 255 were taken from the 1883 New Haven catalog, however, they are imported French clocks and not a product of the New Haven Clock Company. Although New Haven retailed them for a time, they may or may not be marked New Haven.

950

1883

MARBLE, No. 1151.

15 Day, Striking the Full and Half-hour on a Cathedral Gong.
Height, 8½ inches. Width at base, 13 inches.
Black Case, Engraved and Gilt. Real Bronze Inlayings.
Visible Escapement.
White or Black Marble Dial, Gilt Center.

951

1883

MARBLE, No. 1016.

15 Day, Striking the Full and Half-hour on a Cathedral Gong.
Height, 11¾ inches. Width at base, 7¾ inches.
Black Case, Front Inlaid with different colored Marble.
Sunk Center Dial and Visible Escapement.

953

952

1883

1883

954

1883

MARBLE SET, No. 1037.

15 Day, Striking the Full and Half-hour on a Cathedral Gong.
Height, 14½ inches. Width at base, 14½ inches. Height of Side Pieces, 7¾ inches.
Black Case, Inlaid with Hand-painted Porcelain Tablets. Black Marble Dial, with Gilt Figures and Ornamented Porcelain Center.
Side Pieces made and Ornamented with designs in the same style as the clock.

MANTEL CLOCKS - MARBLE CASES

956

1883

MARBLE, No. 1051.

15 Day, Striking the Full and Half-hour on a Cathedral Gong.
Height, 9 inches. Width, 12 inches.
Black Case, Engraved and Gilt, Inlaid with Colored Marble.
Fancy Dial. Gilt Sunk Center.

955

1883

MARBLE, No. 1019.

15 Day, Striking the Full and Half-hour on a Cathedral Gong.
Height, 11½ inches.
Width at base, 9¼ inches.
Black Case, Engraved and Gilt, Inlaid with different colored Marble.
Sunk Center Dial and Visible Escapement.

957

1883

MARBLE, No. 1008.

15 Day, Striking the Full and Half-hour on a Cathedral Gong.
Height, 11½ inches.
Width at base, 9¼ inches.
Black Case, Engraved and Gilt, Inlaid with different colored Marble.

958

1883

959

1883

960

1883

MARBLE SET, No. 1006.

15 Day, Striking the Full and Half-hour on a Cathedral Gong.
Height, including Urn, 15½ inches. Width at base, 10 inches. Height of Coupes, 7½ inches.
With or without Top and Side Pieces.
Black Case, Engraved and Gilt.

MANTEL CLOCKS - MARBLE CASES

962

961

1883

MARBLE, No. 1100.

15 Day, Striking the Full and Half-hour on a Cathedral Gong.
Height, 9 inches. Width, 19 inches.
Black Case, Engraved and Gilt, Inlaid with different colored Marble.
Sunk Center Dial and Visible Escapement.

1883

MARBLE, No. 1001.

8 Day Time-piece, and 15 Day, Striking the Full and Half-hour.
Height, including Figure, 18 inches. Width at base, 9 inches.
With or without Figure.
Black Case, Engraved and Gilt.

963

1883

MARBLE, No. 1086.

15 Day, Striking the Full and Half-hour on a Bell or Cathedral Gong.

Height, 8 inches. Width at base, 14½ inches.

Black Case, Engraved and Gilt, Inlaid with different colored Marble.

964

1883

MARBLE, No. 1005.

15 Day, Striking the Full and Half-hour on a Cathedral Gong.

Height, 12 inches. Width at base, 11 inches.

Black Case, Engraved and Gilt.

Corners and Inlayings of different colored Marble.

The marble clocks on pages 247 to 255 were taken from the 1883 New Haven catalog, however, they are imported French clocks and not a product of the New Haven Clock Company. Although New Haven retailed them for a time, they may or may not be marked New Haven.

MANTEL CLOCKS - MARBLE CASES

The marble clocks on pages 247 to 255 were taken from the 1883 New Haven catalog, however, they are imported French clocks and not a product of the New Haven Clock Company. Although New Haven retailed them for a time, they may or may not be marked New Haven.

965

1883

MARBLE, No. 1065.

15 Day, Striking the Full and Half-hour on a Cathedral Gong.
Height, 9 inches. Width at base, 14 inches.

Black Case, Engraved and Gilt, Inlaid with different colored Marble.
Sunk Center Dial and Visible Escapement.

966

1883

MARBLE, No. 1149.

15 Day, Striking the Full and Half-hour on a Cathedral Gong.
Height, 8¼ inches. Width at base, 13¼ inches.

Black Case, Engraved and Gilt, Inlaid with different colored Marble.

Black Marble Dial, Fancy Sunk Center.

967

1883

MARBLE, No. 1026.

15 Day, Striking the Full and Half-hour on a Cathedral Gong.
Height, 12 inches. Width at base, 8½ inches.
Black Case, Ornamented Black Marble Dial, with Gilt Figures.

With or without Side Pieces, same style as shown on pages 247 and 252

968

1883

MARBLE, No. 1150.

15 Day Striking the Full and Half-hour on a Cathedral Gong.
Height, 14½ inches. Length, 9 inches.

Black Case, Engraved and Gilt. Real Bronze Inlayings.

Fancy Marble Dial.

MANTEL CLOCKS - MARBLE CASES

969

1883

MARBLE, No. 1078.

15 Day, Striking the Full and Half-hour on a Cathedral Gong.
Height, 9 inches. Width at base, 14 inches.

Black Case, Engraved and Gilt, Inlaid with different colored Marble
Gilt Sunk Center Dial and Visible Escapement.

970

1883

MARBLE No. 1012.

15 Day, Striking the Full and Half-hour on a Bell or Cathedral Gong.
Height, 12 inches. Width at base, 9½ inches.
Black Case, Engraved and Gilt, Inlaid with different colored Marble.

971

1883

972

1883

973

1883

MARBLE SET, No. 1011.

15 Day, Striking the Full and Half-hour on a Bell or Cathedral Gong.
Bronze Top and Side Pieces.
Height, with Top Piece, 21 inches. Width at base, 10 inches. Height of Side Pieces, 16 inches.
Black Case, Engraved and Gilt, Inlaid with different colored Marble, with or without Top and Side Pieces.

MANTEL CLOCKS - MARBLE CASES

The marble clocks on pages 247 to 255 were taken from the 1883 New Haven catalog, however, they are imported French clocks and not a product of the New Haven Clock Company. Although New Haven retailed them for a time, they may or may not be marked New Haven.

974

1883

MARBLE No. 1010.

15 Day, Striking the Full and Half-hour on a Cathedral Gong.
Either Visible Escapement or Plain Dial.
Height, 11½ inches. Width at base, 10¼ inches.
Black Case, Engraved and Gilt, Inlaid with different colored Marble.

975

1883

MARBLE, No. 1152.

15 Day, Striking the Full and Half-hour on a Cathedral Gong.
Height, 15 inches. Width at base, 17 inches.

Black Case, Engraved and Gilt. Real Bronze Inlayings.
Visible Escapement or Plain Dial.
White or Black Marble Dial, Gilt Sunk Center.

976

1883

977

1883

978

1883

MARBLE SET, No. 1035.

15 Day, Striking the Full and Half-hour on a Cathedral Gong.
Height, 13 inches. Width at base, 8 inches. Height of Side Pieces, 7½ inches.
Black Case, Inlaid with Hand-painted Porcelain Tablets. Black Marble Dial, with Gilt Figures and Ornamented Porcelain Center.
Side Pieces made and Ornamented with designs in the same style as the clock.

MANTEL CLOCKS - MARBLE CASES

979

1883

MARBLE, No. 1032.

15 Day, Striking the Full and Half-hour on a Cathedral Gong.

Height, 14½ inches. Width at base, 7½ inches.

Black Case, Inlaid with Hand-painted Porcelain Tablets.

Black Marble Dial, with Gilt Figures and Ornamented Porcelain Center.

980

1883

MARBLE, No. 1033.

15 Day, Striking the Full and Half-hour on a Cathedral Gong.

Height, 17 inches. Width at base, 13 inches.

Black Case, Engraved and Gilt, Inlaid with different colored Marble.

Black or White Marble Dial, with Gilt Figures and Sunk Center.

981

1883

MARBLE, No. 1031.

15 Day, Striking the Full and Half-hour on a Cathedral Gong.

Height, 15½ inches. Width at base, 12 inches.

Black Case, Engraved and Gilt. Side Pillars of different colored Marble.

Black Marble Dial, with Gilt Figures and Sunk Center.

982

1883

MARBLE, No. 1034.

15 Day, Striking the Full and Half-hour on a Cathedral Gong.

Height, 12¾ inches. Width at base, 7¾ inches.

Black Case, Inlaid with Hand-painted Porcelain Tablets.

Black Marble Dial, with Gilt Figures and Ornamented Porcelain Center.

MANTEL CLOCKS - MARBLE CASES

The marble clocks on pages 247 to 255 were taken from the 1883 New Haven catalog, however, they are imported French clocks and not a product of the New Haven Clock Company. Although New Haven retailed them for a time, they may or may not be marked New Haven.

983

1883

MARBLE, No. 1154.

15 Day, Striking the Full and Half-hour on a Cathedral Gong.
Height, 12½ inches. Width at base, 18 inches.

Black Case, Engraved and Gilt, Inlaid with different colored Marble.
Marble Pillars, Fancy Dial, Gilt Sunk Center.

984

1883

MARBLE AND BRONZE, No. 1027.

15 Day, Striking the Full and Half-hour on a Cathedral Gong.
Height, with Figure, 24 inches. Width at base, 14 inches.
Black Case, Inlaid with different colored Marble. Bronze Top Piece and Ornaments.
Bronze and Marble Side Pieces to match. Height, 16½ inches.

985

1883

MARBLE, No. 1050.

15 Day, Striking the Full and Half-hour on a Cathedral Gong.
Height, 8¾ inches. Width at base, 13¼ inches.

Black Case, Engraved and Gilt, Inlaid with different colored Marble.
Sunk Center Dial and Visible Escapement.

MANTEL CLOCKS - MARBLE CASES

The year entered beside each clock is the year of the catalog from which the illustration was taken. Some clocks may have been produced a few years earlier.

986

1883

MARBLE, No. 1054.

15 Day, Striking the Full and Half-hour on a Cathedral Gong.
Height, 12 inches. Width at base, 9 inches.
Black Case, Engraved and Gilt, Inlaid with different colored Marble. Fancy Dial.

987

1883

MARBLE, No. 1144.

15 Day, Striking the Full and Half-hour on a Cathedral Gong.
Height, 17½ inches. Width at base, 9¼ inches.

Black Case, Engraved and Gilt. Real Bronze Inlayings.
White or Black Marble Dial. Gilt Sunk Center.

988

1883

MARBLE, No. 1076.

15 Day, Striking the Full and Half-hour on a Cathedral Gong.
Height, 9 inches. Width at base, 16 inches.

Black Case, Engraved and Gilt, Inlaid with different colored Marble.
Sunk Center Dial and Visible Escapement.

989

1883

MARBLE, No. 1153.

15 Day, Striking the Full and Half-hour on a Cathedral Gong.
Height, 13 inches. Width at Base, 10 inches.

Black Case, Engraved and Gilt. Real Bronze Inlayings.
Visible Escapement.
White or Black Marble Dial. Gilt Sunk Center.

The marble clocks on pages 247 to 255 were taken from the 1883 New Haven catalog, however, they are imported French clocks and not a product of the New Haven Clock Company. Although New Haven retailed them for a time, they may or may not be marked New Haven.

MANTEL CLOCKS - MARBLE CASES

990

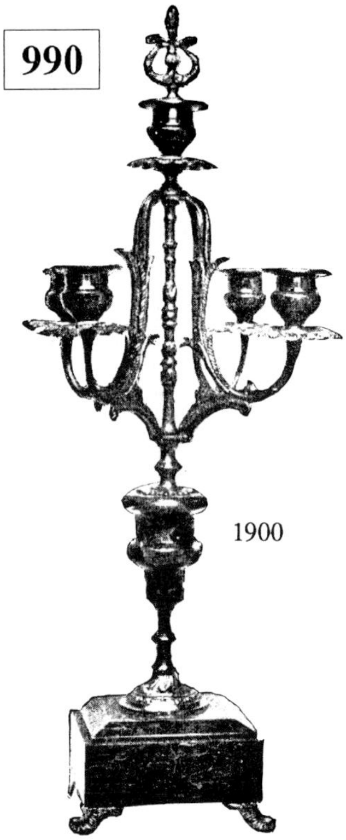

1900

991

1900

MARSTON

992

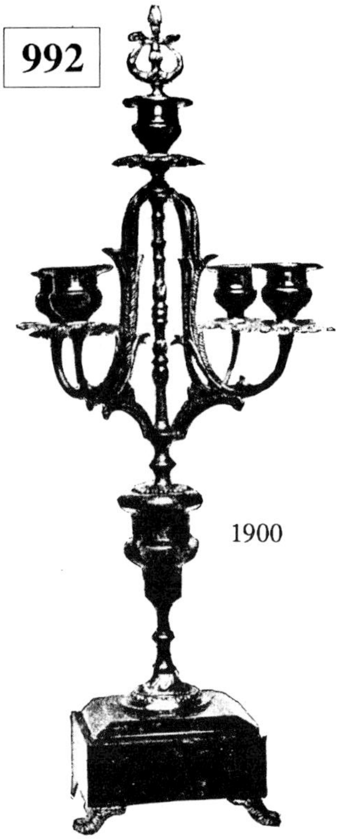

1900

Height of Clock, 16½ inches. Width, 9½ inches. Height of Candelabra, 18 inches. Width, 9½ inches.
Four-inch Ivorized Porcelain Dial, Gilt Center, Empire Sash, Heavy Beveled Glass, Eight-day, Half-hour Strike, Cathedral Gong.
List price, Set, $44.50

993

1900

994

1900

MANDAN

995

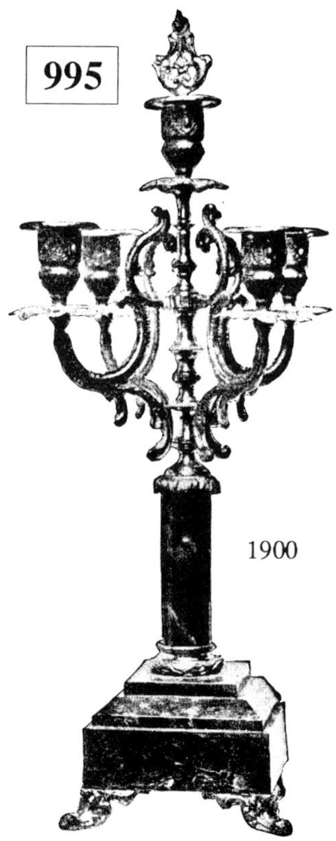

1900

Height of Clock, 18½ inches. Width, 9¾ inches. Height of Candelabra, 18 inches. Width, 5½ inches.
Four-inch Ivorized Porcelain Dial, Gilt Center, Empire Sash, Heavy Beveled Glass, Eight-day, Half-hour Strike, Cathedral Gong.
List price, Set, $66.75

MANTEL CLOCKS - MARBLE CASES

996

1890

MUGWUMP
Base, 14 inches. Height, 9½ inches.
15 Day, Striking Hours and Half-hours
on Cathedral Gong.. $32.50

MUGWUMP

997

1890

KISMET
Base, 14½ inches. Height, 9¾ inches.
15 Day, Striking Hours and Half-hours
on Cathedral Gong.. $25.00

KISMET

998

1890

LORNA
Base, 17 inches. Height, 9½ inches.
15 Day, Striking Hours and Half-hours
on Cathedral Gong.. $40.00

LORNA

MANTEL CLOCKS - MARBLE CASES

999

1890

NAPOLEON

NAPOLEON

Base, 14½ inches. Height, 11 inches.
15 Day, Striking Hours and Half-hours
on Cathedral Gong.. $45.00

1000

NATALIE

Base, 16½ inches. Height, 9 inches.
15 Day, Striking Hours and Half-hours
on Cathedral Gong.. $42.50

1890

NATALIE

1001

1890

MONARCH

MONARCH

Base, 14 inches. Height, 10¼ inches.
15 Day, Striking Hours and Half-hours
on Cathedral Gong.. $46.50

MANTEL CLOCKS - MARBLE CASES

1002

1890

PEERLESS

PEERLESS

Base, 14 inches. Height, 9 inches.
15 Day, Striking Hours and Half-hours
on Cathedral Gong.. $29.00

1003

MARJORIE

Base, 15 inches. Height, 9½ inches.
15 Day, Striking Hours and Half-hours
on Cathedral Gong.. $37.00

1890

MARJORIE

1004

1890

MODJESKA

MODJESKA

Base, 17½ inches. Height, 9½ inches.
15 Day, Striking Hours and Half-hours
on Cathedral Gong.. $41.75

MANTEL CLOCKS - MARBLE CASES

The year entered beside each clock is the year of the catalog from which the illustration was taken. Some clocks may have been produced a few years earlier.

1005

1890

MAGIC

Base, 8½ inches. Height, 9¼ inches.
15 Day, Striking Hours and Half-hours
on Cathedral Gong.. $20.50

1006

1890

QUI VIVE

Base, 12¼ inches. Height, 10 inches.
15 Day, Striking Hours and Half-hours
on Cathedral Gong.. $28.00

1007

1890

PRISCILLA

Base, 19½ inches. Height, 9 ½ inches.
15 Day, Striking Hours and Half-hours on Cathedral Gong.. $45.00

MANTEL CLOCKS - MARBLE CASES

1008

MERMAID

Base, 15½ inches. Height, 11 inches.
15 Day, Striking Hours and Half-hours
on Cathedral Gong.. $40.50

1890

MERMAID

1009

MISTLETOE

Base, 14¼ inches. Height, 9¼ inches.
15 Day, Striking Hours and Half-hours
on Cathedral Gong.. $29.50

1890

MISTLETOE

1010

MINEVRA

Base, 16 inches. Height, 9 inches.
15 Day, Striking Hours and Half-hours
on Cathedral Gong.. $35.50

1890

MINEVRA

MANTEL CLOCKS - MARBLE CASES

1011

NIRVANA
Base, 10 inches. Height, 16 ¾ inches.
15 Day, Striking Hours and Half-hours
on Cathedral Gong.. $36.00

1890

NIRVANA

1012

1890

MYSTERY
Base, 15½ inches. Height, 8 ¾ inches.
15 Day, Striking Hours and Half-hours
on Cathedral Gong.. $44.50

MYSTERY

1013

1890

NAUTILUS
Base, 13 inches. Height, 9 inches.
15 Day, Striking Hours and Half-hours
on Cathedral Gong.. $27.50

1014

1890

NEBULA
Base, 13 inches. Height, 9 inches.
15 Day, Striking Hours and Half-hours
on Cathedral Gong.. $27.00

MANTEL CLOCKS - MARBLE CASES

1015

NYMPH
Base, 14 inches. Height, 9 inches.
15 Day, Striking Hours and Half-hours
on Cathedral Gong.. $38.50

1890

NYMPH

1016

1890

PHANTOM
Base, 12¼ inches. Height, 9¾ inches.
15 Day, Striking Hours and Half-hours
on Cathedral Gong.. $25.00

PHANTOM

1017

PAROLE
Base, 17 inches. Height, 9 inches.
15 Day, Striking Hours and Half-hours
on Cathedral Gong.. $39.00

1890

PAROLE

MANTEL CLOCKS - MARBLOID CASES

MARBLOID CLOCKS.

Each pattern is made in five colors.

MALACHITE.—Rich Green. NUMIDIAN—Delicate Pink.
LAPIS LAZULI—Soft Blue. CRYSTALITE—Blueish Drab.
AGATE—Varying Brown.

1018

1895

ILLYRIA.

Gold Ornaments.

Height, 10 inches; Dial, 5 inches.
8 Day, Half-Hour Strike, Cathedral Gong, List, $16.80
8 " " " " V. E., List, 17.30

1019

1895

PARIAN—Novelty.

One Solid Piece.

Height, 5 inches; Dial 2 inches.
1 Day Time, - - List, $4.20

1020

1895

POLIGNY.

Height, 10½ inches; Width, 10 inches; 5 inch Porcelain Dial.
8 Day, Half-Hour Strike, Cathedral Gong, - List, $50.00
Visible Escapement, - - - - List, .50

The year entered beside each clock is the year of the catalog from which the illustration was taken. Some clocks may have been produced a few years earlier.

MANTEL CLOCKS - MARBLOID CASES

MARBLOID CLOCKS.

Each Pattern is made in Five Colors:

Agate—Varying Brown.

Numidian—Delicate Pink.
Crystallite—Blueish Drab.

Malachite—Rich Green.
Lapis-Lazuli—Soft Blue.

1021

1895

PATRICIAN.
Gold Ornaments.

Height 10¼ inches. Dial 5 inches.
8 Day. Half-Hour Strike, Cathedral Gong, List, $16.80
8 " " " " " V.E., List, 17.30

1022

1895

IVACA.
Gold Ornaments.

Height 10 inches. Dial 5 inches.
8 Day, Half-Hour Strike, Cathedral Gong, List, $16.80
8 " " " " " V.E., List, 17.30

1023

1895

IONE.
Gold Ornaments.

Height 10 inches. Dial 5 inches.
8 Day, Half-Hour Strike, Cathedral Gong, List, $16.80
8 " " " " " V.E., List, 17.30

1024

1895

NUBIAN.
Gold Ornaments.

Height 11½ inches. Dial 5 inches.
8 Day, Half-Hour Strike, Cathedral Gong, List, $19.60
8 " " " " " V.E., List, 20.10

MANTEL CLOCKS - METAL CASES

1025 **1026** **1027**

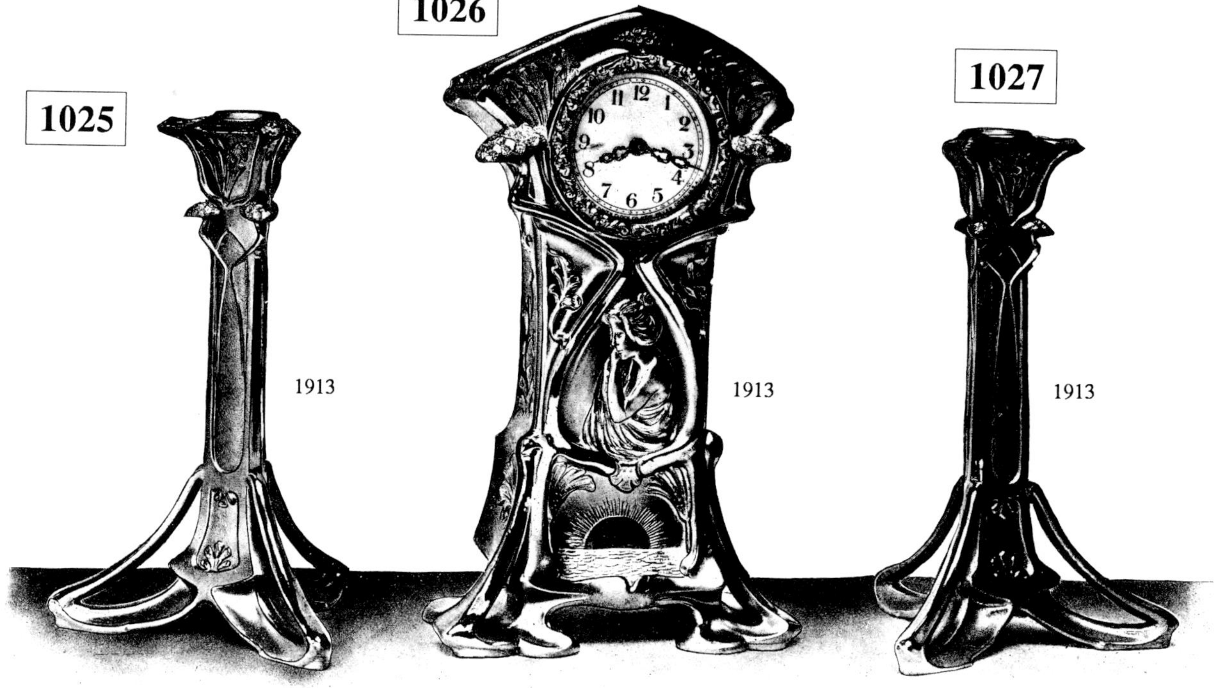

CLOCK. LIST, $11.00
HEIGHT, 10¾ INCHES. WIDTH, 6¼ INCHES.

ANNECY SET
COMPLETE, LIST, $17.50

CANDLESTICKS. PER PAIR, LIST, $6.50
HEIGHT, 9 INCHES. WIDTH, 5¾ INCHES.

1028 **1029** **1030**

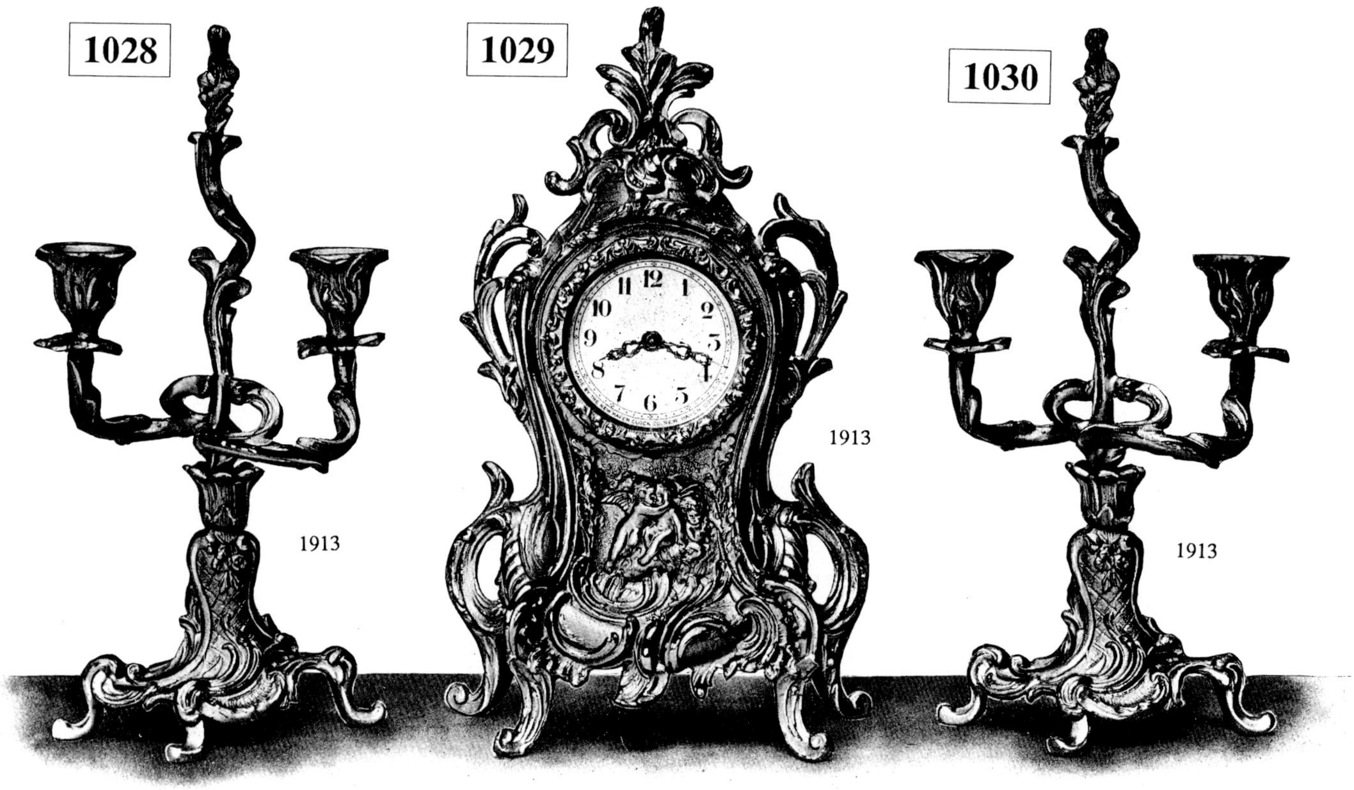

CLOCK. LIST, $10.90
HEIGHT, 11 INCHES. WIDTH, 6½ INCHES.

VERGENNES SET
COMPLETE, LIST, $18.15

CANDELABRA. PER PAIR, LIST, $7.25
HEIGHT, 10½ INCHES. WIDTH, 5½ INCHES.

MANTEL CLOCKS - METAL CASES

1031

COROLAN

French Satin Gold or Bronze Finish.
Height, 8½ inches. Width, 5 inches.
One or Eight-day Time with Two-inch Ivorized Porcelain Dial.

List price, One-day Time, Satin Gold, $ 9.75
" " One-day Time, Bronze, 8.80
" " Eight-day Time, Satin Gold, 11.25
" " Eight-day Time, Bronze, 10.30

Fitted to order only, with Jeweled Sash in Brilliant, Ruby, Emerald, Amethyst, Sapphire, Pearl and Turquoise or alternate stones. Extra list, $2.10

1900

COROLAN

1032 **1033** **1034**

1913 1913 1913

FALAISE SET

CLOCK. LIST, $61.65
HEIGHT, 28 INCHES. WIDTH, 15¾ INCHES.

COMPLETE, LIST, $123.30

CANDELABRA. LIST, PER PAIR, $61.65
HEIGHT, 28½ INCHES. WIDTH, 12½ INCHES.

MANTEL CLOCKS - METAL CASES

CRITERION SET

Best French Satin Gold Finish.

Height of Clock, 14¾ inches.
Width, 6½ inches.

Height of Candelabra, 9¾ inches.

Eight-day Time with Three-inch, Ivorized and Decorated Porcelain Dial.

List price for set, complete, $28.50

Fitted to order only with Jeweled Sash in Brilliant, Ruby, Emerald, Amethyst, Pearl and Turquoise, Sapphire, or alternate stones at extra list of $2.10

1035

1900

1036

1900

CRITERION

1037

1900

1038

1900

1039

1900

LAKESIDE SET

1040

1900

HEIGHT, CLOCK....................18⅝ INCHES
HEIGHT, CANDELABRUM........19¼ INCHES
WIDTH, CLOCK.......................10⅝ INCHES
WIDTH, CANDELABRUM............7¾ INCHES
DIAL, PORCELAIN........................3 INCHES

FINISH, ORMOLU GOLD PLATE, DECORATED PORCELAIN PANELS
MOVEMENT, 8 DAY HALF-HOUR STRIKE, CATHEDRAL GONG
PRICE, CLOCK.......................................LIST, $32.00
PRICE, CANDELABRA, PER PAIR............LIST, 32.00
PRICE, SET ....LIST, 64.00

1911

MANTEL CLOCKS - METAL CASES

1041

1911

1042

1911

1043

1911

LENORE SET

HEIGHT, CLOCK ....15 INCHES
HEIGHT, CANDELABRUM........16½ INCHES
WIDTH, CLOCK ....9 INCHES
WIDTH, CANDELABRUM ....7¾ INCHES
DIAL, PORCELAIN ....3 INCHES

FINISH, ORMOLU GOLD PLATE, DECORATED PORCELAIN PANELS
MOVEMENT, 8 DAY HALF-HOUR STRIKE, CATHEDRAL GONG
PRICE, CLOCK ....LIST, $24.00
PRICE, CANDELABRA, PER PAIR...........LIST, 24.00
PRICE, SET ....LIST, 48.00

1044

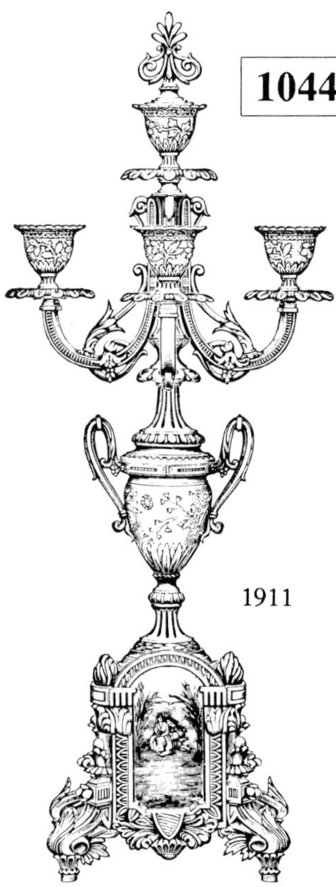

1911

1045

1911

1046

1911

LYON SET

FINISH, ORMOLU GOLD PLATE, DECORATED PORCELAIN PANELS
MOVEMENT, 8 DAY HALF-HOUR STRIKE, CATHEDRAL GONG
PRICE, CLOCK.......................................LIST, $25.40
PRICE, CANDELABRA, PER PAIR............LIST, 25.40
PRICE, SET...LIST, 50.80

HEIGHT, CLOCK.....................16½ INCHES
HEIGHT, CANDELABRUM........19¼ INCHES
WIDTH, CLOCK..........................9 INCHES
WIDTH, CANDELABRUM...........7¼ INCHES
DIAL, PORCELAIN.........................3 INCHES

MANTEL CLOCKS - METAL CASES

1047

1911

1048

1911

ARNOLD SET

1049

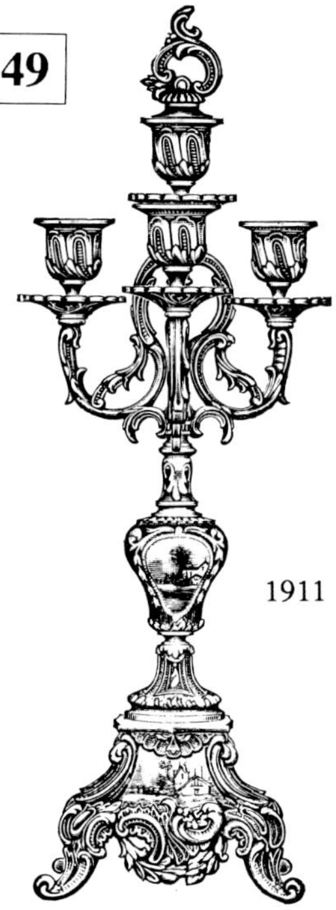

1911

HEIGHT, CLOCK ....13 7/8 INCHES
HEIGHT, CANDELABRUM ....16½ INCHES
WIDTH, CLOCK ....9 INCHES
WIDTH, CANDELABRUM ....6½ INCHES
DIAL, PORCELAIN ....3 INCHES

FINISH, ORMOLU GOLD PLATE, DECORATED PORCELAIN PANELS
MOVEMENT, 8 DAY HALF-HOUR STRIKE, CATHEDRAL GONG
PRICE, CLOCK..LIST, $21.75
PRICE, CANDELABRA, PER PAIR.............LIST, 21.75
PRICE, SET ....LIST, 43.50

1050

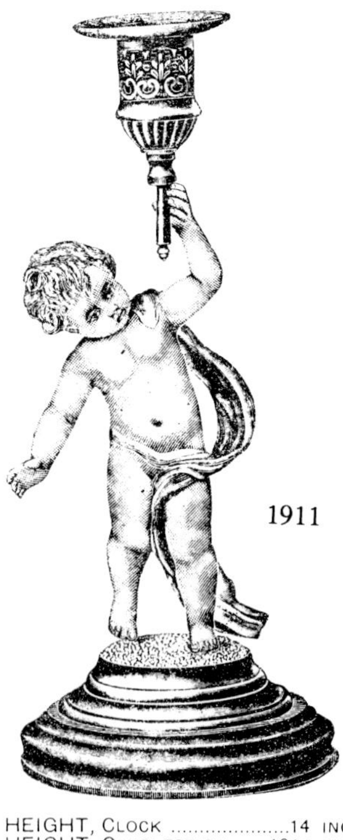

1911

1051

1911

CUPID SET

1052

1911

HEIGHT, CLOCK ....14 INCHES
HEIGHT, CANDLESTICK ......10¼ INCHES
WIDTH, CLOCK ....6 INCHES
WIDTH, CANDLESTICK ....4¾ INCHES
DIAL, PORCELAIN ....2 INCHES

FINISH, ORMOLU GOLD PLATE
MOVEMENT, 1 DAY TIME PIECE
PRICE, CLOCK ....LIST, $7.25
PRICE, CANDLESTICKS, PER PAIR ....LIST, 5.15
PRICE, SET ....LIST, 12.40

MANTEL CLOCKS - METAL CASES

1053

1911

1054

1911

1055

1911

PAVIA SET

DIAL, PORCELAIN.......................2 INCHES
FINISH, ORMOLU GOLD PLATE

HEIGHT, CLOCK ....10¼ INCHES
HEIGHT, CANDLESTICK ....8½ INCHES
WIDTH, CLOCK ....5¼ INCHES
WIDTH, CANDLESTICK ....3¼ INCHES

MOVEMENT, 1 DAY TIME PIECE
PRICE, CLOCK ....LIST. $8.75
PRICE, CANDLESTICKS, PER PAIR ....LIST. 4.40
PRICE, SET ....LIST. 13.15

1056

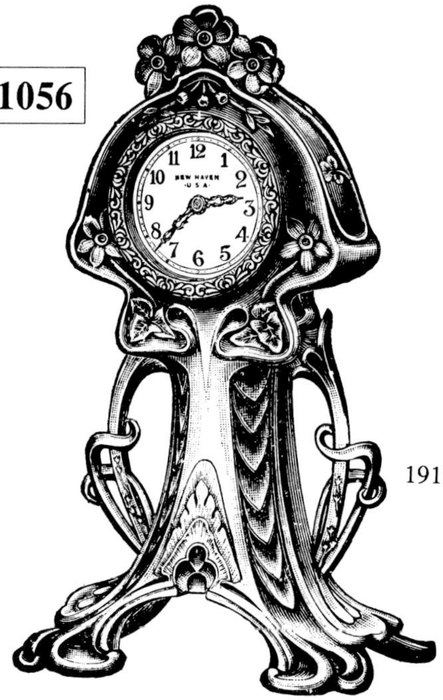

1911

FURNESS

HEIGHT ....11½ INCHES
WIDTH ....7 INCHES
DIAL, PORCELAIN ......3 INCHES
FINISH, ORMOLU GOLD PLATE
MOVEMENT, FINE QUICK-BEAT 8 DAY TIME PIECE
PRICE ....LIST, $10.15

1057

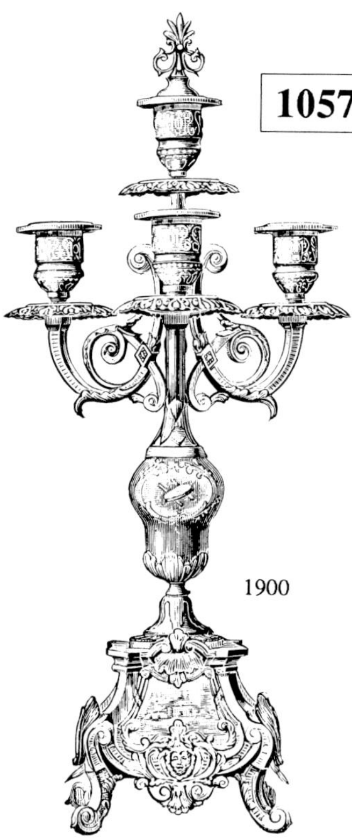

1900

1058

1900

1059

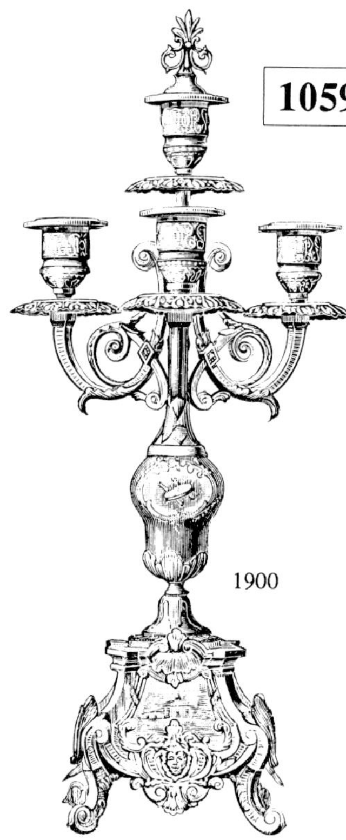

1900

SICILIAN SET

Fitted with Eight-day, Half-hour Strike, Cathedral Gong. Three-inch Decorated Porcelain Dial.
Best French Satin Gold Finish with Decorated Porcelain Panels.
Height of Clock, 16½ inches. Width, 8 inches. Height of Candelabra, 18 inches.
List price for set complete, $50.00

MANTEL CLOCKS - METAL CASES

ROCHELLE SET

French Satin Gold or Bronze Finish.

Height (Clock), 11 inches.
Height (Candlesticks), 8¾ inches.
Width, 6½ inches.

One or Eight-day Time Piece with Two-inch Ivorized Porcelain Dial.

FOR CLOCK ONLY.

	List price.
One-day Time, Satin Gold,	$ 9.00
One-day Time, Bronze,	8.10
Eight-day Time, Satin Gold,	10.50
Eight-day Time, Bronze,	9.60

	List price.
Candlesticks, Satin Gold, .	$ 4.60
Candlesticks, Bronze, .	4.15

Fitted with Jeweled Sash in Brilliant, Ruby, Emerald, Amethyst, Pearl and Turquoise, Sapphire or alternate stones. Extra list, . $ 2.10

1060

1900

1061

1900

1062

1900

ROCHELLE

List price, Complete Set, One-day Time, Satin Gold, $13.60
" " Complete Set, One-day Time, Bronze, 12.25
" " Complete Set, Eight-day Time, Satin Gold, 15.10
" " Complete Set, Eight-day Time, Bronze, 13.75

1063

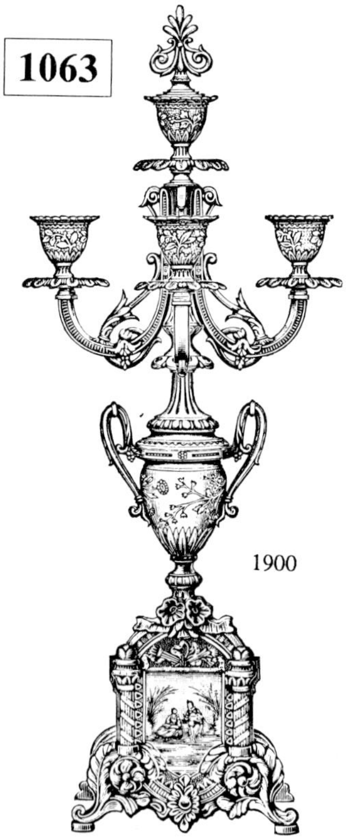

1900

1064

1900

1065

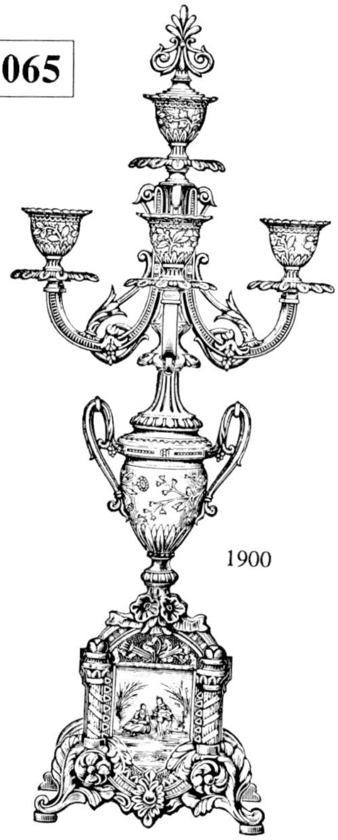

1900

LARCHMONT SET

Height of Clock, 16 inches. Width, 9 inches. Height of Candelabra, 18¾ inches.
Three-inch Decorated Porcelain Dial. Best French Satin Gold. Decorated Porcelain Panels.
Fitted with Eight-day, Half-hour Strike, Cathedral Gong.
List price for set, complete, $62.50

MANTEL CLOCKS - METAL CASES

1066

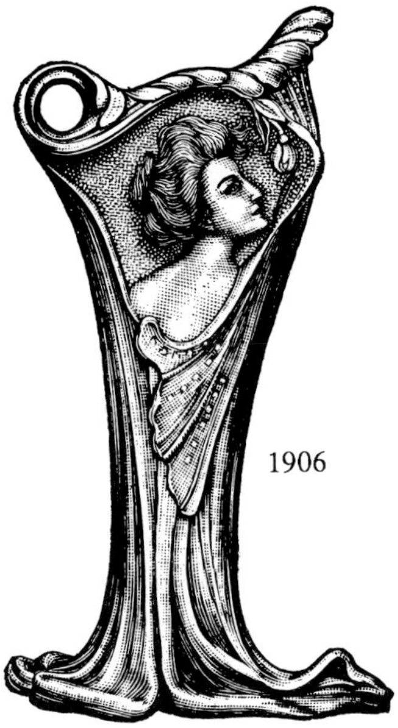

1906

1067

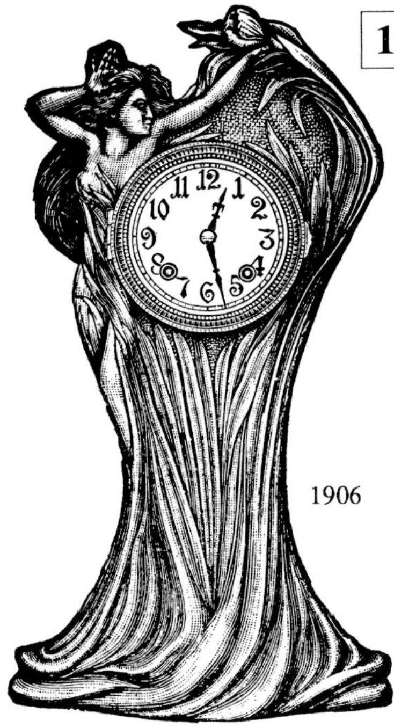

1906

1068

1906

LIBITINA SET

Clock, Height, 17 ¼ inches. Width, 9 ¼ inches. Vases, Height, 14 ¾ inches.
4-inch Porcelain Dial, with beveled Glass. Ormolu Gold Plated Case.
8 Day, Half-hour Strike, Cathedral Gong. Clock List, $26.50. Vases, per pair, List, $29.25. Set Complete, List, $55.75

1069

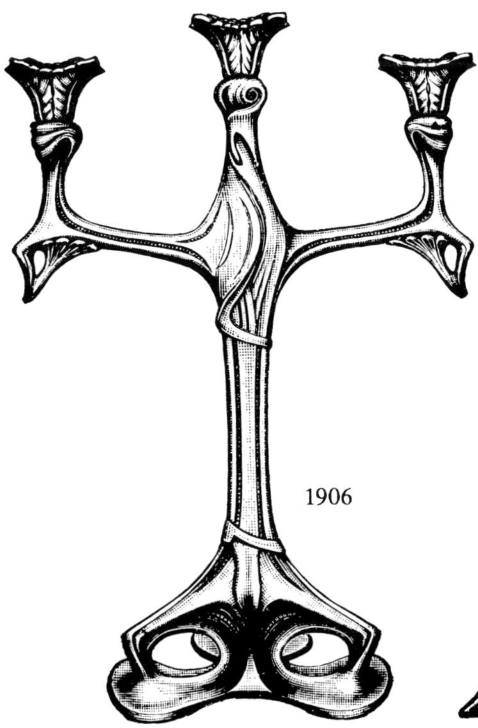

1906

1070

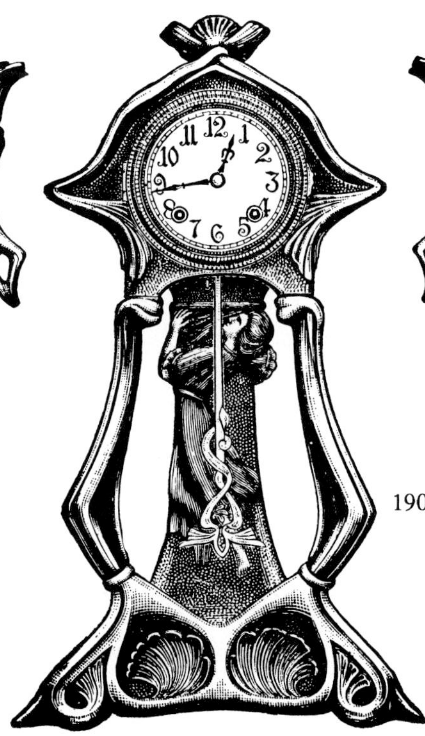

1906

1071

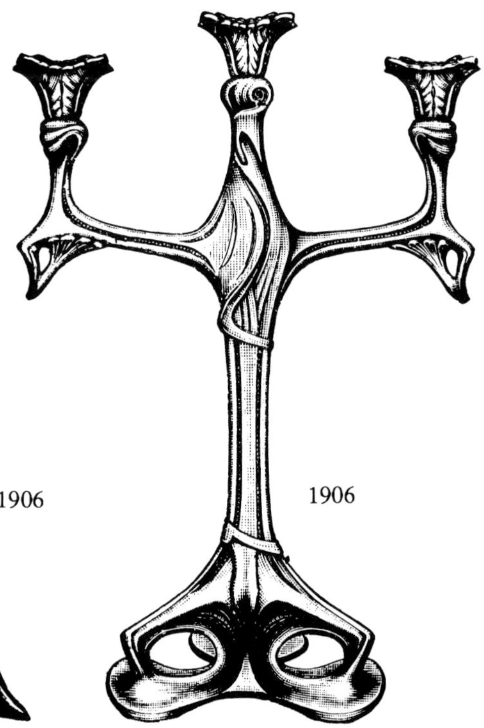

1906

ISERE SET

Clock, Height, 18 ¼ inches. Width, 19 ¼ inches. Candelabra, Height, 18 ¾ inches.
4-inch Porcelain Dial, with beveled Glass. Ormolu Gold Plated Case. Visible Pendulum.
8 Day, Half-hour Strike, Cathedral Gong. Clock List, $29.25. Candelabra, per pair, List, $29.25. Set Complete, List, $58.50

MANTEL CLOCKS - METAL CASES

1072

1880

ANTIQUE BRASS, No. 2

Height, 13½ inches. Dial, 4 inches.
8 Day Plain Dial, Striking on Gong.
8 Day Visible Escapement, Striking on Gong.

1073

ANTIQUE BRASS, No. 2
{William Mittelstadt Collection}

1074 **1075** **1076**

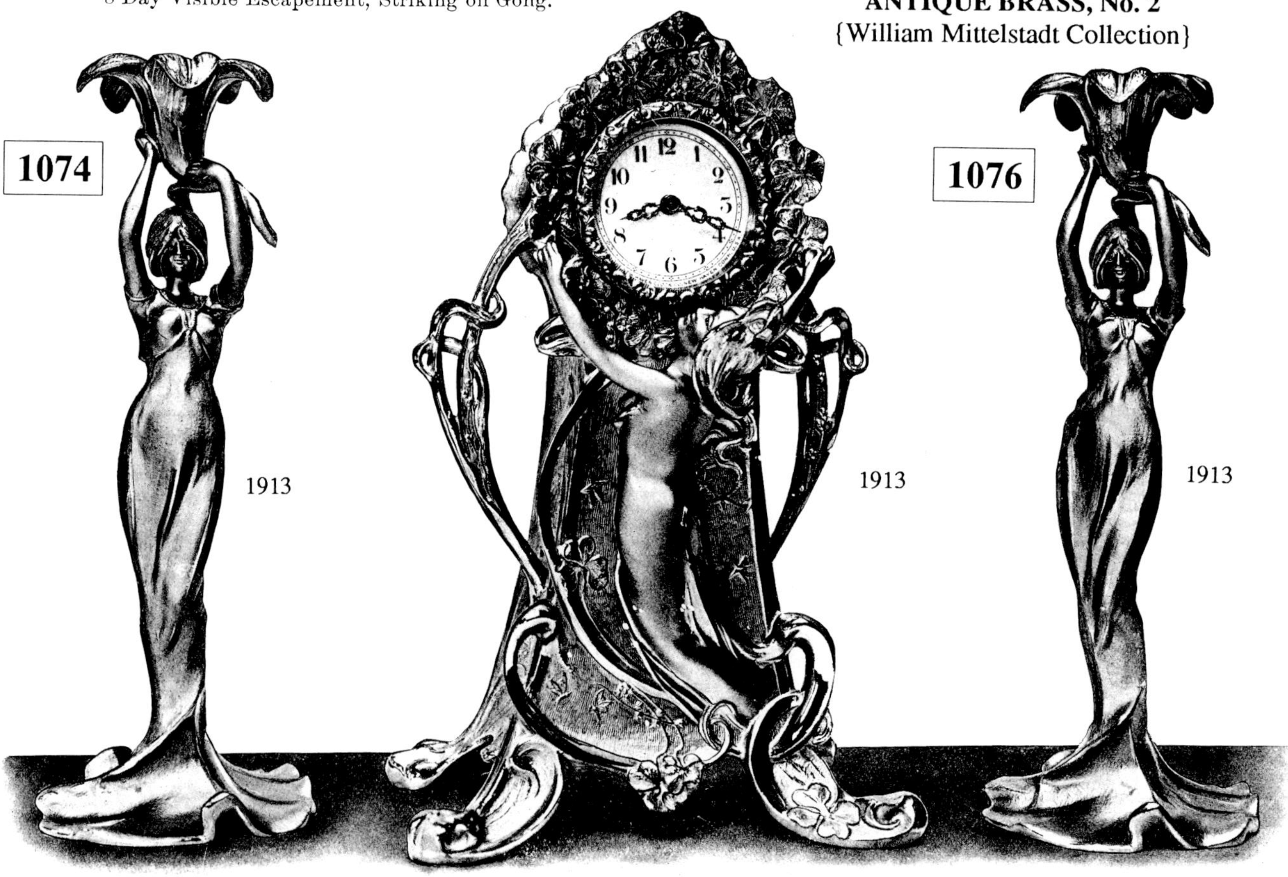

1913 1913 1913

CLOCK. LIST, $18.15
HEIGHT, 12 INCHES. WIDTH, 8 INCHES.

TROUVILLE SET
COMPLETE, LIST, $29.75

CANDLESTICKS. PER PAIR, LIST, $11.60
HEIGHT, 12 INCHES. WIDTH, 4¾ INCHES.

MANTEL CLOCKS - METAL CASES

EIGHT-DAY HALF-HOUR STRIKE, CATHEDRAL GONG, CLOCK SETS.
PORCELAIN DIALS. ORMOLU GOLD PLATE FINISH. BEVELED GLASS

1077

1078

1079

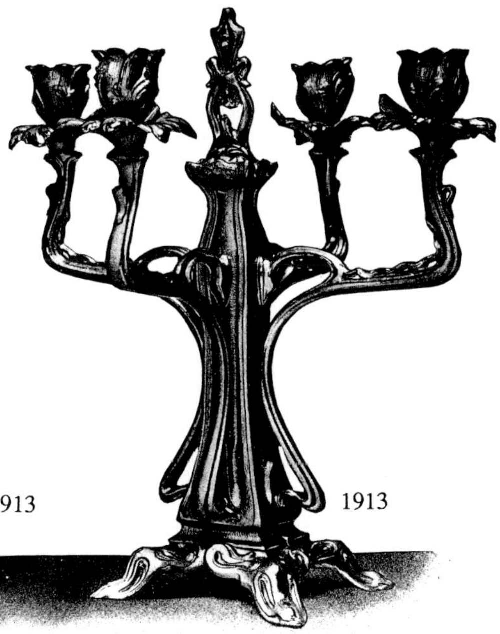

CLOCK. LIST, $18.85
HEIGHT, 15 INCHES. WIDTH, 7¾ INCHES.
DIAL, PORCELAIN 3½ INCHES.

TRUXTON SET
COMPLETE, LIST, $48.00

CANDELABRA. PER PAIR, LIST, $29.15
HEIGHT, 13 INCHES. WIDTH, 11 INCHES.

1080

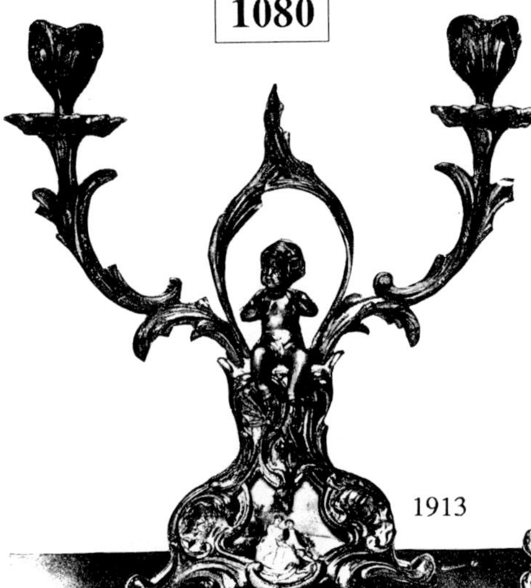

1081

1082

CLOCK. LIST, $25.40
HEIGHT, 15½ INCHES. WIDTH, 10⅜ INCHES.
DIAL, PORCELAIN, 4½ INCHES.

ROSEMONT SET
COMPLETE, LIST, $50.80

CANDELABRA. PER PAIR, LIST, $25.40
HEIGHT, 11 INCHES. WIDTH, 10 INCHES.
DECORATED PORCELAIN PANELS.

MANTEL CLOCKS - METAL CASES

1083

1880

ANTIQUE BRASS, No. 1

Height, 10 inches. Dial, 3 inches.
1 Day Time, Alarm. 1 Day Strike, Alarm.
1 Day ½ Hour Strike.

1084

1895

SPARTAN.
Antique Brass or Barbedienne, Ivorized Dial.

Height to top of Figure, 24 inches; Dial, 5 inches.
8 Day, Half-Hour Strike, Cathedral Gong, - List, $70.00

1085

UNO

LIST, $10.15
HEIGHT, 6½ INCHES. WIDTH, 6 INCHES.
DIAL, 3½ INCHES.
FINISHES, FRENCH BRONZE, OLD BRASS,
GILT, SILVER OR GUN METAL.

1086

ANTIQUE BRASS

Height, 15 ½ inches. Width, 8 inches.
Two-inch Cartouche Dial. Ormolu Gold
Plated Case. Two Springs, Eight Day Time.
{William Mittelstadt Collection}

ANTIQUE BRASS

MANTEL CLOCKS - METAL CASES

1087

1885

No. 1009.
CAST BRASS, HAND FINISHED.

No. 1009.
CAST BRASS, HAND FINISHED.
Height, 13 inches.
Dial, 4 inches.
8 Day Time.........$23 50

1088

1906

LIBOURNE
Height, 10½ inches. Width, 5¼ inches. Three-inch Porcelain Dial. Ormolu Gold Plated Case. One Day Time. List price, $10.50.

1089

1906

ALLIER
Height, 11 ½ inches. Width, 6 inches. Two-inch White Dial. Ormolu Gold Plated Case. One Day Time. List price, $8.35

1090

1900

PLEVNA
Height, 11¾ inches. Width, 8½ inches.
Decorated Ivorized Porcelain Dial, 3 inches.
French Satin Gold Finish.

List price, One-day Time, $11.70
" " Eight-day Time, 13.20

Fitted with Jeweled Sash in Brilliant, Ruby, Emerald, Amethyst, Sapphire, Pearl and Turquoise, or alternate stones.
Extra list, $2.10

1091

1885

No. 1023.
CAST BRASS, HAND FINISHED.
Height, 13½ inches. Dial, 4 inches.

Code Word.	Price.
(Tempo) 8 Day Time......	$17 00

MANTEL CLOCKS - METAL CASES

1092

1917

FALMOUTH

Height, - - - - - 11½ Inches
Width, - - - - - 8½ Inches

1093

1889

HILDA.

Height, 12¼ inches. Dial, 4½ inches.

SOLID CAST BRASS FRAME, WITH RICHLY ORNAMENTED SILVER PANELS, HAND MODELED.

8 Day, Striking Hour and Half-hour on a Cathedral Gong. . $37.50

1094

1886

No. 1019.

SOLID CAST BRASS FRAME, WITH RICHLY ORNAMENTED SILVER PANELS, HAND MODELED.

Height, 9 inches. Breadth, 19¼ inches. Dial, 3¼ inches.

8 Day, Striking Hour and Half-hour on a Cathedral Gong................$55.00

MANTEL CLOCKS - METAL CASES

The year entered beside each clock is the year of the catalog from which the illustration was taken. Some clocks may have been produced a few years earlier.

1095

FLORES
Height, 15 inches. Width, 8 inches.
4-inch Porcelain Dial. Ormolu Gold Plated Case, Gilt Hands. 8 Day Half-hour Strike, Cathedral Gong, List, $18.10

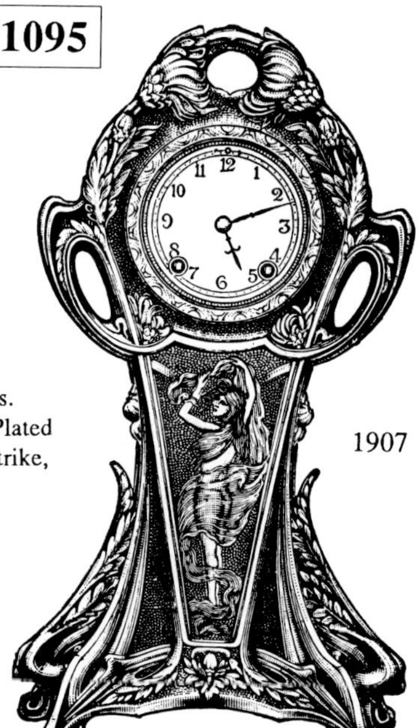

1907

FLORES

1096

1886

LAUREL.

Height, 27 inches. Dial, 8 inches.

FINISHED IN BRONZE AND SILVER AND GOLD.

15 Day Time, Bronze $22.50
15 " " Silver and Gold. .. 25.50

1097

1900

ORPHEUS

French Satin Gold Finish, with Black or Green Enameled Iron Base.

Height, 16 inches.
Width, 12½ inches.

Eight-day, Half-hour Strike, Cathedral Gong, Empire Sash and Heavy Beveled Glass, Four-inch Ivorized Porcelain Dial with Gilt Center, or Decorated Porcelain Dial.

List price, Satin Gold, $36.00
" " Bronze, 30.00

MANTEL CLOCKS - METAL CASES

1099

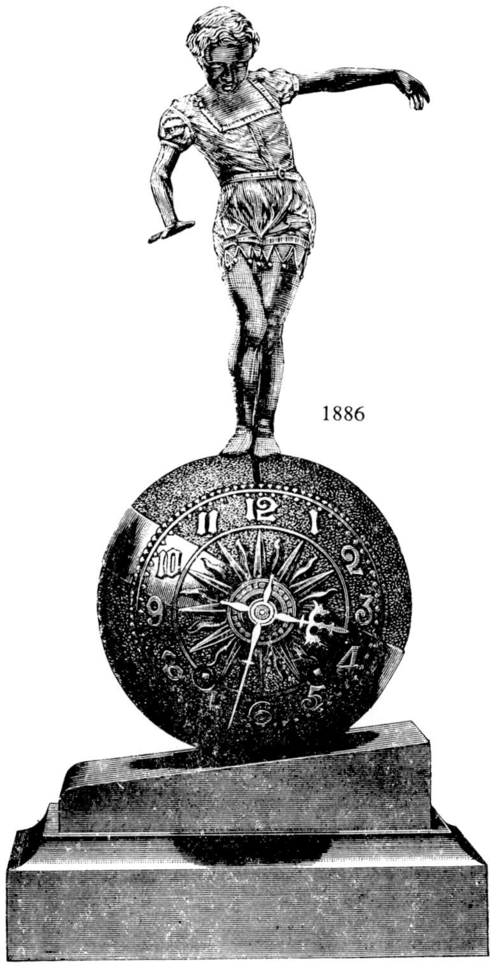

1886

ACROBAT.

CAST IN BRASS, BARBEDIENNE FINISH.

Height, 16¼ inches. Dial, 4 inches.

8 Day Bell Strike $45.00

1098

1886

No. 26.

BRASS OR NICKEL FINISH.

Height, 13½ inches. Dial, 5½ inches.

8 Day Gong Strike, Visible Escapement, French Sash, Porcelain Dial, with Figure ..$26.00
8 " " " " " " " " " without " 24.00

Also made with Gilt Sunk Center, $1.00 list extra.

MANTEL CLOCKS - METAL CASES

1100

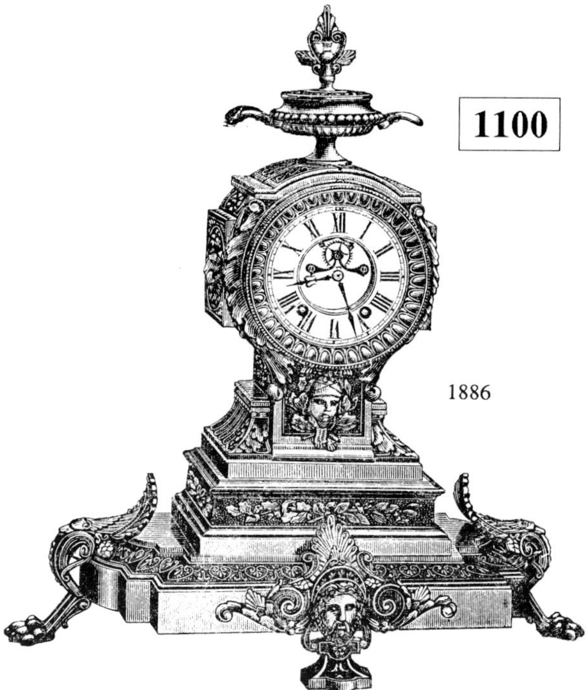

1886

LOUIS
Height, 20 inches. Width, 4 ½ inches.
8 Day Strike, Cathedral Gomg, Visible
Escapement, List, $30.00

1101

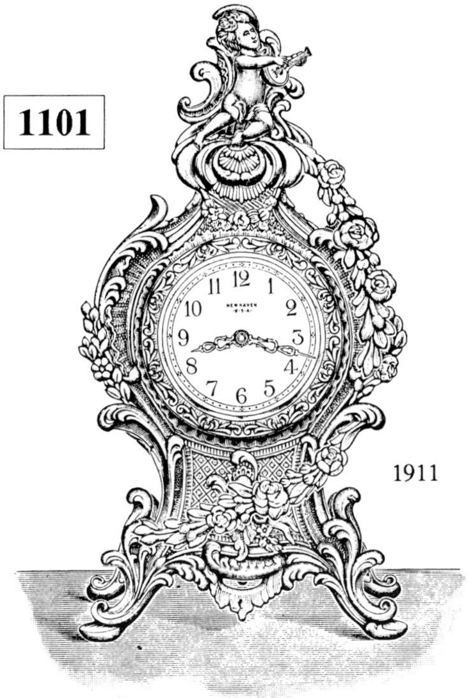

1911

HAMPDEN

HEIGHT ....10¼ INCHES
WIDTH ....5¼ INCHES
DIAL, PORCELAIN ......3 INCHES
FINISH, ORMOLU GOLD PLATE
MOVEMENT, FINE QUICK-BEAT 8 DAY TIME PIECE
PRICE ....LIST, $10.15

1102

1906

SERPENTINE
Height, 13 ¼ inches. Width, 5 ¾ inches.
3-inch Porcelain Dial. Ormolu Gold Plated
Case, Cast Sash. 8 Day Time, List, $13,90

1103

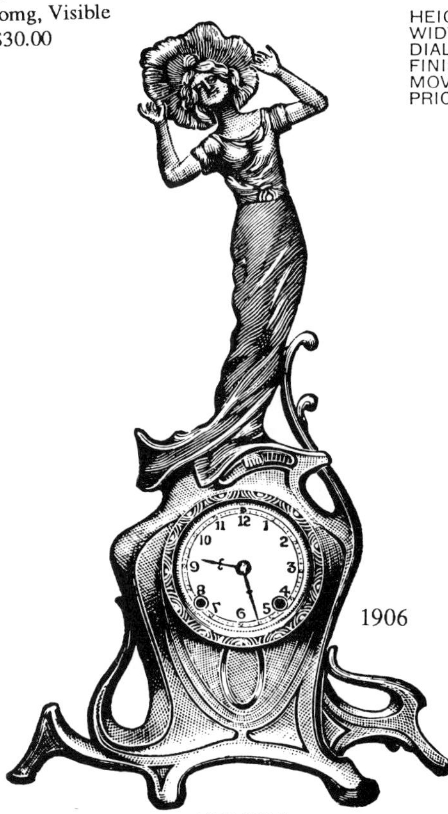

1906

AMANDA
Height, 19 ¼ inches. Width, 11 inches.
4-inch Porcelain Dial. Ormolu Gold Plated or
Antique Verde Case. 8 Day, Half-hour Strike,
Cathedral Gong, List, $21.50

1104

1906

ALTAIR
Height, 12 ½ inches. Width, 7 inches.
3-inch Porcelain Dial. Ormolu Gold Plated
Case, Cast Sash. Decorated Porcelain Panel.
8 Day Time, List, $15.30

MANTEL CLOCKS - MUSICAL CLOCKS

1105

1895

LOHENGRIN.

Musical Clock.

Height, 13⅝ inches; Width, 19¼ inches; Fancy Gilt or Plain White, 6 inch Dial.

8 Day, Half-Hour Strike, Cathedral Gong, Movement with Musical Attachment, List, $27

This clock contains, in addition to the features common to all of our Enameled Black Woods, a device, by which at the half-hour the strike movement, instead of delivering a solitary blow upon the same gong which sounds the hours, plays a tune upon the music box. The mechanism operating this box runs eight days with one winding. The music box furnishes two different tunes, playing one of them alternately at each half-hour upon this device. An application for a patent is now pending. The advantage of this musical attachment in helping the wakeful to tell time in the night season, especially between midnight and 2 A. M. needs only to be suggested.

MANTEL CLOCKS - TAMBOUR CLOCKS

TAMBOURS

Solid Mahogany

Movements are Rack and Snail, 8-Day Half Hour Strike, Cathedral Gong. Plates and Wheels Polished

1106

1917

TAMBOUR No. 2
Height, 11½ Inches Width, 24 Inches
Dial, Silver, 7½ Inches
Cast Sash, Beveled Glass

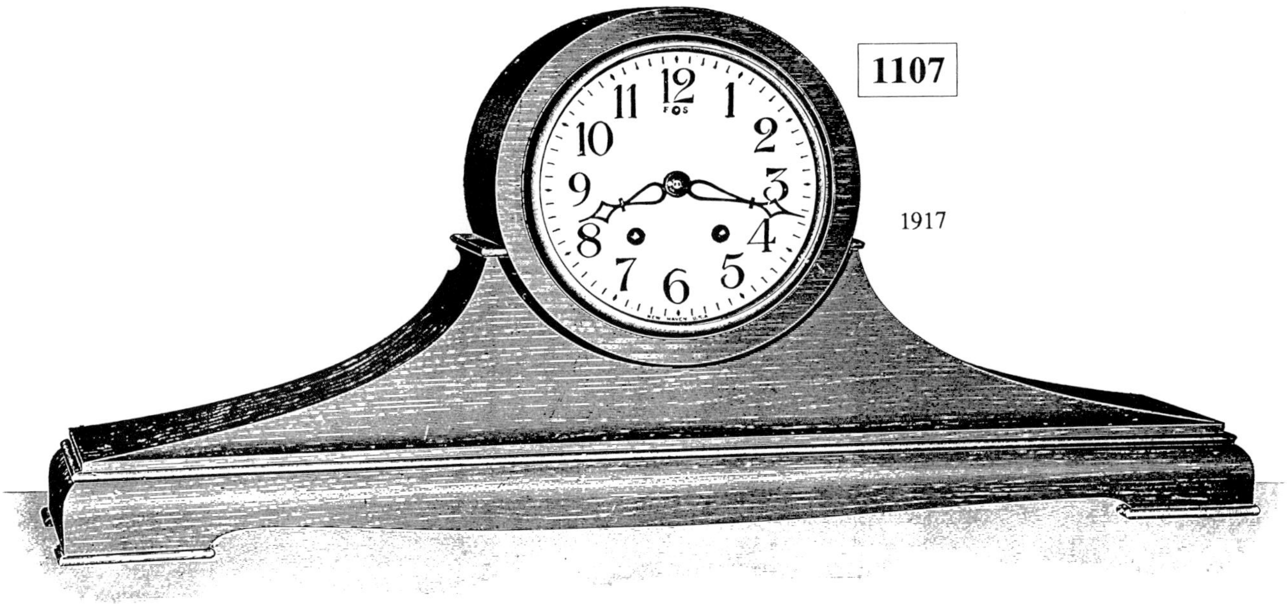

1107

1917

TAMBOUR No. 15
Height, 12⅝ Inches Width, 29½ Inches
Dial, Silver, 7½ Inches
Cast Sash, Beveled Glass

Above clocks furnished in Mahogany, Brown Mahogany, Circassian Walnut Finish, Golden and Fumed Oak, at same list

MANTEL CLOCKS - TAMBOUR CLOCKS

1108

1917

TAMBOUR No. 9
Height, 10½ Inches Width, 19¼ Inches
Dial, Silver, 7½ Inches
Cast Sash, Beveled Glass

1109

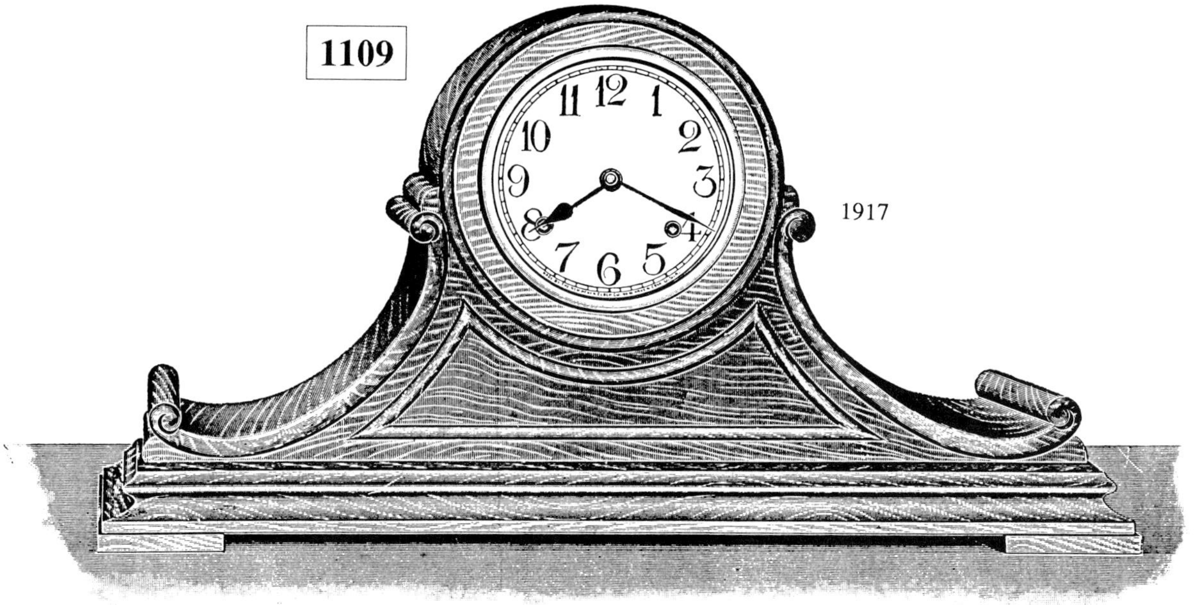

1917

TAMBOUR No. 1
Height, 10¾ Inches Width, 21½ Inches
Dial, Porcelain, 7 Inches
Cast Sash, Beveled Bowed Glass
Also fitted with Silver Dial at Extra List

Movements are Rack and Snail, 8-Day Half Hour Strike, Cathedral Gong. Plates and Wheels Polished

MANTEL CLOCKS - TAMBOUR CLOCKS

1110

TAMBOUR No. 7
Height, 9⅞ Inches Width, 19 Inches
Dial, Porcelain, 5 Inches
Cast Sash, Beveled Bowed Glass

1917

1111

TAMBOUR No. 3
Height, 8 Inches Width, 15½ Inches
Dial, Porcelain, 5 Inches
Cast Sash, Beveled Bowed Glass

1917

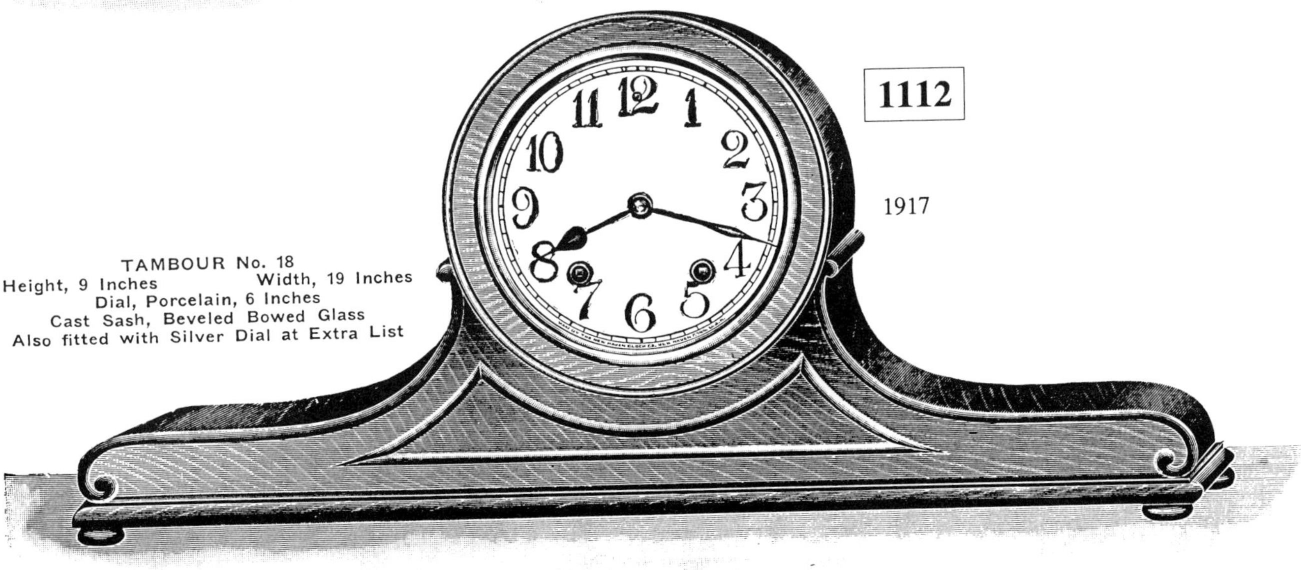

1112

TAMBOUR No. 18
Height, 9 Inches Width, 19 Inches
Dial, Porcelain, 6 Inches
Cast Sash, Beveled Bowed Glass
Also fitted with Silver Dial at Extra List

1917

MANTEL CLOCKS - TAMBOUR CLOCKS

1113

1917

1114

1917

TAMBOUR No. 17
8-Day Half Hour Strike
Cathedral Gong
Height, $9\frac{5}{8}$ Inches Width, $15\frac{7}{8}$ Inches
6 Inch Metal Dial, Bowed Glass
Mahogany Finish
Fitted with Porcelain Dial at Extra List

TAMBOUR No. 14
Height, 9 Inches Width, $19\frac{1}{2}$ Inches
Dial, Porcelain, 6 Inches
Cast Sash, Beveled Bowed Glass
Also fitted with Silver Dial at Extra List

1115

TAMBOUR No. 6
Height, $11\frac{7}{8}$ Inches Width, $24\frac{5}{8}$ Inches
Dial, Porcelain, 7 Inches
Cast Sash, Beveled Bowed Glass
Also fitted with Silver Dial at Extra List

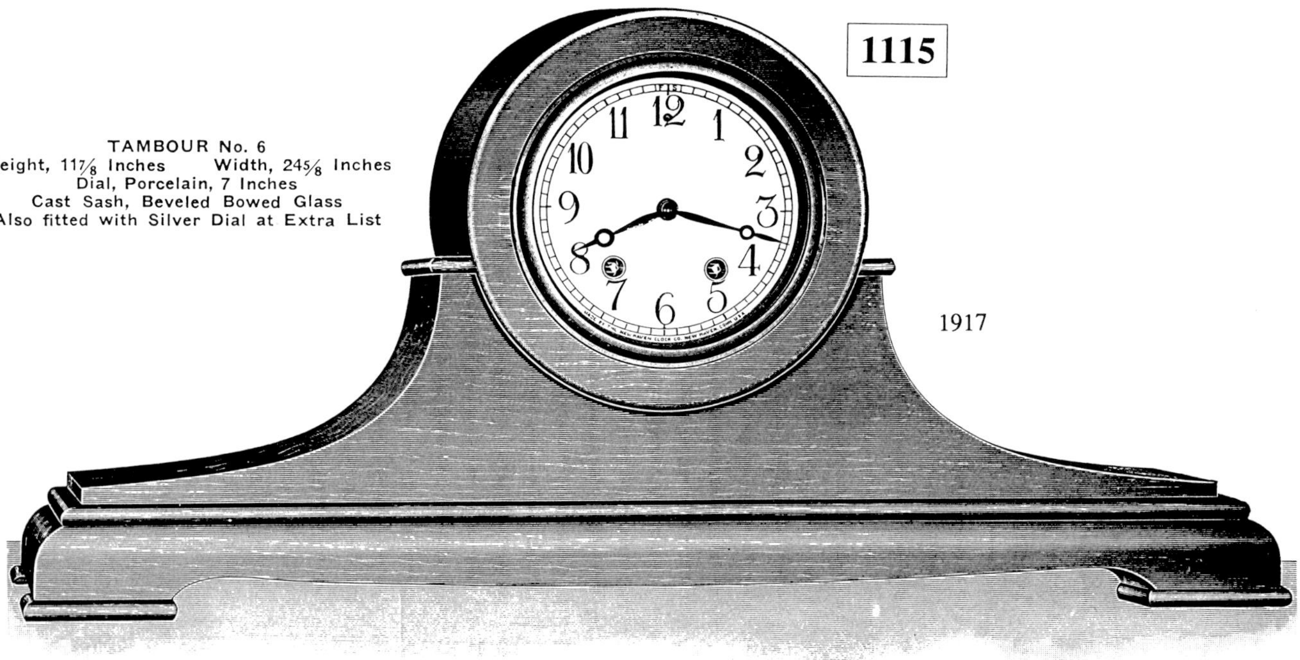

1917

MANTEL CLOCKS - TAMBOUR CLOCKS

The year entered beside each clock is the year of the catalog from which the illustration was taken. Some clocks may have been produced a few years earlier.

1116

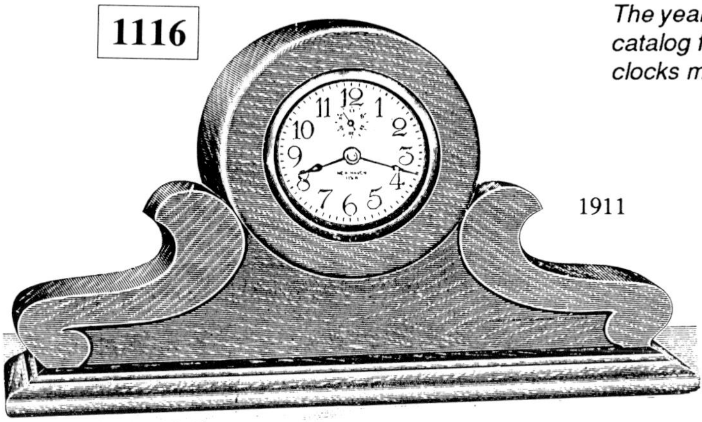

1911

LUZON

HEIGHT ....6 3/8 INCHES
WIDTH ....11 5/8 INCHES
DIAL, PORCELAIN ....3 INCHES
CASE, SOLID MAHOGANY
MOVEMENT, 8 DAY TIME PIECE
PRICE ....LIST, $8.75

1117

1917

QUEDA

Height, - - - - - - - 6½ Inches
Width, - - - - - - - 11¾ Inches

SOLID MAHOGANY

8-Day Lever Time Movements.

3½-inch Porcelain Dials. Beveled Glass

1118

1917

TAMBOUR No. 16
8-Day Half Hour Strike
Cathedral Gong
Height, 9 Inches Width, 18 Inches
6 Inch Metal Dial, Bowed Glass
Mahogany Finish, or Fumed or Golden Oak
Fitted with Porcelain Dial at Extra List

1119

1911

LINTON

HEIGHT ....7 INCHES
WIDTH ....12 5/8 INCHES
DIAL, PORCELAIN ....3 INCHES
CASE, SOLID MAHOGANY
MOVEMENT, 8 DAY TIME PIECE
PRICE ....LIST, $8.75

MANTEL CLOCKS - TAMBOUR CLOCKS

1120

1911

LANSING

HEIGHT ....6½ INCHES
WIDTH ....11½ INCHES
DIAL, PORCELAIN ....3 INCHES
CASE, SOLID MAHOGANY
MOVEMENT, 8 DAY TIME PIECE
PRICE ....LIST, $8.00

1121

1917

TAMBOUR No. 11
8-Day Half Hour Strike
Cathedral Gong
Height, 10¼ Inches Width, 20½ Inches
6 Inch Metal Dial, Bowed Glass
Mahogany Finish
Fitted with Porcelain Dial at Extra List

1122

1911

LUCIFER

HEIGHT ....5¼ INCHES
WIDTH ....10 INCHES
DIAL, PORCELAIN ....3 INCHES
CASE, SOLID MAHOGANY
MOVEMENT, 8 DAY TIME PIECE
PRICE ....LIST, $7.25

1123

1917

TAMBOUR No. 13
8-Day Half Hour Strike
Cathedral Gong
Height, 10 Inches Width, 22 Inches
6 Inch Metal Dial, Bowed Glass
Mahogany Finish
Fitted with Porcelain Dial at Extra List

MANTEL CLOCKS - TAMBOUR CLOCKS

1124

1913

EIGHT-DAY, HALF-HOUR STRIKE, CATHEDRAL GONG.
MOVEMENT, RACK-AND-SNAIL,
HEAVY BRASS AND HIGHLY POLISHED PLATES AND WHEELS.
PORCELAIN DIAL, FRENCH CAST SASH, BEVELED BOWED GLASS.
SOLID MAHOGANY CASES

QUINBY
LIST, $13.00
HEIGHT, 7¼ INCHES. WIDTH, 13¼ INCHES.

1125

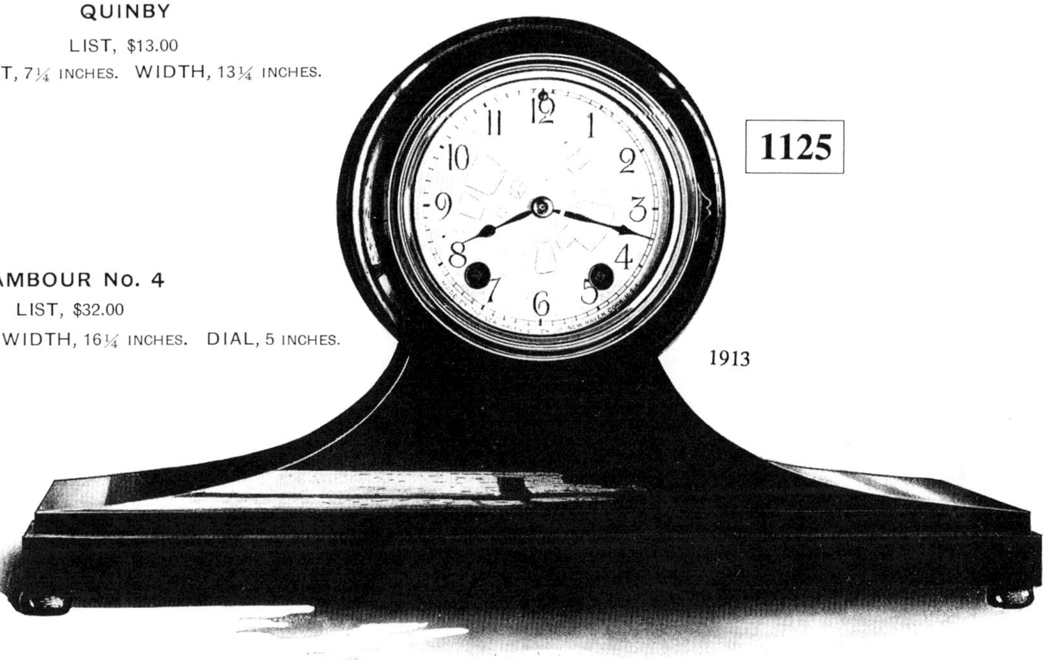

1913

TAMBOUR No. 4
LIST, $32.00
HEIGHT, 9½ INCHES. WIDTH, 16¼ INCHES. DIAL, 5 INCHES.

1126

1913

TAMBOUR No. 5
LIST, $34.75
HEIGHT, 10¾ INCHES. WIDTH, 19¼ INCHES. DIAL, 6 INCHES.

MANTEL CLOCKS - TAMBOUR CLOCKS

1127

1913

EIGHT-DAY HALF-HOUR STRIKE, CATHEDRAL GONG.
MOVEMENT, RACK-AND-SNAIL,
HEAVY BRASS AND HIGHLY POLISHED PLATES AND WHEELS.
PORCELAIN DIAL, FRENCH CAST SASH, BEVELED BOWED GLASS.
SOLID MAHOGANY CASES

QUEBEC
LIST, $10.15
HEIGHT, 6½ INCHES. WIDTH, 12¾ INCHES.

1128

1913

QUESTA
LIST, $10.90
HEIGHT, 6⅜ INCHES. WIDTH, 12 INCHES.

1129

1913

QUIRK
LIST, $10.90
HEIGHT, 7¼ INCHES. WIDTH, 14 INCHES.

1130

1913

TAMBOUR No. 8
LIST, $27.50
HEIGHT, 8¼ INCHES WIDTH, 16⅜ INCHES. DIAL, 4 INCHES.

1131

1913

QUINCY
LIST, $10.90
HEIGHT, 5½ INCHES. WIDTH, 10½ INCHES.

MANTEL CLOCKS - TAMBOUR CLOCKS

THESE CLOCKS ARE FITTED WITH ARABIC DIALS UNLESS ROMAN IS SPECIFIED.

FITTED 1-DAY TIMEPIECE, LIST, $1.50 LESS.
FITTED 1-DAY INTERMITTENT ALARM, LIST, SAME AS 8-DAY TIMEPIECE.

1132

1913

HANFORD
LIST, $8.00.
HEIGHT, 4¾ INCHES. WIDTH, 9¾ INCHES.

1133

1913

ROMERO
LIST, $8.00
HEIGHT, 4¾ INCHES. WIDTH, 9¾ INCHES.

1134

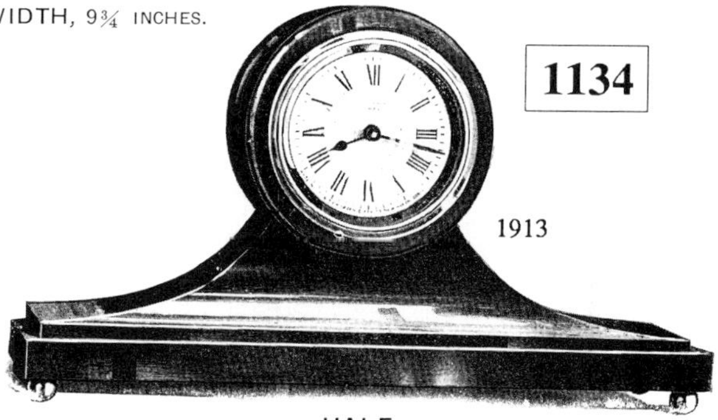

1913

HALE
LIST, $8.00
HEIGHT, 4¾ INCHES. WIDTH, 8⅜ INCHES.

1135

1913

HARPER
LIST $8.75
HEIGHT, 4½ INCHES. WIDTH 8 INCHES.

1136

1913

QUITO
LIST, $10.15
HEIGHT, 6 INCHES. WIDTH, 11¾ INCHES.

1137

1913

QUAYLE
LIST, $10.15
HEIGHT, 6½ INCHES. WIDTH, 12¾ INCHES.

1138

1913

RADWAY
LIST, $7.25
HEIGHT, 4½ INCHES. WIDTH 6½ INCHES.

MANTEL CLOCKS - TILE CASES

These clocks, manufactured by New Haven Clock Company with "Low Tile" cases are considered extremely rare. Clock collectors, American Pottery Enthusiasts as well as Art Potery Collectors all desire these fine clocks.

1139

1895

ILIUM

Height, 14 inches. Width, 11 inches.

OAK FRAME, WITH CHELSEA TILE PANELS.

8-Day, Half-hour Strike, Cathedral Gong, List, $15.00

1140

1885

No. 1017

Height, 12 inches. Breadth, 9 inches. Dial, 4 ¼ inches.

RICHLY MODELED, CAST BRASS FRAME WITH PANELS OF CHELSEA TILES.

Eight-Day Striking Hour and Half-hour on a Cathedral Gong........$55.00

1141

No. 1017

Eight-Day, striking hour and half-hour on Cathedral Gong with visible escapement. Cast brass frame with panels of Chelsea tiles.

{Private Collection}

1142

Back view of **No. 1017**

{Private Collection}

MANTEL CLOCKS - TILE CASES

1143

Rear view of **No. 1017** eight-day movement.
{Private Collection}

1144

Side view of **No. 1017**
{Private Collection}

1145

Tile inscription inside reads:

J & J. G. LOW
PATENT
ART TILE WORKS
CHELSEA
MASS. U.S.A.
COPYRIGHT 1884
by J. G. & J. E. LOW

See page 416 number 1694
for novelty Low Tile clock.

1146

Top view of **No. 1017**
{Private Collection}

MANTEL CLOCKS - WOOD CASES

1147

1880

ANDES.

Spring.

Height, 17 inches.
1 Day Strike.
8 Day Strike.
Also with Alarm.

1148

1889

Height, 17 inches. Dial, 6 inches.

ECLIPSE.

ROSEWOOD.

1 Day Strike................. $5.85
1 Day Strike, Alarm.......... 6.45
8 Day Strike................. 7.35
8 Day Strike, Alarm.......... 7.95

1149

1880

NO NAME.

Spring.

Height, 9¼ inches.

1 Day Time.

1 Day Time, Alarm.

1150

1880

RELIABLE.

Walnut.

Height, 9¼ inches.

1 Day Time.

1 Day Time, Alarm.

1151

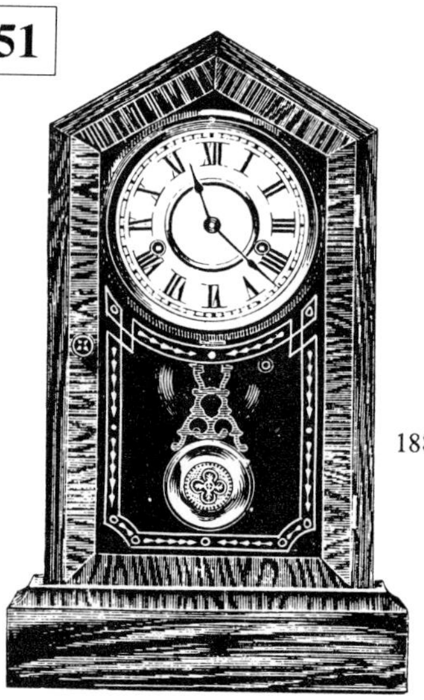

1880

PYRAMID.

Spring.

Height, 17 inches.
1 Day Strike.
8 Day Strike.
Also with Alarm.

1152

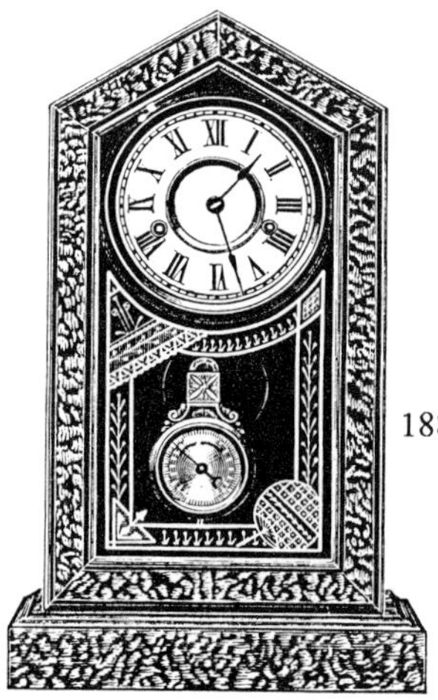

1885

MOTTLED TUSCAN.

MOTTLED ASH.

Height, 16 inches. Dial, 5 inches.

1 Day Strike. 8 Day Strike.

Also with Alarm.

Finished with Gilt Edges.

1153

1886

Height, 10¾ inches. Dial, 4 inches.

ROCKET.

CHESTNUT OR IMITATION ZEBRA.

1 Day Time............. $1.80
1 Day Time, Alarm...... 2.40

MANTEL CLOCKS - WOOD CASES

1154

1885

GEM COTTAGE.

WALNUT VENEERED AND SOLID CHESTNUT.

Height, 14 inches. Dial, 6 inches.

1 Day Strike	....	$3 50
1 " " Alarm	....	4 10
8 " "	....	5 00
8 " " Alarm	....	5 60

1155

1878

GOTHIC GEM, V. P.

1 Day. Strike. Spring.
8 " " "

Hight, 18 inches.

1156

1880

GOTHIC GEM.

Spring.

Height, 17¾ inches.
1 Day Strike.
8 Day Strike.
Also with Alarm.

1157

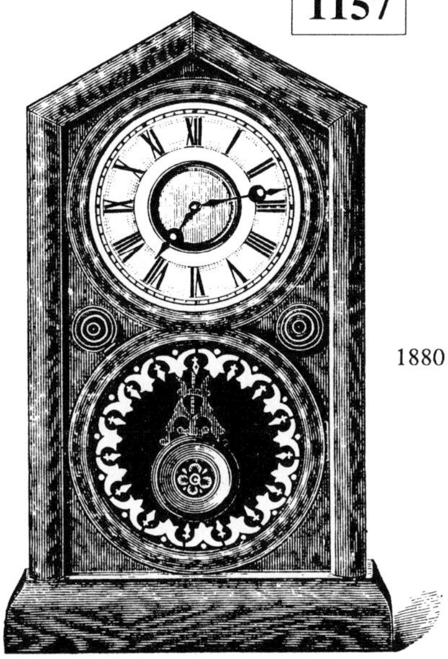

1880

GOTHIC GEM, D. R.

Spring.

Height, 18 inches.
1 Day Strike.
8 Day Strike.
Also with Alarm.

1158

1878

GOTHIC GEM.

1 Day. Strike. Spring.
8 " " "

Hight, 17 1-2 inches.

1159

1880

GOTHIC GEM, V. P.

Spring.

Height, 17¾ inches.
1 Day Strike.
8 Day Strike.
Also with Alarm.

MANTEL CLOCKS - WOOD CASES

1160

1880

TUSCAN.

Spring.

Height, $14\frac{1}{2}$ inches.

1 Day Time.

1 Day Time, Alarm.

1161

1880

SMALL TUSCAN, EXTRA.

Spring.

Height, 16 inches.

1 Day Strike.

8 Day Strike.

Also with Alarm.

1162

1880

TUSCAN.

Spring.

Height, $17\frac{3}{4}$ inches.

1 Day Strike.

8 Day Strike.

Also with Alarm.

1163

1880

TUSCAN, V. P.

Spring.

Height, $17\frac{3}{4}$ inches.

1 Day Strike.

8 Day Strike.

Also with Alarm.

1164

1880

TUSCAN, D. R.

Spring.

Height, 18 inches.

1 Day Strike.

8 Day Strike.

Also with Alarm.

Also made with Gilt Sash.

1165

1880

COUPON PREFERRED.

Spring.

Height, $15\frac{1}{2}$ inches.

1 Day Strike.

8 Day Strike.

Also with Alarm.

MANTEL CLOCKS - WOOD CASES

The year entered beside each clock is the year of the catalog from which the illustration was taken. Some clocks may have been produced a few years earlier.

1166

1886

Height, 13 inches. Dial, 5 inches.

OCTAGON ROCKET.

IMITATION ZEBRA.

1 Day Time.................... $1.60
1 " " Alarm............ 2.20

1167

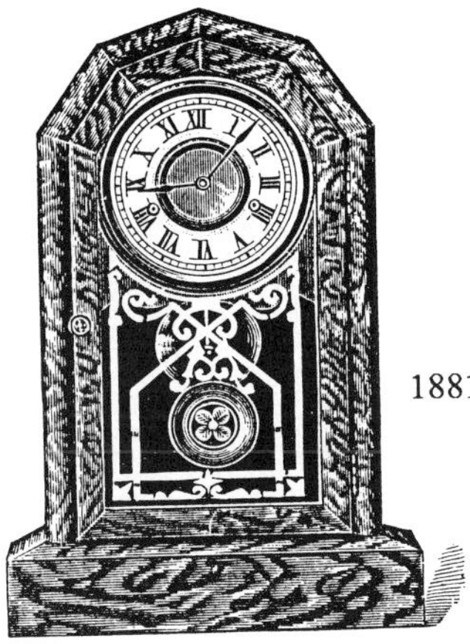

1881

ALPS.

WALNUT VENEER.

Height, 17 inches. Dial, 5 inches.
1 Day Strike. 8 Day Strike.
Also with Alarm.

1168

1886

Height, 17½ inches. Dial, 5 inches.

OCTAGON GEM, EXTRA.

ROSEWOOD.

1 Day Strike.................. $5.10
1 " " Alarm.............. 5.70
8 " " 6.60
8 " " Alarm.............. 7.20

1169

1886

Height, 17½ inches. Dial, 5 inches.

OCTAGON GEM.

ROSEWOOD.

1 Day Strike.................. $4.85
1 " " Alarm.............. 5.45
8 " " 6.35
8 " " Alarm.............. 6.95

1170

1886

Height, 17½ inches. Dial, 5 inches.

OCTAGON GEM, GILT.

ROSEWOOD.

1 Day Strike.................. $5.10
1 " " Alarm.............. 5.70
8 " " 6.60
8 " " Alarm.............. 7.20

1171

1886

Height, 17½ inches. Dial, 5 inches.

OCTAGON GEM, EXTRA GILT.

ROSEWOOD.

1 Day Strike.................. $5.35
1 " " Alarm.............. 5.95
8 " " 6.85
8 " " Alarm.............. 7.45

MANTEL CLOCKS - WOOD CASES

1172

DUCHESS

Flame Mahogany case.
Height, 18 inches. 1 Day. Strike. Spring.
{Jack Wyer Collection}

1173

1878

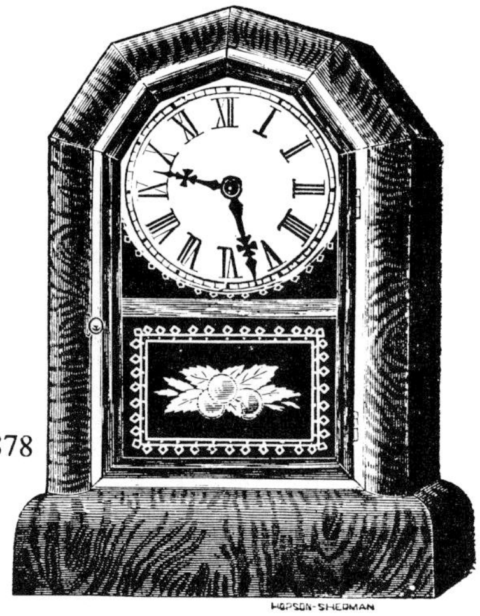

OCTAGON PRIZE

Height, 15¼ inches. 1 Day. Strike. Spring.
8 Day. Strike. Spring.

1174

1878

GEM

Height, 15 inches. 1 Day. Strike. Spring.
8 Day. Strike. Spring.

1175

Close-up of **Duchess** 30-Hour Time and Strike Movement.
{Jack Wyer Collection}

1176

OCTAGON PRIZE, V. P.

Height, 18 inches. 1 Day. Strike. Spring.
8 Day. Strike. Spring.

MANTEL CLOCKS - WOOD CASES

1177

1886

No. 254.

ROSEWOOD, ZEBRA, OR MAHOGANY.

Height, 14 inches. Dial, 5 inches.

1 Day Strike	$3.90
1 " " Alarm	4.50

1178

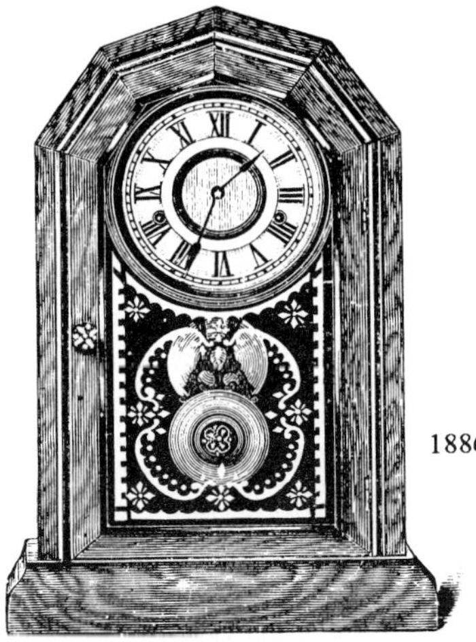

1886

Height, 16 inches. Dial, 5 inches.

SMALL OCTAGON PRIZE, EXTRA.

ROSEWOOD.

1 Day Strike	$4.65
1 " " Alarm	5.25
8 " "	6.15
8 " " Alarm	6.75

1179

1886

No. 254.

ZEBRA OR MAHOGANY.

Height, 18 inches. Dial, 6 inches.

8 Day Strike	$5.40
8 " " Alarm	6.00

1180

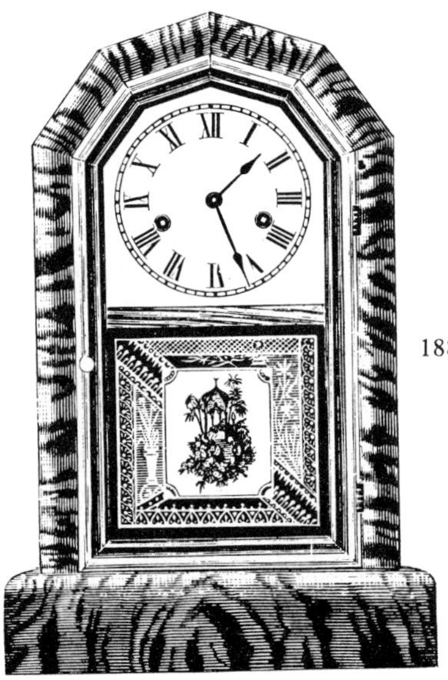

1880

OCTAGON PRIZE.

Height, 17¼ inches.
1 Day Strike.
8 Day Strike.
Also with Alarm.

1181

1889

Height, 17¾ inches. Dial, 6 inches.

OCTAGON PRIZE, V. P.

ROSEWOOD OR ZEBRA.

1 Day Strike	$5.40
1 Day Strike, Alarm	6.00
8 Day Strike	7.00
8 Day Strike, Alarm	7.60

1182

1880

OCTAGON PRIZE.

Height, 14½ inches.

1 Day Time.

1 Day Time, Alarm.

MANTEL CLOCKS - WOOD CASES

1183

1900

SHARP GOTHIC — "A"

Rosewood, Zebra or Mahogany.

Height, 21 inches. Dial, 6 inches.

List price,	One-day Strike,		$4.30
"	"	One-day Strike, Alarm,	4.80
"	"	Eight-day Strike,	5.30
"	"	Eight-day Strike, Alarm,	5.80

1184

1886

Height, 17¾ inches. Dial, 5 inches.

No. 251.

ZEBRA, VENEERED.

1	Day Time		$2.60
1	" "	Alarm	3.20
1	" Strike		3.90
1	" "	Alarm	4.50
8	" "		5.40
8	" "	Alarm	6.00

1185

1900

SHARP GOTHIC—"G"

Rosewood, Zebra or Mahogany.

Height, 21 inches. Dial, 6 inches.

List price,	One-day Strike,		$4.65
"	"	One-day Strike, Alarm,	5.15
"	"	Eight-day Strike,	5.65
"	"	Eight-day Strike, Alarm,	6.15

1186

1880

DERBY.

Walnut.

Height, 18½ inches.

Dial, 5½ inches.

1 Day Strike

1 Day Strike, Alarm.

1187

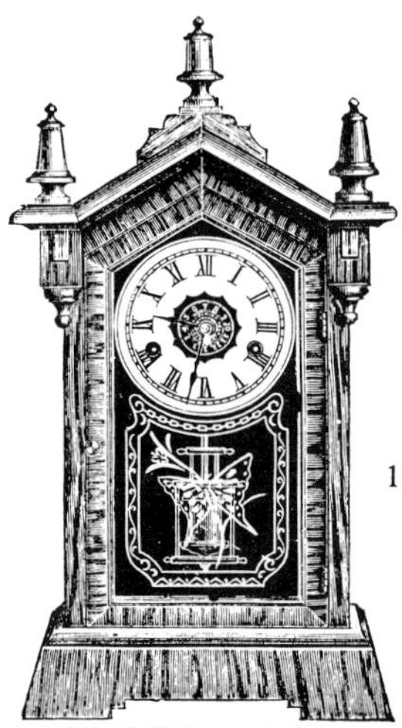

1900

DIADEM — Zebra

Height, 18 inches. Dial, 5 inches.

List price,	One-day Time,		$3.35
"	"	One-day Time, Alarm,	3.85
"	"	One-day Strike,	4.35
"	"	One-day Strike, Alarm,	4.85

1188

1880

DERBY, T. P.

Walnut.

Height, 17 inches.

Dial, 5 inches.

1 Day Time.

1 Day Time, Alarm.

MANTEL CLOCKS - WOOD CASES

1189

SMALL GOTHIC
Flame Mahogany case. 30-Hour Time and Alarm.
Height, 14½ inches. Dial, 4 inches.
{Jack Wyer Collection}

1190

Dial removed showing **Small Gothic** 30-Hour Time and Alarm Movement and Label.
{Jack Wyer Collection}

1191

Dial removed showing **Gothic** 30-Hour Time and Strike Movement and Label.
{Jack Wyer Collection}

1192

GOTHIC
Mahogany case. 30-Hour Time and Strike.
Height, 19½ inches. Dial, 5 inches.
{Jack Wyer Collection}

MANTEL CLOCKS - WOOD CASES

1193

1886

Height, 21 inches. Dial, 6 inches.

SHARP GOTHIC, V. P.

ROSEWOOD, ZEBRA, OR MAHOGANY.

1 Day Strike	$4.65
1 " " Alarm	5.25
8 " "	6.15
8 " " Alarm	6.75

1194

1880

SMALL GOTHIC, V. P.

Spring.

Height, 16 inches.

1 Day Time.

1 Day Time, Alarm.

1195

1886

Height, 15½ inches. Dial, 5 inches.

SMALL GOTHIC.

ROSEWOOD OR ZEBRA.

1 Day Time	$2.60
1 " " Alarm	3.20
1 " Strike	3.75
1 " " Alarm	4.35

1196

1886

Height, 16 inches.

SMALL GOTHIC, V. P.

ROSEWOOD OR ZEBRA.

1 Day Time	$2.95
1 " " Alarm	3.55
1 " Strike	4.10
1 " " Alarm	4.70

1197

1886

Height, 16 inches.

SMALL GOTHIC, V. P.

ROSEWOOD OR ZEBRA.

1 Day Time	$3.15
1 Day Time, Alarm	3.75
1 Day Strike	4.45
1 Day Strike, Alarm	5.05

1198

1886

Height, 15½ inches.

SMALL GOTHIC.

ROSEWOOD OR ZEBRA.

1 Day Time	$2.60
1 " " Alarm	3.20

MANTEL CLOCKS - WOOD CASES

1199

1880

GEM COTTAGE.

Height, 9 inches.

1 Day Time.

1 Day Time, Alarm.

1200

RATTLER. 1880

Spring.

Height, 10¾ inches.

1 Day Time.

1 Day Time, Alarm.

1201

COTTAGE, No. 2.

Spring.

Height, 12 inches.

1 Day Time.

1 Day Strike.

Also with Alarm.

1880

1202

1889

Height, 12 inches. Dial, 6 inches.

COTTAGE, No. 2.

ZEBRA OR MAHOGANY.

1 Day Time	$1.90
1 Day Time, Alarm	2.50
1 Day Strike	3.15
1 Day Strike, Alarm	3.75

1203

1878

ROSE GILT

Height, 9½ inches.
1 Day, Time, Spring.
Above furnished either Square or Octagon Top, as desired, at same price.

The year entered beside each clock is the year of the catalog from which the illustration was taken. Some clocks may have been produced a few years earlier.

1204

1880

COTTAGE, EXTRA.

Spring.

1 Day Time. Height, 12 inches.

1 Day Strike, } Height, 13 inches.
8 Day Strike, }

Also with Alarm.

1205

1880

AUBURN.

Spring.

Height, 14 inches.
1 Day Strike.
8 Day Strike.
Also with Alarm.

1206

1889

Height, 14 inches. Dial, 6 inches.

COTTAGE (Extra).

ZEBRA OR MAHOGANY.

1 Day Strike	$3.70
1 Day Strike, Alarm	4.30
8 Day Strike	5.25
8 Day Strike, Alarm	5.85

MANTEL CLOCKS - WOOD CASES

1207

COTTAGE
Jerome & Co.
1 Day Time and Alarm.
{Morgan Jay Taylor Collection}

1208

Close-up of **Jerome & Co.** Label.
{Morgan Jay Taylor Collection}

1209

1878

PRIZE.

1 Day. Strike. Spring.
8 " " "

Hight, 14 inches.

1210

Dial removed showing **Jerome & Co.** 30-Hour Time and Alarm Movement.
{Morgan Jay Taylor Collection}

PLEASE NOTE:

Clocks marked "Jerome & Co." or having this trademark were manufactured by the New Haven Clock Company, not by the earlier firm of Chauncey Jerome. "Jerome & Co." was a trademark on many New Haven models in the 1880's.

MANTEL CLOCKS - WOOD CASES

The year entered beside each clock is the year of the catalog from which the illustration was taken. Some clocks may have been produced a few years earlier.

1211

Close-up of **Column** 30-hour
Time and Strike Movement.
{Jack Wyer Collection}

1212

1878

COLUMN
Height, 16 inches.
1 Day. Strike. Spring.
8 Day. Strike. Spring.

1213

COLUMN
Mahogany case. Height, 16 inches.
1 Day. Strike. Spring.
{Jack Wyer Collection}

MANTEL CLOCKS - WOOD CASES

1214

1886

Height, 17¼ inches. Dial, 6 inches.

GUIDE, GILT PILLAR.

FINISHED WITH GILT OR ROSEWOOD COLUMNS.

1 Day Strike $6.45
1 " " Alarm 7.05
8 " " 7.95
8 " " Alarm 8.55

1215

1880

GUIDE, V. P.

Spring.

Height, 17¼ inches.
1 Day Strike.
8 Day Strike.
Also with Alarm.
Also with Gilt Columns.

1216

1886

Height, 17¼ in. Dial, 6 inches.

GUIDE, FULL GILT.

FINISHED WITH FULL GILT COLUMNS AND MOULDING.

1 Day Strike $7.45
1 " " Alarm 8.05
8 " " 8.95
8 " " Alarm 9.55

1217

1880

GUIDE.

Spring.

Height, 17¼ inches.
1 Day Strike.
8 Day Strike.
Also with Alarm.
Also with Gilt Columns.

1218

1886

ROUND GOTHIC, V. P.

ROSEWOOD OR ZEBRA.

Height, 17¼ inches. Dial, 6 inches.

1 Day Strike $5.25
1 " " Alarm 5.85
8 " " 6.75
8 " " Alarm 7.35

1219

1880

ROUND GOTHIC, EXTRA.

Spring.

Height, 17¼ inches.
1 Day Strike.
8 Day Strike.
Also with Alarm.

MANTEL CLOCKS - WOOD CASES

1220

1881

O. G. (WEIGHT. NO. 1 OR NO. 2.)

MAHOGANY OR ZEBRA.

Height, 26 inches. Dial, 8 inches.
1 Day Strike. 1 Day Strike, Alarm.

1221

1881

O. O. G. EXTRA V. P. (WEIGHT.)

ROSEWOOD.

Height, 21½ inches. Dial, 6 inches

1 Day Strike. 1 Day Strike, Alarm.

Smallest Rosewood Weight Clock made.

1222

1881

O. O. G. (WEIGHT.)

MAHOGANY OR ZEBRA.

Height, 26 inches. Dial, 8 inches.
1 Day Strike. 1 Day Strike, Alarm.

1223

1881

O. G. OR O. O. G. (WEIGHT.)

MAHOGANY OR ZEBRA.

Height, 30 inches. Dial, 9 inches.
8 Day Strike. 8 Day Strike, Alarm.

MANTEL CLOCKS - WOOD CASES

1224

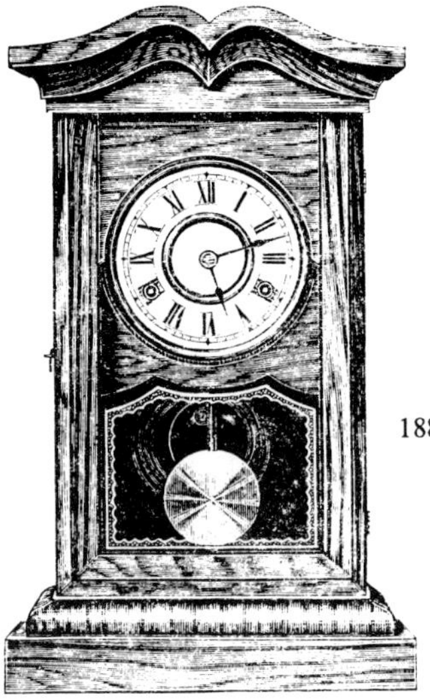

1886

Height, 18 inches. Dial, 5 inches.

No. 751.

WALNUT OR MAHOGANY VENEERED.

8 Day Strike........................$6.35
8 " " Alarm........................6.95

1225

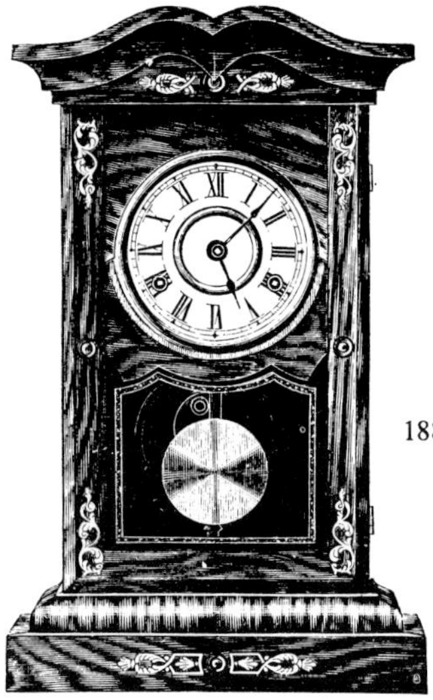

1886

Height, 18 inches. Dial, 5 inches.

No. 751.

WITH BRASS ORNAMENTS.

WALNUT OR MAHOGANY VENEERED.

8 Day Strike........................$6.70
8 " " Alarm........................7.30

1226

1886

Height, 19½ inches. Dial, 6 inches.

No. 752.

WALNUT OR MAHOGANY VENEERED.

8 Day Strike........................$6.50
8 " " Alarm........................7.10

1227

1886

Height, 19½ inches. Dial, 6 inches.

No. 752.

WITH BRASS ORNAMENTS.

WALNUT OR MAHOGANY, VENEERED.

8 Day Strike........................$6.85
8 " " Alarm........................7.45

1228

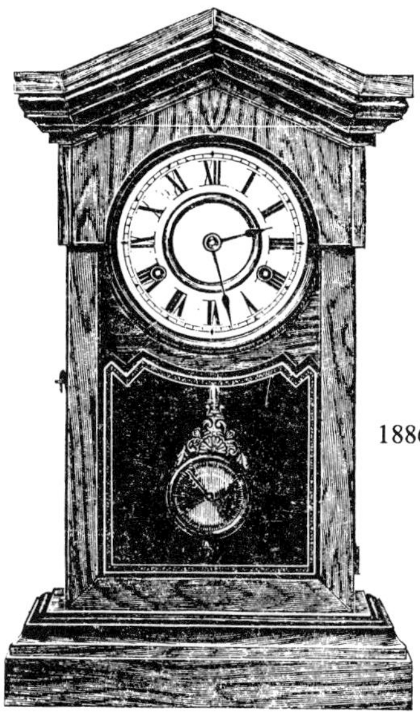

1886

Height, 19 inches. Dial, 5 inches.

No. 753.

WALNUT OR MAHOGANY VENEERED.

8 Day Strike........................$6.95
8 " " Alarm........................7.55

1229

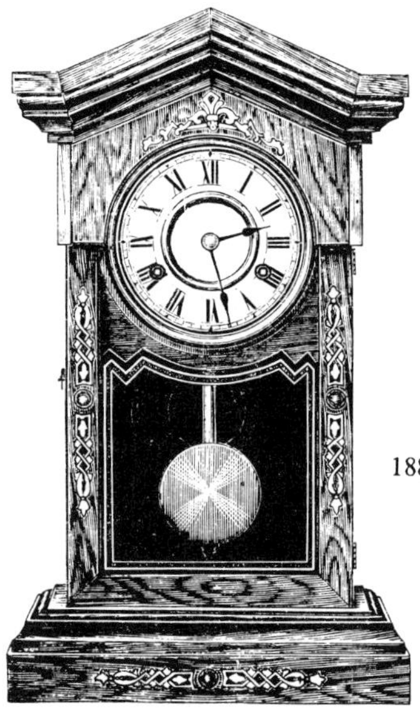

1886

Height, 19 inches. Dial, 5 inches.

No. 753.

WITH BRASS ORNAMENTS.

WALNUT OR MAHOGANY, VENEERED.

8 Day Strike........................$7.30
8 " " Alarm........................7.90

MANTEL CLOCKS - WOOD CASES

1230

1880

O. B. TIME.

BLACK MOULDING.

Height, 10 inches. Dial, 5 inches.

1 Day Time. 1 Day Time, Alarm.

Black Finish, ornamented with Gilt.

1231

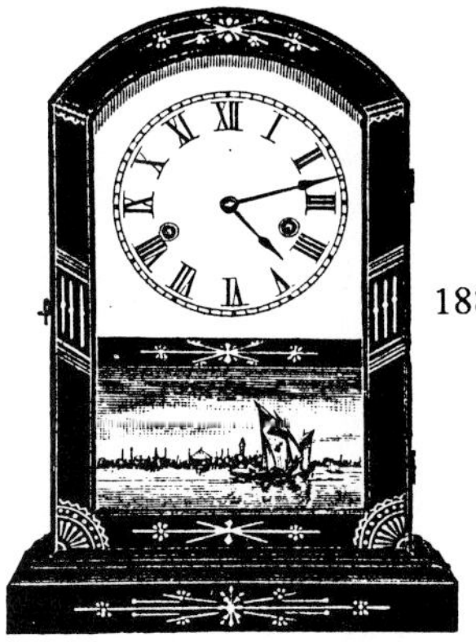

1881

O. B.

BLACK MOULDING.

Height, 14 inches. Dial, 6 inches.

1 Day Strike. 8 Day Strike.

Also with Alarm.

Black Finish, Ornamented with Gilt.

1232

1881

STERLING.

BLACK MOULDING (LEVER).

Height, 9 inches. Dial, 4 inches.

8 Day Time. 1 Day Strike.

Black Finish, with White Carvings.

1233

1881

ARGYLE TIME.

BLACK MOULDING.

Height, 13 inches. Dial, 4 inches.

1 Day Time. 1 Day Time, Alarm.

Black Finish, with White Carvings, and Japanese Tablets, showing Pendulum Ball.

The year entered beside each clock is the year of the catalog from which the illustration was taken. Some clocks may have been produced a few years earlier.

1234

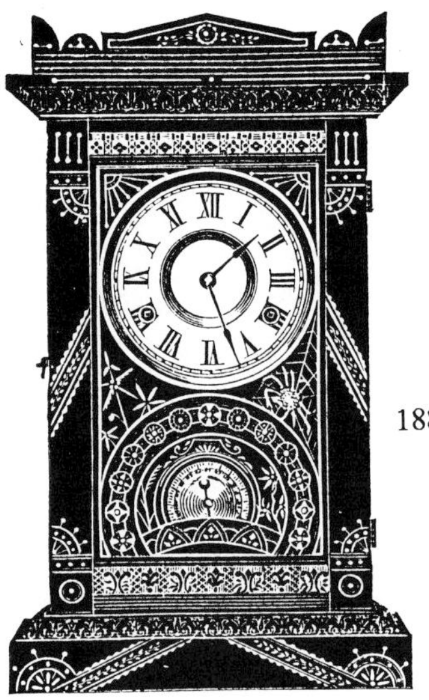

1881

CONWAY.

BLACK MOULDING

Height, 16½ inches. Dial, 5 inches.

1 Day Strike. 8 Day Strike. Also with Alarm.

Black Finish, with White Carvings, Fancy Tablet, showing Pendulum Ball.

1235

1881

WARWICK.

BLACK MOULDING.

Height, 17 inches. Dial, 8 inches.

1 Day Strike. 8 Day Strike. Also with Alarm.

Extra Width, Black Finish, with White Carvings.

1236

1881

ARGYLE.

BLACK MOULDING.

Height, 17 inches. Dial, 5 inches.

1 Day Strike. 8 Day Strike. Also with Alarm.

Black Finish, with White Carvings, Japanese Tablets, showing Pendulum Ball.

MANTEL CLOCKS - WOOD CASES

1237

1886

No. 750.

WALNUT OR MAHOGANY VENEERED.

Height, $19\frac{1}{4}$ inches. Dial, 5 inches.

8 Day Strike.....................$6.15
8 " " Alarm.............. 6.75

1238

1886

No. 750.

WITH BRASS ORNAMENTS.

WALNUT OR MAHOGANY VENEERED.

Height, $19\frac{1}{4}$ inches. Dial, 5 inches.

8 Day Strike...............$6.50
8 " " Alarm.......... 7.10

1239

1880

RARUS.

Walnut.

Height, $17\frac{1}{2}$ inches.

Dial, $5\frac{1}{2}$ inches.

1 Day Strike.

1 Day Strike, Alarm.

1240

1886

Height, $19\frac{1}{8}$ inches. Dial, 5 inches.

No. 754.

WALNUT OR MAHOGANY VENEERED.

8 Day Strike...................$6.75
8 " " Alarm............. 7.35

1241

1886

Height, $19\frac{1}{2}$ inches. Dial, 5 inches.

No. 754.

WITH BRASS ORNAMENTS.

WALNUT OR MAHOGANY VENEERED.

8 Day Strike...............$7.10
8 " " Alarm.......... 7.70

1242

1880

RARUS, T. P.

Walnut.

Height, $15\frac{1}{2}$ inches.

Dial, 5 inches.

1 Day Time.

1 Day Time, Alarm.

MANTEL CLOCKS - WOOD CASES

1243

1895

OCCIDENTAL.

Oak with Silver Ornaments. Walnut with Gilt Ornaments.

Height, 24 inches; Dial, 6 inches; Mirror on Sides.

8 Day Strike,	List, $9.00
8 Day Strike, Alarm,	List, 9.50
8 Day Gong,	List, 9.40
8 Day Gong, Alarm,	List, 9.90

1244

1911

OCCIDENTAL

HEIGHT ....24 INCHES
WIDTH ....16½ INCHES
DIAL ....6 INCHES
CASE, OAK OR WALNUT
PACKED, 1 IN A BOX
PRICE, 8 DAY HALF-HOUR STRIKE....LIST, $10.40
PRICE, 8 DAY HALF-HOUR STRIKE,
CATHEDRAL GONG....................LIST, 10.85
PRICE, WITH ALARM............LIST EXTRA, .45

OCCIDENTAL.

1245

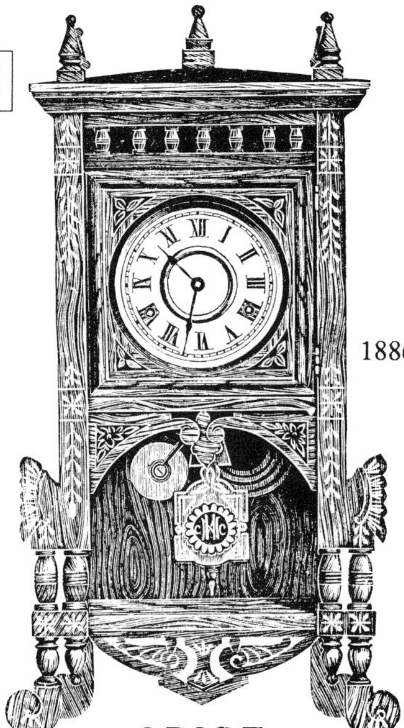

1886

ADICE.

WALNUT CABINET CLOCK.

(Oven) 8 Day Strike $9.40
(Play) 8 " " Cathedral Gong 9.90

This Clock is novel in construction, the lower half being open. The movement is fully protected by boxing underneath.

1246

1880

TYNE.

Walnut.

Height, 19 inches.
Dial, 5½ inches.
8 Day Strike.
8 Day Strike, Alarm.

1247

1911

OCCIDENTAL EXTRA

HEIGHT ....23¼ INCHES
WIDTH ....16¾ INCHES
DIAL ....6 INCHES
CASE, OAK OR WALNUT
PACKED, 1 IN A BOX
PRICE, 8 DAY HALF-HOUR STRIKE....LIST, $10.00
PRICE, 8 DAY HALF-HOUR STRIKE,
CATHEDRAL GONG ....LIST, 10.45
PRICE, WITH ALARM...........LIST EXTRA, .45

MANTEL CLOCKS - WOOD CASES

1248

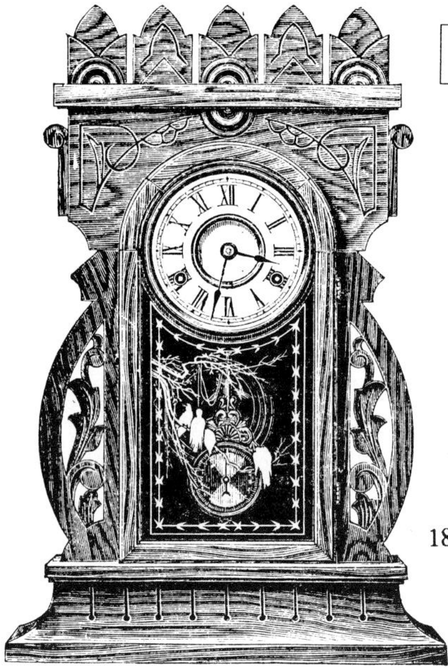

1886

Height, 20½ inches. Dial, 5 inches.

No. 502.

BLACK WALNUT.

8 Day Strike	$7.20
8 " " Alarm	7.80
8 " " Cathedral Gong	7.70
8 " " " " Alarm	8.30

1249

ODER.

1880

Walnut.

Height, 20½ inches.
Dial, 6 inches.
8 Day Strike.
8 Day Strike, Alarm.

1250

1886

Height, 21 inches. Dial, 5 inches.

No. 501.

BLACK WALNUT.

8 Day Strike	$7.05
8 " " Alarm	7.65
8 " " Cathedral Gong	7.55
8 " " " " Alarm	8.15

1251

1900

OAKDALE

Oak with Silver Ornaments. Walnut with Gilt Ornaments.
Height, 24 inches. Width, 15¼ inches. Dial, 6 inches.

List price, Eight-day, Strike,	$8.65
" " Eight-day, Strike Alarm,	9.15
" " Eight-day, Gong,	9.05
" " Eight-day, Gong Alarm,	9.55

1252

1880

APOLLO.

Walnut.

Height, 20 inches.
Dial, 5½ inches.
8 Day Strike.
8 Day Strike, Alarm.

MANTEL CLOCKS - WOOD CASES

1253

MAIL.

IMITATION WALNUT.

Height, 16 inches.

Dial, 5 inches.

1 Day Strike $3.75
1 " " Alarm ... 4.35

1886

MAIL.

1254

No. 509.

BLACK WALNUT.

Height, 19½ inches. Dial, 5 inches.

	Price.
8 Day Strike	$5.75
8 " " Alarm	6.35
8 " " Cathedral Gong	6.25
8 " " " " Alarm..	6.85

1886

No. 509.

1255

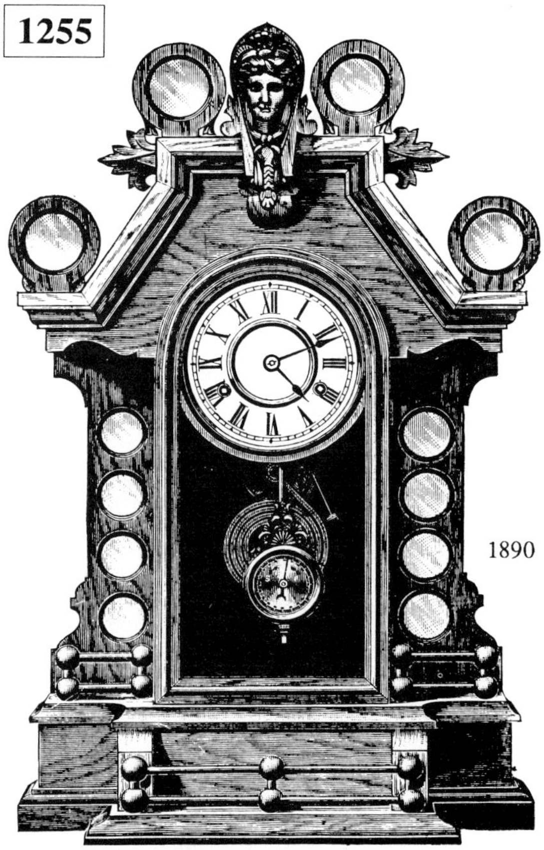

1890

Height, 26 inches. Dial, 6 inches.

COUNTESS.

BLACK WALNUT.

8 Day Strike..............................	$11.00
8 Day Strike, Alarm..................	11.60
8 Day Strike, Cathedral Gong..........	11.50
8 Day Strike, Cathedral Gong, Alarm...	12.10

The above Clock has 12 Mirrors and Solid Metal Head.

1256

1880

CHAMPION.

Walnut.

Height, 25 inches. 8 Day Strike.

Dial, 6 inches. 8 Day Strike, Alarm.

MANTEL CLOCKS - WOOD CASES

1257

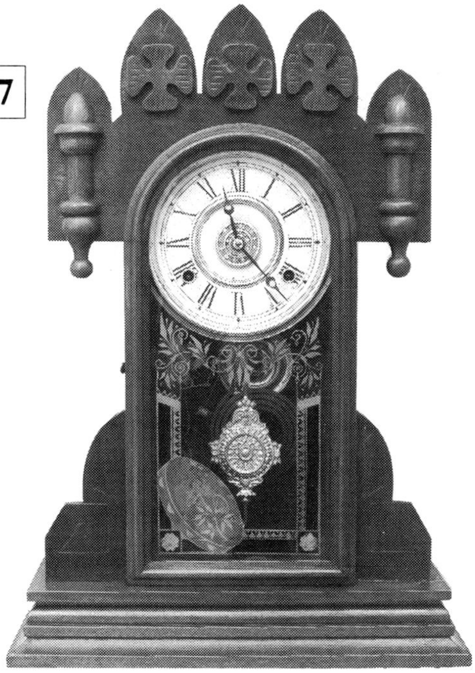

SIGMA
Walnut case. 8-Day Time, Strike and Alarm.
Height, 21 inches. Width, 15 inches. Dial, 6 inches.
{The Millie & Jerry Maltz Collection}
{Photo courtesy of Jack Wyer}

1258

GINGERBREAD
Oak case. 8-Day Time, Strike and Alarm.
Height, 24½ inches. Dial, 5 inches.
{Jack Wyer Collection}

1259

Close-up of **Sigma** Time, Strike and Alarm movement signed New Haven Clock Company, with two patent dates Feb. 11, 1879 and April 1884.
{The Millie & Jerry Maltz Collection}
{Photo courtesy of Jack Wyer}

1260

Dial removed showing **Gingerbread** 8-Day Time, Strike and Alarm Movement.
{Jack Wyer Collection}

MANTEL CLOCKS - WOOD CASES

1261

Elbe Patent Dial Pendulum Ball.
For 1883 catalog illustration see #1563 page 401.
{Jack Wyer Collection}

1262

ELBE
Walnut case. 8-Day Time, Strike and Alarm.
Height, 24 inches. Dial, 6 inches.
{Jack Wyer Collection}

1263

Elbe Striking Alarm Label.
{Jack Wyer Collection}

1264

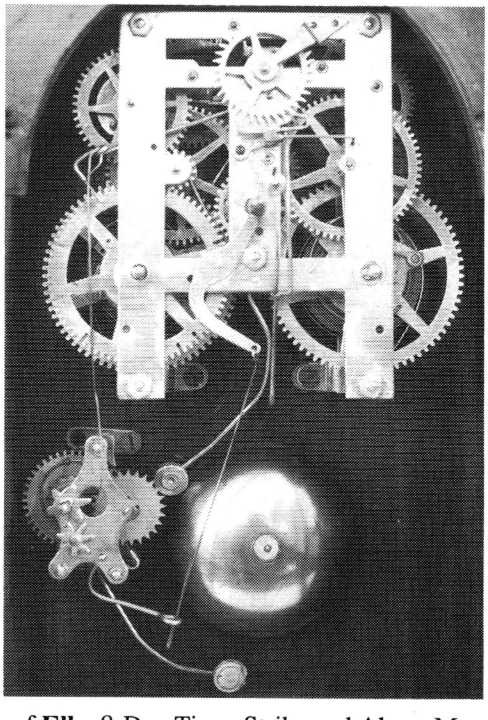

Close-up of **Elbe** 8-Day Time, Strike and Alarm Movement.
{Jack Wyer Collection}

MANTEL CLOCKS - WOOD CASES

1265

1880

RHINE.

1266

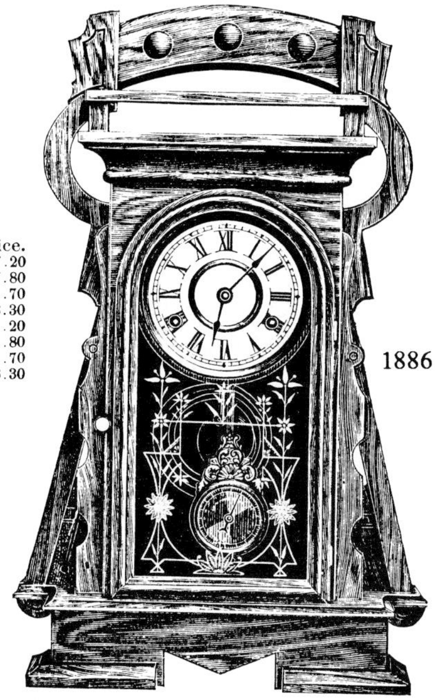

1886

No. 510.

No. 510.

BLACK WALNUT OR CHERRY.

Height, 22 inches. Dial, 5 inches.

	Price.
8 Day Strike, Black Walnut	$7.20
8 " " Alarm, Black Walnut	7.80
8 " " Cath. Gong, Black Walnut	7.70
8 " " " " Alarm, B'k Wal.	8.30
8 " " Cherry	7.20
8 " " Alarm, Cherry	7.80
8 " " Cathedral Gong, Cherry	7.70
8 " " " " Alm. Cherry.	8.30

RHINE.

Walnut.

Height, 21 inches.
Dial, 6 inches.
8 Day Strike.
8 Day Strike, Alarm.

1267

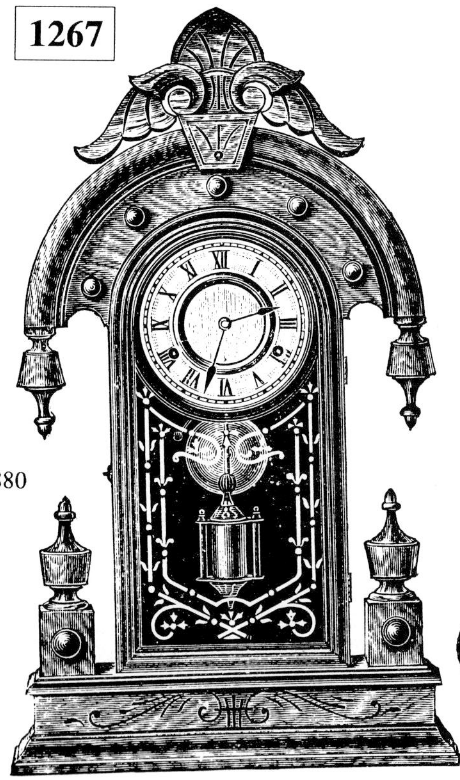

1880

1268

1886

1269

1880

ELBE.

Walnut.

Height, 24 inches.
Dial, 6 inches.
8 Day Strike.
8 Day Strike, Alarm.

DRAVE.

BLACK WALNUT.

Height, 20 inches.

Dial, 5 inches.

8 Day Strike	$7.20
8 " " Alarm	7.80
8 " " Cathedral Gong	7.70
8 " " " " Alarm	8.30

MOSELLE. Walnut.

Height, 26 inches.
Dial, 6 inches.
8 Day Strike. 8 Day Strike, Alarm.

MANTEL CLOCKS - WOOD CASES

1270

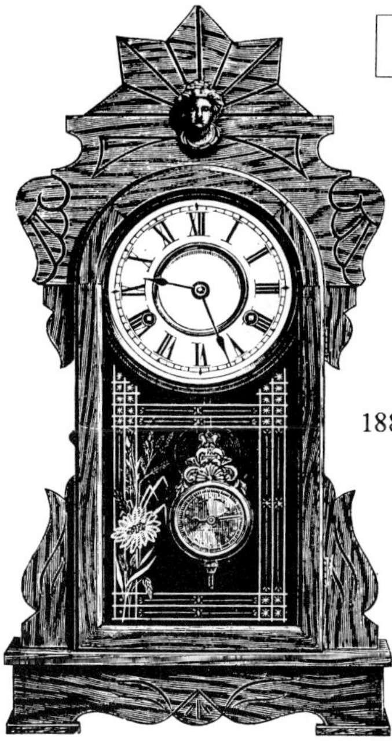

1886

1271

1886

No. 515.

BLACK WALNUT.

Height, 19¾ inches. Dial, 5 inches.

8 Day Strike $6.20
8 " " Alarm 6.80
8 " " Cathedral Gong........ 6.70
8 " " " " Alarm.. 7.30

No. 517.

IMITATION BLACK WALNUT.

Height, 22¼ inches. Dial, 6 inches.

8 Day Strike $5.60
8 " " Alarm 6.20
8 " " Cathedral Gong....... 6.10
8 " " " " Alarm. 6.70

1272

1890

Height, 24 inches. Dial, 6 inches.

CORSAIR.

BLACK WALNUT.

8 Day Strike.. $10.00
8 Day Strike, Alarm.. 10.60
8 Day Strike, Cathedral Gong............................ 10.50
8 Day Strike, Cathedral Gong, Alarm.................... 11.10

above Clock has ¾ in. Beveled Glass in Door and Solid Metal Ornaments.

1273

1881

LENA.

WALNUT.

Height, 35½ inches. Dial, 8 inches.

8 Day Time. 8 Day Strike. 8 Day Strike, Alarm.

MANTEL CLOCKS - WOOD CASES

1274

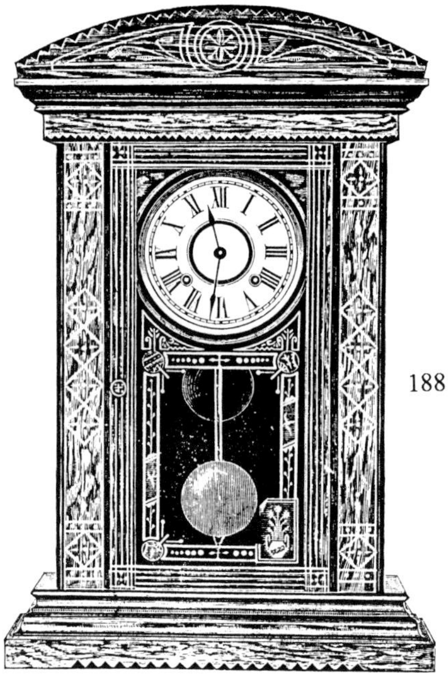

1881

GANGES.
WALNUT (WEIGHT)

Height, 23½ inches. Dial, 6 inches.
1 Day Strike. 1 Day Strike, Alarm.

1275

1881

INDUS.
WALNUT (WEIGHT).

Height, 23 inches. Dial, 6 inches.
1 Day Strike. 1 Day Strike, Alarm.

These are the smallest weight Clocks made.

1276

INDUS
Walnut case. 30-Hour Weight Driven Movement.
Height, 24 inches. Dial, 5 inches.
{Jack Wyer Collection}

1277

Close-up of **Indus** 30-Hour Weight Driven Movement.
{Jack Wyer Collection}

MANTEL CLOCKS - WOOD CASES

1278

1883

MAAS
WALNUT
Height, 23 inches. Dial 6 inches.
8 Day Strike. 8 Day Strike, Alarm.

1279

1883

BOYNE
ASH
Height, 23 inches. Dial 6 inches.
8 Day Strike. 8 Day Strike, Alarm.

1280

1881

LIFFEY
WALNUT
Height, 22½ inches. Dial 6 inches.
8 Day Strike. 8 Day Strike, Alarm.

1281

1883

DOURO
WALNUT
Height, 23 inches. Dial 6 inches.
8 Day Strike. 8 Day Strike, Alarm.

1282

1881

AVON
WALNUT
Height, 21 inches. Dial 5 inches.
8 Day Strike. 8 Day Strike, Alarm.

MANTEL CLOCKS - WOOD CASES

1283

1883

THEISS.

WALNUT.

Height, 19½ inches. Dial, 5 inches.

1 Day Strike. 1 Day Strike, Alarm.

1284

ARNO. 1883

WALNUT.

Height, 19¾ inches. Dial, 5 inches.

8 Day Strike. 8 Day Strike, Alarm.

1285

1883

DESNA.

WALNUT.

Height, 20½ inches. Dial, 5 inches.

8 Day Strike. 8 Day Strike, Alarm.

1286

1883

DORA.

WALNUT.

Height, 20 inches. Dial, 5 inches.

8 Day Strike. 8 Day Strike, Alarm.

1287

1883

EBRO.

ASH.

Height, 21 inches. Dial, 5 inches.

8 Day Strike. 8 Day Strike, Alarm.

MANTEL CLOCKS - WOOD CASES

1288

1881

TRENT.

WALNUT.

Height, 20 inches. Dial, 5 inches.
8 Day Strike. 8 Day Strike, Alarm.

1289

1890

Height, 22 inches. Dial, 6 inches.

COQUETTE.

BLACK WALNUT.

8 Day Strike........................\$7.50
8 Day Strike, Alarm.................. 8.10
8 Day Strike, Cathedral Gong........ 8.00
8 Day Strike, Cathedral Gong, Alarm. 8.60

1290

1885

No. 518.

BLACK WALNUT.

Height, 21¼ inches. Dial, 6 inches.

1 Day Strike.............. ....\$5 20
1 " " Alarm................ 5 80

1291

1881

RIGA.

WALNUT.

Height, 21½ inches. Dial, 5 inches.
1 Day Strike. 1 Day Strike, Alarm.

1292

1881

MERSEY.

WALNUT.

Height, 21 inches. Dial, 5 inches.
1 Day Strike. 1 Day Strike, Alarm.

1293

1881

DUNA.

WALNUT.

Height, 22 inches. Dial, 6 inches.
8 Day Strike. 8 Day Strike, Alarm.

MANTEL CLOCKS - WOOD CASES

1294

1886

No. 511

1295

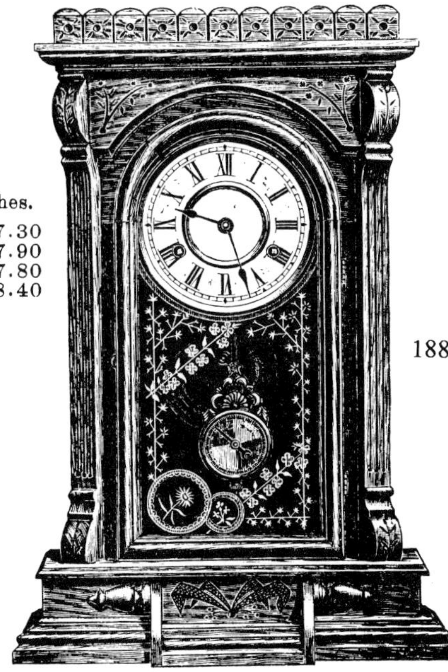

1886

No. 512

No. 512.

BLACK WALNUT.

Height, 21½ inches. Dial, 6 inches.

8 Day Strike $7.30
8 " " Alarm................ 7.90
8 " " Cathedral Gong...... 7.80
8 " " " " Alarm. 8.40

No. 511.

BLACK WALNUT.

Height, 18¾ inches. Dial, 5 inches.

8 Day Strike $5.00
8 " " Alarm................. 5.60
8 " " Cathedral Gong....... 5.50
8 " " " " Alarm. 6.10

1296

1890

Height, 24 inches. Dial, 6 inches.

PARISIAN.

BLACK WALNUT.

8 Day Strike.......................$7.20
8 Day Strike, Alarm............... 7.80
8 Day Strike, Cathedral Gong...... 7.70
8 Day Strike, Cathedral Gong, Al'm. 8.30

1297

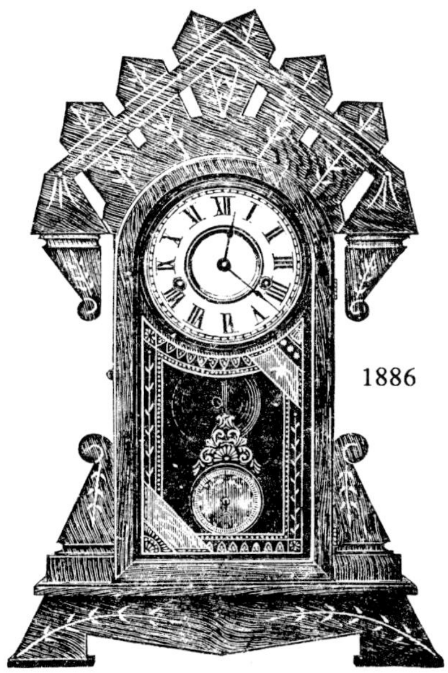

1886

ARAL.

BLACK WALNUT.

Height, 20¼ inches.

Dial, 5 inches.

1 Day Strike.........$5.20
1 " " Alarm... 5.80

1298

1880

DANUBE.

Walnut.

Height, 24 inches.
Dial, 6 inches.
8 Day Strike.
8 Day Strike, Alarm.

MANTEL CLOCKS - WOOD CASES

1299

1886

No. 522.

BLACK WALNUT.

Height, 20 inches. Dial, 6 inches.

1 Day Strike............$4.30
1 " " Alarm...... 4.90

1300

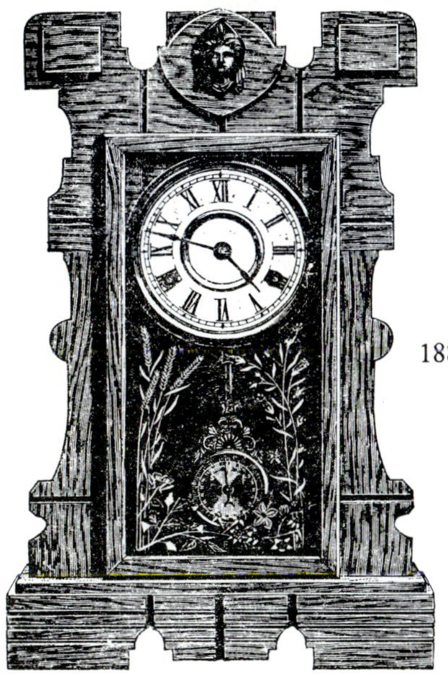

1886

Height, 20 inches. Dial, 5 inches.

No. 513.

BLACK WALNUT.

8 Day Strike.............................$5.25
8 " " Alarm.................... 5.85
8 " " Cathedral Gong........ 5.75
8 " " " " Alarm.. 6.35

1301

1886

Height, 21 inches. Dial, 5 inches.

No. 514.

BLACK WALNUT.

1 Day Strike.........$4.60
1 " " Alarm... 5.20

1302

1886

Height, 24 inches. Dial, 6 inches.

No. 503.

BLACK WALNUT.

8 Day Strike.............................$7.20
8 " " Alarm..................... 7.80
8 " " Cathedral Gong....... 7.70
8 " " " " Alarm. 8.30

1303

1886

HUMBER.

BLACK WALNUT.

Height, 21 inches. Dial, 6 inches.

8 Day Strike.............................$7.35
8 " " Alarm..................... 7.95
8 " " Cathedral Gong....... 7.85
8 " " " " Alarm. 8.45

1304

1880

NECKAR.

Walnut.

Height, 24 inches.
Dial, 6 inches.
8 Day Strike.
8 Day Strike, Alarm.

MANTEL CLOCKS - WOOD CASES

1305

1886

PRUTH.

BLACK WALNUT.

Height, 21¾ inches. Dial, 5 inches.

8 Day Strike	$7.15
8 " " Alarm	7.75
8 " " Cathedral Gong	7.65
8 " " " " Alarm.	8.25

1306

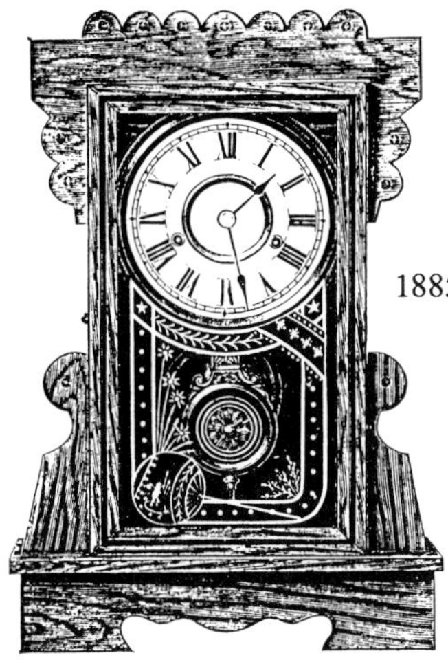

1885

Height, 17½ inches. Dial, 6 inches.

No. 520.

BLACK WALNUT.

1 Day Strike	$4 50
1 " " Alarm	5 10

1307

1880

NEVA.

Walnut.

Height, 19 inches.
Dial, 5½ inches.
8 Day Strike.
8 Day Strike, Alarm.

1308

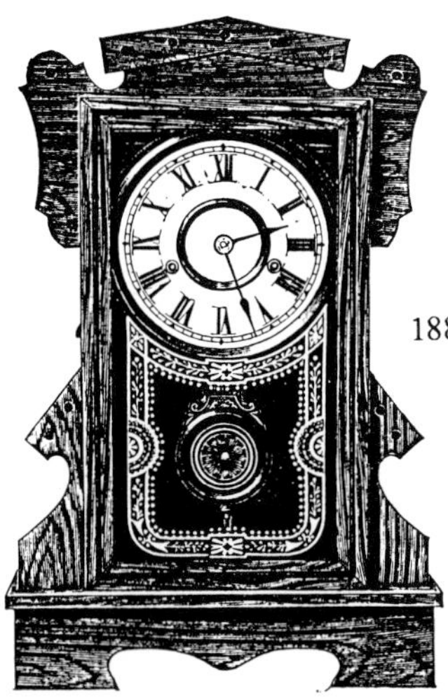

1885

Height, 17¾ inches, Dial, 6 inches.

No. 519.

BLACK WALNUT.

1 Day Strike	$4 50
1 " " Alarm	5 10

1309

1890

Height, 20¾ inches. Dial, 6 inches.

CALUMET.

BLACK WALNUT.

1 Day Strike	$5.00
1 Day Strike, Alarm	5.60

1310

1885

Height, 19 inches. Dial, 5 inches.

OUSE.

BLACK WALNUT.

1 Day Strike	$4 75
1 " " Alarm	5 35

MANTEL CLOCKS - WOOD CASES

1311

1886

Height, 20¼ inches. Dial, 5 inches.

AYR.

BLACK WALNUT.

8 Day Strike	$7.40
8 " " Alarm	8.00
8 " " Cathedral Gong	7.90
8 " " " " Alarm	8.50

1312

1885

Height, 20¼ inches. Dial, 5 inches.

AYR.

BLACK WALNUT,

8 Day Strike	$7 40
8 " " Alarm	8 00
8 " " Cathedral Gong	7 90
8 " " " " Alarm	8 50

1313

1880

WESER.

Walnut.

Height, 20½ inches.
Dial, 6 inches.
8 Day Strike.
8 Day Strike, Alarm.

1314

1885

Height, 25 inches; Dial, 6 inches.

ENCHANTRESS

Oak or Walnut.

8 Day Strike,	List, $8.00
8 Day Strike, Alarm,	List, 8.50
8 Day Gong,	List, 8.40
8 Day Gong, Alarm,	List, 8.90

1315

1890

Height, 23½ inches. Dial, 6 inches.

NEREID.

BLACK WALNUT.

8 Day Strike	$5.75
8 Day Strike, Alarm	6.35
8 Day Strike, Cathedral Gong	6.25
8 Day Strike, Cathedral Gong, Al'm	6.85

MANTEL CLOCKS - WOOD CASES

1316

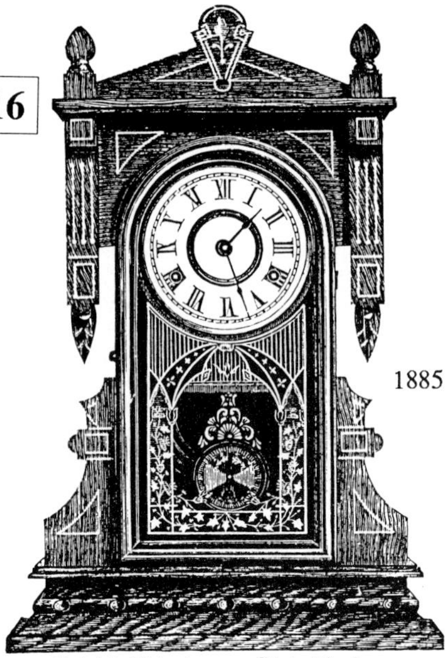

1885

Height, 20 inches. Dial, 5 inches.

KAMA.

BLACK WALNUT.

8 Day Strike.......................$7 20
8 " " Alarm....... ... . 7 80
8 " " Cathedral Gong....... 7 70
8 " " " " Alarm.. 8 30

1317

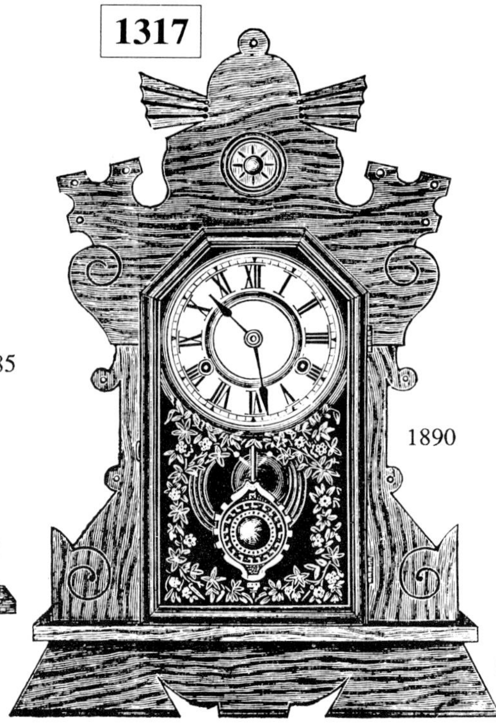

1890

Height, 19½ inches. Dial, 5 inches.

GRAYLING.

BLACK WALNUT.

8 Day Strike...........................$5.50
8 Day Strike, Alarm.................... 6.10
8 Day Strike, Cathedral Gong.......... 6.00
8 Day Strike, Cathedral Gong, Alarm... 6.60

1318

1885

Height, 22 inches. Dial, 6 inches.

TENNESSEE.

ASH.

8 Day Strike.........................$7 35
8 " " Alarm................. 7 95
8 " " Cathedral Gong........ 7 85
8 " " " " Alarm.. 8 45

1319

1885

Height, 23 inches. Dial, 5½ inches.

CLYDE.

BLACK WALNUT.

8 Day Strike....... ....$6 60
8 " " Alarm...... .... 7 20
8 " " Cathedral Gong.. .... 7 10
8 " " " " Alarm. 7 70

1320

1886

No. 504.

BLACK WALNUT.

Height, 20½ inches. Dial, 5 inches.

8 Day Strike......................$6.00
8 " " Alarm................ 6.60
8 " " Cathedral Gong....... 6.50
8 " " " " Alarm. 7.10

1321

1885

Height, 21 inches. Dial, 6 inches.

OSAGE.

ASH.

8 Day Strike.............. ....$7 15
8 " " Alarm................. 7 75
8 " " Cathedral Gong........ 7 65
8 " " " " Alarm.. 8 25

MANTEL CLOCKS - WOOD CASES

1322

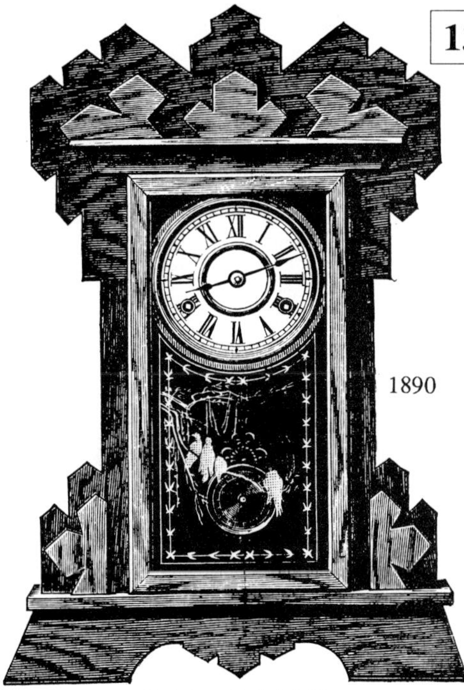

1890

CELESTE.

BLACK WALNUT.

Height, 21 inches. Dial, 5 inches.

8 Day Strike	$5.75
8 Day Strike, Alarm	6.35
8 Day Strike, Cathedral Gong	6.25
8 Day Strike, Cathedral Gong, Al'm.	6.85

1323

1890

CARMEN.

BLACK WALNUT.

Height, 21½ inches. Dial, 6 inches.

8 Day Strike	$5.50
8 Day Strike, Alarm	6.10
8 Day Strike, Cathedral Gong	6.00
8 Day Strike, Cathedral Gong, Al'm.	6.60

1324

1885

Height, 21¼ inches. Dial, 5 inches.

No. 506.

IMITATION BLACK WALNUT OR CHESTNUT.

8 Day Strike, Walnut	$5 25
8 " " Alarm, Walnut	5 85
8 " " Cathedral Gong, Walnut	5 75
8 " " " " Alarm. Walnut.	6 35
8 " " Chestnut	5 25
8 " " Alarm, Chestnut	5 85
8 " " Cathedral Gong	5 75
8 " " " " Alarm, Chestnut	6 35

1325

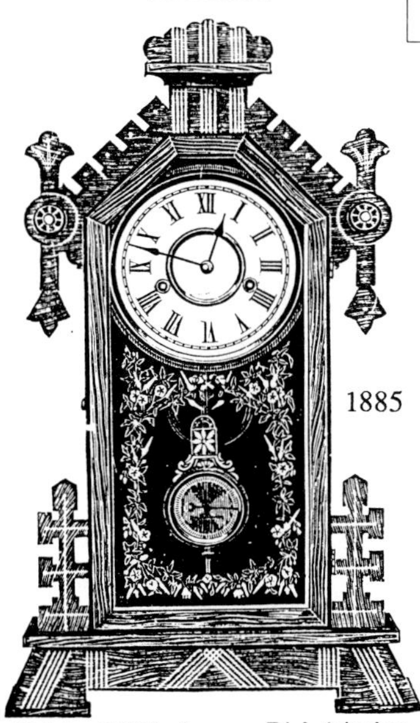

1885

Height, 21¼ inches. Dial, 6 inches.

EUPHRATES.

BLACK WALNUT.

8 Day Strike	$6 20
8 " " Alarm	6 80
8 " " Cathedral Gong	6 70
8 " " " " Alarm	7 30

1326

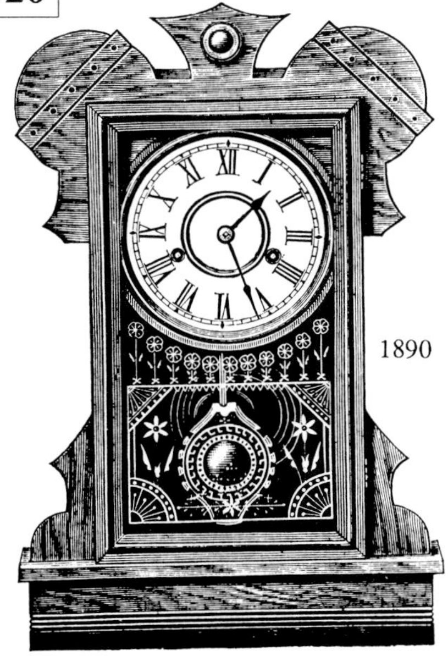

1890

Height, 18 inches. Dial, 6 inches.

CLOCHETTE.

BLACK WALNUT.

1 Day Strike	$4.00
1 Day Strike, Alarm	4.60

1327

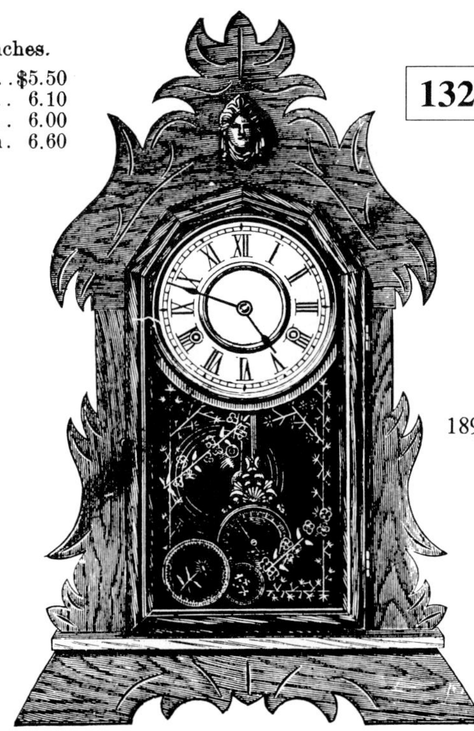

1890

Height, 22 inches. Dial, 6 inches.

CECILE.

BLACK WALNUT.

8 Day Strike	$5.50
8 Day Strike, Alarm	6 10
8 Day Strike, Cathedral Gong	6.00
8 Day Strike, Cathedral Gong, Al'm.	6.60

MANTEL CLOCKS - WOOD CASES

1328

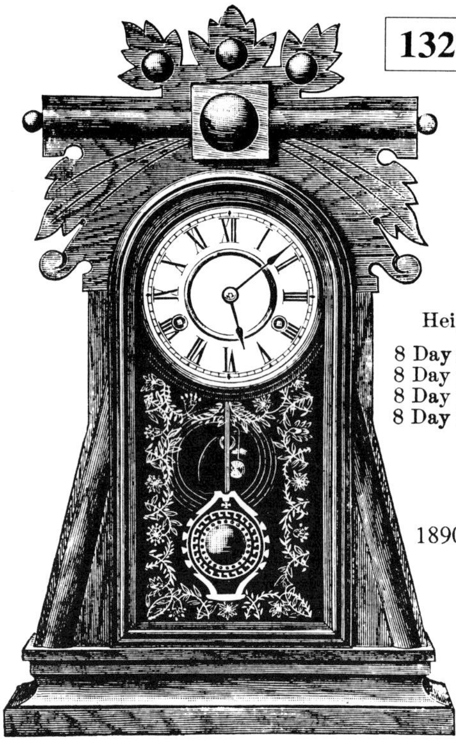

1890

IREX.

BLACK WALNUT.

Height, 23½ inches. Dial, 6 inches.

8 Day Strike	$6.25
8 Day Strike, Alarm	6.85
8 Day Strike, Cathedral Gong	6.75
8 Day Strike, Cathedral Gong, Al'm.	7.35

1329

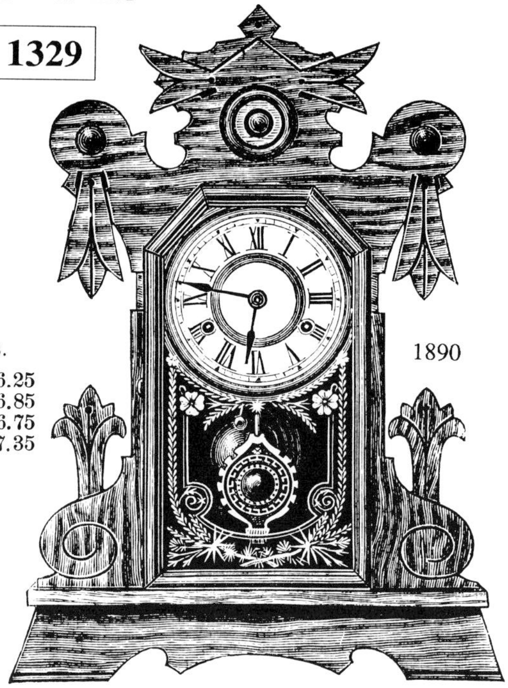

1890

RAMBLER.

BLACK WALNUT.

Height, 19½ inches. Dial, 5 inches.

8 Day Strike	$5.50
8 Day Strike, Alarm	6.10
8 Day Strike, Cathedral Gong	6.00
8 Day Strike, Cathedral Gong, Alarm	6.60

1330

1890

VAMPIRE.

BLACK WALNUT.

Height, 21½ inches. Dial, 6 inches.

8 Day Strike	$6.75
8 Day Strike, Alarm	7.35
8 Day Strike, Cathedral Gong	7.25
8 Day Strike, Cathedral Gong, Al'm..	7.85

1331

1890

ELMA.

BLACK WALNUT.

Height, 21 inches. Dial, 6 inches.

8 Day Strike	$6.50
8 Day Strike, Alarm	7.10
8 Day Strike, Cathedral Gong	7.00
8 Day Strike, Cathedral Gong, Al'm..	7.60

MANTEL CLOCKS - WOOD CASES

1332

1886

Height, 21 inches. Dial, 5 inches.

DEE.

BLACK WALNUT.

8 Day Strike........................$7.50
8 " " Alarm 8.10
8 " " Cathedral Gong 8.00
8 " " " " Alarm. 8.60

1333

1885

Height, 20 inches. Dial, 5 inches.

No. 505.

IMITATION WALNUT OR ASH.

1 Day Strike, Walnut.......$4 65
1 " " Alarm Walnut. 5 15
1 " " Ash........... 4 55
1 " " Alarm, Ash... 5 15

1334

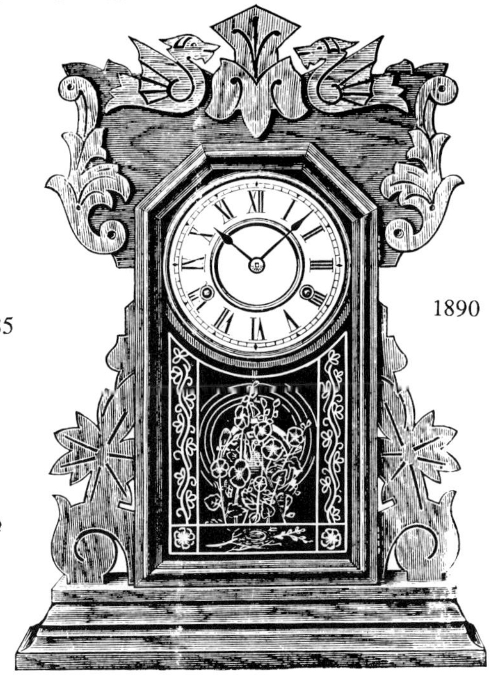

1890

Height, 20½ inches. Dial, 6 inches.

CLARITA.

BLACK WALNUT.

8 Day Strike........................$6.75
8 Day Strike, Alarm................ 7.35
8 Day Strike, Cathedral Gong....... 7.25
8 Day Strike, Cathedral Gong, Al'm. 7.85

1335

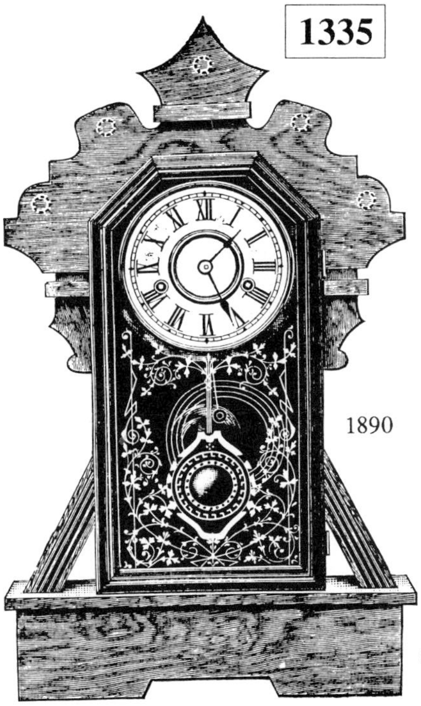

1890

CHRISTINE.

BLACK WALNUT.

Height, 20¾ inches Dial, 5 inches.

1 Day Strike...................$4.25
1 Day Strike, Alarm 4.85

1336

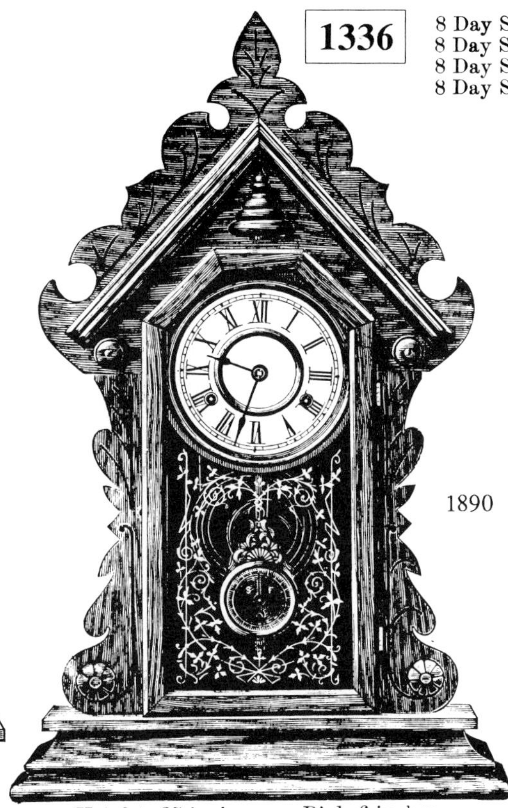

1890

Height, 25 inches. Dial, 6 inches.

CINDERELLA.

BLACK WALNUT.

8 Day Strike......................$6.25
8 Day Strike, Alarm............... 6.85
8 Day Strike, Cathedral Gong..... 6.75
8 Day Strike, Cathedral Gong, Al'm. 7.35

1337

1880

DON.

Walnut.

Height, 18 inches.

Dial, 5½ inches.

1 Day Strike.

1 Day Strike, Alarm.

MANTEL CLOCKS - WOOD CASES

1338

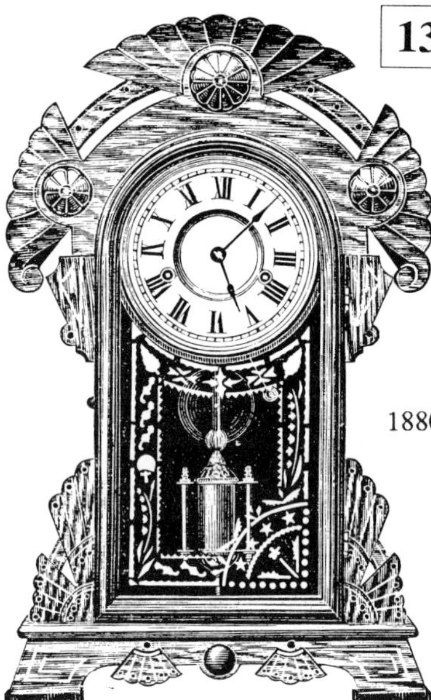

1880

TAGUS.

Walnut.

Height, 19 inches.
Dial, 5½ inches.
1 Day Strike.
1 Day Strike, Alarm.

1339

1880

TIBER.

Walnut.

Height, 19 inches.
Dial, 5½ inches.
1 Day Strike.
1 Day Strike, Alarm.

1340

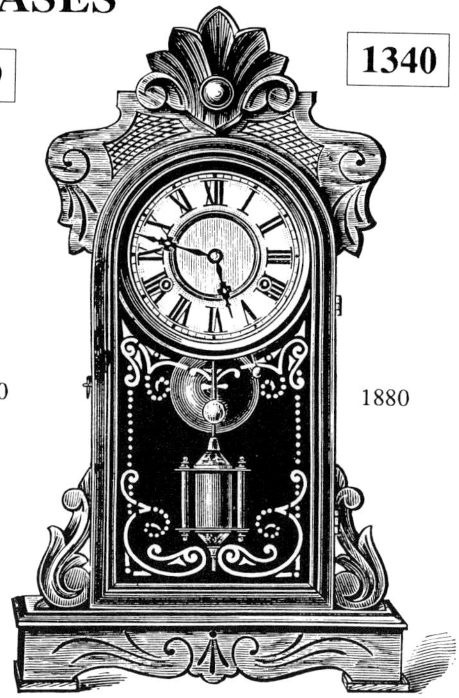

1880

SABINE.

Walnut.

Height, 21½ inches.
Dial, 6 inches.
1 Day Strike.
1 Day Strike, Alarm.

1341

1880

THAMES.

Walnut.

Height, 23 inches.
Dial, 6 inches.
8 Day Strike.
8 Day Strike, Alarm.

1342

1880

SEINE.

Walnut.

Height, 23 inches.
Dial, 6 inches.
1 Day Strike.
1 Day Strike, Alarm.

1343

1880

SHANNON.

Walnut.

Height, 18 inches.
Dial, 5½ inches.
1 Day Strike.
1 Day Strike, Alarm.

MANTEL CLOCKS - WOOD CASES

1344

1880

SEVERN.

Walnut.
Height, 19 inches.
Dial, 5½ inches.
1 Day Strike.
1 Day Strike, Alarm.

1345

1890

TITANIA

TITANIA.

BLACK WALNUT.

Height, 24 inches. Dial, 6 inches.

8 Day Strike........................$6.00
8 Day Strike, Alarm................ 6.60
8 Day Strike, Cathedral Gong........ 6.50
8 Day Strike, Cathedral Gong, Al'm.. 7.10

1346

1890

Height, 22¾ inches. Dial, 6 inches.

SEMINOLE.

BLACK WALNUT.

8 Day Strike........................$7.00
8 Day Strike, Alarm................ 7.60
8 Day Strike, Cathedral Gong........ 7.50
8 Day Strike, Cathedral Gong, Al'm... 8.10

1347

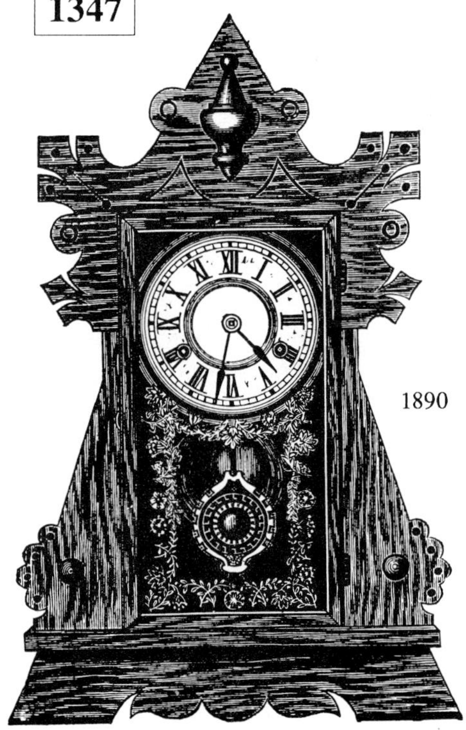

1890

Height, 22½ inches. Dial, 5 inches.

LIRIS.

BLACK WALNUT.

1 Day Strike.......... ....$4.75
1 Day Strike, Alarm................ 5.35

MANTEL CLOCKS - WOOD CASES

1348

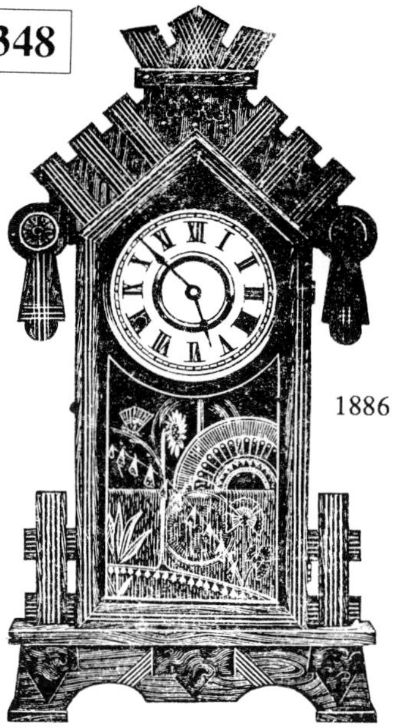

1886

Height, 21 inches. Dial, 5 inches.

TEES.

BLACK WALNUT.

1 Day Strike $4.75
1 " " Alarm 5.35

1349

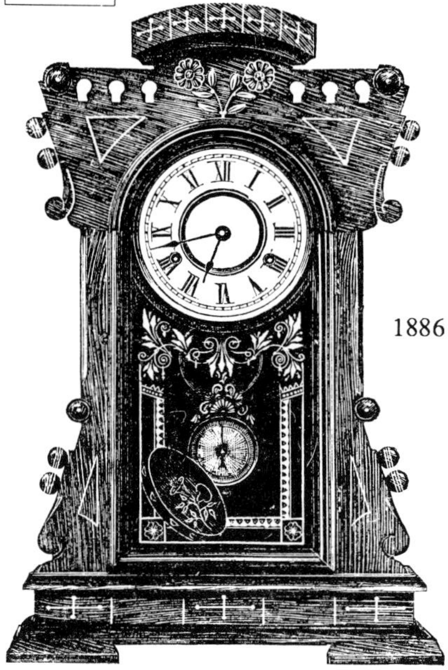

1886

Height, 21¼ inches. Dial, 6 inches.

TAMAR.

BLACK WALNUT.

8 Day Strike $7.30
8 " " Alarm 7.90
8 " " Cathedral Gong 7.80
8 " " " " Alarm . . 8.40

1350

1886

Height, 22½ inches. Dial, 6 inches.

TIGRIS.

BLACK WALNUT.

1 Day Strike $5.00
1 " " Alarm 5.60

1351

1890

Height, 20½ inches. Dial, 6 inches.

VOLANTE.

BLACK WALNUT.

8 Day Strike $6.75
8 Day Strike, Alarm 7.35
8 Day Strike, Cathedral Gong 7.25
8 Day Strike, Cathedral Gong, Al'm. 7.85

1352

1890

Height, 21½ inches. Dial, 6 inches.

VINDEX.

BLACK WALNUT.

8 Day Strike $6.75
8 Day Strike, Alarm 7.35
8 Day Strike, Cathedral Gong 7.25
8 Day Strike, Cathedral Gong, Al'm. 7.85

MANTEL CLOCKS - WOOD CASES

1353

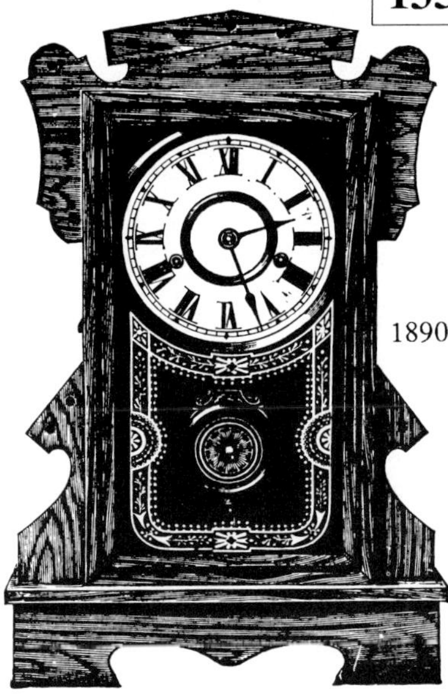

1890

Height, 17¾ inches. Dial, 6 inches.

BONITA.

BLACK WALNUT.

1 Day Strike..........$4.50
1 Day Strike, Alarm... 5.10

1354

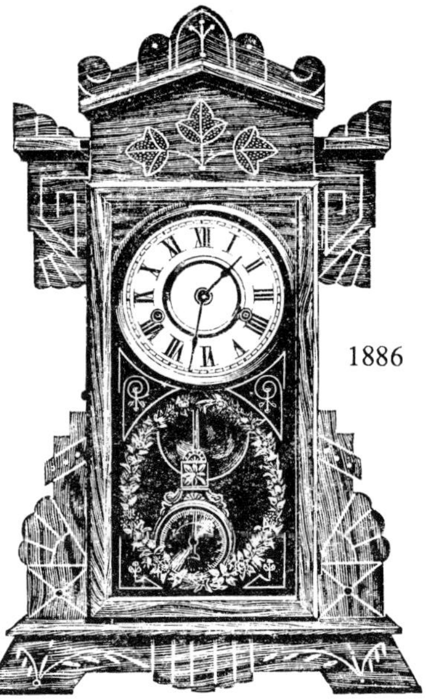

1886

Height, 20¾ inches. Dial, 5 inches.

SANTEE.

ASH.

1 Day Strike........$5.25
1 " " Alarm.. 5.85

1355

1890

Height, 17½ inches. Dial, 6 inches.

BOBOLINK.

BLACK WALNUT.

1 Day Strike..........$4.50
1 Day Strike, Alarm... 5.10

1356

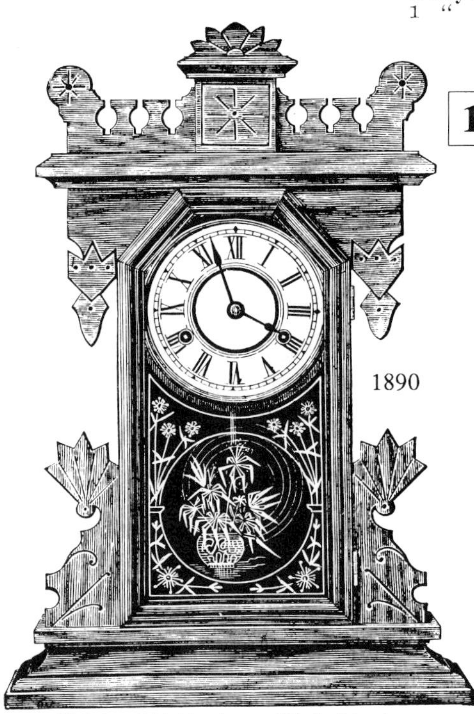

1890

Height, 21¼ inches. Dial, 6 inches.

YARANA.

BLACK WALNUT.

8 Day Strike........................$6.25
8 Day Strike, Alarm................ 6.85
8 Day Strike, Cathedral Gong....... 6.75
8 Day Strike, Cathedral Gong, Al'm.. 7.35

1357

1890

Height, 23 inches. Dial, 6 inches.

TOMAHAWK.

BLACK WALNUT.

8 Day Strike.......................$6.00
8 Day Strike, Alarm................ 6.60
8 Day Strike, Cathedral Gong........ 6.50
8 Day Strike, Cathedral Gong, Al'm.. 7.10

MANTEL CLOCKS - WOOD CASES

1358

1890

Height, 20½ inches. Dial, 5 inches.

IVY.

BLACK WALNUT.

1 Day Strike............$4.80
1 Day Strike, Alarm.... 5.40

1359

VOLGA. 1880

Walnut.
Height, 21½ inches.
Dial, 6 inches.
8 Day Strike.
8 Day Strike, Alarm.

1360

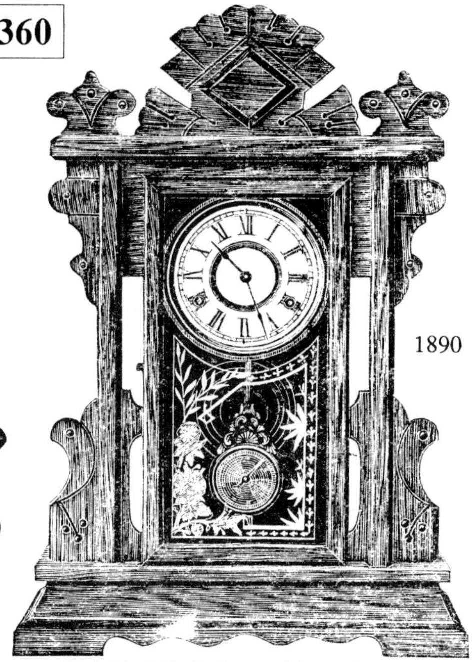

1890

Height, 22¼ inches. Dial, 5 inches.

No. 508.

BLACK WALNUT.

8 Day Strike $6.50
8 " " Alarm 7.10
8 " " Cathedral Gong 7.00
8 " " " " Alarm. 7.60

1361

1890

Height, 21½ inches. Dial, 6 inches.

BOREAS.

BLACK WALNUT.

1 Day Strike.........................$4.75
1 Day Strike, Alarm................ 5.35

1362

1890

Height, 21 inches. Dial, 6 inches

ZINGARA.

BLACK WALNUT.

8 Day Strike.........................$6.50
8 Day Strike, Alarm................ 7.10
8 Day Strike, Cathedral Gong........ 7.00
8 Day Strike, Cathedral Gong, Al'm... 7.60

MANTEL CLOCKS - WOOD CASES

1363

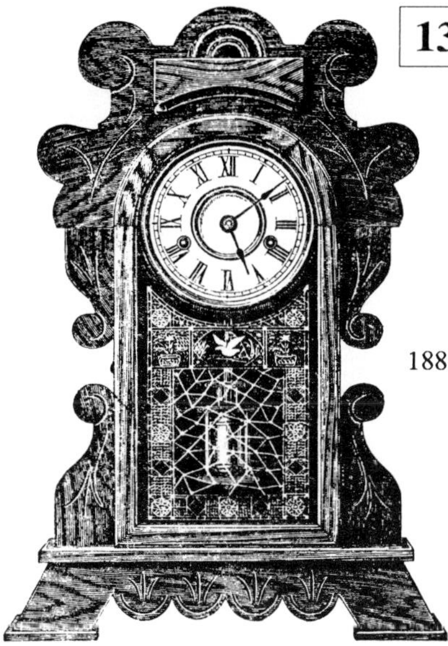

1886

Height, 20 inches. Dial, 5 inches.

No. 505.

IMITATION WALNUT.

1 Day Strike............$4.65
1 " " Alarm..... 5.25

1364

1886

Height, 19¾ inches. Dial, 6 inches.

No. 521.

BLACK WALNUT.

1 Day Strike..........$4.50
1 " " Alarm.... 5.10

1365

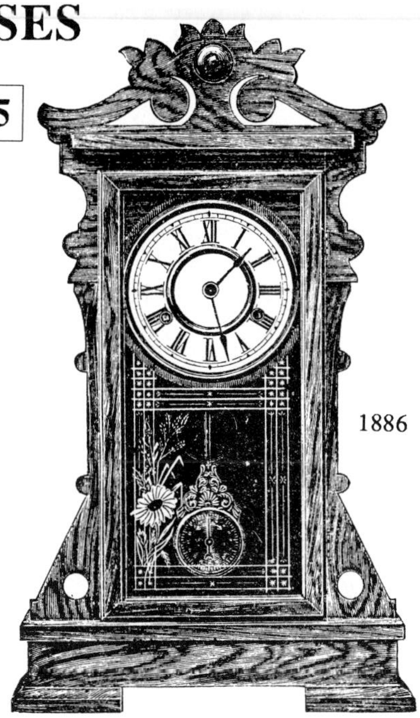

1886

Height, 22¾ inches. Dial, 6 inches.

No. 523.

BLACK WALNUT.

8 Day Strike $6.25
8 " " Alarm.................. 6.85
8 " " Cathedral Gong...... 6.75
8 " " " " Alarm.. 7.35

1366

1890

Height, 23½ inches. Dial, 6 inches.

SHAMROCK.

BLACK WALNUT.

8 Day Strike........................$6.00
8 Day Strike, Alarm................ 6.60
8 Day Strike, Cathedral Gong....... 6.50
8 Day Strike, Cathedral Gong, Al'm.. 7.10

1367

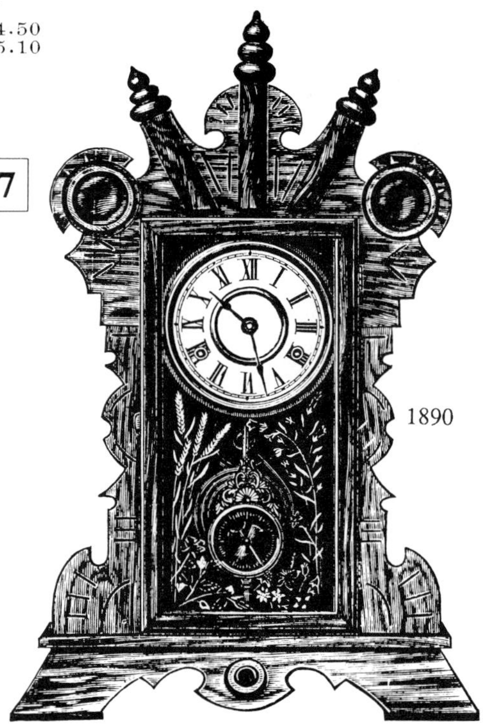

1890

Height, 22½ inches. Dial, 5 inches.

CHARMER.

BLACK WALNUT.

8 Day Strike.......................$5.75
8 Day Strike, Alarm.............. 6.35
8 Day Strike, Cathedral Gong..... 6.25
8 Day Strike, Cathedral Gong, Al'm 6.85

MANTEL CLOCKS - WOOD CASES

1367-A

1895

PARTHIAN.
Oak or Walnut.
Height, 24 inches; Dial, 6 inches.
With Thermometer, Barometer and Level.

8 Day Strike, -	List, $6.25
8 Day Strike, Alarm,	List, 6.75
8 Day Gong, -	List, 6.65
8 Day Gong, Alarm,	List, 7.15

1368

1887

No. 529.
BLACK WALNUT.
Height, 21½ inches. Dial, 6 inches.
8 day Strike. 8 day Strike, Alarm.

1369

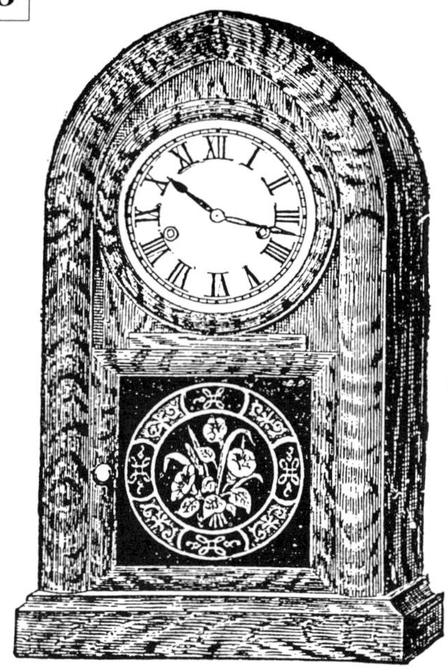

1885

ROUND GOTHIC.
ROSEWOOD, ZEBRA OR MAHOGANY.
Height, 19¼ inches. Dial, 6 inches.

1 Day Strike	$4 90
1 " " Alarm	5 50
8 " "	6 40
8 " " Alarm	7 00

1370

1911

1371

1911

1372

1911

GENEVA SET

HEIGHT, CLOCK ....9¾ INCHES
HEIGHT, CANDLESTICKS ....6⅞ INCHES
WIDTH, CLOCK ....4⅜ INCHES
WIDTH, CANDLESTICKS ....2½ INCHES
DIAL, PORCELAIN ....2 INCHES, FRENCH CAST SASH, WITH HEAVY BEVELED BOWED GLASS, FANCY BRASS HANDS
CASE, MARBLEIZED MAHOGANY FINISH, WITH BRASS FINISHED ORNAMENTS
MOVEMENT, 8 DAY TIME PIECE
PRICE ....LIST, $23.20

MANTEL CLOCKS - WOOD CASES

1373

1895

PRESIDENT.

1 Day Strike, - - -	List, $4 40
1 Day Strike, Alarm, -	List, 4 90
8 Day, Half-Hour Strike, - -	List, 5 00
8 Day, Half-Hour Strike, Alarm,	List, 5.50
8 Day, Half-Hour Strike, Gong,	List, 5.40
8 Day, Half-Hour Strike, Gong Alarm,	List, 5.90

1374

1895

SECRETARY.

1 Day Strike, - - -	List, $4.40
1 Day Strike, Alarm, - -	List, 4.90
8 Day, Half-Hour Strike, - -	List, 5.00
8 Day, Half-Hour Strike, Alarm,	List, 5.50
8 Day, Half-Hour Strike, Gong,	List, 5.40
8 Day, Half-Hour Strike, Gong Alarm,	List, 5.90

1375

1895

GOVERNOR.

1 Day Strike, - - -	List, $4 40
1 Day Strike, Alarm, - -	List, 4.90
8 Day, Half-Hour Strike, - -	List, 5.00
8 Day, Half-Hour Strike, Alarm,	List, 5.50
8 Day, Half-Hour Strike, Gong,	List, 5.40
8 Day, Half-Hour Strike, Gong Alarm,	List, 5.90

Interchangeable Tops.
23 inches High.

STATESMAN LINE.
Oak or Walnut.

Interchangeable Tops.
6 inch Dials.

1376

1895

SENATOR.

1 Day Strike, - - -	List, $4.40
1 Day Strike, Alarm, - -	List, 4.90
8 Day, Half-Hour Strike, - -	List, 5.00
8 Day, Half-Hour Strike, Alarm,	List, 5.50
8 Day, Half-Hour Strike, Gong, -	List, 5.40
8 Day, Half-Hour Strike, Gong Alarm,	List, 5.90

1377

1895

LOBBYIST.

1 Day Strike, - - -	List, $4.40
1 Day Strike, Alarm, - -	List, 4.90
8 Day, Half-Hour Strike, -	List, 5.00
8 Day, Half-Hour Strike, Alarm, -	List, 5.50
8 Day, Half-Hour Strike, Gong,	List, 5.40
8 Day, Half-Hour Strike, Gong Alarm,	List, 5.90

1378

1895

AMBASSADOR.

1 Day Strike, - -	List, $4.40
1 Day Strike, Alarm, - -	List, 4.90
8 Day, Half-Hour Strike, -	List, 5.00
8 Day, Half-Hour Strike, Alarm, -	List, 5.50
8 Day, Half-Hour Strike, Gong,	List, 5.40
8 Day, Half-Hour Strike, Gong Alarm,	List, 5.90

MANTEL CLOCKS - WOOD CASES

1379

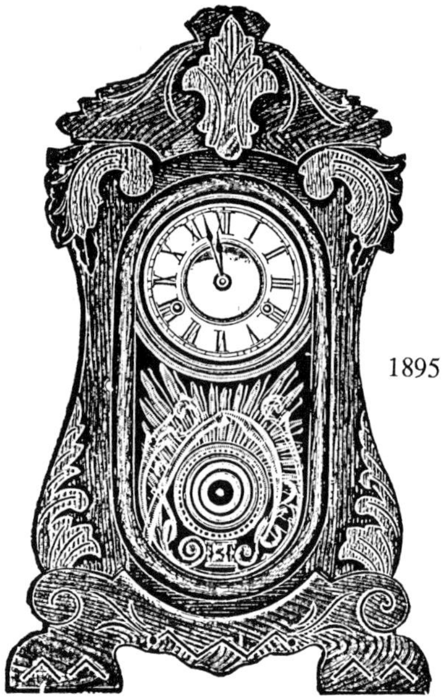

1895

GENERAL.

8 Day, Half Hour Strike. ....List, $5.25
8 Day, Half Hour Strike, Alarm.......List, 5.75
8 Day, Half Hour Strike, Gong........Lisi, 5.65
8 Day, Half Hour Strike, Gong Alarm.List, 6.15

1380

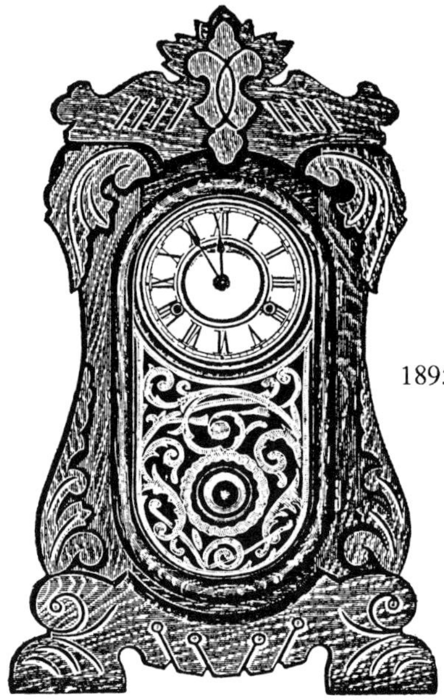

1895

COLONEL.

8 Day, Half Hour Strike..............List, $5.25
8 Day, Half Hour Strike, Alarm.......List, 5.75
8 Day, Half Hour Strike, Gong........List, 5.65
8 Day, Half Hour Strike, Gong Alarm..List, 6.15

1381

1895

MAJOR.

8 Day, Half Hour Strike..............List, $5.25
8 Day, Half Hour Strike, Alarm. ... List, 5.75
8 Day, Half Hour Strike, Gong........List, 5.65
8 Day, Half Hour Strike, Gong Alarm.List, 6 15

Interchangeable Tops. MILITARY LINE. Interchangeable Tops.

23 inches high. Oak or Walnut. 6 inch Dials.

1382

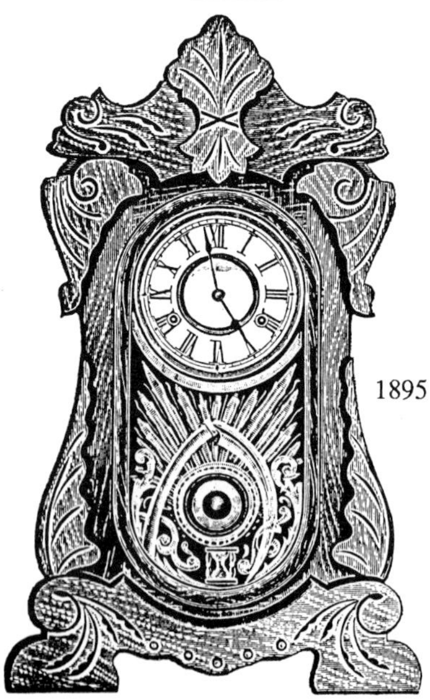

1895

CAPTAIN.

8 Day, Half Hour Strike..............List, $5.25
8 Day, Half Hour Strike, Alarm.......List, 5.75
8 Day, Half Hour Strike, Gong........List, 5.65
8 Day, Half Hour Strike, Gong Alarm.List, 6.15

1383

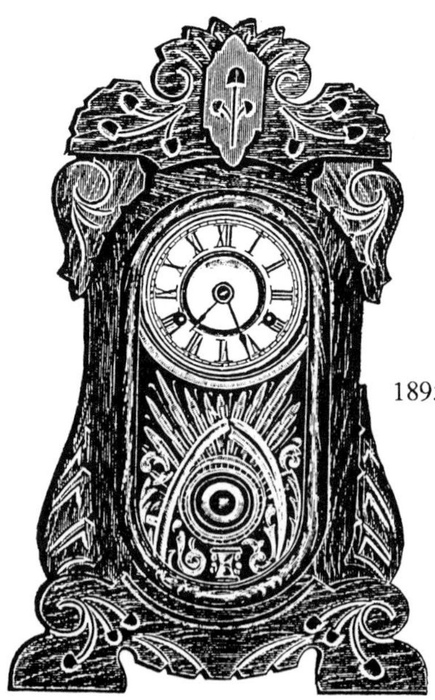

1895

LIEUTENANT.

8 Day, Half Hour Strike..............List, $5.25
8 Day, Half Hour Strike, Alarm. .....List, 5.75
8 Day, Half Hour Strike, Gong........List, 5.65
8 Day, Half Hour Strike, Gong Alarm.List, 6.15

1384

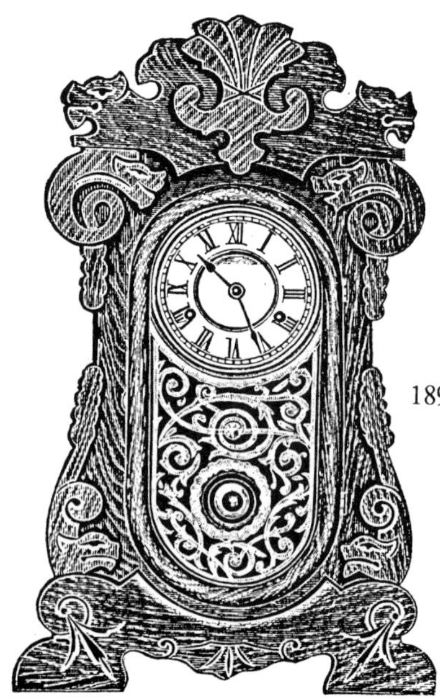

1895

SERGEANT.

8 Day, Half Hour Strike.. . ...List, $5.25
8 Day, Half Hour Strike, Alarm.......List, 5.75
8 Day, Half Hour Strike, Gong........List, 5.65
8 Day, Half Hour Strike, Gong Alarm.List, 6.15

MANTEL CLOCKS - WOOD CASES

1385

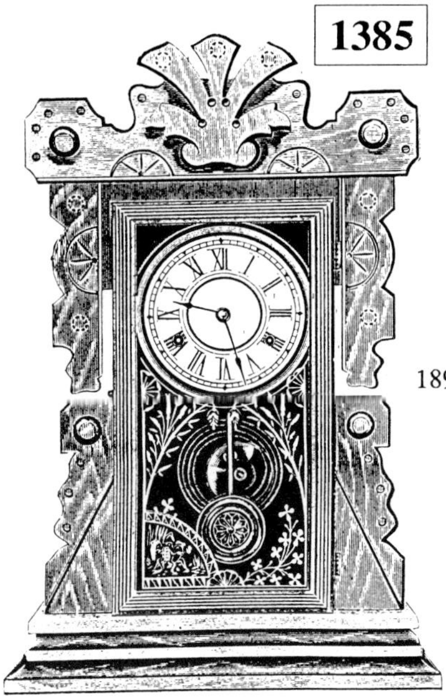

1895

1386

1895

1387

1895

MARSHAL.

8 Day, Half-Hour Strike, - - List, $4.75
8 Day, Half-Hour Strike, Alarm, List, 5.25
8 Day, Half-Hour Strike, Gong, - List, 5.15
8 Day, Half-Hour Strike, Gong Alarm, List, 5.65

ALDERMAN.

8 Day, Half-Hour Strike, - - List, $4.75
8 Day, Half-Hour Strike, Alarm, List, 5.25
8 Day, Half-Hour Strike, Gong, - List, 5.15
8 Day, Half-Hour Strike, Gong Alarm, List, 5.65

RECORDER.

8 Day, Half-Hour Strike, - - List, $4.75
8 Day, Half-Hour Strike, Alarm, List, 5.25
8 Day, Half-Hour Strike, Gong, - List, 5.15
8 Day, Half-Hour Strike, Gong Alarm, List, 5.65

HUSTLER LINE.

Interchangeable Tops. 23 inches High. Oak or Walnut. 6-inch Dial. Interchangeable Tops.

1388

1895

1389

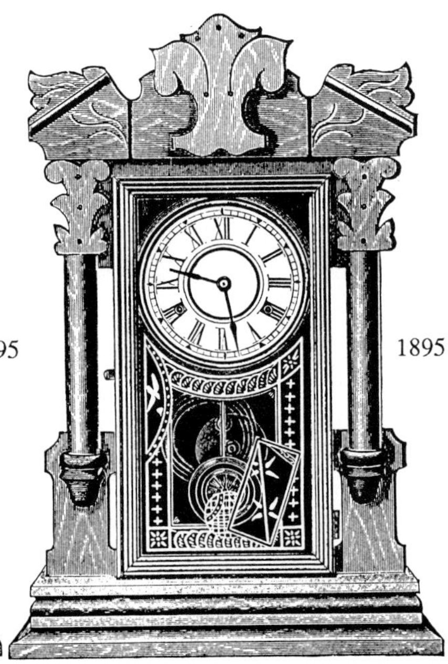

1895

1390

1895

MAYOR.

8 Day, Half-Hour Strike, - - List, $4.75
8 Day, Half-Hour Strike, Alarm, List, 5.25
8 Day, Half-Hour Strike, Gong, - List, 5.15
8 Day, Half-Hour Strike, Gong Alarm, List, 5.65

COUNCILMAN.

8 Day, Half-Hour Strike, - - List, $4.75
8 Day, Half-Hour Strike, Alarm, List, 5.25
8 Day, Half-Hour Strike, Gong, - List, 5.15
8 Day, Half-Hour Strike, Gong Alarm, List, 5.65

JANITOR.

8 Day, Half-Hour Strike, - - List, $4.75
8 Day, Half-Hour Strike, Alarm, List, 5.25
8 Day, Half-Hour Strike, Gong, - List, 5.15
8 Day, Half-Hour Strike, Gong Alarm, List, 5.65

MANTEL CLOCKS - WOOD CASES

1391

1895

NUTWOOD.

8 Day, Half-Hour Strike, - - List, $4.50
8 Day, Half-Hour Strike, Alarm, - List, 5.00
8 Day, Half-Hour Strike, Gong, - List, 4.90
8 Day, Half-Hour Strike, Gong Alarm, List, 5.40

1392

1895

NANCY.

8 Day, Half-Hour Strike, - - List, $4.50
8 Day, Half-Hour Strike, Alarm, - List, 5.00
8 Day, Half-Hour Strike, Gong, - List, 4.90
8 Day, Half-Hour Strike, Gong Alarm, List, 5.40

1393

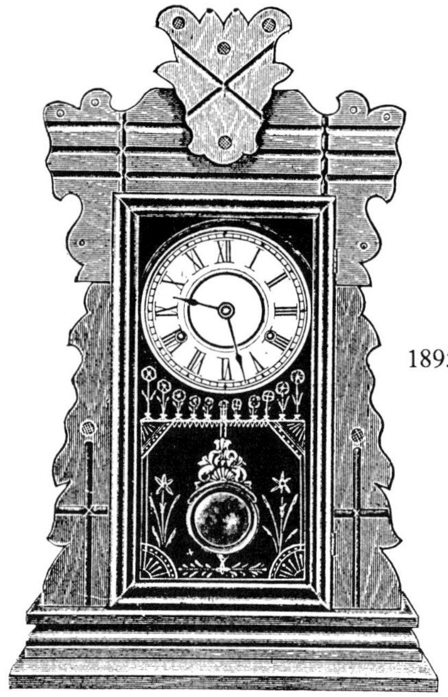

1895

SUNOL.

8 Day, Half-Hour Strike, - - List, $4.50
8 Day, Half-Hour Strike, Alarm, - List, 5.00
8 Day, Half-Hour Strike, Gong, - List, 4.90
8 Day, Half-Hour Strike, Gong Alarm, List, 5.40

PRIZE WINNER LINE.

23 inches High, Oak or Walnut, 6 inch Dial.

1394

1895

STAMBOUL.

8 Day, Half-Hour Strike, - - List, $4.50
8 Day, Half-Hour Strike, Alarm, - List, 5.00
8 Day, Half-Hour Strike, Gong, - List, 4.90
8 Day, Half-Hour Strike, Gong Alarm, List, 5.40

1395

1895

ARION.

8 Day, Half-Hour Strike, - - List, $4.50
8 Day, Half-Hour Strike, Alarm, - List, 5.00
8 Day, Half-Hour Strike, Gong, - List, 4.90
8 Day, Half-Hour Strike, Gong Alarm, List, 5.40

1396

1895

DIRECT.

8 Day, Half-Hour Strike, - - List, $4.50
8 Day, Half-Hour Strike, Alarm, - List, 5.00
8 Day, Half-Hour Strike, Gong, - List, 4.90
8 Day, Half-Hour Strike, Gong Alarm, List, 5.40

MANTEL CLOCKS - WOOD CASES

ROMAN LINE.

Oak or Walnut.

22 inches High; 6 inch Dial.

1397

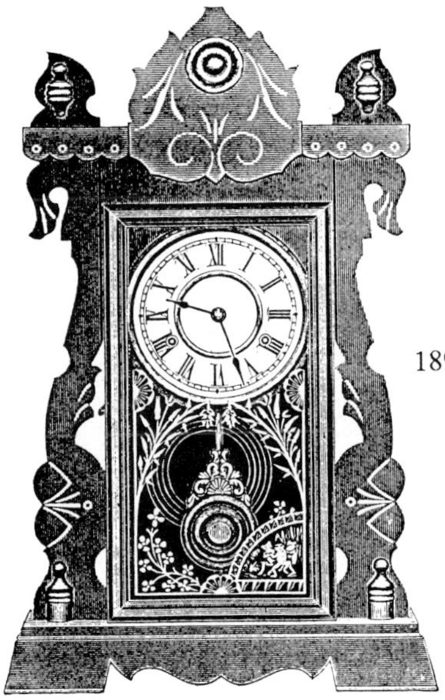

1895

METELLUS.

8 Day, Half-Hour Strike............List, $4.75
8 Day, Half-Hour Strike, Alarm.......List, 5.25
8 Day, Half-Hour Strike, Gong.......List, 5.15
8 Day, Half-Hour Strike, Gong Alarm..List, 5.65

1398

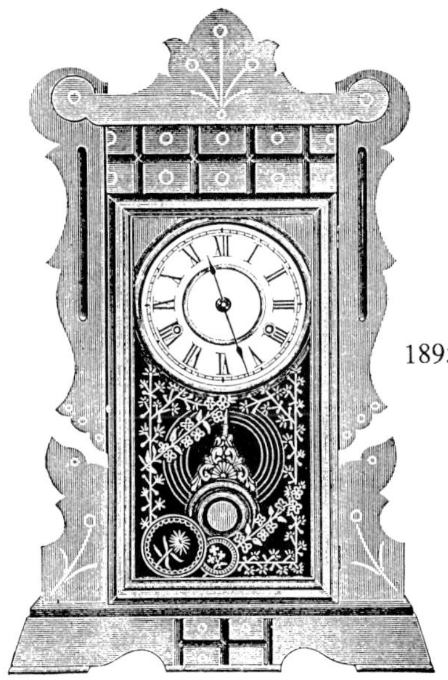

1895

TARQUIN.

8 Day, Half-Hour Strike............List, $4.75
8 Day, Half-Hour Strike, Alarm.......List, 5.25
8 Day, Half-Hour Strike, Gong.......List, 5.15
8 Day, Half-Hour Strike, Gong Alarm..List, 5.65

1399

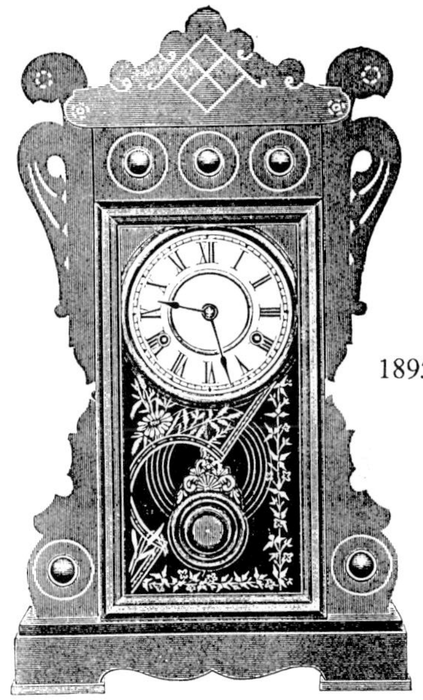

1895

JUVENAL.

8 Day, Half-Hour Strike............List, $4.75
8 Day, Half-Hour Strike, Alarm.......List, 5.25
8 Day, Half-Hour Strike, Gong.......List, 5.15
8 Day, Half-Hour Strike, Gong Alarm..List, 5.65

1400

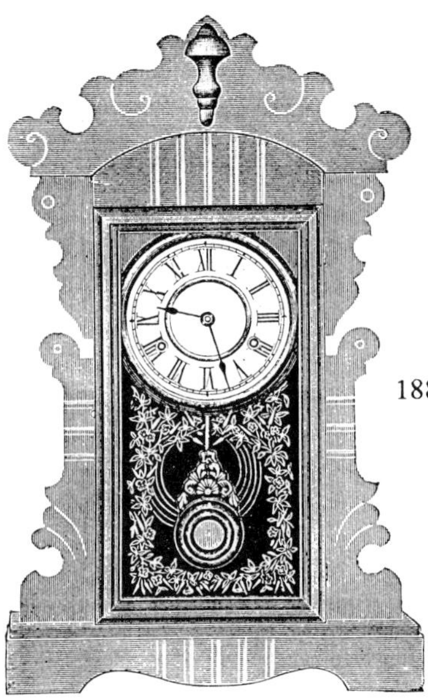

1885

AGRIPPA.

8 Day, Half-Hour Strike............List, $4.75
8 Day, Half-Hour Strike, Alarm.......List, 5.25
8 Day, Half-Hour Strike, Gong.......List, 5.15
8 Day, Half-Hour Strike, Gong Alarm..List, 5.65

1401

1885

GRACCHUS.

8 Day, Half-Hour Strike............List, $4.75
8 Day, Half-Hour Strike, Alarm.......List, 5.25
8 Day, Half-Hour Strike, Gong.......List, 5.15
8 Day, Half Hour Strike, Gong Alarm..List, 5.65

1402

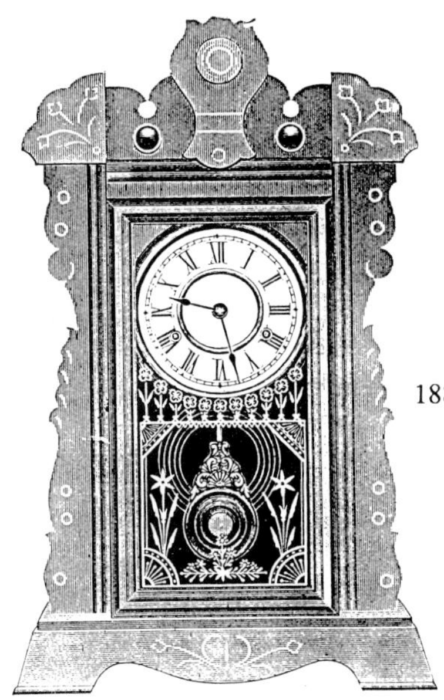

1885

REGULUS.

8 Day, Half-Hour Strike............List, $4.75
8 Day, Half-Hour Srtike, Alarm.......List, 5.25
8 Day, Half-Hour Strike, Gong.......List, 5.15
8 Day, Half-Hour Strike, Gong Alarm..List, 5.65

MANTEL CLOCKS - WOOD CASES

1403

1900

A

1404

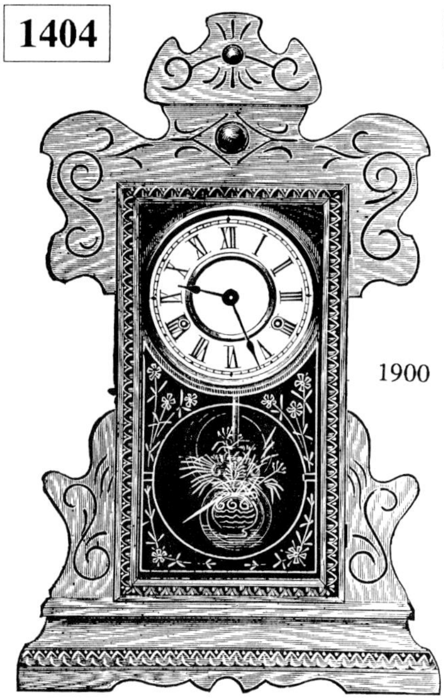

1900

B

1405

1900

C

PICKET LINE—Oak or Walnut.

Height, 22 inches. Width, 13½ inches. Dial, 6 inches.

List price,	Eight-day, Half-hour Strike,	$ 4.00
" "	Eight-day, Half-hour Strike, Alarm,	4.50
" "	Eight-day, Half-hour Strike, Gong,	4.40
" "	Eight-day, Half-hour Strike, Gong Alarm,	4.90

1406

1900

A

1407

1900

B

1408

1900

C

PATROL LINE—Oak or Walnut.

Packed assorted six in a Case. Height, 22¾ inches. Width, 14¼ inches. Dial, 6 inches.

List price,	Eight-day, Half-hour Strike,	$ 4.25
" "	Eight-day, Half-hour Strike, Alarm,	4.75
" "	Eight-day, Half-hour Strike, Gong,	4.65
" "	Eight-day, Half-hour Strike, Gong Alarm,	5.15

MANTEL CLOCKS - WOOD CASES

1409

1890

Height, 20½ inches. Dial, 6 inches.

MAYFLOWER.

BLACK WALNUT.

8 Day Strike........................$6.50
8 Day Strike, Alarm................. 7.10
8 Day Strike, Cathedral Gong....... 7.00
8 Day Strike, Cathedral Gong, Al'm.. 7.60

1410

CYGNET.

BLACK WALNUT.

1890

Height, 23½ inches. Dial, 6 inches.

8 Day Strike.......................$7.00
8 Day Strike, Alarm................ 7.60
8 Day Strike, Cathedral Gong...... 7.50
8 Day Strike, Cathedral Gong, Al'm. 8.10

1411

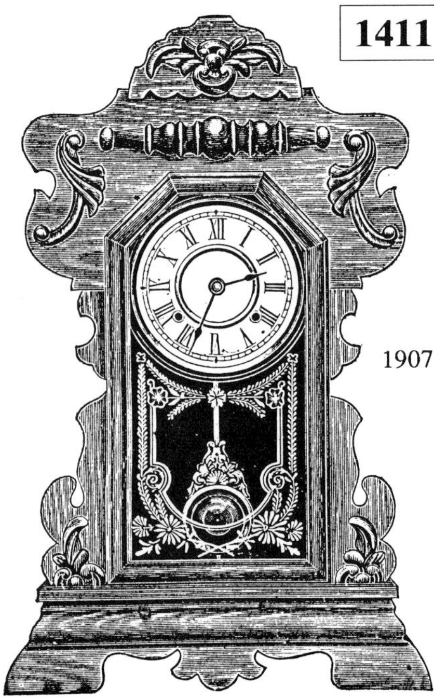

1907

A

1412

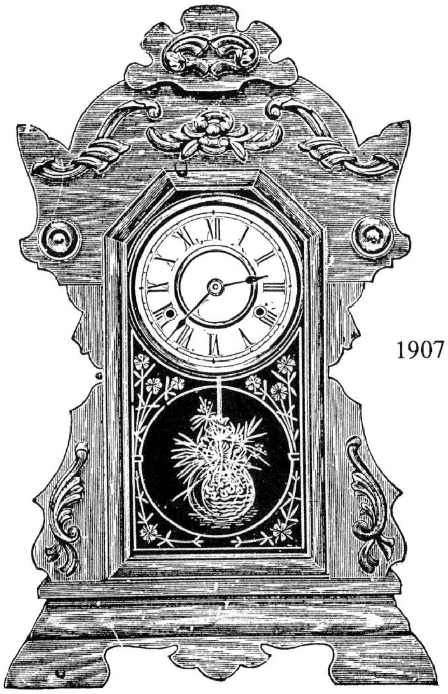

1907

B

1413

1907

C

MERSEY LINE - OAK OR WALNUT.

Height, 23 inches. Width, 14 inches. Dial, 6 inches.

Eight-Day, Half-hour Strike, List Price...$4.40 Eight-Day, Half-hour Gong, List Price...$4.80
Eight-Day, Half-hour Strike, Alarm, List Price...$4.90 Eight-Day, Half-hour Gong, Alarm, List Price...$5.30

MANTEL CLOCKS - WOOD CASES

1414

1911

A

1415

1911

B

1416

1911

C

GYPSY LINE

HEIGHT ....18¾ INCHES
WIDTH ....9¾ INCHES
DIAL ....6 INCHES
CASE, SOLID OAK
PACKED, ASSORTED 6 IN A BOX
PRICE, 8 DAY HALF-HOUR STRIKE ....LIST, $4.35
PRICE, 8 DAY HALF-HOUR STRIKE,
CATHEDRAL GONG ....LIST, 4.80
PRICE, WITH ALARM ....LIST EXTRA, .45

1417

1907

A

1418

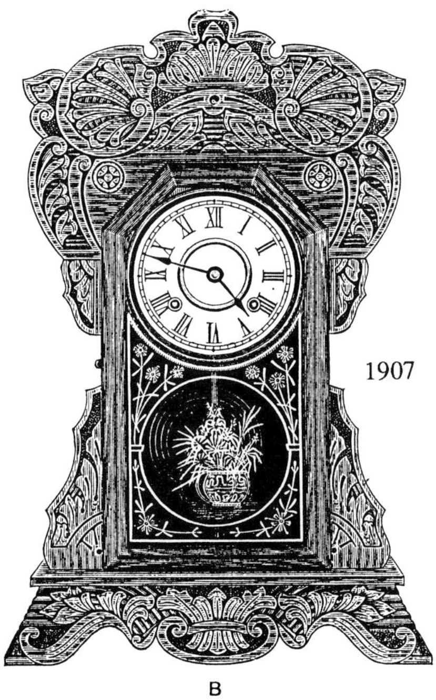

1907

B

1419

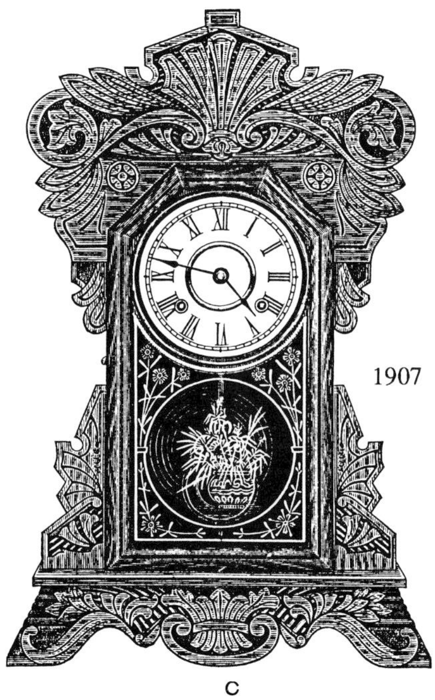

1907

C

CAMDEN LINE—All Oak, Assorted Patterns

Height, 22 inches. Width, 14 inches. Dial, 6 inches.

Eight-day, Half-hour Strike ....List Price, $4.25
Eight-day, Half-hour Strike, Alarm ... " " 4.75
Eight-day, Half-hour Gong...........List Price, $4.65
Eight-day, Half-hour Gong, Alarm .... " " 5.15

MANTEL CLOCKS - WOOD CASES

1420

1886

No. 525.

BLACK WALNUT.

Height, 23½ inches. Dial, 6 inches.

8 Day Strike		$6.50
8 " " Alarm		7.10
8 " " Cathedral Gong		7.00
8 " " " " Alarm		7.60

1421

1881

LOIRE.

WALNUT.

Height, 20 inches. Dial, 5 inches

8 Day Strike. 8 Day Strike, Alarm.

1422

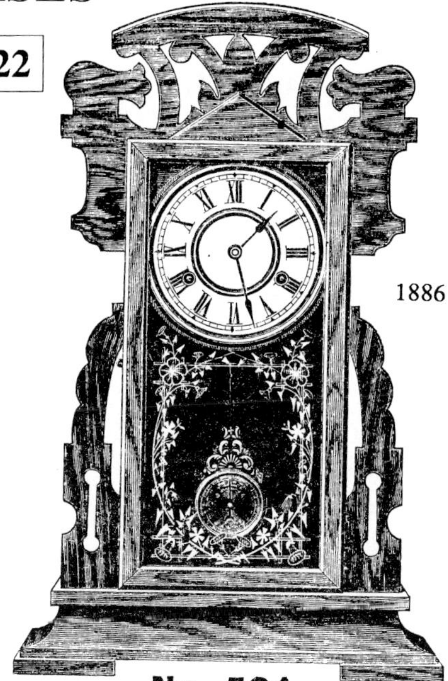

1886

No. 524.

BLACK WALNUT.

Height, 22 inches. Dial, 6 inches.

8 Day Strike		$6.35
8 " " Alarm		6.95
8 " " Cathedral Gong		6.85
8 " " " " Alarm		7.45

1423

1900

A

1424

1900

B

1425

1900

C

METAL TRIMMED MANTELS

BRAZILIAN LINE

Oak with Cast Metal Trimmings. Half-hour Strike on Wire Bell or Cathedral Gong.

Height, 22¼ inches. Width, 15 inches. Dial, 6 inches.

List price, Single Clock, Eight-day Strike,	each,	$5.00
" " " " Eight-day Strike, Alarm,	"	5.50
" " " " Eight-day Gong,	"	5.40
" " " " Eight-day Gong, Alarm,	"	5.90

MANTEL CLOCKS - WOOD CASES

1426

1900

A

1427

1900

B

1428

1900

C

MAINE LINE—Oak or Walnut.

Height, 23½ inches. Width, 15 inches.

Dial, 6 inches.

List price, Eight-day, Half-hour Strike, ...		$ 4.50
" " Eight-day, Half-hour Strike, Alarm, . .		5.00
" " Eight-day, Half-hour Strike, Gong, . .		4.90
" " Eight-day, Half-hour Strike, Gong Alarm, .		5.40

1429

1900

DEE

1430

1900

DOON

1431

1900

DON

"D" LINE—Oak or Walnut.

Packed assorted six in a Case. Height, 24⅜ inches. Width, 15⅝ inches. Dial, 6 inches.

List price, Eight-day, Half-hour Strike, ...	$ 5.25
" ". Eight-day, Half-hour Strike, Alarm, ...	5.75
" " Eight-day, Half-hour Strike, Gong, ...	5.65
" " Eight-day, Half-hour Strike, Gong Alarm, ...	6.15

MANTEL CLOCKS - WOOD CASES

The year entered beside each clock is the year of the catalog from which the illustration was taken. Some clocks may have been produced a few years earlier.

1432

1907

A

NORWICH LINE—Oak or Walnut, Assorted Patterns

Height, 25 inches. Width, 15½ inches. Dial, 6 inches.

Eight-day, Half-hour Strike . . . . List Price, $4.50
Eight-day, Half-hour Strike, Alarm . . . " " 5.00

Eight-day, Half-hour Gong. . . . . List Price, $4.90
Eight-day, Half-hour Gong, Alarm . . . . " " 5.40

SOLD IN CASE LOTS AT THE FOLLOWING PRICES:

Six Clocks (two of each pattern), Eight-day, Half-hour Strike	List Price,	$23.00
" " " " " " " " Alarm . .	" "	26.00
" " " " " " " Gong	" "	25.40
" " " " " " " " Alarm . .	" "	28.40

1434

1433

1907

B

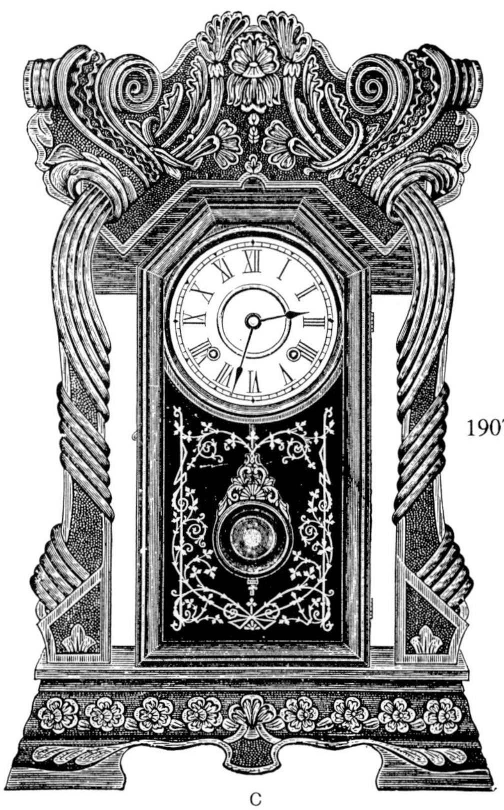

1907

C

MANTEL CLOCKS - WOOD CASES

1435 22 inches High. DANDY LINE. Oak or Walnut. 6 inch Dial. **1436-A**

1895

A

8 Day Strike, - - List, $4.00
8 Day Strike, Alarm, - - List, 4.50

1436

1895

B

8 Day Strike, - List, $4.00
8 Day Strike, Alarm, - - List, 4.50

1895

C

8 Day Strike, - - List, $4.00
8 Day Strike, Alarm, - - List, 4.50

1437

1911

A

1438

B

1911

MERCHANTS LINE

HEIGHT ......22¾ INCHES
WIDTH ....15 INCHES
DIAL ....6 INCHES
CASE, OAK OR DARK WOOD
PACKED, ASSORTED 6 IN A BOX
PRICE, 8 DAY HALF-HOUR STRIKE.....LIST, $4.35
PRICE, 8 DAY HALF-HOUR STRIKE,
CATHEDRAL GONG......................LIST, 4.80
PRICE, WITH ALARM.............LIST EXTRA, .45

1439

1911

C

MANTEL CLOCKS - WOOD CASES

1439-A

1913

MIRROR

LIST, $4.65

HAS MIRRORS IN SIDE PANELS.
HEIGHT, 24¼ INCHES. WIDTH, 16½ INCHES.

BOUDOIR CLOCKS

With

Twelve Day

Movement

1440

SQUARE BOUDOIR
HEIGHT, 7¾ inches WIDTH, 7¼ inches
CASE, Antique Mahogany with Dark Panel
Rich 3½ inch Gold Dial

1441

GOTHIC BOUDOIR
HEIGHT, 7¾ inches WIDTH, 7¼ inches
CASE, Antique Mahogany with Dark Panels
Beautiful Gold Dial 3½ inches wide

1442

OVAL BOUDOIR
HEIGHT, 7¾ inches WIDTH, 7¼ inches
CASE, Antique Mahogany with Dark Panels
Fancy Gold Dial, 3½ inches in width

MISSION CLOCKS

The year entered beside each clock is the year of the catalog from which the illustration was taken. Some clocks may have been produced a few years earlier.

1443

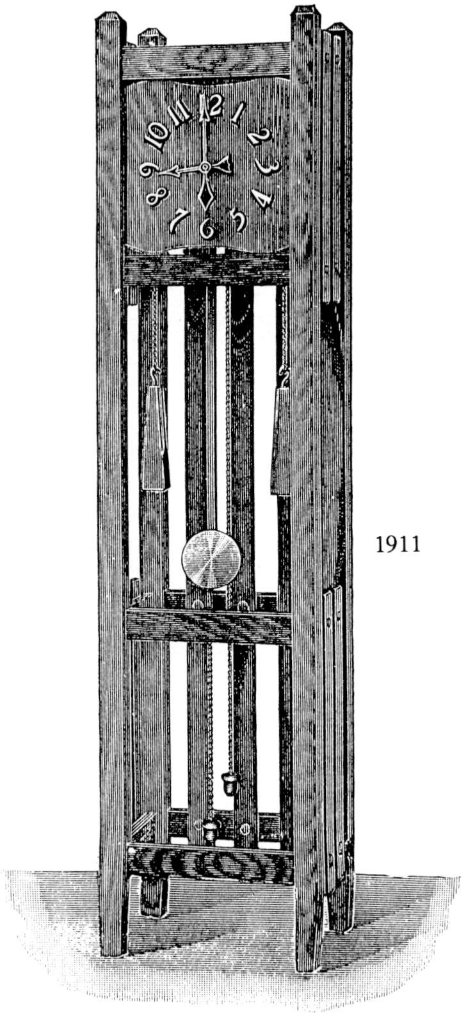

1911

DUTCH No. 20

HEIGHT ....78 INCHES
WIDTH ....18½ INCHES
DIAL, WOOD ....12 INCHES, CAST HANDS AND NUMERALS
CASE, SOLID OAK
FINISH, FLEMISH
MOVEMENT, 8 DAY WEIGHT, HALF-HOUR STRIKE, CATHEDRAL GONG
PRICE ....LIST, $43.50

1444

1906

DUTCH No. 45

Height, 68½ inches. Width, 15½ inches.
Solid Oak, Flemish Finish.
Stamped Brass Hands and Numerals.
Eight-day, Half-hour Strike, Cathedral Gong, Spring Movement.

List Price, $18.10

1445

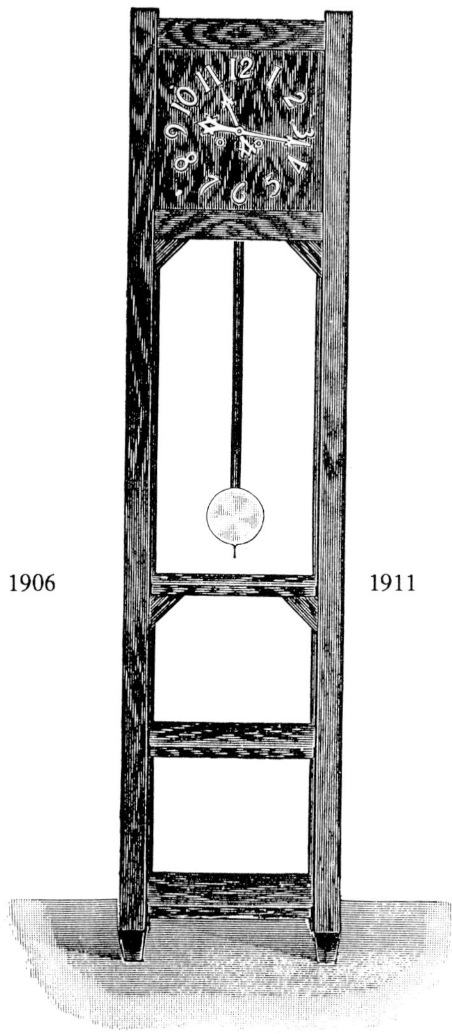

1911

DUTCH No. 40

HEIGHT ....77½ INCHES
WIDTH ....17½ INCHES
DIAL, WOOD ....12 INCHES, CAST HANDS AND NUMERALS
CASE, SOLID OAK
FINISH, FLEMISH
MOVEMENT, 8 DAY SPRING, HALF-HOUR STRIKE, CATHEDRAL GONG
PRICE ....LIST, $27.60

MISSION CLOCKS

1446

1906

DULCE

Height, 24 inches. Width, 16 inches.
12 inch Wood Dial. Solid Oak, Flemish Finish.
Cast Brass Hands and Numerals.

Eight-day Time . . . . List Price, $6.45
Eight-day Strike . . . . " " 7.45

1447

1911

LA CRUZ

HEIGHT ....22½ INCHES
WIDTH ......16½ INCHES
DIAL, WOOD....8 INCHES, CAST NUMERALS AND FANCY BRASS HANDS
CASE, SOLID OAK WITH STAINED GLASS
FINISH, FLEMISH
PACKED, 6 IN A BOX
MOVEMENT, 8 DAY HALF-HOUR STRIKE, CATHEDRAL GONG
PRICE....LIST, $7.25

1448

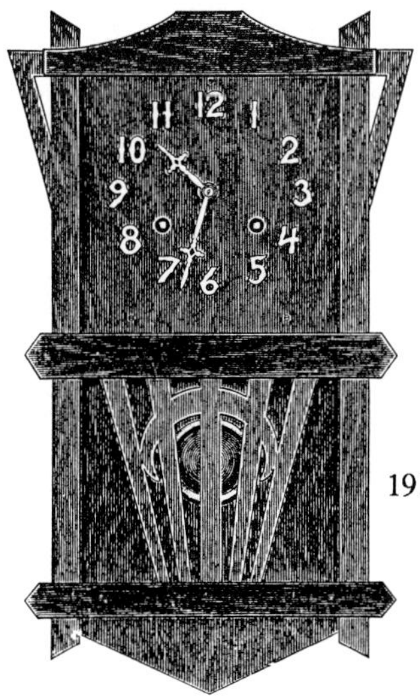

1911

SAN LUCAS

HEIGHT ....20¼ INCHES
WIDTH ......11½ INCHES
DIAL, WOOD....6 INCHES, CAST NUMERALS AND FANCY BRASS HANDS
CASE, SOLID OAK
FINISH, FLEMISH
PACKED, 6 IN A BOX
MOVEMENT, 8 DAY HALF-HOUR STRIKE, CATHEDRAL GONG
PRICE....LIST, $4.35

1449

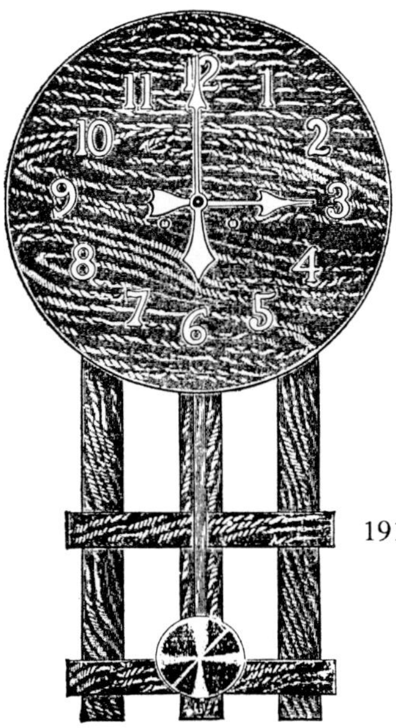

1911

GORDON

HEIGHT ....32 INCHES
WIDTH ....16 INCHES
DIAL, WOOD ....12 INCHES, BRASS NUMERALS AND FANCY HANDS
CASE, SOLID OAK
FINISH, FLEMISH
PRICE, 8 DAY HALF-HOUR STRIKE, CATHEDRAL GONG ......LIST, $6.50

1450

1906

DURANCE HANGING

DURANCE HANGING

Height, 37½ inches. Width, 15¾ inches.
Solid Oak, Flemish Finish.
Cast Brass Hands and Numerals.

Eight-day Time . . . . List Price, $21.50
Eight-day, Strike, Gong . . " " 23.00
Thirty-day Time . . . . " " 23.50

MISSION CLOCKS

MISSION MANTEL CLOCKS. 8-DAY HALF-HOUR STRIKE,
CATHEDRAL GONG. SOLID OAK CASES, FLEMISH FINISH. 6-INCH WOOD DIALS.
RAISED CAST NUMERALS AND FANCY HANDS. PACKED 6 IN A BOX.

1451

1913

SAN RAFAEL
LIST, $4.35
HEIGHT, 20½ INCHES. WIDTH, 12 INCHES.

1452

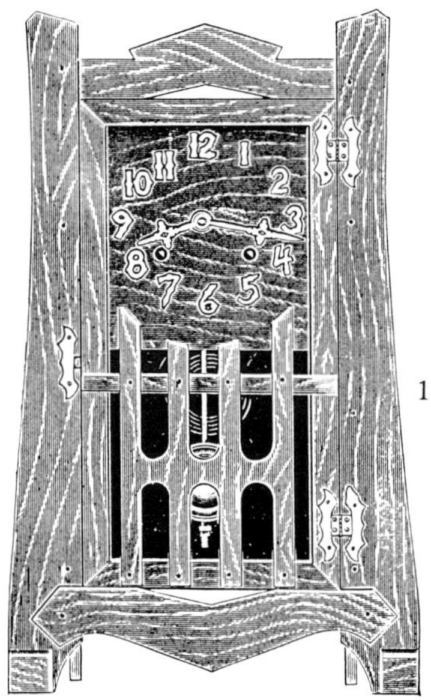

1913

SAN JOSE
LIST, $4.35
HEIGHT, 20½ INCHES. WIDTH, 12 INCHES.

1453

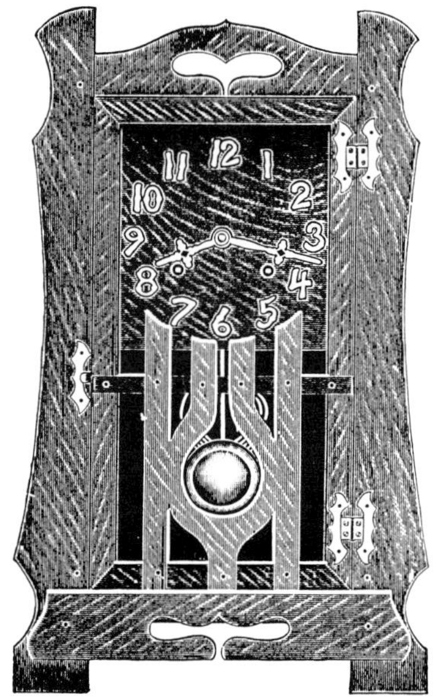

1913

SAN MARTIN
LIST, $4.35
HEIGHT, 20½ INCHES. WIDTH, 12 INCHES.

1454

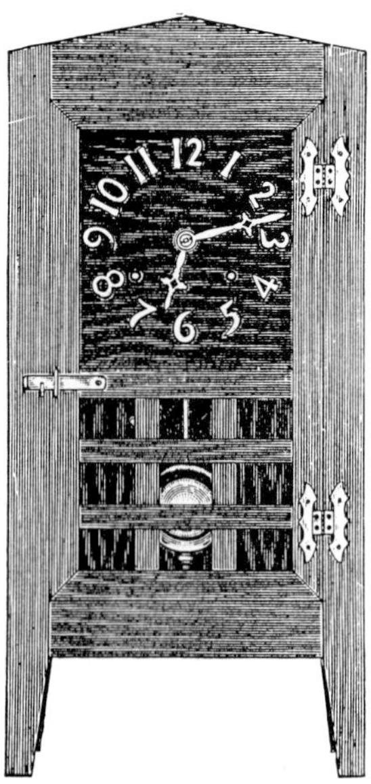

1913

SAN MATEO
LIST, $5.80
HEIGHT, 20½ INCHES. WIDTH, 9½ INCHES

1455

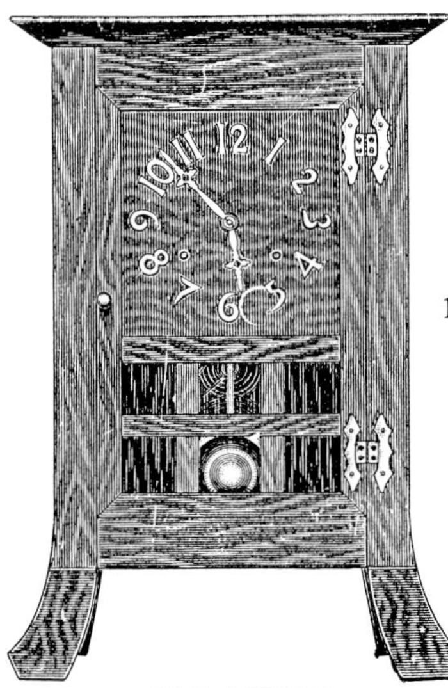

1913

SAN PEDRO
LIST, $5.80
HEIGHT, 19 INCHES. WIDTH, 12½ INCHES.

1456

1913

SANTA CLARA
LIST, $5.80
HEIGHT, 20¼ INCHES. WIDTH, 11¼ INCHES.

MISSION CLOCKS

MISSION MANTEL CLOCKS. 8-DAY HALF-HOUR STRIKE,
CATHEDRAL GONG. SOLID OAK CASES, FLEMISH FINISH. 6-INCH WOOD DIALS.
RAISED CAST NUMERALS AND FANCY HANDS. PACKED 6 IN A BOX.

1457

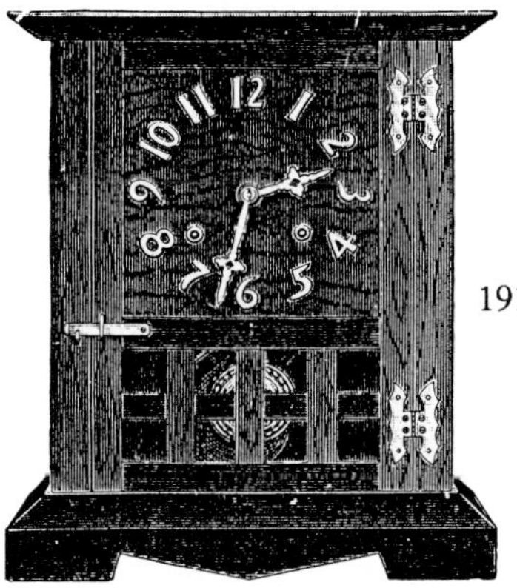

1913

LOS ALAMOS
LIST. $5.80
HEIGHT, 14¼ INCHES. WIDTH, 11¾ INCHES.

1458

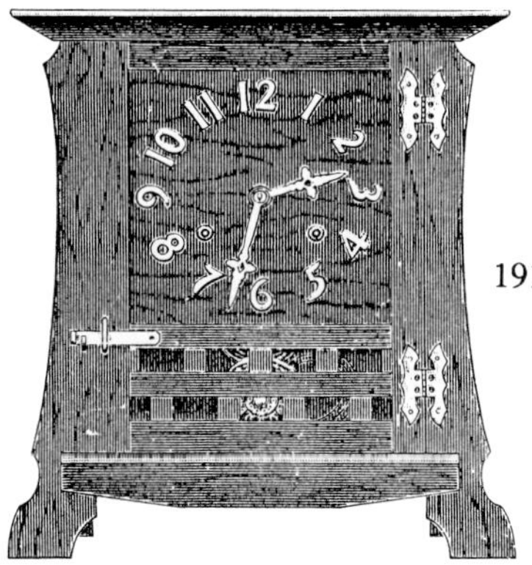

1913

LOS SANTOS
LIST, $5.80
HEIGHT, 13¾ INCHES. WIDTH, 12¼ INCHES.

1459

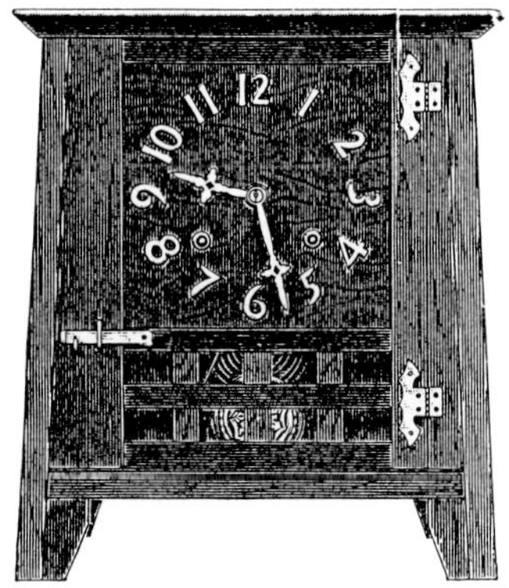

1913

LOS BARRIOS
LIST, $5.80
HEIGHT, 13½ INCHES. WIDTH, 11½ INCHES.

The year entered beside each clock is the year of the catalog from which the illustration was taken. Some clocks may have been produced a few years earlier.

1460

1906

LA LANZA
Mantel Weathered Oak Mission Clock. Eight-inch Dial, with Brass Numerals and Hands. Height, 22 inches. Width, 13 ½ inches. Eight-Day, Half-hour Strike on Cathedral Gong. List price, $10.50

1461

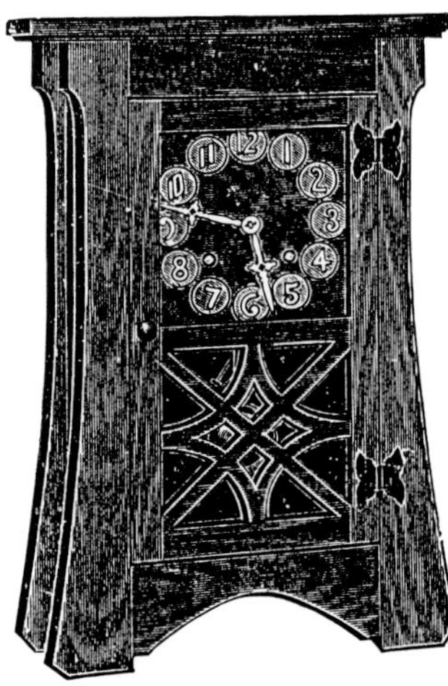

1911

SAN LUIS

HEIGHT ....20 INCHES
WIDTH ....13 INCHES
DIAL, WOOD....6 INCHES, WHITE FIGURES ON BACKGROUND OF ARTISTIC COLOR, FANCY BRASS HANDS
CASE, SOLID OAK
FINISH, FLEMISH
PACKED, 6 IN A BOX
MOVEMENT, 8 DAY HALF-HOUR STRIKE, CATHEDRAL GONG
PRICE....LIST, $6.50

1462

1907

SAN CARLOS
Mantel Weathered Oak Mission Clock. Wooden Dial, with white figures on a background of dark brown. Fancy Brass Hands. Height, 20 inches. Width, 9 inches. Dial, 6 inches. List price, $5.55

MISSION CLOCKS

1463

1907

SAN PABLO

Mantel Weathered Oak Mission Clock. Wooden Dial, with white figures on a background of dark brown. Fancy Brass Hands. Height, 20 inches. Width, 11 inches. Dial, 6 inches. List price, $5.55

1464

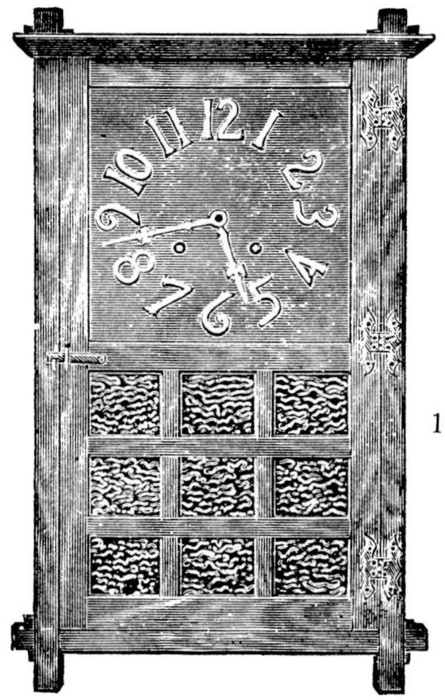

1907

LA PLATA

Mantel Weathered Oak Mission Clock. 8 inch Dial, with Brass Numerals and Hands. Height, 22 ¾ inches. Width, 13 ¾ inches. 8 Day, Half-hour Strike on Cathedral Gong. List price, $10.50

1465

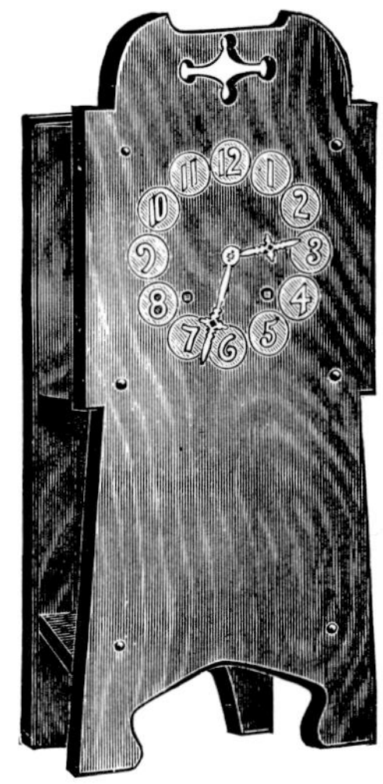

1907

SAN JUAN

Mantel Weathered Oak Mission Clock. Wooden Dial, with white figures on a background of dark brown. Fancy Brass Hands. Height, 20 ½ inches. Width, 9 ½ inches. Dial, 6 inches. List price, $5.55

1466

1907

A

1467

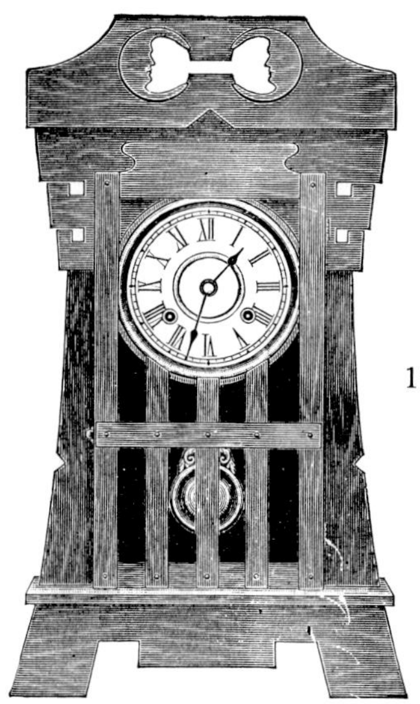

1907

B

1468

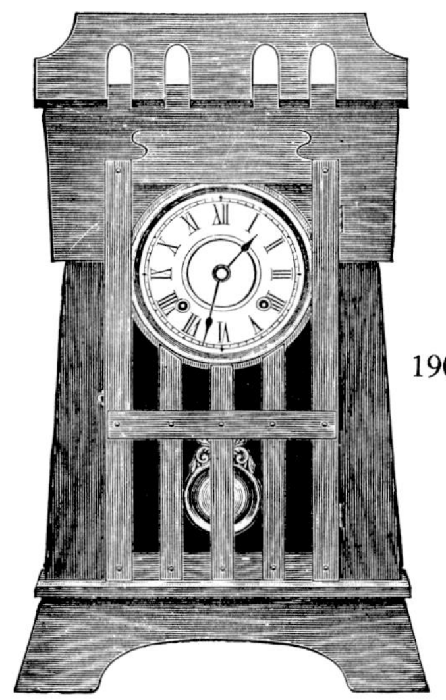

1907

C

SANCHO LINE - SOLID OAK, FLEMISH FINISH.

Height, 22 ½ inches. Width, 13 inches. Dial, 6 inches.

Eight-Day, Half-hour Strike, List Price...$4.00 Eight-Day, Half-hour Gong, List Price...$4.40

Eight-Day, Half-hour Strike, Alarm, List Price...$4.50 Eight-Day, Half-hour Gong, Alarm, List Price...$4.90

MISSION CLOCKS

1469

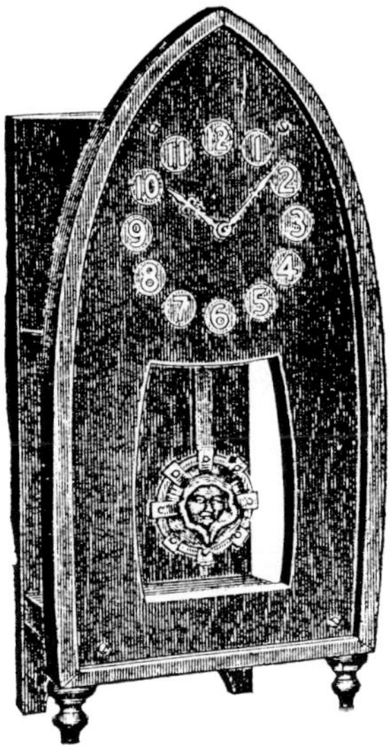

1907

MORO.
No. 6229 Height 13 inches, width 6¼ inches, dial 4 inches, mahogany case, each......................$4 70

1470

1907

VEGA.
No. 6230 Height 13 inches, width 6 inches, dial 4 inches, mahogany case, each......................$4 70

1471

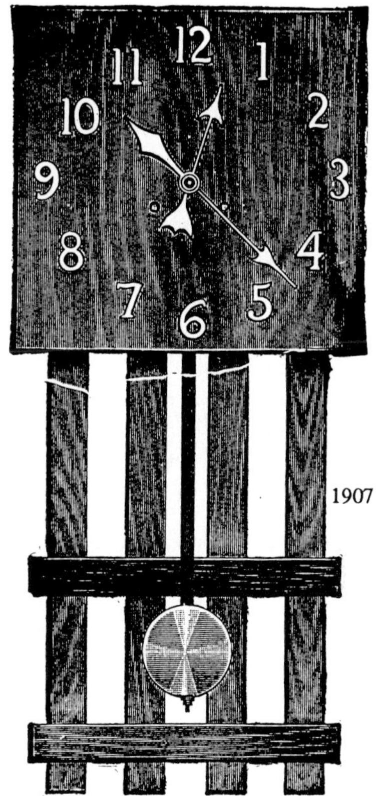

1907

No. B5084. $6.50
Height, 30 inches; width, 13 inches; 12-inch wood dial with brass numerals and fancy hands.
Solid oak case. Flemish finish.
8-day, half-hour strike. Cathedral gong.
(New Haven.)

1472

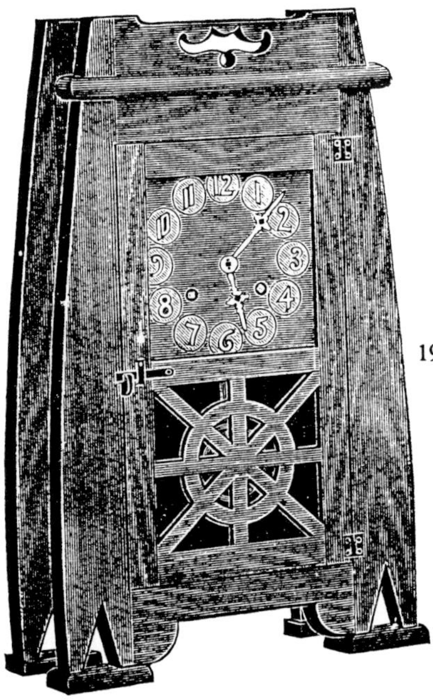

1907

SAN BENITO.
No. 6234 Mantel Weathered Oak Mission Clock, wooden dial with white figures on a background of cobalt blue, fancy brass hands, height 21 inches, width 13½ inches, dial 6 each.................. ....$8 35

1473

1907

No. B5085. SAN FELIPE. $4.35
Weathered oak.
Height, 20¼ inches; dial, 6 inches.
Brass hands and numerals.
8-day, half-hour strike. Cathedral gong.
(New Haven.)

1474

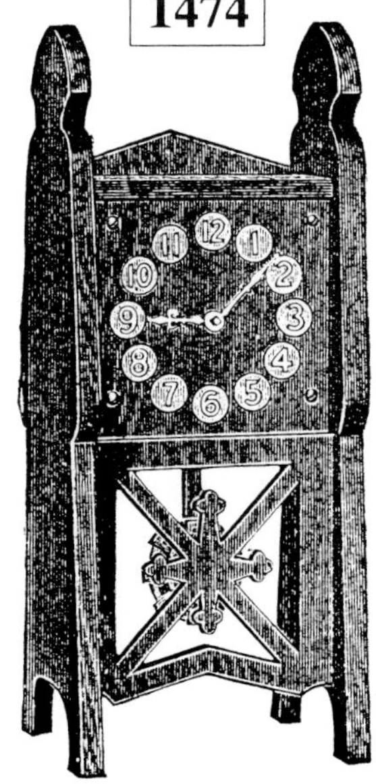

1907

PALOS.
No. 6231 Height 13½ inches, width 5½ inches, dial 4 inches, mission oak case, each......................$4 70

MISSION CLOCKS

1475

1907

TOSCA.
No. 6232 Height 12½ inches, width 6½ inches, dial 4 inches, mahogany case, each........................$5 70

1476

1907

No. 7252 (Also called **MINCH**)
Solid Oak case. Flemish Finish. Fitted with pendulum movement, with brass hands and numerals. Height, 12 inches. Width, 4½ inches. Dial, 3 inches.
One-day Time..............................$5.35

1477

1907

No. 7216 (Also called **LISO**)
7216
Height 8½ inches, Width 4½ inches, Dial 2 inches, The Case is Solid Oak, Flemish Finish, Wood Dial with White Figures Painted on a Dark Blue Background, Fitted with Lever Movement, One-Day Time,
$3 50

1478

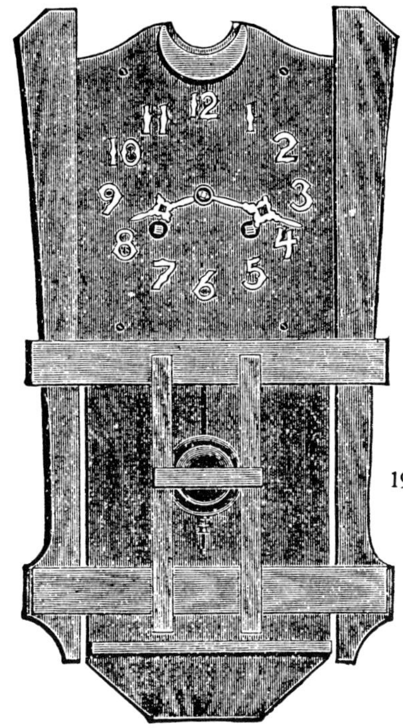

1907

7211
Weathered Oak Mission Hanging Clock, Height 20¾ inches, Width 10¾ inches, Dial 6 inches,
$6 00 each.

1479

1907

No. 7250 (Also called **MINAS**)
Solid Oak case. Flemish Finish. Fitted with pendulum movement, with brass hands and numerals. Height, 12 inches. Width, 4½ inches. Dial, 3 inches.
One-day Time..............................$5.35

1480

1907

No. 7251 (Also called **MINHO**)
Solid Oak case. Flemish Finish. Fitted with pendulum movement, with brass hands and numerals. Height, 12 inches. Width, 4½ inches. Dial, 3 inches.
One-day Time..............................$5.35

1481

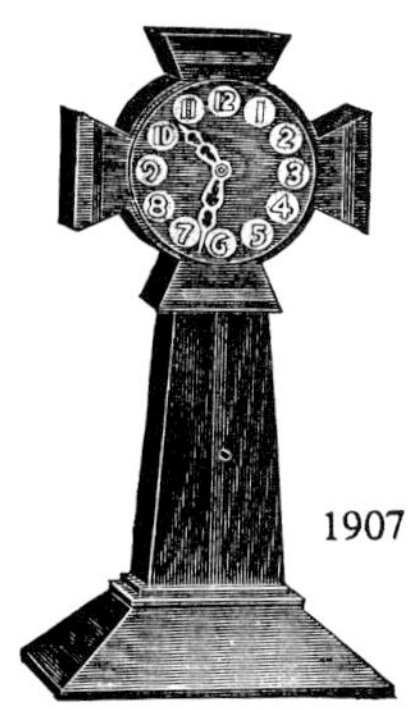

1907

No. 7213 (Also called **LUTE**)
7213
Height 9 inches, Width 4½ inches, Dial, 2 inches. The case is Solid Oak, Flemish Finish, Wood Dial with White Figures Painted on a Dark Blue Background, Fitted with Lever Movement.
$3 50

1482

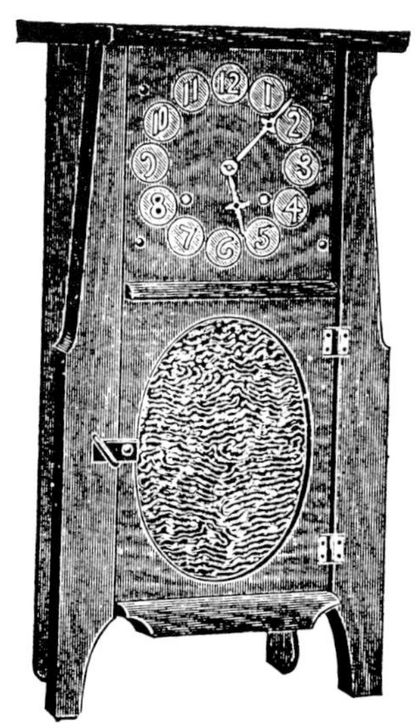

1906

SAN MARCOS

Mantel Weathered Oak Mission Clock.
Wooden Dial, with white figures on a background of dark brown.
Fancy Brass Hands.
Height, 21 inches. Width, 10½ inches.
Dial, 6 inches.

List Price, $6.25

MOVEMENT ILLUSTRATIONS

1483

1883

No. 0 MOVEMENT
(Weight)

Sweep Second Regulator Movement. Dead-Beat Escapement. Retaining Power.
Cut and Polished Hard Steel Pinions. Heavy Engraved Frames. Screw Pillars.
A superior Regulator Movement for Watchmakers' use.

MOVEMENT ILLUSTRATIONS

The year entered beside each movement is the year of the catalog from which the illustration was taken. Some movements may have been produced a few years earlier.

1484

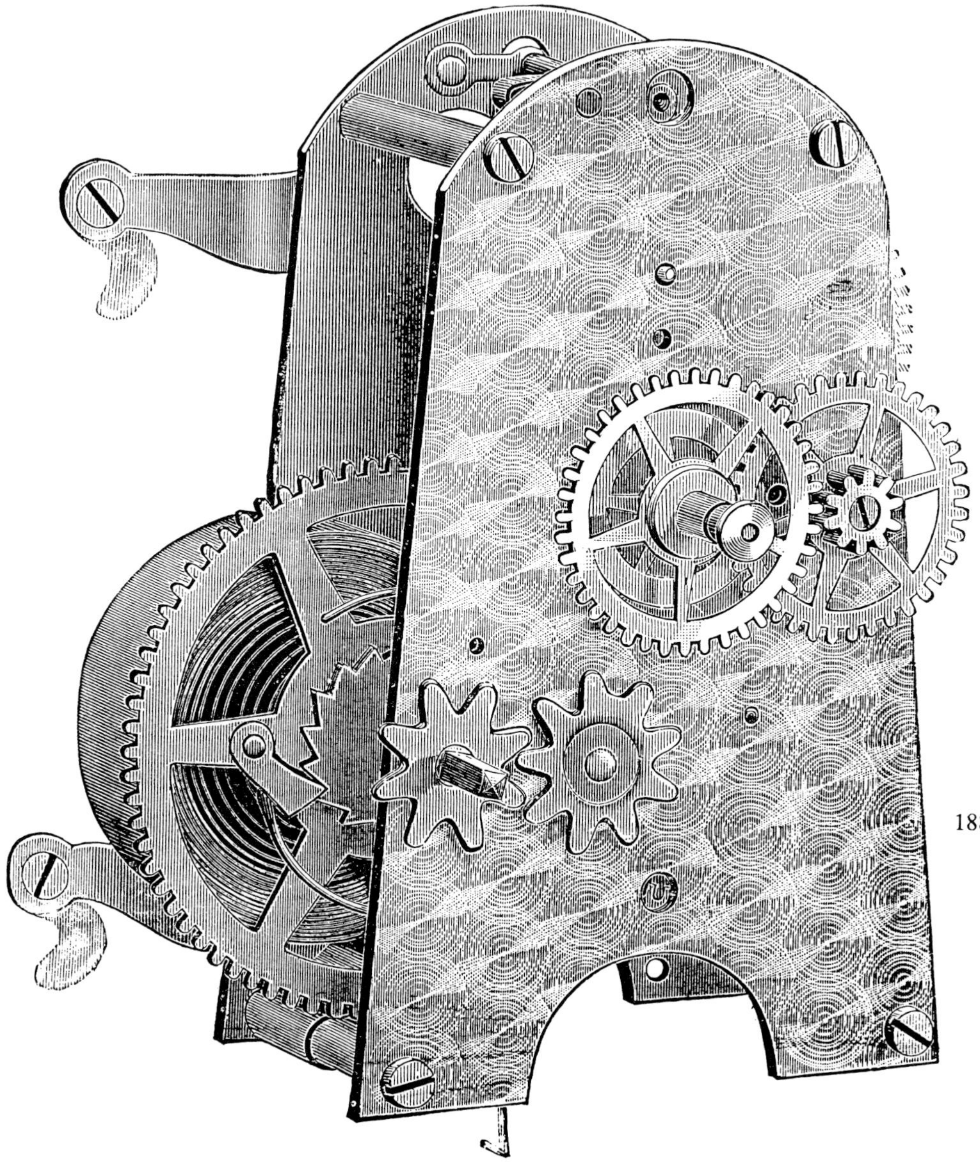

1883

No. 1 MOVEMENT
(Spring)

Solid Engraved Frames. Heavy Wheels. Screw Pillars. Stop Work.
A very accurate running Movement, and is superior for Office and Railroad use.

MOVEMENT ILLUSTRATIONS

1485

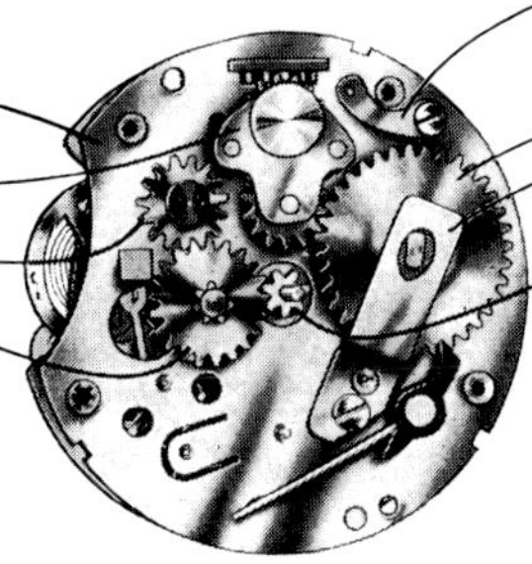

FRONT VIEW

1486

REAR VIEW

No. 50
Small Watch Movement
2-16-1927

MOVEMENT ILLUSTRATIONS

1487

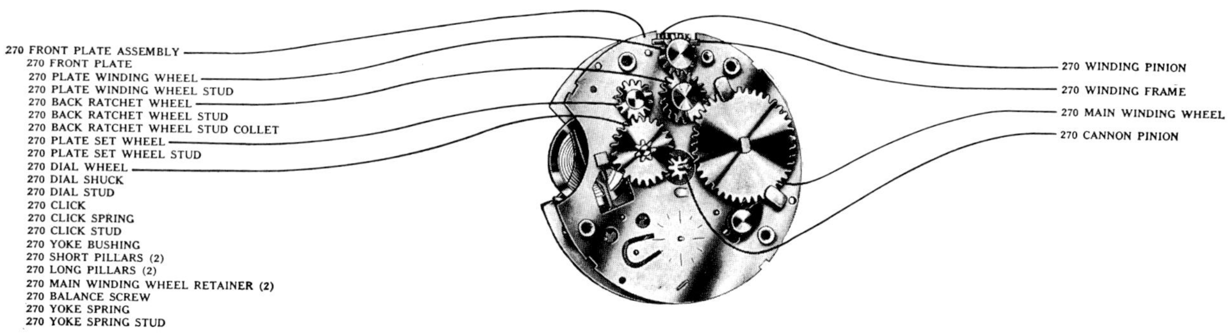

FRONT VIEW

1488

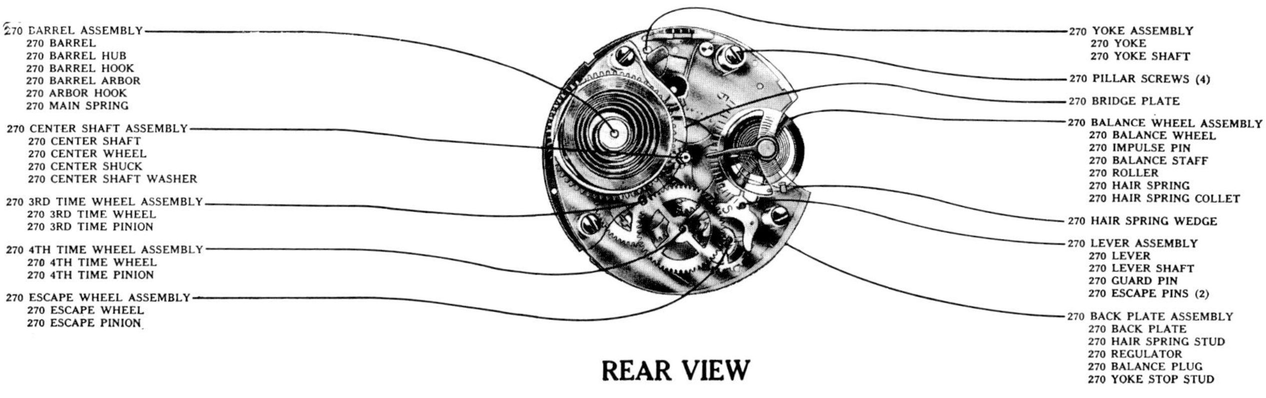

REAR VIEW

No. 270
Large Watch Movement
2-16-1927

MOVEMENT ILLUSTRATIONS

1489

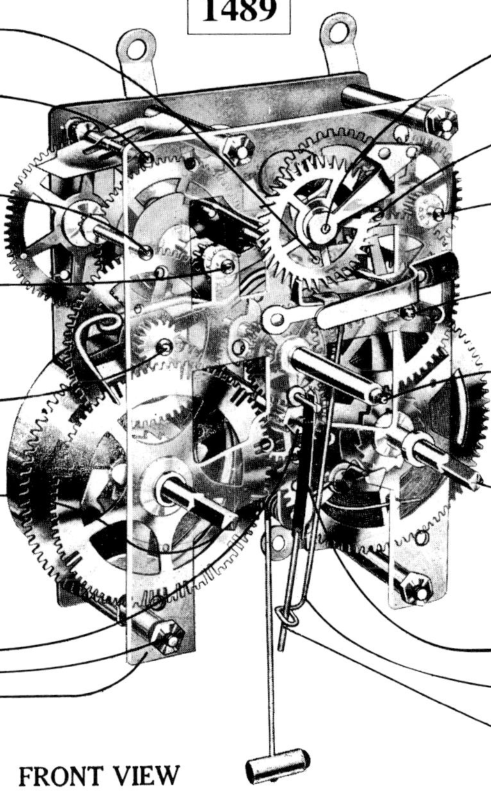

FRONT VIEW

1490

REAR VIEW

No. 303
8-Day Pendulum Time and Strike Movement
1-4-1928

MOVEMENT ILLUSTRATIONS

1491

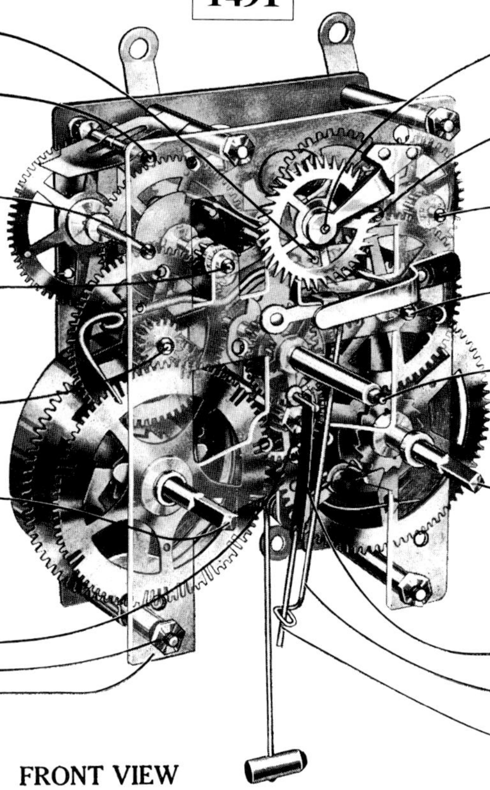

FRONT VIEW

1492

REAR VIEW

No. 304

8-Day Pendulum Time and Strike Movement

1-4-1928

MOVEMENT ILLUSTRATIONS

1493

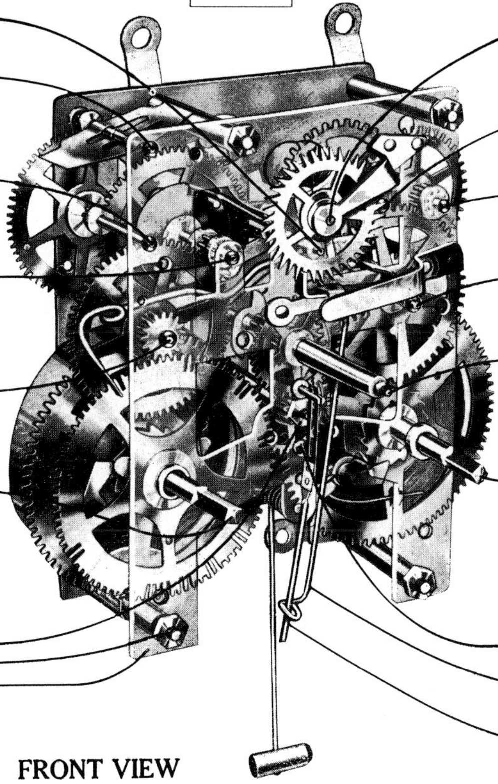

FRONT VIEW

1494

REAR VIEW

No. 307

8-Day Pendulum Time and Strike Movement

1-4-1928

MOVEMENT ILLUSTRATIONS

1495

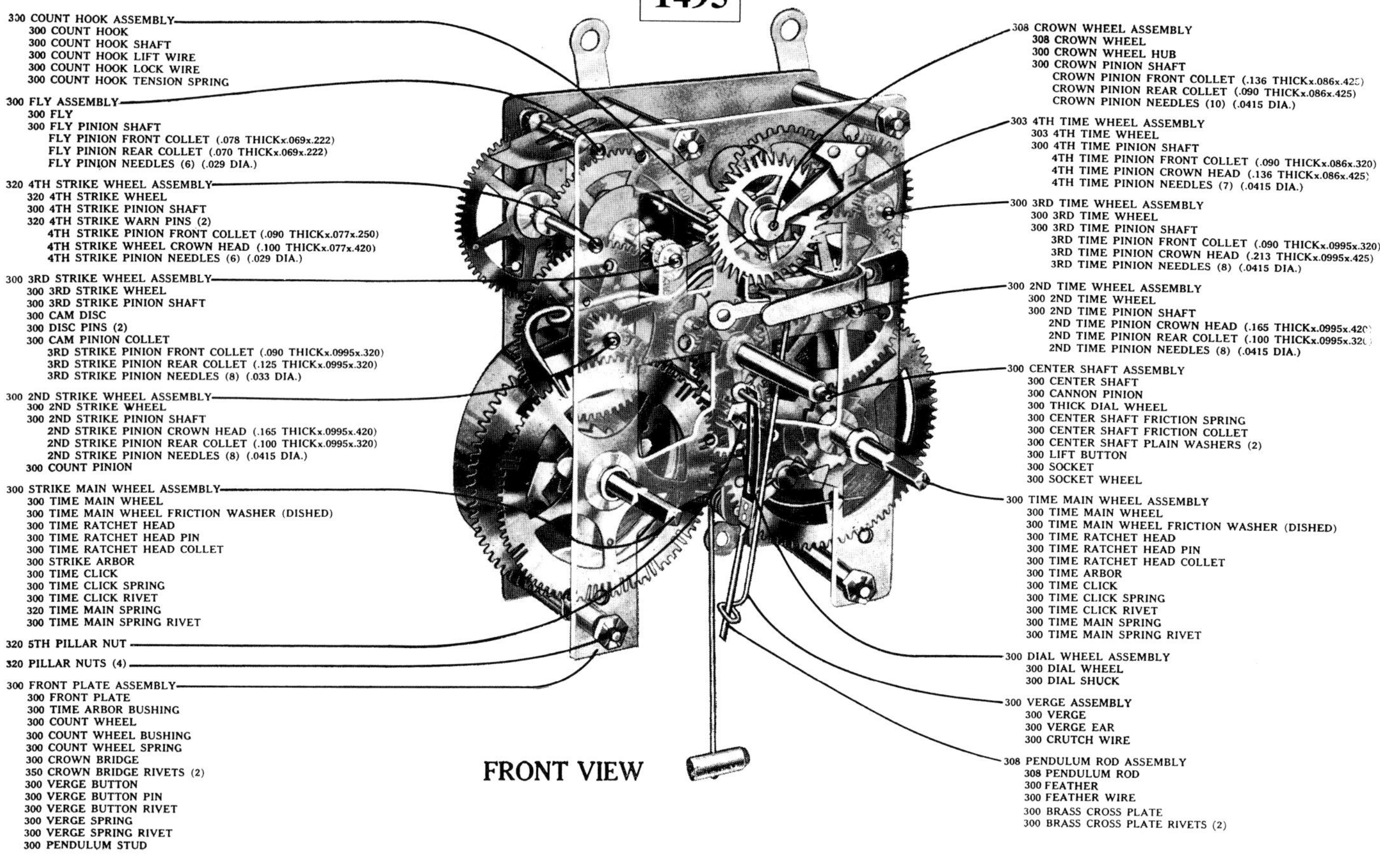

FRONT VIEW

1496

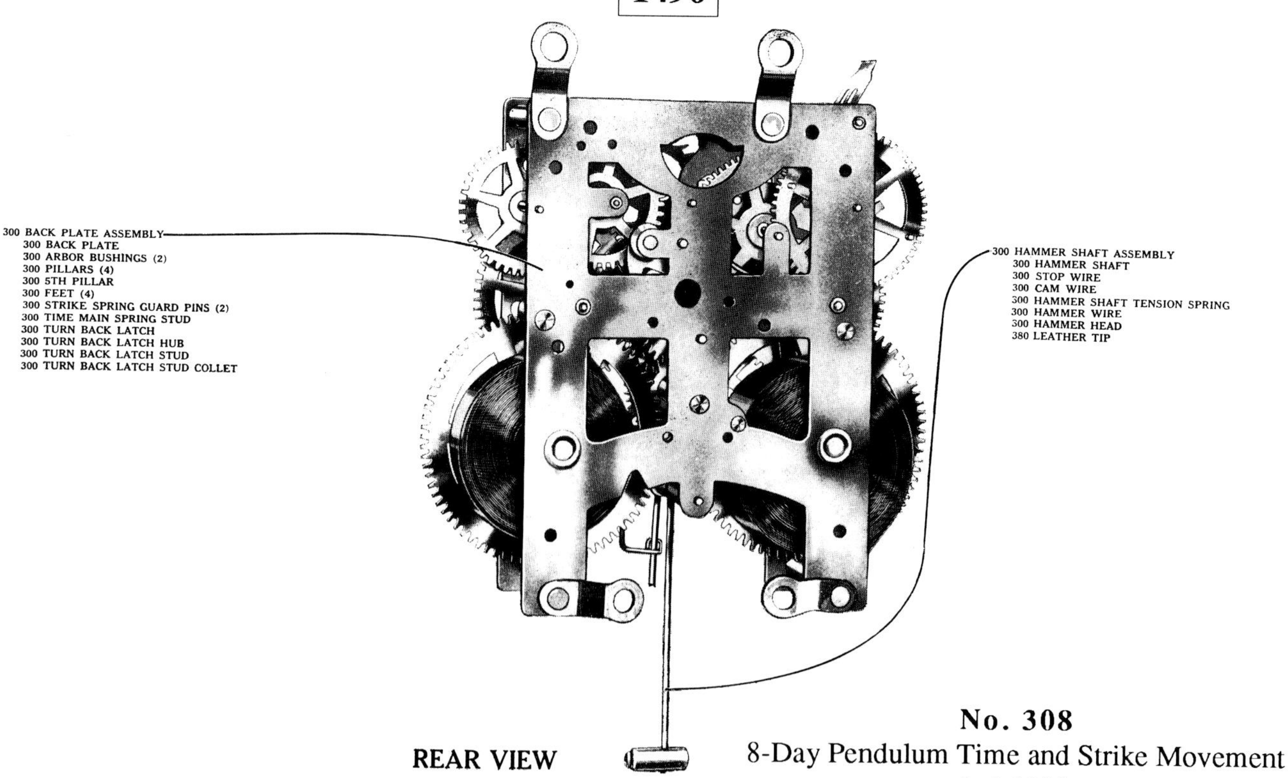

REAR VIEW

No. 308
8-Day Pendulum Time and Strike Movement
1-4-1928

MOVEMENT ILLUSTRATIONS

1497

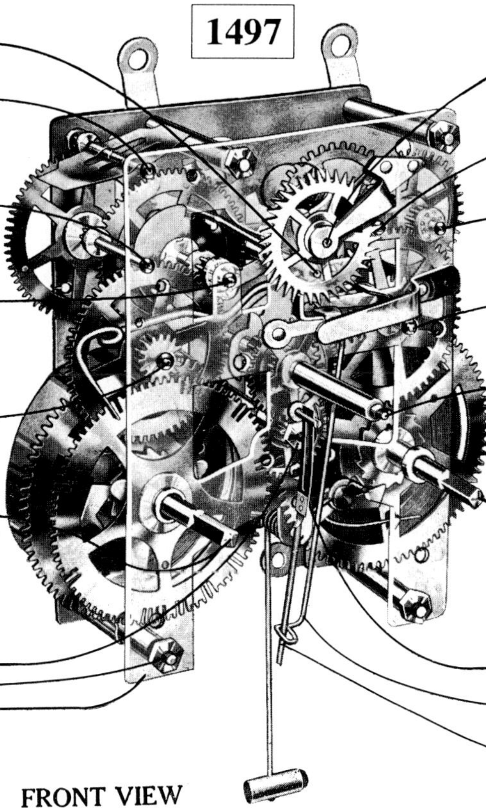

FRONT VIEW

1498

REAR VIEW

No. 309
8-Day Pendulum Time and Strike Movement
1-4-1928

MOVEMENT ILLUSTRATIONS

1499

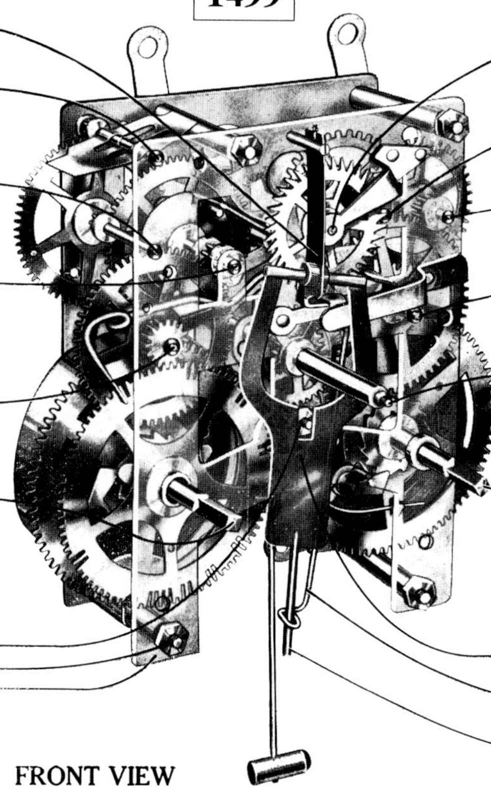

FRONT VIEW

1500

REAR VIEW

No. 310
8-Day Pendulum Time and Strike Movement
1-4-1928

MOVEMENT ILLUSTRATIONS

1501

300 COUNT HOOK ASSEMBLY
- 300 COUNT HOOK
- 300 COUNT HOOK SHAFT
- 300 COUNT HOOK LIFT WIRE
- 300 COUNT HOOK LOCK WIRE
- 300 COUNT HOOK TENSION SPRING

300 FLY ASSEMBLY
- 300 FLY
- 300 FLY PINION SHAFT
  - FLY PINION FRONT COLLET (.078 THICKx.069x.222)
  - FLY PINION REAR COLLET (.070 THICKx.069x.222)
  - FLY PINION NEEDLES (6) (.029 DIA.)

320 4TH STRIKE WHEEL ASSEMBLY
- 320 4TH STRIKE WHEEL
- 300 4TH STRIKE PINION SHAFT
- 320 4TH STRIKE WARN PINS (2)
  - 4TH STRIKE PINION FRONT COLLET (.090 THICKx.077x.250)
  - 4TH STRIKE WHEEL CROWN HEAD (.100 THICKx.077x.420)
  - 4TH STRIKE PINION NEEDLES (6) (.029 DIA.)

300 3RD STRIKE WHEEL ASSEMBLY
- 300 3RD STRIKE WHEEL
- 300 3RD STRIKE PINION SHAFT
- 300 CAM DISC
- 300 DISC PINS (2)
- 300 CAM PINION COLLET
  - 3RD STRIKE PINION FRONT COLLET (.090 THICKx.0995x.320)
  - 3RD STRIKE PINION REAR COLLET (.125 THICKx.0995x.320)
  - 3RD STRIKE PINION NEEDLES (8) (.033 DIA.)

300 2ND STRIKE WHEEL ASSEMBLY
- 300 2ND STRIKE WHEEL
- 300 2ND STRIKE PINION SHAFT
  - 2ND STRIKE PINION CROWN HEAD (.165 THICKx.0995x.420)
  - 2ND STRIKE PINION REAR COLLET (.100 THICKx.0995x.320)
  - 2ND STRIKE PINION NEEDLES (8) (.0415 DIA.)
- 300 COUNT PINION

300 STRIKE MAIN WHEEL ASSEMBLY
- 300 TIME MAIN WHEEL
- 300 TIME MAIN WHEEL FRICTION WASHER (DISHED)
- 300 TIME RATCHET HEAD
- 300 TIME RATCHET HEAD PIN
- 300 TIME RATCHET HEAD COLLET
- 300 STRIKE ARBOR
- 300 TIME CLICK
- 300 TIME CLICK SPRING
- 300 TIME CLICK RIVET
- 320 TIME MAIN SPRING
- 300 TIME MAIN SPRING RIVET

320 5TH PILLAR NUT

320 PILLAR NUTS (4)

300 FRONT PLATE ASSEMBLY
- 300 FRONT PLATE
- 300 TIME ARBOR BUSHING
- 300 COUNT WHEEL
- 300 COUNT WHEEL BUSHING
- 300 COUNT WHEEL SPRING
- 300 CROWN BRIDGE
- 350 CROWN BRIDGE RIVETS (2)
- 300 VERGE BUTTON
- 300 VERGE BUTTON PIN
- 300 VERGE BUTTON RIVET
- 300 VERGE SPRING
- 300 VERGE SPRING RIVET
- 300 PENDULUM STUD

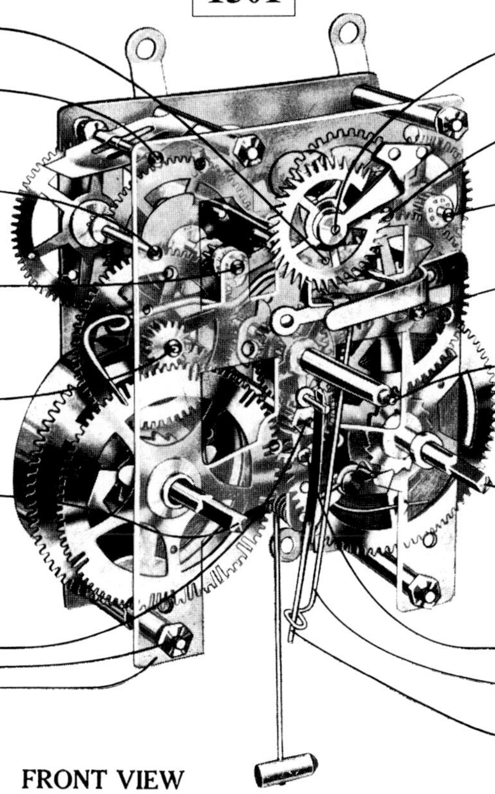

FRONT VIEW

311 CROWN WHEEL ASSEMBLY
- 311 CROWN WHEEL
- 300 CROWN WHEEL HUB
- 300 CROWN PINION SHAFT
  - CROWN PINION FRONT COLLET (.136 THICKx.086x.362)
  - CROWN PINION REAR COLLET (.090 THICKx.086x.362)
  - CROWN PINION NEEDLES (8) (.0415 DIA.)

310 4TH TIME WHEEL ASSEMBLY
- 310 4TH TIME WHEEL
- 300 4TH TIME PINION SHAFT
  - 4TH TIME PINION FRONT COLLET (.090 THICKx.086x.320)
  - 4TH TIME PINION CROWN HEAD (.096 THICKx.086x.425)
  - 4TH TIME PINION NEEDLES (8) (.0415 DIA.)

310 3RD TIME WHEEL ASSEMBLY
- 310 4TH TIME WHEEL
- 300 3RD TIME PINION SHAFT
  - 3RD TIME PINION FRONT COLLET (.090 THICKx.0995x.320)
  - 3RD TIME PINION CROWN HEAD (.213 THICKx.0995x.425)
  - 3RD TIME PINION NEEDLES (8) (.0415 DIA.)

300 2ND TIME WHEEL ASSEMBLY
- 300 2ND TIME WHEEL
- 300 2ND TIME PINION SHAFT
  - 2ND TIME PINION CROWN HEAD (.165 THICKx.0995x.420)
  - 2ND TIME PINION REAR COLLET (.100 THICKx.0995x.320)
  - 2ND TIME PINION NEEDLES (8) (.0415 DIA.)

300 CENTER SHAFT ASSEMBLY
- 300 CENTER SHAFT
- 300 CANNON PINION
- 300 THICK DIAL WHEEL
- 300 CENTER SHAFT FRICTION SPRING
- 300 CENTER SHAFT FRICTION COLLET
- 300 CENTER SHAFT PLAIN WASHERS (2)
- 300 LIFT BUTTON
- 300 SOCKET
- 300 SOCKET WHEEL

300 TIME MAIN WHEEL ASSEMBLY
- 300 TIME MAIN WHEEL
- 300 TIME MAIN WHEEL FRICTION WASHER (DISHED)
- 300 TIME RATCHET HEAD
- 300 TIME RATCHET HEAD PIN
- 300 TIME RATCHET HEAD COLLET
- 300 TIME ARBOR
- 300 TIME CLICK
- 300 TIME CLICK SPRING
- 300 TIME CLICK RIVET
- 300 TIME MAIN SPRING
- 300 TIME MAIN SPRING RIVET

300 DIAL WHEEL ASSEMBLY
- 300 DIAL WHEEL
- 300 DIAL SHUCK

300 VERGE ASSEMBLY
- 300 VERGE
- 300 VERGE EAR
- 300 CRUTCH WIRE

311 PENDULUM ROD ASSEMBLY
- 311 PENDULUM ROD
- 300 FEATHER
- 300 FEATHER WIRE

1502

300 BACK PLATE ASSEMBLY
- 300 BACK PLATE
- 300 ARBOR BUSHINGS (2)
- 300 PILLARS (4)
- 300 5TH PILLAR
- 300 FEET (4)
- 300 STRIKE SPRING GUARD PINS (2)
- 300 TIME MAIN SPRING STUD
- 300 TURN BACK LATCH
- 300 TURN BACK LATCH HUB
- 300 TURN BACK LATCH STUD
- 300 TURN BACK LATCH STUD COLLET

REAR VIEW

300 HAMMER SHAFT ASSEMBLY
- 300 HAMMER SHAFT
- 300 STOP WIRE
- 300 CAM WIRE
- 300 HAMMER SHAFT TENSION SPRING
- 300 HAMMER WIRE
- 300 HAMMER HEAD
- 380 LEATHER TIP

No. 311

8-Day Pendulum Time and Strike Movement

1-4-1928

MOVEMENT ILLUSTRATIONS

1503

300 COUNT HOOK ASSEMBLY
300 COUNT HOOK
300 COUNT HOOK SHAFT
300 COUNT HOOK LIFT WIRE
300 COUNT HOOK LOCK WIRE
300 COUNT HOOK TENSION SPRING

300 FLY ASSEMBLY
300 FLY
300 FLY PINION SHAFT
FLY PINION FRONT COLLET (.078 THICKx.069x.222)
FLY PINION REAR COLLET (.070 THICKx.069x.222)
FLY PINION NEEDLES (6) (.029 DIA.)

320 4TH STRIKE WHEEL ASSEMBLY
320 4TH STRIKE WHEEL
300 4TH STRIKE PINION SHAFT
320 4TH STRIKE WARN PINS (2)
4TH STRIKE PINION REAR COLLET (.090 THICKx.077x.250)
4TH STRIKE WHEEL CROWN HEAD (.100 THICKx.077x.420)
4TH STRIKE PINION NEEDLES (6) (.029 DIA.)

300 3RD STRIKE WHEEL ASSEMBLY
300 3RD STRIKE WHEEL
300 3RD STRIKE PINION SHAFT
300 CAM DISC
300 DISC PINS (2)
300 CAM PINION COLLET
3RD STRIKE PINION FRONT COLLET (.090 THICKx.0995x.320)
3RD STRIKE PINION REAR COLLET (.125 THICKx.0995x.320)
3RD STRIKE PINION NEEDLES (8) (.033 DIA.)

300 2ND STRIKE WHEEL ASSEMBLY
300 2ND STRIKE WHEEL
300 2ND STRIKE PINION SHAFT
2ND STRIKE PINION CROWN HEAD (.165 THICKx.0995x.420)
2ND STRIKE PINION REAR COLLET (.100 THICKx.0995x.320)
2ND STRIKE PINION NEEDLES (8) (.0415 DIA.)
300 COUNT PINION

300 STRIKE MAIN WHEEL ASSEMBLY
300 TIME MAIN WHEEL
300 TIME MAIN WHEEL FRICTION WASHER (DISHED)
300 TIME RATCHET HEAD
300 TIME RATCHET HEAD PIN
300 TIME RATCHET HEAD COLLET
300 STRIKE ARBOR
300 TIME CLICK
300 TIME CLICK SPRING
300 TIME CLICK RIVET
320 TIME MAIN SPRING
300 TIME MAIN SPRING RIVET

320 5TH PILLAR NUT

320 PILLAR NUTS (4)

300 FRONT PLATE ASSEMBLY
300 FRONT PLATE
300 TIME ARBOR BUSHING
300 COUNT WHEEL
300 COUNT WHEEL BUSHING
300 COUNT WHEEL SPRING
300 CROWN BRIDGE
350 CROWN BRIDGE RIVETS (2)
300 VERGE BUTTON
300 VERGE BUTTON PIN
300 VERGE BUTTON RIVET
300 VERGE SPRING
300 VERGE SPRING RIVET
300 PENDULUM STUD

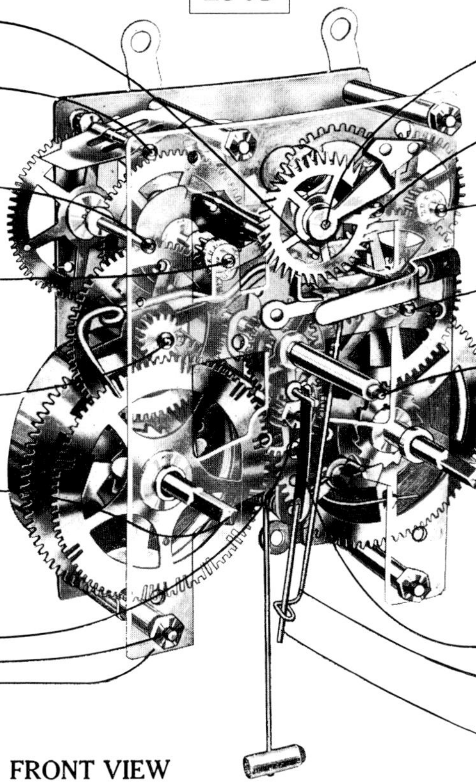

FRONT VIEW

312 CROWN WHEEL ASSEMBLY
312 CROWN WHEEL
300 CROWN WHEEL HUB
300 CROWN PINION SHAFT
CROWN PINION SHAFT COLLET (.136 THICKx.086x.362)
CROWN PINION REAR COLLET (.090 THICKx.086x.362)
CROWN PINION NEEDLES (8) (.0415 DIA.)

303 4TH TIME WHEEL ASSEMBLY
303 4TH TIME WHEEL
300 4TH TIME PINION SHAFT
4TH TIME PINION FRONT COLLET (.090 THICKx.086x.320)
4TH TIME PINION CROWN HEAD (.136 THICKx.086x.425)
4TH TIME PINION NEEDLES (7) (.0415 DIA.)

300 3RD TIME WHEEL ASSEMBLY
300 3RD TIME WHEEL
300 3RD TIME PINION SHAFT
3RD TIME PINION FRONT COLLET (.090 THICKx.0995x.320)
3RD TIME PINION CROWN HEAD (.213 THICKx.0995x.425)
3RD TIME PINION NEEDLES (8) (.0415 DIA.)

300 2ND TIME WHEEL ASSEMBLY
300 2ND TIME WHEEL
300 2ND TIME PINION SHAFT
2ND TIME PINION CROWN HEAD (.165 THICKx.0995x.420)
2ND TIME PINION REAR COLLET (.100 THICKx.0995x.320)
2ND TIME PINION NEEDLES (8) (.0415 DIA.)

300 CENTER SHAFT ASSEMBLY
300 CENTER SHAFT
300 CANNON PINION
300 THICK DIAL WHEEL
300 CENTER SHAFT FRICTION SPRING
300 CENTER SHAFT FRICTION COLLET
300 CENTER SHAFT PLAIN WASHERS (2)
300 LIFT BUTTON
300 SOCKET
300 SOCKET WHEEL

300 TIME MAIN WHEEL ASSEMBLY
300 TIME MAIN WHEEL
300 TIME MAIN WHEEL FRICTION WASHER (DISHED)
300 TIME RATCHET HEAD
300 TIME RATCHET HEAD PIN
300 TIME RATCHET HEAD COLLET
300 TIME ARBOR
300 TIME CLICK
300 TIME CLICK SPRING
300 TIME CLICK RIVET
300 TIME MAIN SPRING
300 TIME MAIN SPRING RIVET

300 DIAL WHEEL ASSEMBLY
300 DIAL WHEEL
300 DIAL SHUCK

300 VERGE ASSEMBLY
300 VERGE
300 VERGE EAR
300 CRUTCH WIRE

312 PENDULUM ROD ASSEMBLY
312 PENDULUM ROD
300 FEATHER
300 FEATHER WIRE

1504

300 BACK PLATE ASSEMBLY
300 BACK PLATE
300 ARBOR BUSHINGS (2)
300 PILLARS (4)
300 5TH PILLAR
300 FEET (4)
300 STRIKE SPRING GUARD PINS (2)
300 TIME MAIN SPRING STUD
300 TURN BACK LATCH
300 TURN BACK LATCH HUB
300 TURN BACK LATCH STUD
300 TURN BACK LATCH STUD COLLET

REAR VIEW

300 HAMMER SHAFT ASSEMBLY
300 HAMMER SHAFT
300 STOP WIRE
300 CAM WIRE
300 HAMMER SHAFT TENSION SPRING
300 HAMMER WIRE
300 HAMMER HEAD
380 LEATHER TIP

No. 312
8-Day Pendulum Time and Strike Movement
1-4-1928

MOVEMENT ILLUSTRATIONS

1505

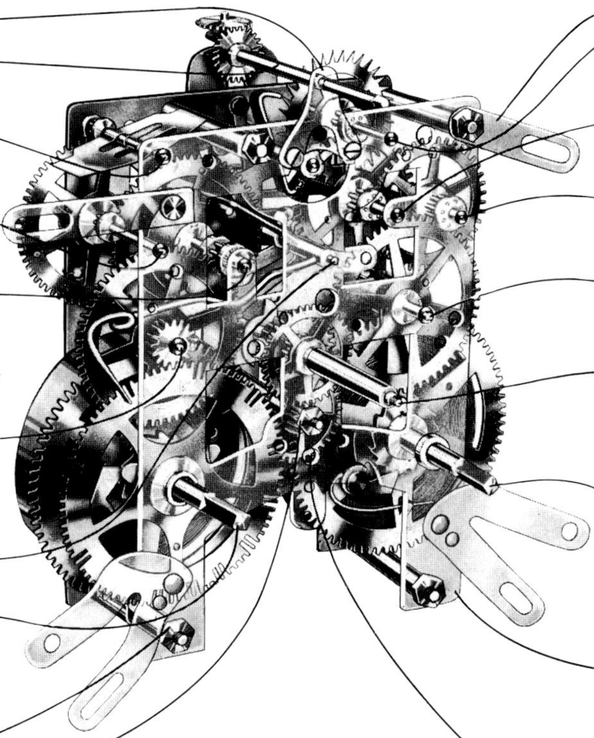

FRONT VIEW

1506

REAR VIEW

No. 320
8-Day Pendulum Time and Strike Movement
2-16-1927

MOVEMENT ILLUSTRATIONS

1507

320 FRONT REGULATOR BRIDGE

320 REGULATOR SHAFT ASSEMBLY
- 320 REGULATOR SHAFT
- 320 REGULATOR ADJUSTING WHEEL
- 320 REGULATOR ADJUSTING WHEEL HUB

320 FLY ASSEMBLY
- 320 FLY
- 300 FLY PINION SHAFT
  - FLY PINION FRONT COLLET (.078 THICKx.069x.222)
  - FLY PINION REAR COLLET (.078 THICKx.069x.222)
  - FLY PINION NEEDLES (6) (.033 DIA.)

320 4TH STRIKE WHEEL ASSEMBLY
- 320 4TH STRIKE WHEEL
- 300 4TH STRIKE PINION SHAFT
- 320 4TH STRIKE WARN PINS (2)
  - 4TH STRIKE WHEEL CROWN HEAD (.100 THICKx.077x.420)
  - 4TH STRIKE PINION REAR COLLET (.090 THICKx.077x.250)
  - 4TH STRIKE PINION NEEDLES (6) (.029 DIA.)

320 3RD STRIKE WHEEL ASSEMBLY
- 300 3RD STRIKE WHEEL
- 300 3RD STRIKE PINION SHAFT
- 300 CAM DISC
- 300 CAM DISC PINS (2)
- 300 CAM PINION COLLET
  - 3RD STRIKE PINION FRONT COLLET (.090 THICKx.0995x.320)
  - 3RD STRIKE PINION REAR COLLET (.125 THICKx.0995x.320)
  - 3RD STRIKE PINION NEEDLES (8) (.033 DIA.)

300 2ND STRIKE WHEEL ASSEMBLY
- 300 2ND STRIKE WHEEL
- 300 2ND STRIKE PINION SHAFT
  - 2ND STRIKE PINION CROWN HEAD (.165 THICKx.0995x.420)
  - 2ND STRIKE PINION REAR COLLET (.100 THICKx.0995x.320)
  - 2ND STRIKE PINION NEEDLES (8) (.0415 DIA.)
- 300 COUNT PINION

300 COUNT HOOK ASSEMBLY
- 300 COUNT HOOK
- 300 COUNT HOOK SHAFT
- 300 COUNT HOOK LIFT WIRE
- 300 COUNT HOOK LOCK WIRE
- 300 COUNT HOOK TENSION SPRING

320 STRIKE MAIN WHEEL ASSEMBLY
- 300 TIME MAIN WHEEL
- 300 TIME MAIN WHEEL FRICTION WASHED (DISHED)
- 300 TIME RATCHET HEAD
- 300 TIME RATCHET HEAD PIN
- 300 TIME RATCHET HEAD COLLET
- 320 STRIKE ARBOR
- 300 TIME CLICK
- 300 TIME CLICK SPRING
- 300 TIME CLICK RIVET
- 320 TIME MAIN SPRING
- 300 TIME MAIN SPRING RIVET

320 PILLAR NUTS (4)

320 5TH PILLAR NUT

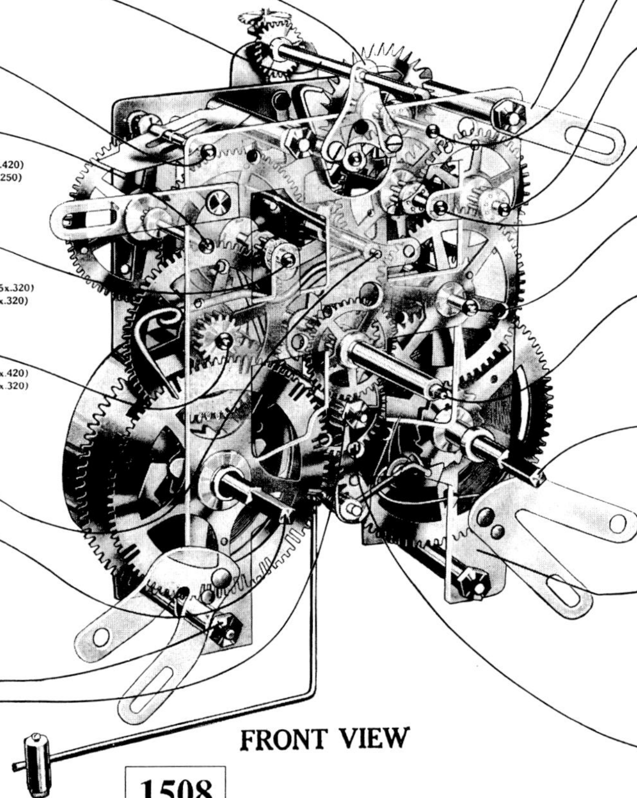

FRONT VIEW

320 SHORT SINGLE FOOT

700 CROWN WHEEL ASSEMBLY
- 700 CROWN WHEEL
- 700 CROWN PINION SHAFT
  - CROWN PINION FRONT COLLET (.090 THICK x.086x.320)
  - CROWN PINION CROWN HEAD (.125 THICK x.086x.420)
  - CROWN PINION NEEDLES (6) (.0415 DIA)

320 4TH TIME WHEEL ASSEMBLY
- 320 4TH TIME WHEEL
- 300 4TH TIME PINION SHAFT
  - 4TH TIME PINION FRONT COLLET (.090 THICKx.086x.320)
  - 4TH TIME PINION CROWN HEAD (.125 THICKx.086x.420)
  - 4TH TIME PINION NEEDLES (7) (.0415 DIA.)

300 3RD TIME WHEEL ASSEMBLY
- 300 3RD TIME WHEEL
- 300 3RD TIME PINION SHAFT
  - 3RD TIME PINION FRONT COLLET (.090 THICKx.0995x.320)
  - 3RD TIME PINION CROWN HEAD (.165 THICKx.0995x.420)
  - 3RD TIME PINION NEEDLES (8) (.0415 DIA.)

300 2ND TIME WHEEL ASSEMBLY
- 300 2ND TIME WHEEL
- 300 2ND TIME PINION SHAFT
  - 2ND TIME PINION CROWN HEAD (.165 THICKx.0995x.420)
  - 2ND TIME PINION REAR COLLET (.100 THICKx.0995x.320)
  - 2ND TIME PINION NEEDLES (8) (.0415 DIA.)

320 CENTER SHAFT ASSEMBLY
- 320 CENTER SHAFT
- 300 CANNON PINION
- 300 THICK DIAL WHEEL
- 300 CENTER SHAFT FRICTION SPRING
- 300 CENTER SHAFT FRICTION COLLET
- 300 CENTER SHAFT PLAIN WASHERS (2)
- 300 LIFT BUTTON
- 320 SOCKET
- 300 SOCKET WHEEL

320 TIME MAIN WHEEL ASSEMBLY
- 300 TIME MAIN WHEEL
- 300 TIME MAIN WHEEL FRICTION WASHER (DISHED)
- 300 TIME RATCHET HEAD
- 300 TIME RATCHET HEAD PIN
- 300 TIME RATCHET HEAD COLLET
- 320 TIME ARBOR
- 300 TIME CLICK
- 300 TIME CLICK SPRING
- 300 TIME CLICK RIVET
- 320 TIME MAIN SPRING
- 300 TIME MAIN SPRING RIVET

320 FRONT PLATE ASSEMBLY
- 320 FRONT PLATE
- 320 LEFT DOUBLE FOOT
- 320 RIGHT DOUBLE FOOT
- 320 LONG SINGLE FOOT
- 300 ARBOR BUSHING
- 300 COUNT WHEEL BUSHING
- 320 FOOT RIVET (LARGE) (3)
- 320 FOOT RIVET (SMALL) (2)
- 380 VERGE BUTTON
- 380 VERGE BUTTON RIVET
- 300 COUNT WHEEL
- 300 COUNT WHEEL SPRING

300 DIAL WHEEL ASSEMBLY
- 300 DIAL WHEEL
- 300 DIAL SHUCK

1508

320 FEATHER WIRE

700 VERGE ASSEMBLY
- 700 VERGE
- 700 VERGE SHAFT
- 320 CRUTCH
- 320 CRUTCH HUB

321 BACK PLATE ASSEMBLY
- 300 BACK PLATE
- 300 ARBOR BUSHINGS (2)
- 320 PILLARS (4)
- 300 5TH PILLAR
- 300 STRIKE SPRING GUARD PINS (2)
- 300 TIME MAIN SPRING STUD
- 300 TURN BACK LATCH
- 300 TURN BACK LATCH HUB
- 300 TURN BACK LATCH STUD COLLET
- 300 TURN BACK LATCH STUD
- 380 VERGE BUTTON
- 380 VERGE BUTTON RIVET

320 PENDULUM ROD ASSEMBLY
- 320 PENDULUM ROD
- 320 FEATHER

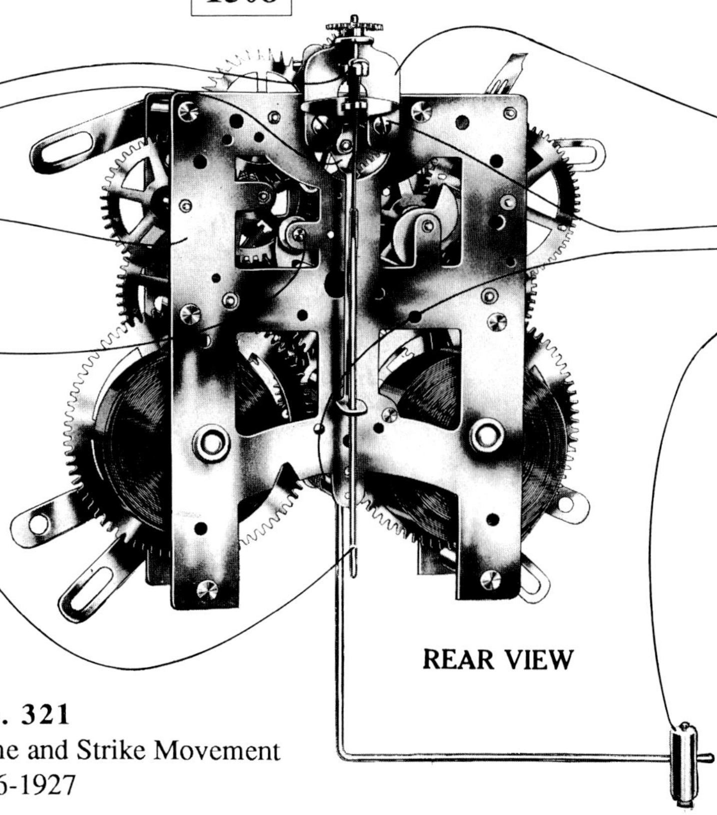

REAR VIEW

320 BACK REGULATOR BRIDGE ASSEMBLY
- 320 BACK REGULATOR BRIDGE
- 320 BACK REGULATOR BRIDGE STUD
- 320 BRIDGE SPRING
- 320 BRIDGE SPRING STUD
- 320 REGULATOR ADJUSTING WHEEL
- 320 REGULATOR ADJUSTING WHEEL HUB
- 320 REGULATOR ADJUSTING SCREW

320 REGULATOR BRIDGE SCREWS (4)

321 HAMMER SHAFT ASSEMBLY
- 300 HAMMER SHAFT
- 300 STOP WIRE
- 300 CAM WIRE
- 300 HAMMER SHAFT TENSION SPRING
- 321 HAMMER WIRE

700 HAMMER HEAD ASSEMBLY
- 700 HAMMER HEAD
- 380 LEATHER TIP
- 700 HAMMER HEAD SCREW

No. 321

8-Day Pendulum Time and Strike Movement

2-16-1927

MOVEMENT ILLUSTRATIONS

The year entered beside each movement is the year of the catalog from which the illustration was taken. Some movements may have been produced a few years earlier.

1509

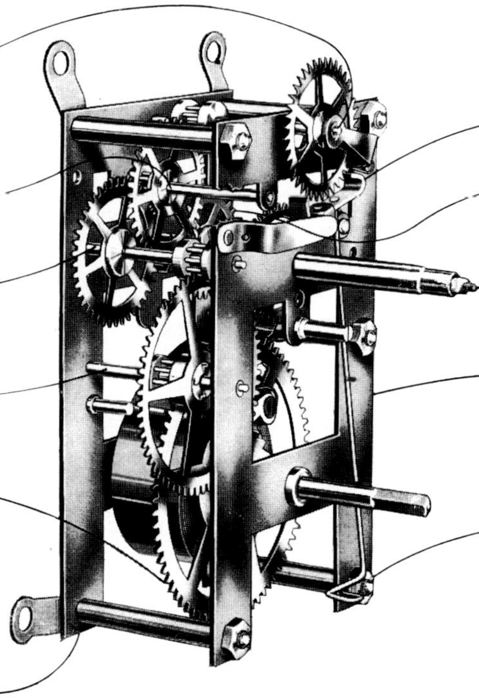

No. 325
8-Day Pendulum Timepiece
11-1-1927

MOVEMENT ILLUSTRATIONS

1510

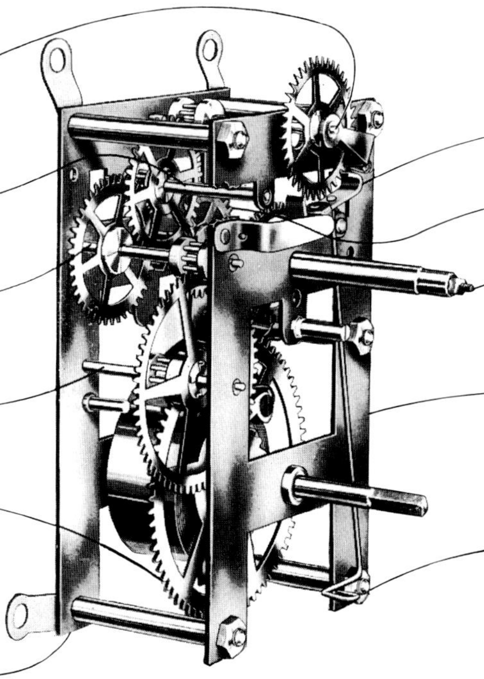

No. 327

8-Day Pendulum Timepiece

11-1-1927

MOVEMENT ILLUSTRATIONS

1511

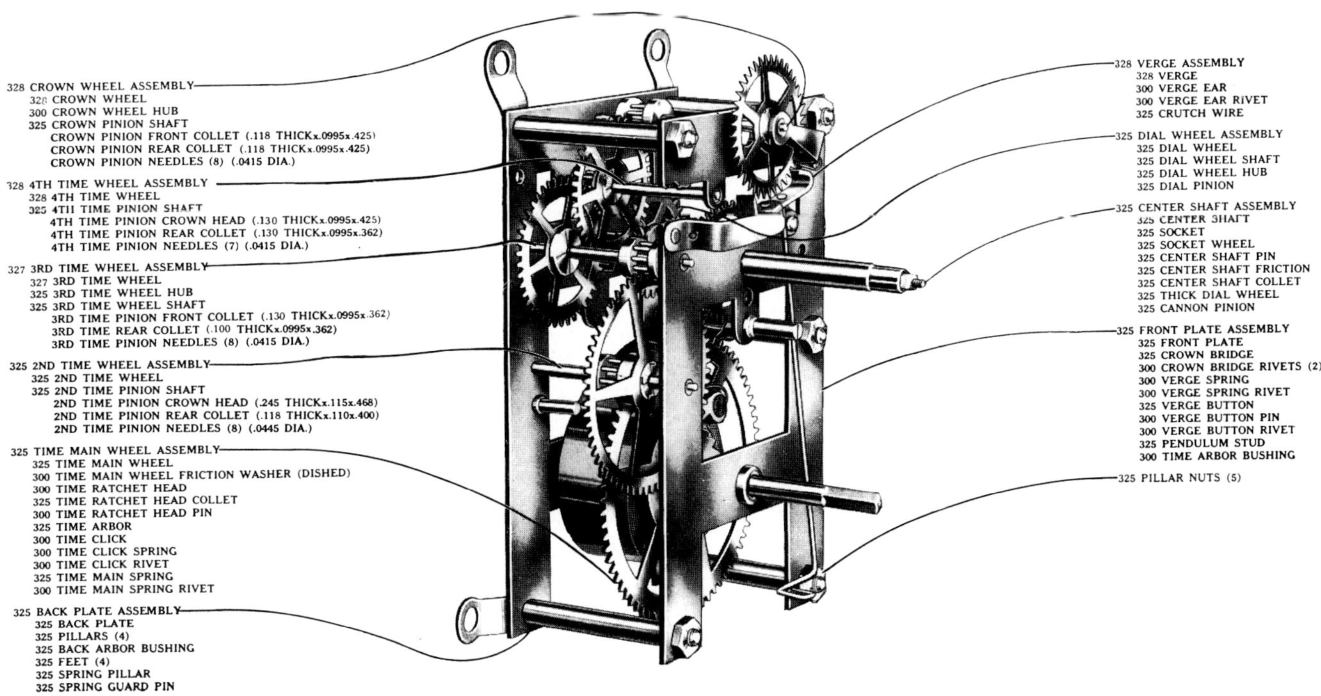

No. 328
8-Day Pendulum Timepiece
11-1-1927

MOVEMENT ILLUSTRATIONS

1512

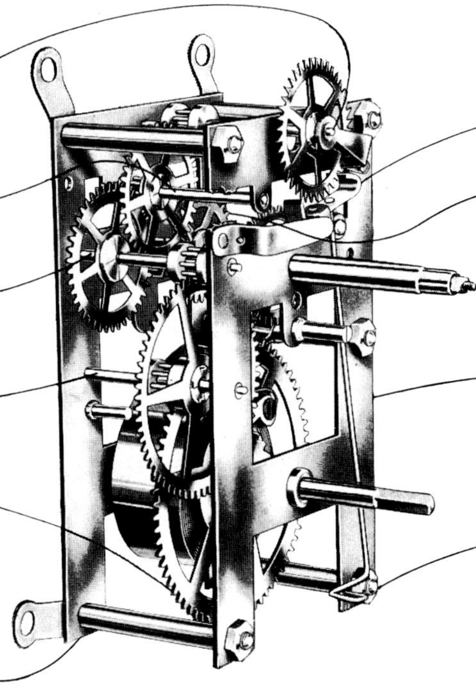

No. 329
8-Day Pendulum Timepiece
12-1-1927

MOVEMENT ILLUSTRATIONS

The year entered beside each movement is the year of the catalog from which the illustration was taken. Some movements may have been produced a few years earlier.

1513

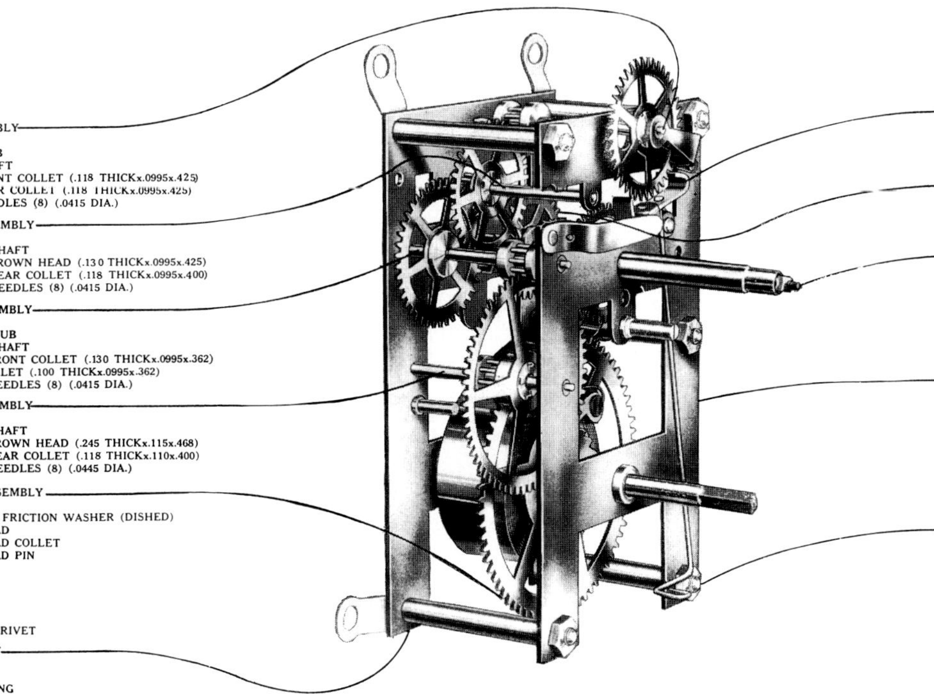

No. 330
8-Day Pendulum Timepiece
11-1-1927

MOVEMENT ILLUSTRATIONS

1514

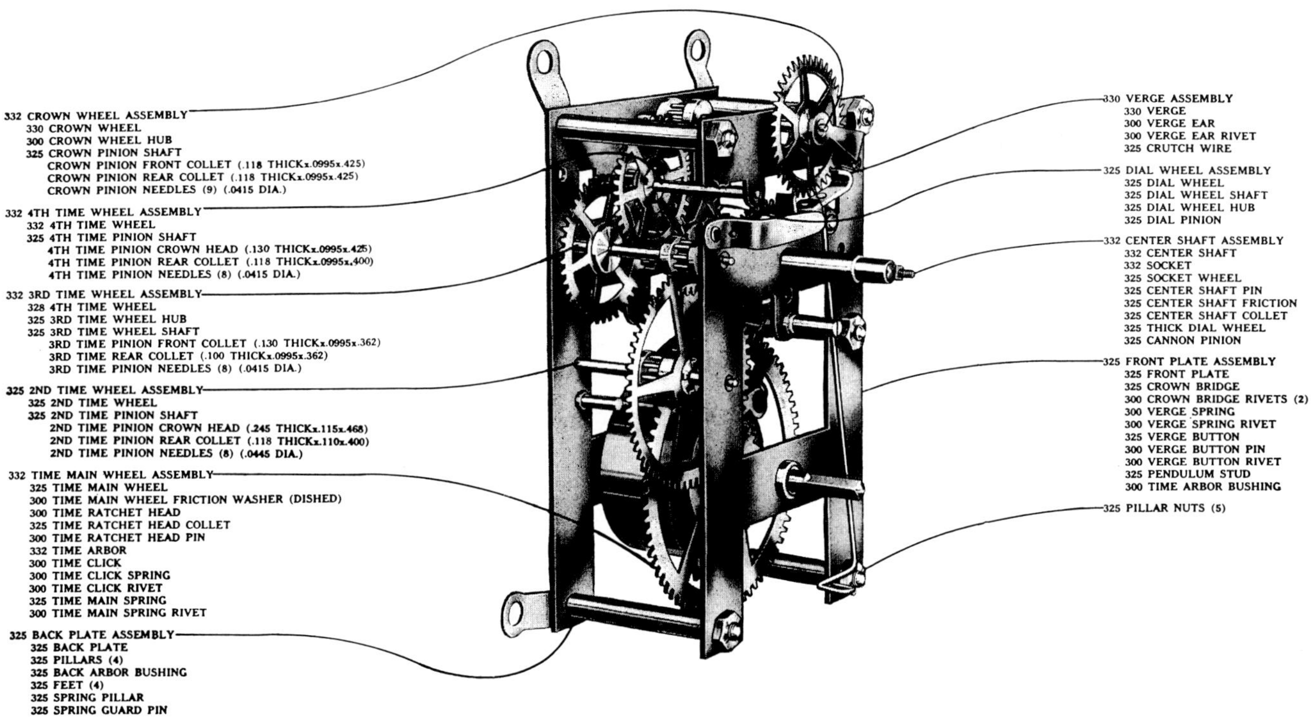

No. 332
8-Day Pendulum Timepiece
12-1-1927

MOVEMENT ILLUSTRATIONS

1515

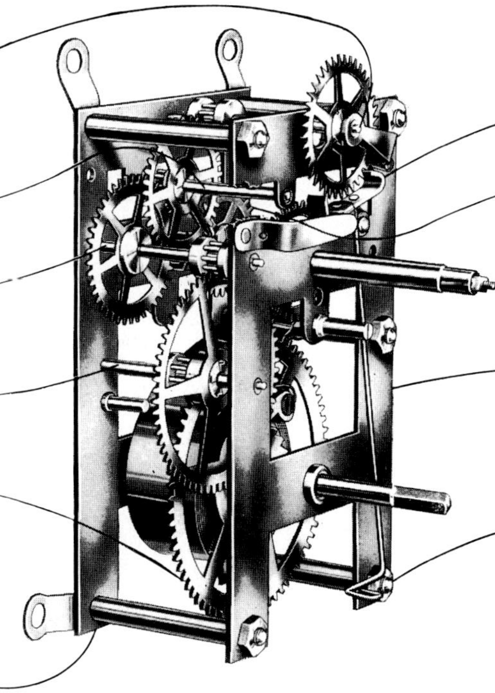

No. 333
8-Day Pendulum Timepiece
11-1-1927

MOVEMENT ILLUSTRATIONS

1516

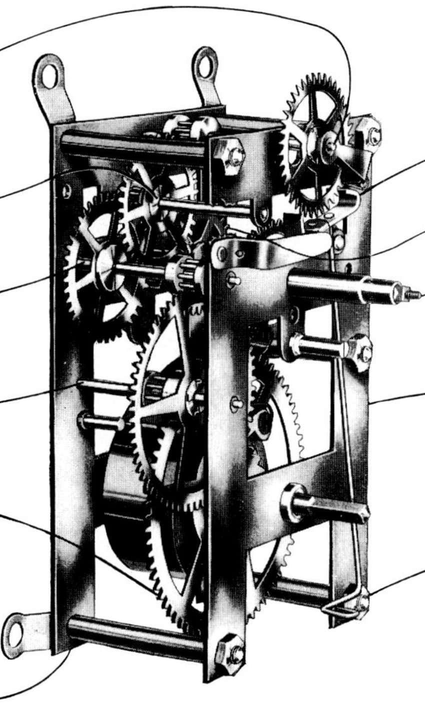

No. 334

8-Day Pendulum Timepiece

11-1-1927

MOVEMENT ILLUSTRATIONS

1517

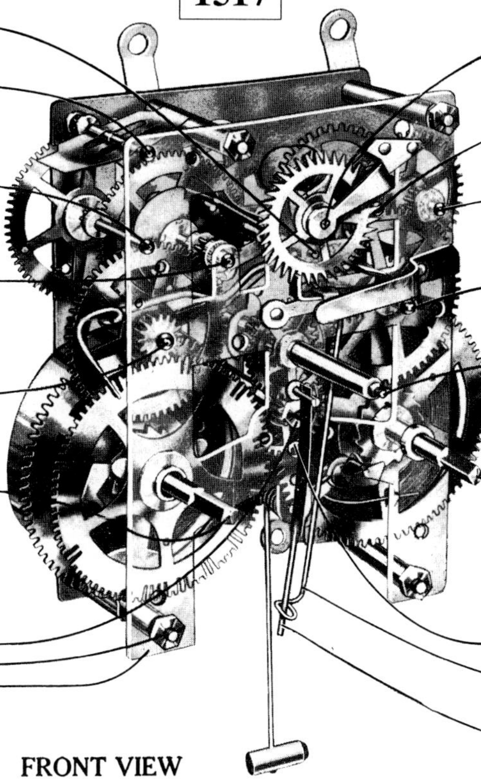

FRONT VIEW

300 COUNT HOOK ASSEMBLY
- 300 COUNT HOOK
- 300 COUNT HOOK SHAFT
- 300 COUNT HOOK LIFT WIRE
- 300 COUNT HOOK LOCK WIRE
- 300 COUNT HOOK TENSION SPRING

300 FLY ASSEMBLY
- 300 FLY
- 300 FLY PINION SHAFT
  - FLY PINION FRONT COLLET (.078 THICKx.069x.222)
  - FLY PINION REAR COLLET (.070 THICKx.069x.222)
  - FLY PINION NEEDLES (6) (.029 DIA.)

320 4TH STRIKE WHEEL ASSEMBLY
- 320 4TH STRIKE WHEEL
- 300 4TH STRIKE PINION SHAFT
- 320 4TH STRIKE WARN PINS (2)
  - 4TH STRIKE WHEEL CROWN HEAD (.100 THICKx.077x.420)
  - 4TH STRIKE PINION REAR COLLET (.090 THICKx.077x.250)
  - 4TH STRIKE PINION NEEDLES (6) (.029 DIA.)

300 3RD STRIKE WHEEL ASSEMBLY
- 300 3RD STRIKE WHEEL
- 300 3RD STRIKE PINION SHAFT
- 300 CAM DISC
- 300 DISC PINS (2)
- 300 CAM PINION COLLET
  - 3RD STRIKE PINION FRONT COLLET (.090 THICKx.0995x.320)
  - 3RD STRIKE PINION REAR COLLET (.125 THICKx.0995x.320)
  - 3RD STRIKE PINION NEEDLES (8) (.033 DIA.)

300 2ND STRIKE WHEEL ASSEMBLY
- 300 2ND STRIKE WHEEL
- 300 2ND STRIKE PINION SHAFT
  - 2ND STRIKE PINION CROWN HEAD (.165 THICKx.0995x.420)
  - 2ND STRIKE PINION REAR COLLET (.100 THICKx.0995x.320)
  - 2ND STRIKE PINION NEEDLES (8) (.0415 DIA.)
- 300 COUNT PINION

300 STRIKE MAIN WHEEL ASSEMBLY
- 300 TIME MAIN WHEEL
- 300 TIME MAIN WHEEL FRICTION WASHER (DISHED)
- 300 TIME RATCHET HEAD
- 300 TIME RATCHET HEAD PIN
- 300 TIME RATCHET HEAD COLLET
- 300 STRIKE ARBOR
- 300 TIME CLICK
- 300 TIME CLICK SPRING
- 300 TIME CLICK RIVET
- 320 TIME MAIN SPRING
- 300 TIME MAIN SPRING RIVET

320 5TH PILLAR NUT

320 PILLAR NUTS (4)

300 FRONT PLATE ASSEMBLY
- 300 FRONT PLATE
- 300 TIME ARBOR BUSHING
- 300 COUNT WHEEL
- 300 COUNT WHEEL BUSHING
- 300 COUNT WHEEL SPRING
- 300 CROWN BRIDGE
- 350 CROWN BRIDGE RIVETS (2)
- 300 VERGE BUTTON
- 300 VERGE BUTTON PIN
- 300 VERGE BUTTON RIVET
- 300 VERGE SPRING
- 300 VERGE SPRING RIVET
- 300 PENDULUM STUD

335 CROWN WHEEL ASSEMBLY
- 335 CROWN WHEEL
- 300 CROWN WHEEL HUB
- 300 CROWN PINION SHAFT
  - CROWN PINION FRONT COLLET (.118 THICKx.086x.425)
  - CROWN PINION REAR COLLET (.118 THICKx.086x.425)
  - CROWN PINION NEEDLES (9) (.0415 DIA.)

310 4TH TIME WHEEL ASSEMBLY
- 310 4TH TIME WHEEL
- 300 4TH TIME PINION SHAFT
  - 4TH TIME PINION FRONT COLLET (.136 THICKx.086x.320)
  - 4TH TIME PINION CROWN HEAD (.090 THICKx.086x.425)
  - 4TH TIME PINION NEEDLES (8) (.0415 DIA.)

310 3RD TIME WHEEL ASSEMBLY
- 310 4TH TIME WHEEL
- 300 3RD TIME PINION SHAFT
  - 3RD TIME PINION FRONT COLLET (.090 THICKx.0995x.320)
  - 3RD TIME PINION CROWN HEAD (.213 THICKx.0995x.425)
  - 3RD TIME PINION NEEDLES (8) (.0415 DIA.)

300 2ND TIME WHEEL ASSEMBLY
- 300 2ND TIME WHEEL
- 300 2ND TIME PINION SHAFT
  - 2ND TIME PINION CROWN HEAD (.165 THICKx.0995x.420)
  - 2ND TIME PINION REAR COLLET (.100 THICKx.0995x.320)
  - 2ND TIME PINION NEEDLES (8) (.0415 DIA.)

300 CENTER SHAFT ASSEMBLY
- 300 CENTER SHAFT
- 300 CANNON PINION
- 300 THICK DIAL WHEEL
- 300 CENTER SHAFT FRICTION SPRING
- 300 CENTER SHAFT FRICTION COLLET
- 300 CENTER SHAFT PLAIN WASHERS (2)
- 300 LIFT BUTTON
- 300 SOCKET
- 300 SOCKET WHEEL

300 TIME MAIN WHEEL ASSEMBLY
- 300 TIME MAIN WHEEL
- 300 TIME MAIN WHEEL FRICTION WASHER (DISHED)
- 300 TIME RATCHET HEAD
- 300 TIME RATCHET HEAD PIN
- 300 TIME RATCHET HEAD COLLET
- 300 TIME ARBOR
- 300 TIME CLICK
- 300 TIME CLICK SPRING
- 300 TIME CLICK RIVET
- 300 TIME MAIN SPRING
- 300 TIME MAIN SPRING RIVET

300 DIAL WHEEL ASSEMBLY
- 300 DIAL WHEEL
- 300 DIAL SHUCK

300 VERGE ASSEMBLY
- 300 VERGE
- 300 VERGE EAR
- 300 CRUTCH WIRE

335 PENDULUM ROD ASSEMBLY
- 335 PENDULUM ROD
- 300 FEATHER
- 300 FEATHER WIRE

1518

REAR VIEW

300 BACK PLATE ASSEMBLY
- 300 BACK PLATE
- 300 ARBOR BUSHINGS (2)
- 300 PILLARS (4)
- 300 5TH PILLAR
- 300 FEET (4)
- 300 STRIKE SPRING GUARD PINS (2)
- 300 TIME MAIN SPRING STUD
- 300 TURN BACK LATCH
- 300 TURN BACK LATCH HUB
- 300 TURN BACK LATCH STUD
- 300 TURN BACK LATCH STUD COLLET

300 HAMMER SHAFT ASSEMBLY
- 300 HAMMER SHAFT
- 300 STOP WIRE
- 300 CAM WIRE
- 300 HAMMER SHAFT TENSION SPRING
- 300 HAMMER WIRE
- 300 HAMMER HEAD
- 380 LEATHER TIP

No. 335
8-Day Pendulum Time and Strike Movement
1-4-1928

MOVEMENT ILLUSTRATIONS

1519

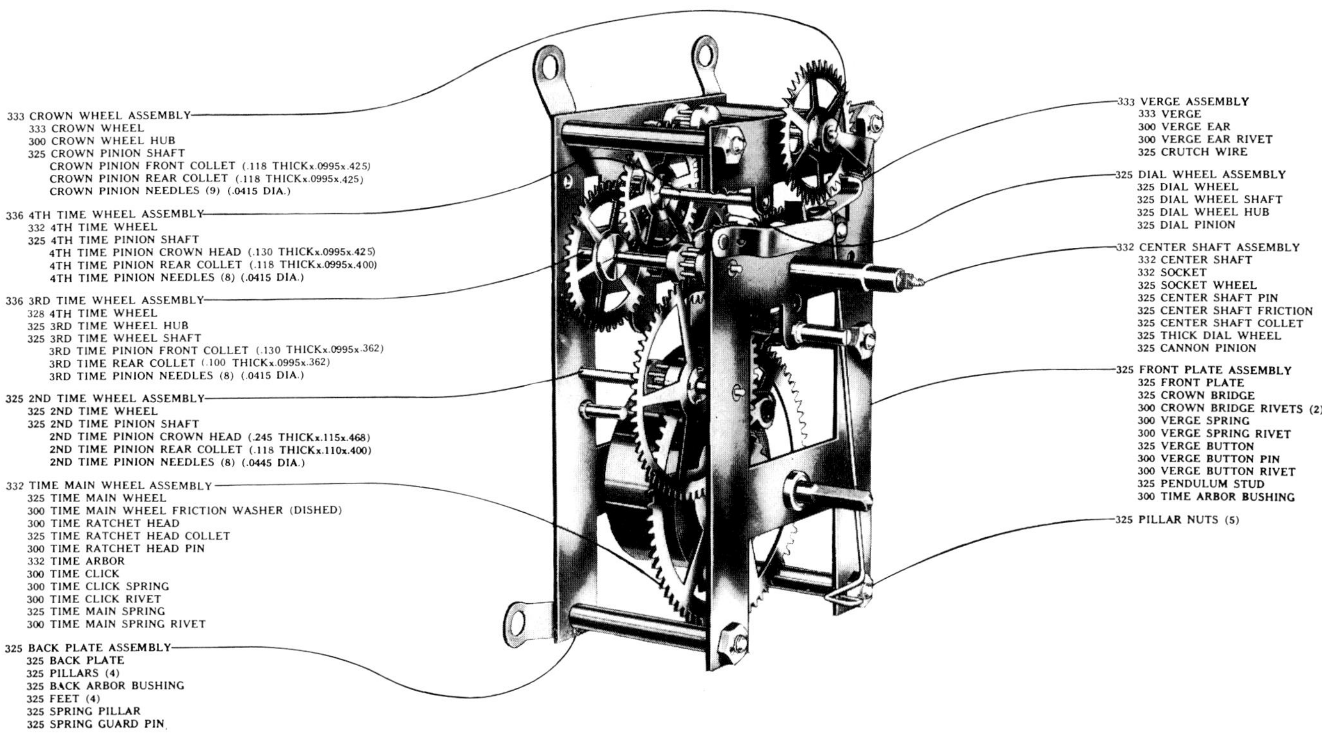

No. 336
8-Day Pendulum Timepiece
11-1-1927

MOVEMENT ILLUSTRATIONS

1520

300 COUNT HOOK ASSEMBLY
- 300 COUNT HOOK
- 300 COUNT HOOK SHAFT
- 300 COUNT HOOK LIFT WIRE
- 300 COUNT HOOK LOCK WIRE
- 300 COUNT HOOK TENSION SPRING

300 FLY ASSEMBLY
- 300 FLY
- 300 FLY PINION SHAFT
- FLY PINION FRONT COLLET (.078 THICKx.069x.222)
- FLY PINION REAR COLLET (.070 THICKx.069x.222)
- FLY PINION NEEDLES (6) (.029 DIA.)

320 4TH STRIKE WHEEL ASSEMBLY
- 320 4TH STRIKE WHEEL
- 300 4TH STRIKE PINION SHAFT
- 320 4TH STRIKE WARN PINS (2)
- 4TH STRIKE PINION FRONT COLLET (.090 THICKx.077x.250)
- 4TH STRIKE WHEEL CROWN HEAD (.100 THICKx.077x.420)
- 4TH STRIKE PINION NEEDLES (6) (.029 DIA.)

300 3RD STRIKE WHEEL ASSEMBLY
- 300 3RD STRIKE WHEEL
- 300 3RD STRIKE PINION SHAFT
- 300 CAM DISC
- 300 DISC PINS (2)
- 300 CAM PINION COLLET
- 3RD STRIKE PINION FRONT COLLET (.090 THICKx.0995x.320)
- 3RD STRIKE PINION REAR COLLET (.125 THICKx.0995x.320)
- 3RD STRIKE PINION NEEDLES (8) (.033 DIA.)

300 2ND STRIKE WHEEL ASSEMBLY
- 300 2ND STRIKE WHEEL
- 300 2ND STRIKE PINION SHAFT
- 2ND STRIKE PINION CROWN HEAD (.165 THICKx.0995x.420)
- 2ND STRIKE PINION REAR COLLET (.100 THICKx.0995x.320)
- 2ND STRIKE PINION NEEDLES (8) (.0415 DIA.)
- 300 COUNT PINION

300 STRIKE MAIN WHEEL ASSEMBLY
- 300 TIME MAIN WHEEL
- 300 TIME MAIN WHEEL FRICTION WASHER (DISHED)
- 300 TIME RATCHET HEAD
- 300 TIME RATCHET HEAD PIN
- 300 TIME RATCHET HEAD COLLET
- 300 STRIKE ARBOR
- 300 TIME CLICK
- 300 TIME CLICK SPRING
- 300 TIME CLICK RIVET
- 320 TIME MAIN SPRING
- 300 TIME MAIN SPRING RIVET

320 5TH PILLAR NUT

320 PILLAR NUTS (4)

300 FRONT PLATE ASSEMBLY
- 300 FRONT PLATE
- 300 TIME ARBOR BUSHING
- 300 COUNT WHEEL
- 300 COUNT WHEEL BUSHING
- 300 COUNT WHEEL SPRING
- 300 CROWN BRIDGE
- 350 CROWN BRIDGE RIVETS (2)
- 300 VERGE BUTTON
- 300 VERGE BUTTON PIN
- 300 VERGE BUTTON RIVET
- 300 VERGE SPRING
- 300 VERGE SPRING RIVET
- 300 PENDULUM STUD

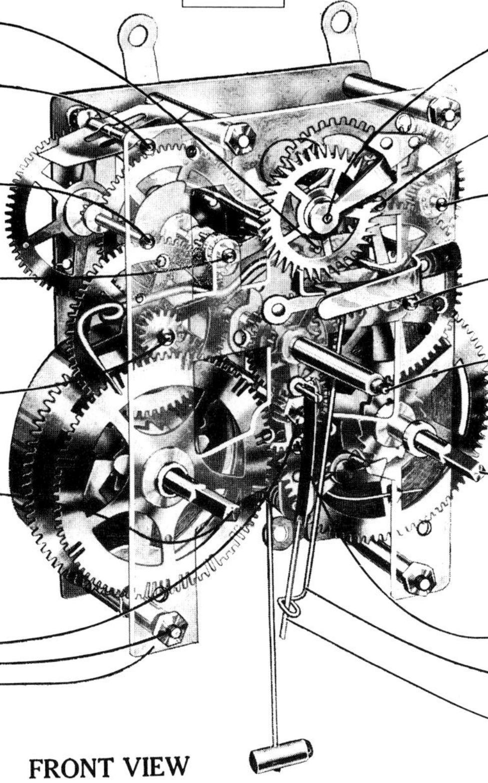

FRONT VIEW

337 CROWN WHEEL ASSEMBLY
- 337 CROWN WHEEL
- 300 CROWN WHEEL HUB
- 300 CROWN PINION SHAFT
- CROWN PINION FRONT COLLET (.136 THICKx.086x.425)
- CROWN PINION REAR COLLET (.090 THICKx.086x.425)
- CROWN PINION NEEDLES (10) (.0415 DIA.)

303 4TH TIME WHEEL ASSEMBLY
- 303 4TH TIME WHEEL
- 300 4TH TIME PINION SHAFT
- 3RD TIME PINION FRONT COLLET (.090 THICKx.086x.320)
- 3RD TIME PINION CROWN HEAD (.136 THICKx.086x.425)
- 4TH TIME PINION NEEDLES (7) (.0415 DIA.)

300 3RD TIME WHEEL ASSEMBLY
- 300 3RD TIME WHEEL
- 300 3RD TIME PINION SHAFT
- 3RD TIME PINION FRONT COLLET (.090 THICKx.0995x.320)
- 3RD TIME PINION CROWN HEAD (.213 THICKx.0995x.425)
- 3RD TIME PINION NEEDLES (8) (.0415 DIA.)

300 2ND TIME WHEEL ASSEMBLY
- 300 2ND TIME WHEEL
- 300 2ND TIME PINION SHAFT
- 2ND TIME PINION CROWN HEAD (.165 THICKx.0995x.420)
- 2ND TIME PINION REAR COLLET (.100 THICKx.0995x.320)
- 2ND TIME PINION NEEDLES (8) (.0415 DIA.)

300 CENTER SHAFT ASSEMBLY
- 300 CENTER SHAFT
- 300 CANNON PINION
- 300 THICK DIAL WHEEL
- 300 CENTER SHAFT FRICTION SPRING
- 300 CENTER SHAFT FRICTION COLLET
- 300 CENTER SHAFT PLAIN WASHERS (2)
- 300 LIFT BUTTON
- 300 SOCKET
- 300 SOCKET WHEEL

300 TIME MAIN WHEEL ASSEMBLY
- 300 TIME MAIN WHEEL
- 300 TIME MAIN WHEEL FRICTION WASHER (DISHED)
- 300 TIME RATCHET HEAD
- 300 TIME RATCHET HEAD PIN
- 300 TIME RATCHET HEAD COLLET
- 300 TIME ARBOR
- 300 TIME CLICK
- 300 TIME CLICK SPRING
- 300 TIME CLICK RIVET
- 300 TIME MAIN SPRING
- 300 TIME MAIN SPRING RIVET

300 DIAL WHEEL ASSEMBLY
- 300 DIAL WHEEL
- 300 DIAL SHUCK

300 VERGE ASSEMBLY
- 300 VERGE
- 300 VERGE EAR
- 300 CRUTCH WIRE

337 PENDULUM ROD ASSEMBLY
- 337 PENDULUM ROD
- 300 FEATHER
- 300 FEATHER WIRE

1521

300 BACK PLATE ASSEMBLY
- 300 BACK PLATE
- 300 ARBOR BUSHINGS (2)
- 300 PILLARS (4)
- 300 5TH PILLAR
- 300 FEET (4)
- 300 STRIKE SPRING GUARD PINS (2)
- 300 TIME MAIN SPRING STUD
- 300 TURN BACK LATCH
- 300 TURN BACK LATCH HUB
- 300 TURN BACK LATCH STUD
- 300 TURN BACK LATCH STUD COLLET

300 HAMMER SHAFT ASSEMBLY
- 300 HAMMER SHAFT
- 300 STOP WIRE
- 300 CAM WIRE
- 300 HAMMER SHAFT TENSION SPRING
- 300 HAMMER WIRE
- 300 HAMMER HEAD
- 380 LEATHER TIP

REAR VIEW

No. 337
8-Day Pendulum Time and Strike Movement
1-4-1928

MOVEMENT ILLUSTRATIONS

1522

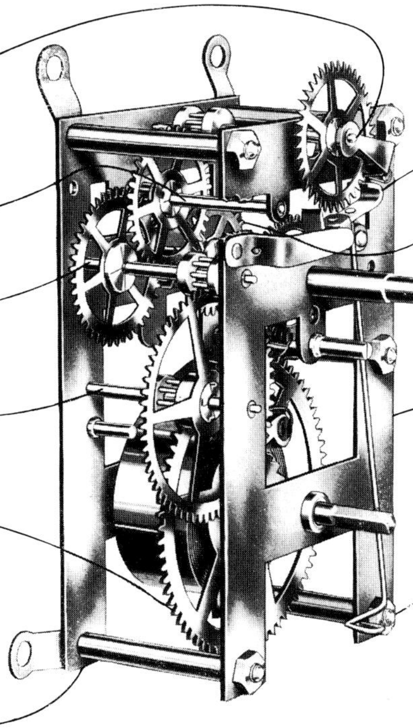

No. 338
8-Day Pendulum Timepiece
11-1-1927

MOVEMENT ILLUSTRATIONS

1523

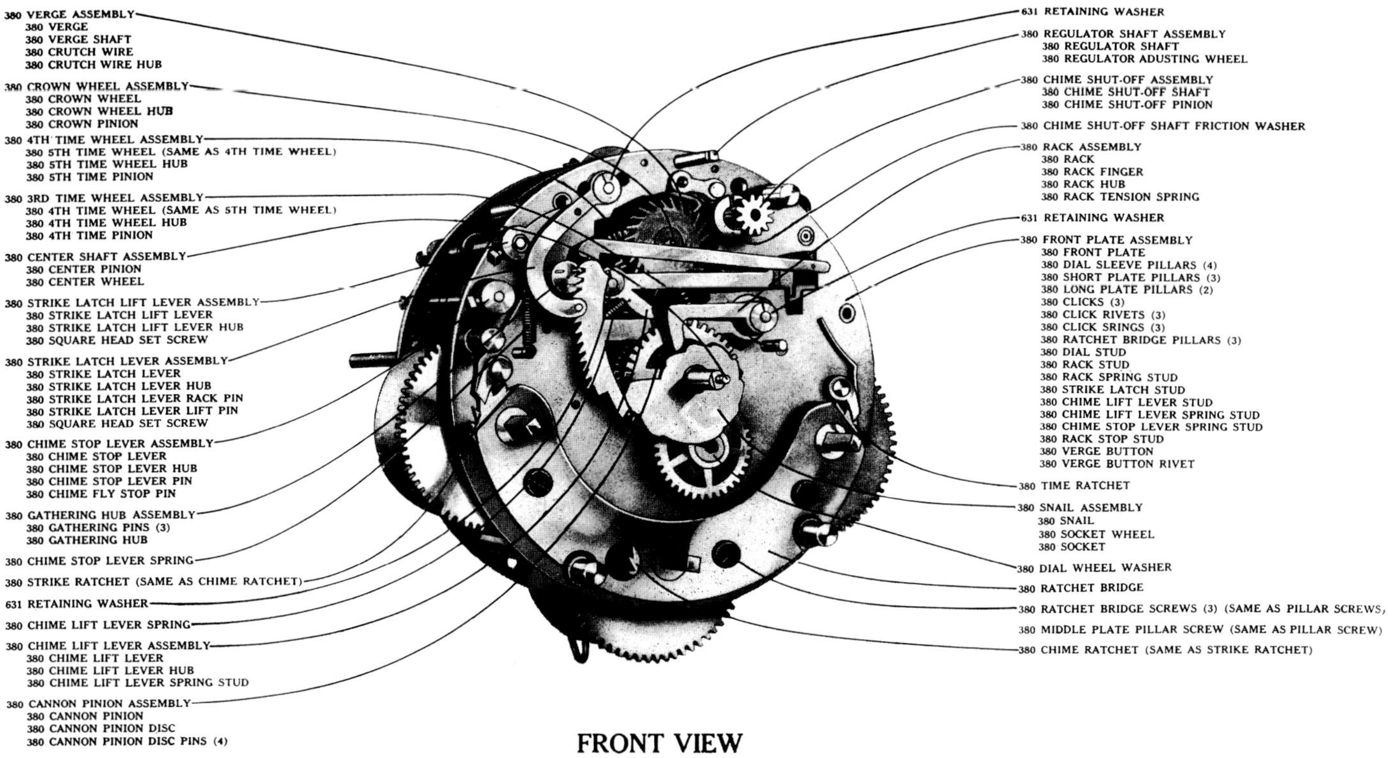

FRONT VIEW

No. 380
8-Day Pendulum Chime (Rack and Snail)
1-4-1928

MOVEMENT ILLUSTRATIONS

1524

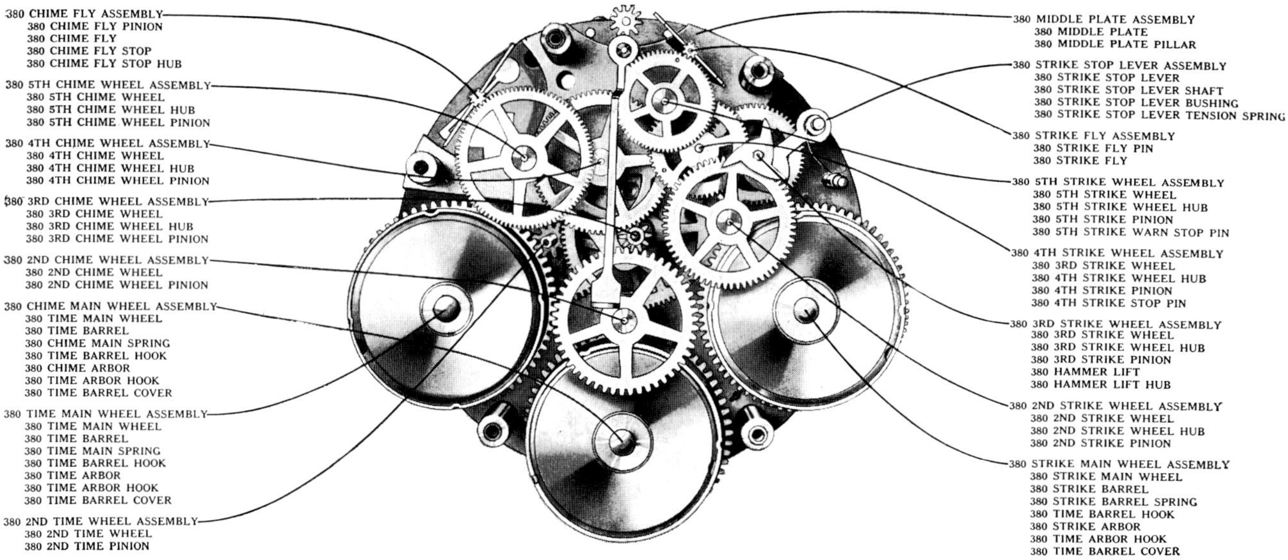

REAR VIEW
(BACK PLATE REMOVED)

1525

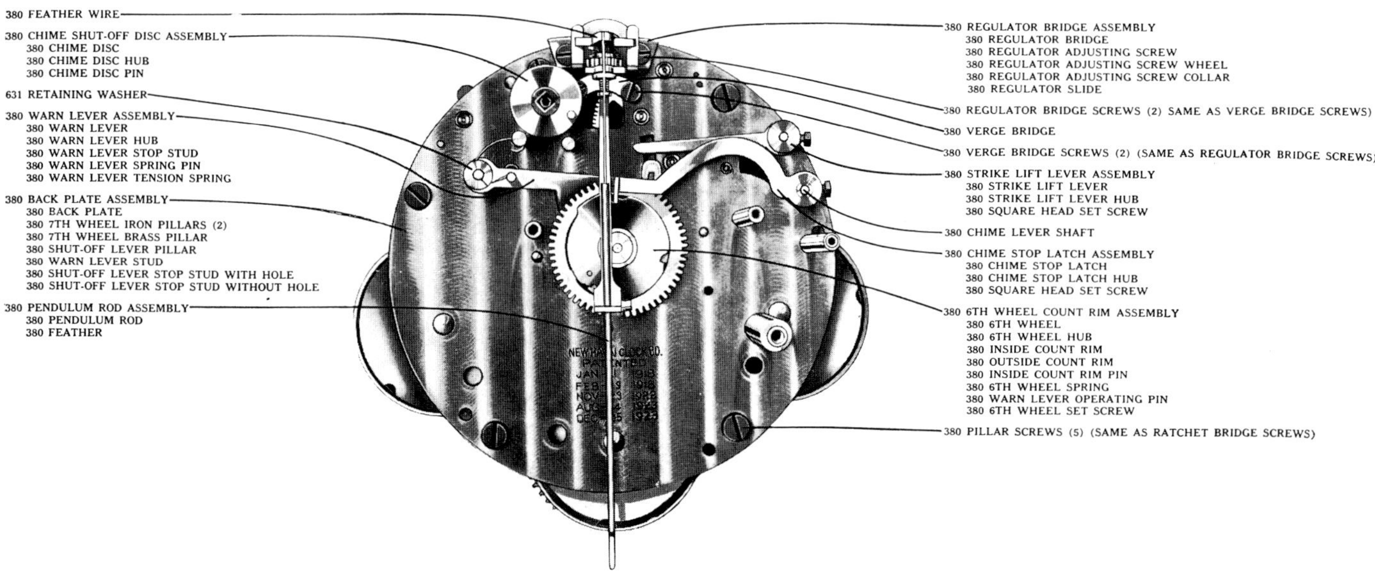

REAR VIEW

No. 380
8-Day Pendulum Chime (Rack and Snail)
12-1-1927

MOVEMENT ILLUSTRATIONS

1526

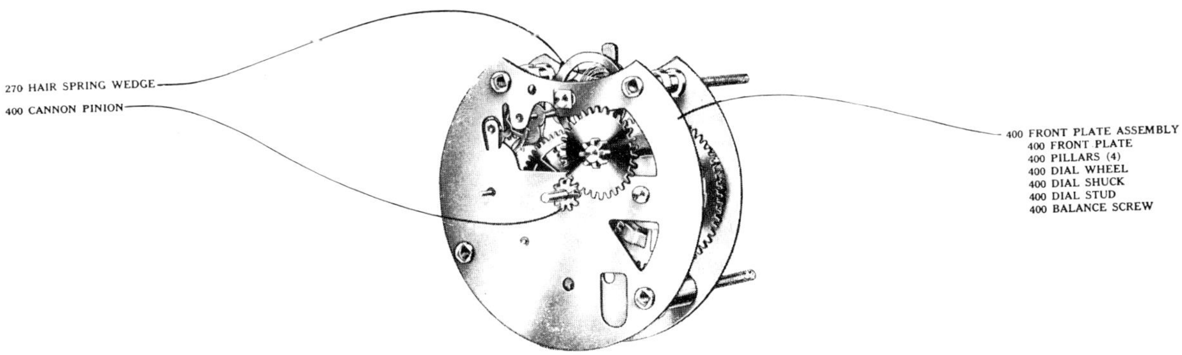

FRONT VIEW

1527

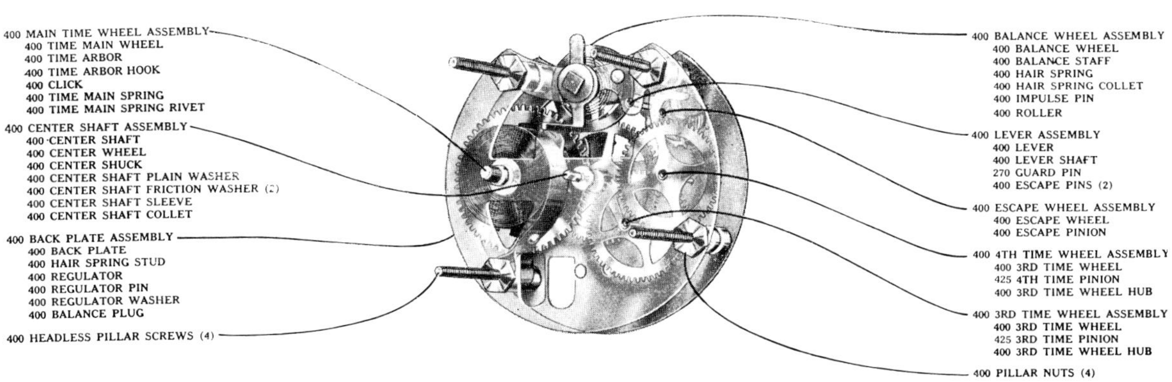

REAR VIEW

No. 400

1-Day Small Lever Timepiece

11-1-1927

MOVEMENT ILLUSTRATIONS

1528

FRONT VIEW

1529

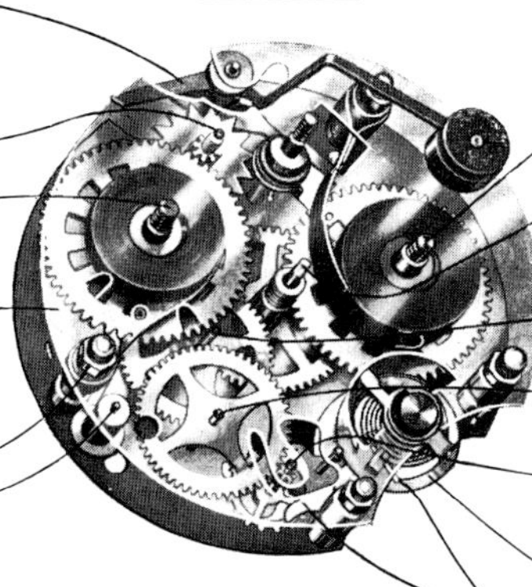

REAR VIEW

No. 426
1-Day Tidy-Tot Alarm Movement
11-1-1927

MOVEMENT ILLUSTRATIONS

1530

FRONT VIEW

1531

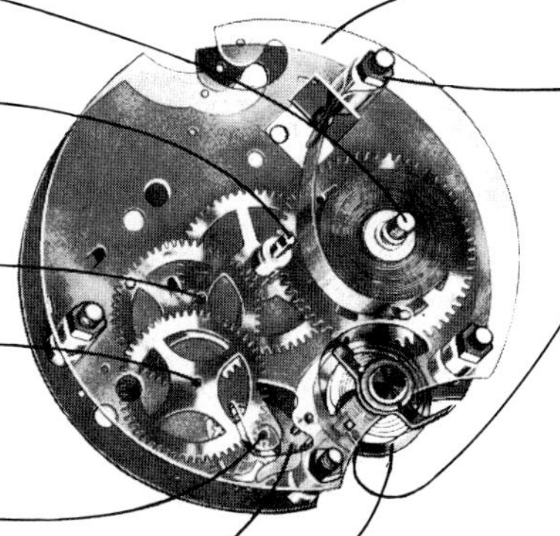

REAR VIEW

No. 434
1-Day Small Slip-in Alarm Movement with Short Back Pillars
11-1-1927

MOVEMENT ILLUSTRATIONS

1532

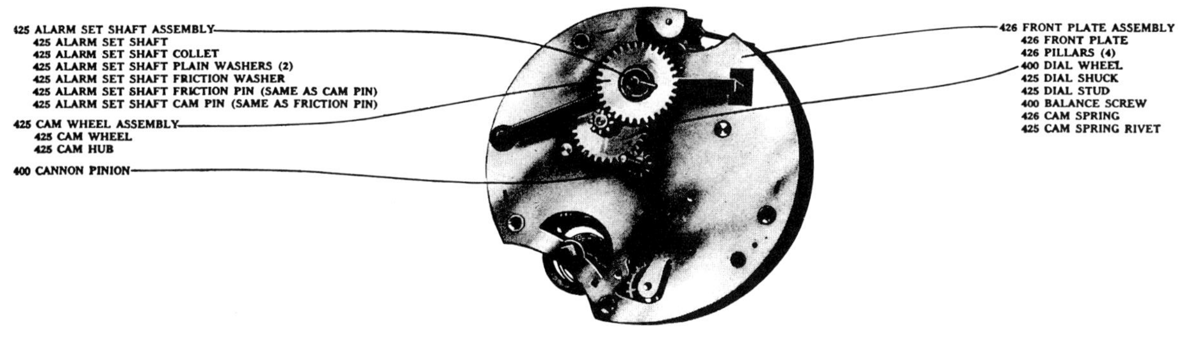

FRONT VIEW

1533

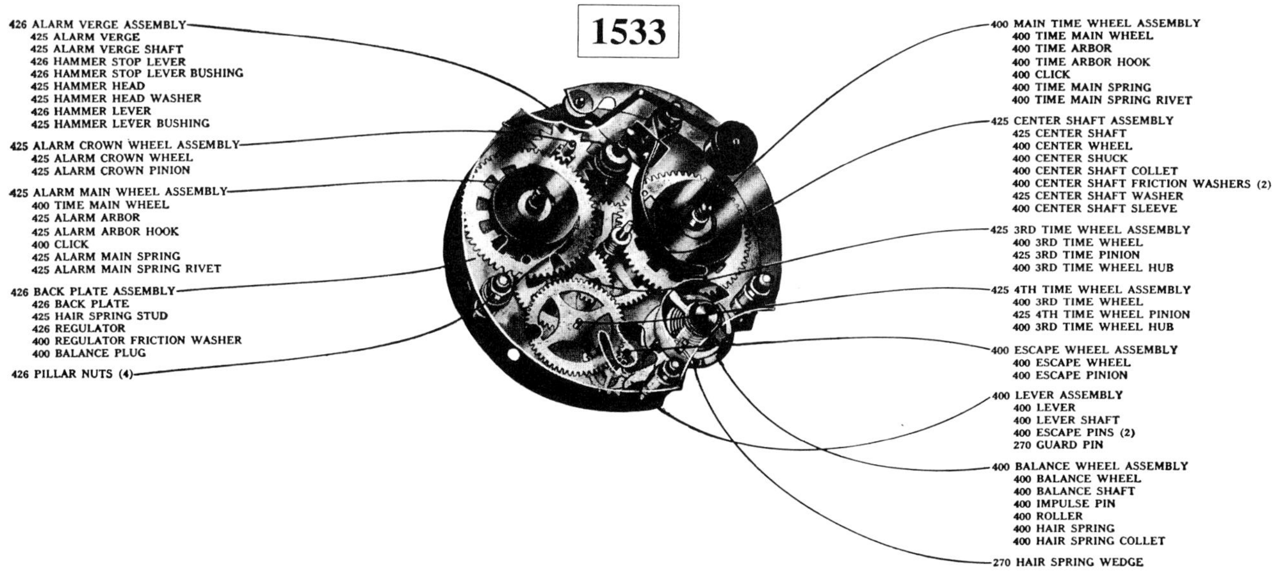

REAR VIEW

No. 435
1-Day Small Slip-in Alarm Movement
with Short Back Pillars
12-1-1927

MOVEMENT ILLUSTRATIONS

1534

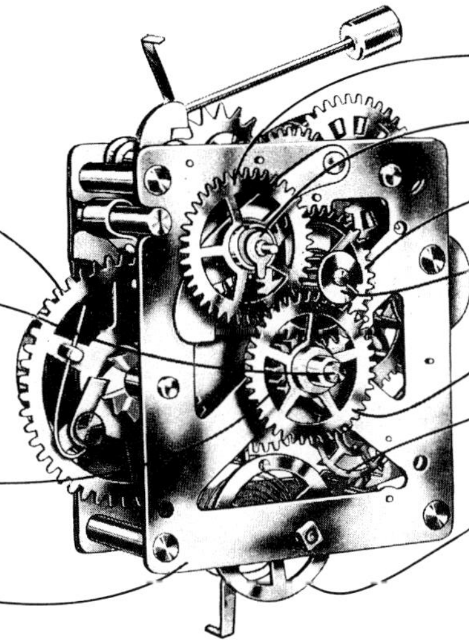

FRONT VIEW

1535

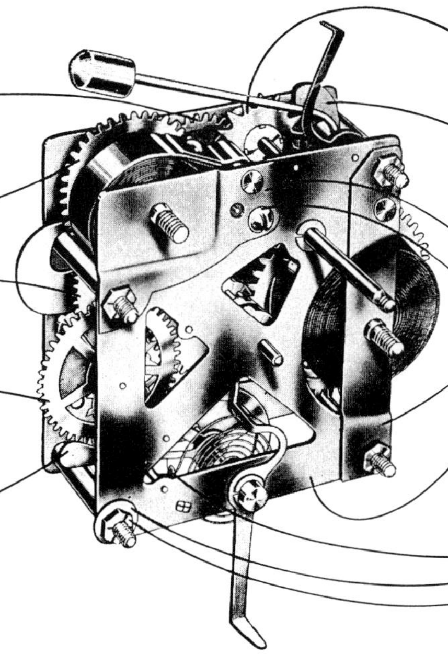

REAR VIEW

No. 505
1-Day Tat-Too Alarm Movement
9-20-1926

MOVEMENT ILLUSTRATIONS

1536

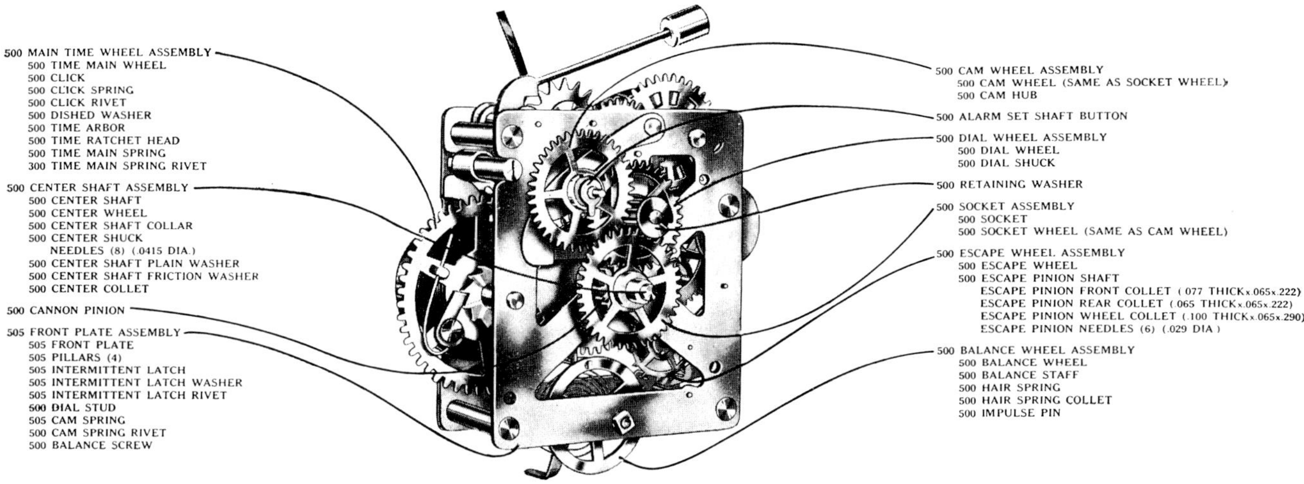

FRONT VIEW

1537

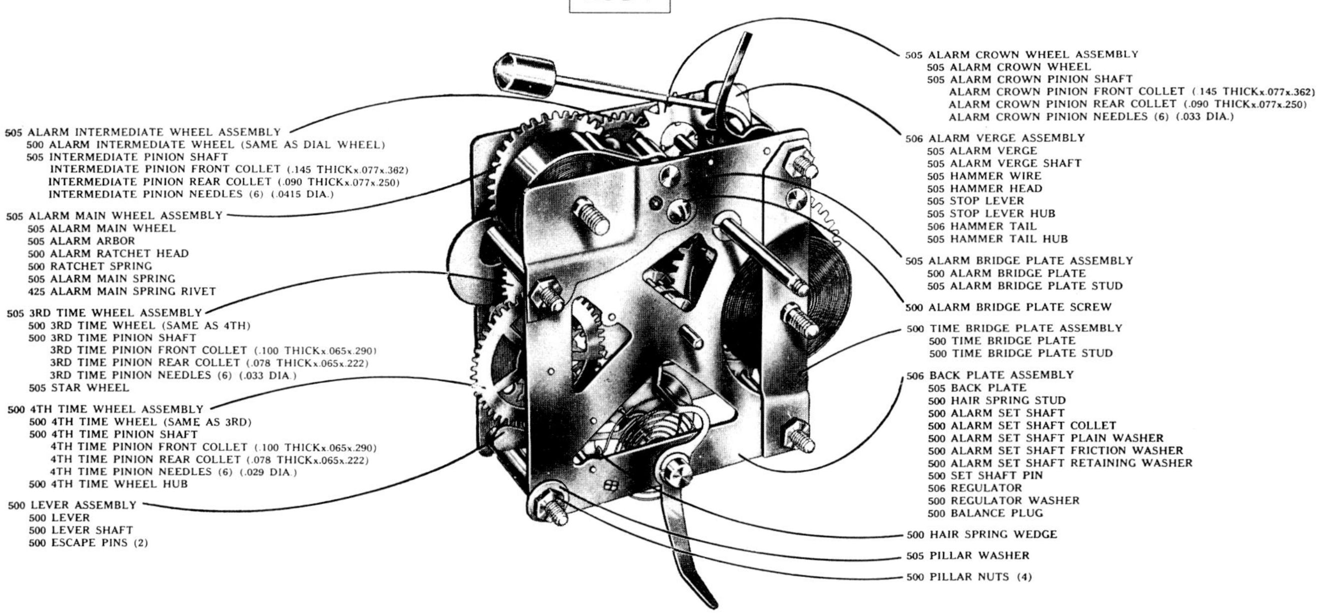

REAR VIEW

No. 506
1-Day Tom-Tom Alarm Movement
9-20-1926

MOVEMENT ILLUSTRATIONS

1538

FRONT VIEW

1539

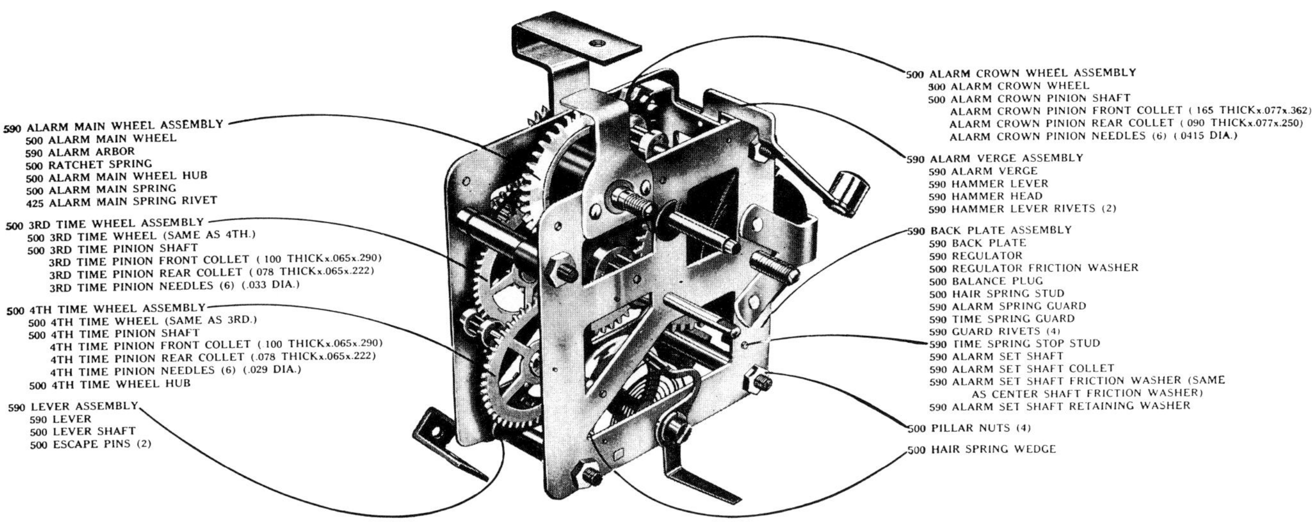

REAR VIEW

No. 590
1-Day Tick-Tock Alarm Movement
9-20-1926

MOVEMENT ILLUSTRATIONS

1540

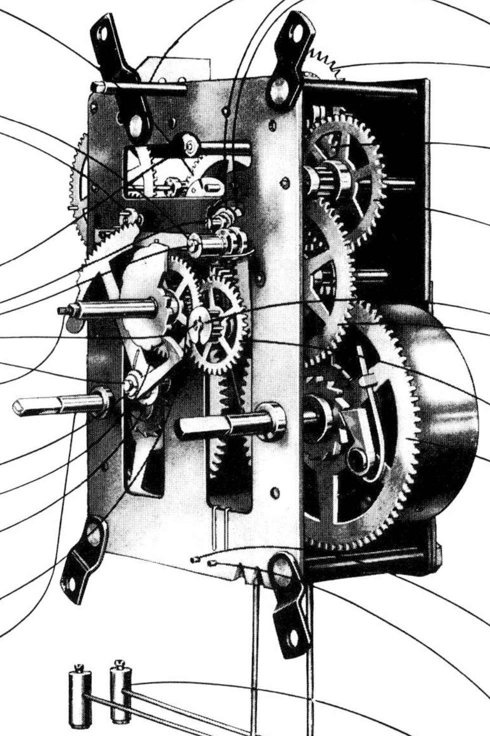

700 3RD STRIKE WHEEL ASSEMBLY
700 3RD STRIKE WHEEL
700 CAM DISC PINS (4)
700 CAM DISC
700 3RD STRIKE PINION SHAFT
3RD STRIKE PINION FRONT COLLET (.090x.086x.285)
3RD STRIKE PINION BACK COLLET (.125x.086x.285)
700 CAM PINION COLLET
3RD STRIKE PINION NEEDLES (8) (.033x.256)

631 RETAINING WASHER

700 GATHERING HUB ASSEMBLY
700 GATHERING HUB
GATHERING HUB PINS (2) (SAME AS 500 ESCAPE PINS)

700 LATCH ASSEMBLY
700 LATCH HUB
700 LATCH
700 WARN PIN (SAME AS LATCH PIN)
700 LATCH PIN (SAME AS WARN PIN)
700 LATCH TENSION WIRE (SAME AS RACK TENSION WIRE)

700 FRONT PLATE ASSEMBLY
700 FRONT PLATE
700 LONG FOOT
700 SHORT FEET (3)
380 VERGE BUTTON RIVET
380 VERGE BUTTON
700 TURN BACK LATCH STUD (SAME AS DIAL STUD)
700 LATCH STUD (SAME AS RACK STUD)
700 DIAL STUD (SAME AS TURN BACK LATCH STUD)
700 RACK STUD (SAME AS LATCH STUD)
300 ARBOR BUSHINGS (2)
700 PLATE PILLARS (4)

700 CENTER SHAFT ASSEMBLY
700 THICK DIAL WHEEL
700 CENTER WASHER (LARGE)
700 LIFT HUB
700 LIFT PINS (2)
700 CENTER SHAFT
REST OF ASSEMBLY SHOWN ON REAR VIEW

631 RETAINING WASHER

700 SNAIL ASSEMBLY
700 SNAIL
500 SOCKET WHEEL
700 SOCKET

700 RACK ASSEMBLY
700 RACK
700 RACK TENSION WIRE (SAME AS LATCH TENSION WIRE)
380 RACK FINGER
700 RACK HUB

700 CANNON PINION

700 STRIKE MAIN WHEEL ASSEMBLY
300 TIME MAIN WHEEL
300 TIME MAIN WHEEL FRICTION WASHER (DISHED)
300 TIME RATCHET HEAD
300 TIME RATCHET HEAD PINS (2)
300 TIME RATCHET HEAD COLLET
700 STRIKE ARBOR
300 TIME CLICK
300 TIME CLICK RIVET
300 TIME CLICK SPRING
700 STRIKE MAIN SPRING
600 TIME BARREL HOOK
700 TIME BARREL

700 REGULATOR SHAFT ASSEMBLY
700 REGULATOR SHAFT
700 REGULATOR WHEEL (SAME AS REGULATOR ADJUSTING SCREW WHEEL)

700 TURN BACK LATCH ASSEMBLY
700 TURN BACK LATCH
700 TURN BACK LATCH HUB

500 RETAINING WASHER

700 CROWN WHEEL ASSEMBLY
700 CROWN WHEEL
CROWN PINION FRONT COLLET (.090x.086x.320)
CROWN PINION CROWN HEAD (.125x.076x.420)
700 CROWN PINION SHAFT
CROWN PINION NEEDLES (7) (.041½x.315)

700 4TH TIME WHEEL ASSEMBLY
700 4TH TIME WHEEL
700 4TH TIME PINION
4TH TIME PINION FRONT COLLET (.090x.086x.320)
4TH TIME PINION CROWN HEAD (.125x.076x.420)
4TH TIME PINION NEEDLES (7) (.041½x.315)

700 3RD TIME WHEEL ASSEMBLY
300 3RD TIME WHEEL
700 3RD TIME PINION SHAFT
3RD TIME PINION FRONT COLLET (.090x.086x.320)
3RD TIME PINION REAR COLLET (320 2ND TIME PINION COLLET)
3RD TIME PINION NEEDLES (8) (.041½x.315)

500 RETAINING WASHER

700 2ND TIME WHEEL ASSEMBLY
700 2ND TIME WHEEL
700 2ND TIME PINION SHAFT
2ND TIME PINION FRONT COLLET (300 3RD STRIKE PINION COLLET)
2ND TIME PINION REAR COLLET (.100x.0995x.362)
2ND TIME PINION NEEDLES (8) (.041½x.315)

700 DIAL WHEEL ASSEMBLY
500 DIAL WHEEL
700 DIAL SHUCK

700 TIME MAIN WHEEL ASSEMBLY
300 TIME MAIN WHEEL
300 TIME MAIN WHEEL FRICTION WASHER (DISHED)
300 TIME RATCHET HEAD
300 TIME RATCHET HEAD PINS (2)
300 TIME RATCHET HEAD COLLET
700 TIME ARBOR
300 TIME CLICK
300 TIME CLICK RIVET
300 TIME CLICK SPRING
700 TIME MAIN SPRING
600 TIME BARREL HOOK
700 TIME BARREL

700 1ST HAMMER SHAFT ASSEMBLY
700 1ST HAMMER TAIL WIRE (SAME AS 2ND)
700 1ST HAMMER TENSION SPRING (SAME AS 2ND)
700 1ST HAMMER SHAFT
700 1ST HAMMER SET WIRE (SAME AS 2ND)
700 1ST HAMMER WIRE (SAME AS 2ND)

700 2ND HAMMER SHAFT ASSEMBLY
700 2ND HAMMER TAIL WIRE (SAME AS 1ST)
700 2ND HAMMER TENSION SPRING (SAME AS 1ST)
700 2ND HAMMER SHAFT
700 2ND HAMMER SET WIRE (SAME AS 1ST)
700 2ND HAMMER WIRE (SAME AS 1ST)

700 HAMMER HEAD ASSEMBLY (2)
700 HAMMER HEADS (2)
700 SET SCREWS (2)
380 LEATHER TIPS (2)

FRONT VIEW

1541

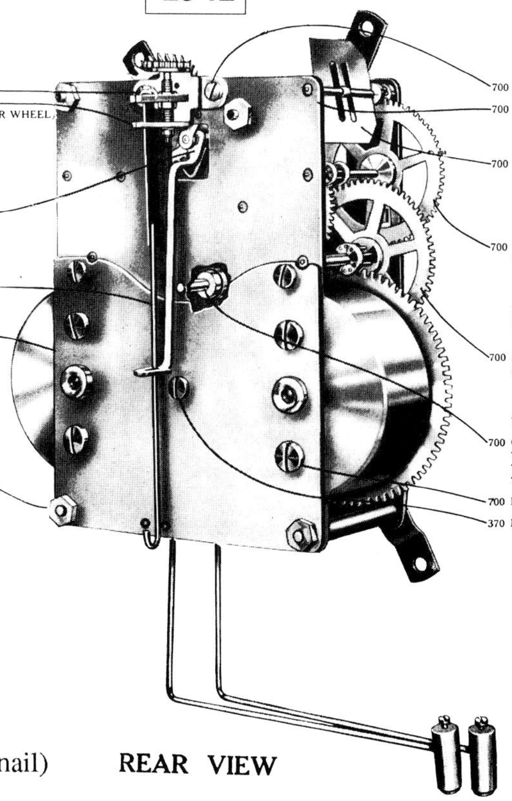

320 FEATHER WIRE

700 REGULATOR BRIDGE ASSEMBLY
700 REGULATOR ADJUSTING SCREW WHEEL (SAME AS REGULATOR WHEEL)
700 REGULATOR ADJUSTING SCREW
700 PENDULUM STUD
700 PENDULUM STUD BUSHING
700 REGULATOR FRICTION WASHER
700 REGULATOR FRICTION COLLET
700 REGULATOR BRIDGE

700 VERGE ASSEMBLY
320 CRUTCH WIRE HUB
700 VERGE SHAFT
700 VERGE
320 CRUTCH WIRE

320 PENDULUM ROD ASSEMBLY
320 FEATHER
320 PENDULUM ROD

700 BARRELPLATE ASSEMBLY
700 BARREL PLATE
700 ARBOR BUSHINGS (2)

320 PILLAR NUTS (4)

700 REGULATOR BRIDGE SCREWS (2)

700 BACK PLATE ASSEMBLY
380 VERGE BUTTON RIVET
380 VERGE BUTTON
700 BACK PLATE

700 FLY ASSEMBLY
700 FLY
700 FLY PINION SHAFT
FLY PINION REAR COLLET (.078x.065x.222)
FLY PINION FRONT COLLET (.078x.065x.222)
FLY PINION NEEDLES (6) (.033x.250)

700 4TH STRIKE WHEEL ASSEMBLY
300 4TH STRIKE WHEEL
4TH STRIKE WHEEL HUB CROWN HEAD (.100x.086x.420)
4TH STRIKE PINION REAR COLLET (.090x.077x.250)
4TH STRIKE PINION NEEDLES (7) (.029x.210)
700 4TH STRIKE PINION SHAFT
320 STRIKE PIN
4TH STRIKE WHEEL FRONT COLLET (.090x.077x.250)

700 2ND STRIKE WHEEL ASSEMBLY
300 2ND STRIKE WHEEL
2ND STRIKE CROWN HEAD (.165x.105x.420)
2ND STRIKE PINION NEEDLES (8) (.041½x.315)
2ND STRIKE PINION REAR COLLET (.100x.0995x.362)
700 2ND STRIKE PINION SHAFT

700 CENTER SHAFT ASSEMBLY (CONTINUED FROM FRONT PLATE)
700 CENTER SHAFT FRICTION WASHER
700 CENTER SHAFT SPLIT FRICTION WASHER
700 CENTER SHAFT FRICTION HUB

700 BARREL SCREWS (4)

370 BARREL PLATE SCREWS (3)

No. 700
8-Day Duo-Strike (Rack and Snail)

REAR VIEW

MOVEMENT ILLUSTRATIONS

1542

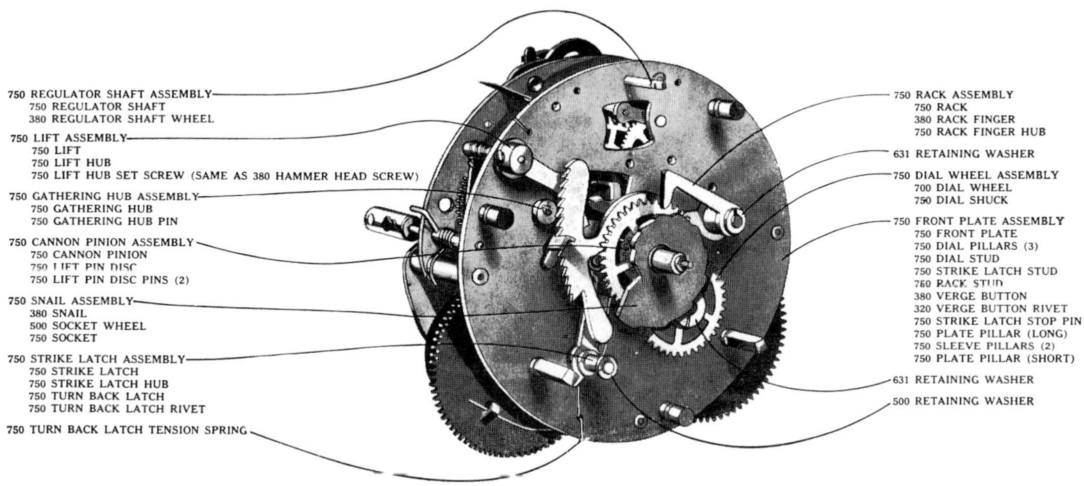

FRONT VIEW

1543

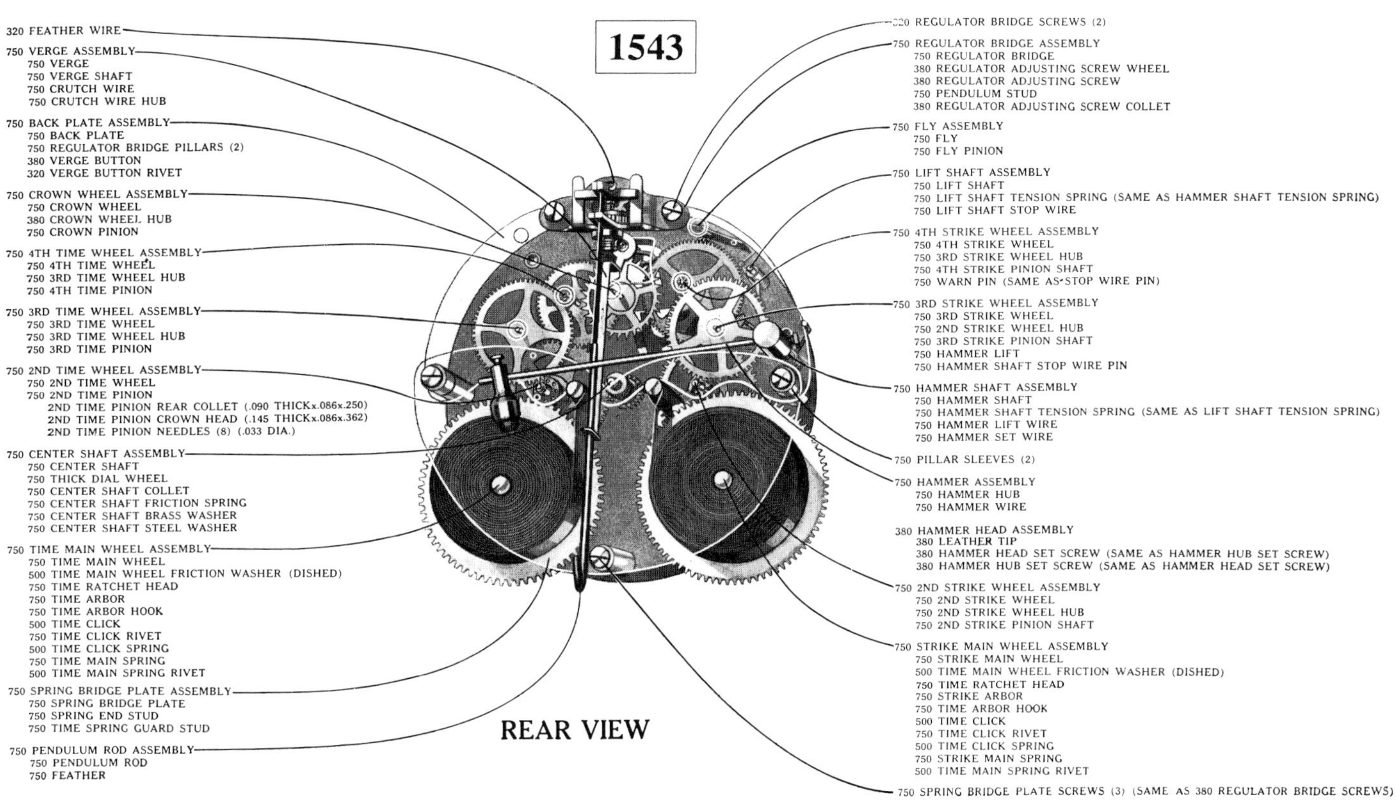

REAR VIEW

No. 750

8-Day Time and Strike (Rack and Snail)

11-1-1927

MOVEMENT ILLUSTRATIONS

The year entered beside each movement is the year of the catalog from which the illustration was taken. Some movements may have been produced a few years earlier.

MOVEMENT FOR 8 DAY MOTOR CLOCKS

1544

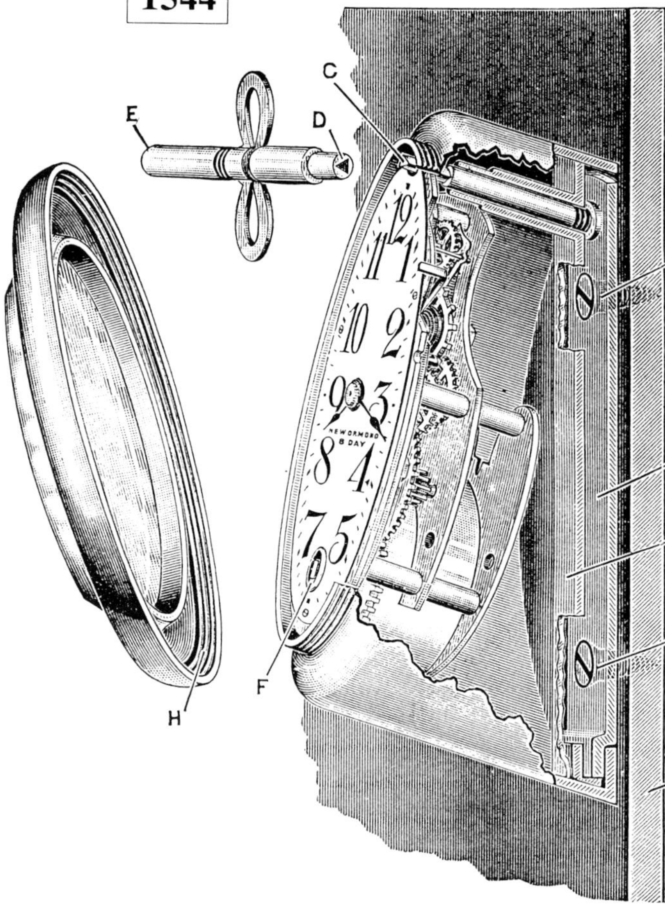

1911

SECTIONAL VIEW
Showing Special Features

A. LOCKING PLATE WHICH SCREWS OR BOLTS TO DASHBOARD AND INTO WHICH CLOCK LOCKS
B. SCREWS OR BOLTS, HOLDING LOCKING PLATE TO DASHBOARD
C. BOLT WITH TRIANGULAR HEAD (REQUIRING SPECIAL KEY) WHICH LOCKS CLOCK TO LOCKING PLATE
D. END OF KEY TO LOCK CLOCK TO LOCKING PLATE
E. END OF KEY FOR WINDING CLOCK
F. WINDING ARBOR
G. BACK OF CLOCK WHICH LOCKS INTO LOCKING PLATE
H. SCREW BEZEL WHICH MAKES CLOCK DUST- AND WATER-PROOF
I. DASHBOARD

OUR NEW 12 DAY MOVEMENT HAS A LONGER, WIDER AND THINNER MAIN SPRING THAN PREVIOUS MOVEMENTS OF THIS CLASS AND A LARGER WINDING BARREL. IT WILL RUN FROM 18 TO 21 DAYS AND KEEP ACCURATE TIME FOR AT LEAST 12 DAYS WITH ONE WINDING. ON OUR 12 DAY RATING THERE IS THEREFORE A LARGE RESERVE POWER.

ALL PINIONS ARE CUT STEEL, WITH SMALL PIVOTS, SMOOTHLY FINISHED TO ELIMINATE FRICTION AND WEAR. THE PLATES AND WHEELS ARE MANUFACTURED OF HEAVY HARD ROLLED BRASS WITH FINELY MADE TOOLS. THE BALANCE SHAFTS ARE HARDENED BY A NEW ELECTRICAL PROCESS AND ARE POLISHED. THE MOVEMENT IS MADE TO SET STANDARDS UNDER CLOSE INSPECTION AND THOROUGHLY TESTED BEFORE AND AFTER FITTING INTO THE CASE. ALL PARTS ARE EASILY ACCESSABLE.

IN PRODUCING THIS MOVEMENT, WE HAVE MADE A DISTINCT ADVANCE IN THE ART OF BUILDING DURABLE AND DEPENDABLE TIME PIECES.

1545

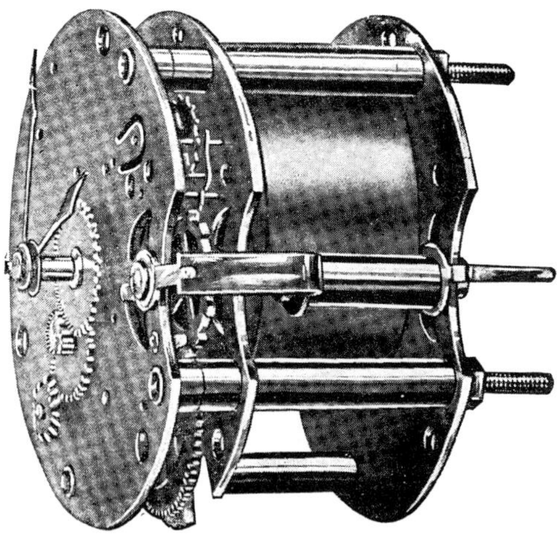

1923

New 12 Day Movement

MOVEMENT ILLUSTRATIONS

1545-A

Front view of **Saturn** eight-day weight driven movement. Catalog illustration #546 page 144. {Private Collection}

1546

THIS MOVEMENT IS USED IN ALL CLOCKS OF THE JUNIOR TATTOO FAMILY. ITS STRONGEST POINTS ARE ACCURATE TIMEKEEPING, GENTLE TICK, INTERMITTENT ALARM, SHUT-OFF SWITCH. IT ALARMS INTERMITTENTLY WITH 20-SECOND INTERVALS FOR 5 MINUTES.

1547

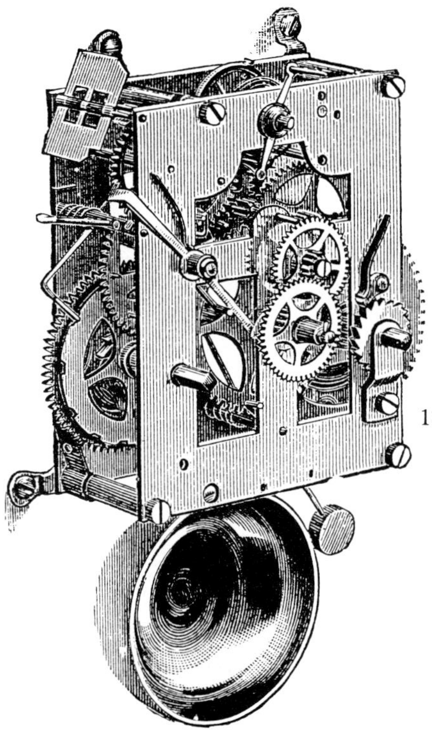

1883

MARINE MOVEMENT

One-Day Strike. Heavy Plates and Wheels. Cut and polished Hard Steel Pinions. Superior make and finish, and excellent Time-keeping qualities.

1548

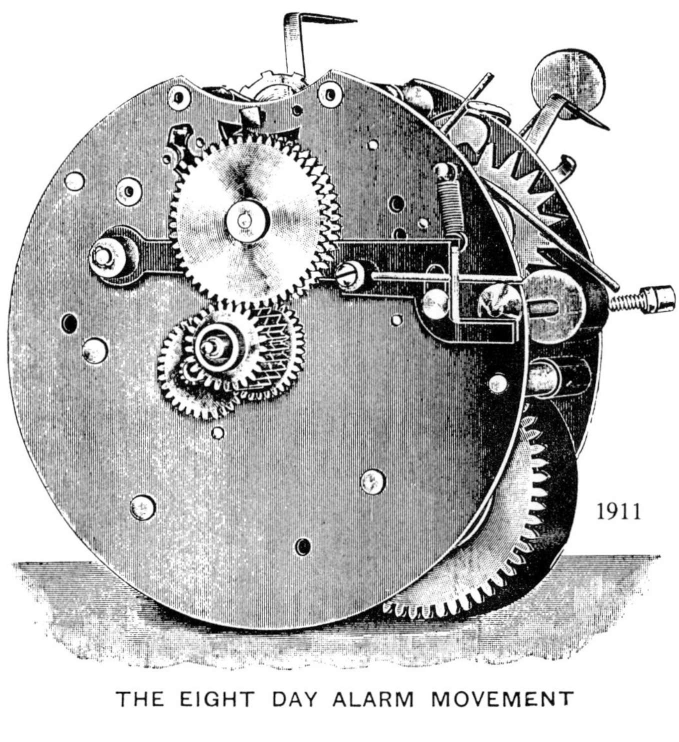

1911

THE EIGHT DAY ALARM MOVEMENT

MOVEMENT ILLUSTRATIONS

1549

The New Haven Clock Co.

Announces

"ELM CITY" ELECTRIC CLOCKS

This is the electrical age. Electricity has revolutionized modern home life and is responsible for many of the comforts of present day civilization. Clockmaking is one of the oldest industries. The requirements for a good clock are very exacting, and the life expected of a good clock is long. Electric clocks are not new, but the production of a popularly-priced, dependable electric clock requires a high degree of skill and long experience in clock manufacture. THE NEW HAVEN CLOCK COMPANY, which has manufactured clocks continuously since 1817, now presents electric clocks of unquestioned reliability, at prices within the reach of all.

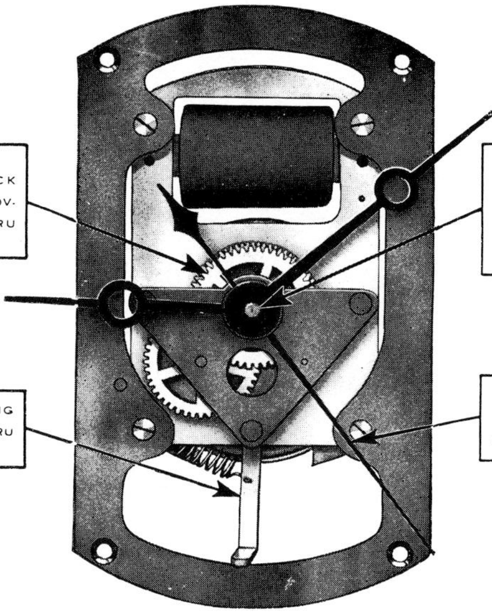

Front View of Movement

Progressive dealers can now meet the demand with the popularly priced "Elm City" Electric Clocks, distributed through legitimate clock channels. Progressive power stations furnish the majority of homes with alternating current. These regular electric vibrations are utilized in Elm City Electric Clocks by a simple, dependable, synchronous motor which operates a dependable clock movement.

All moving parts of the motor are sealed in a metal case, where they turn silently in an oil bath. The motor and field coil may each be quickly detached by removing two screws. The movement is of New Haven design. All mantel and wall clocks are constructed to permit convenient setting and starting from the dial, without disturbing the original position or getting in back of clocks. All clocks are completely equipped with cord and plug, attachable to any convenient outlet.

"Elm City" Electric Clocks are cased for every purpose and are adapted to any room in the home or office. The smaller models are for desk, radio or boudoir table. Cheerfully colored wall models add a decorative touch to kitchen or sun porch. Larger mahogany cases in banjo, tambour and upright designs harmonize with their surroundings in beautifully furnished homes. All wooden cases are durably finished with several coats of hand-rubbed Duco and are solid, without veneer.

1550

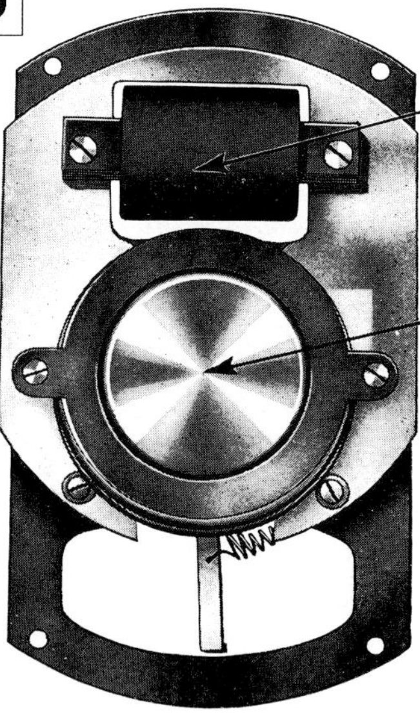

1929

Back View of Movement

METHOD OF OPERATION

1. Plug cord into convenient outlet.
2. Push automatic starting lever across slot and release. The motor will instantly pick up timed impulses from your central station.
3. Set minute and second hands to correct time, forward or backward as with ordinary clock. Adjustments are made from the front on mantel and hanging clocks. Clocks do not have to be turned or lifted from wall.
4. Correct operation of clock is indicated by rotation of a long second hand. Stopping of second hand denotes current interruption. Re-start and re-set as previously indicated.
5. Clock will indicate correct time as long as current supply is continuous.
6. Electric current used is negligible and cost of operating is but a few cents a month.

IMPORTANT

Elm City Electric Clocks are regularly furnished for 110-120 volts, 60 cycle. Clocks for 50 cycle current are furnished on special order, without extra charge. If the generating stations deliver the frequency of current they are supposed to, clocks will keep absolutely accurate time when properly started and set. Progressive stations closely control their frequency. In other cases, the error in time is small and may be corrected as in ordinary clocks.

MOVEMENT ILLUSTRATIONS

SELF STARTING, SLOW SPEED MOTORED ELECTRIC CLOCKS

1551

New, slow speed, Whitehead, 300 R. P. M. self starting motor.

To give satisfactory service over a period of years, electric clocks must approach the mythical perpetual motion machine in durability. Years of experience, careful research, and study have enabled us to produce an outstanding new line of electric clocks. On time controlled circuits, they are as perfect timekeepers as the finest chronometers used for navigation.

We have successfully met five requirements to insure a maximum life and silent operation:

1. Low wattage input to reduce hum and heat. Heat destroys the oil supply and hum results in noisy clocks. The low wattage also results in a very considerable reduction in operating expense.
2. Polished bronze and steel bearings are used in the motor and all other bearings are polished so as to reduce friction in the movement. The fastest moving part is reduced to a speed of only 5 revolutions per second. As clocks run 24 hours a day, it is important to have the parts move slowly, to provide ample lubrication, and to eliminate friction so as to insure long life.
3. The characteristics of these motors insure an ample supply of power and synchronous operation, regardless of voltage fluctuations on the circuits to which they may be attached.
4. The motor and structure are practically inaudible in operation. The slow speed not only eliminates wear and noise but, in addition, it enables us to have the maximum velocity of gear teeth at any part of the train—the fastest moving 32 teeth per second, which is below the level of audibility of 40 vibrations per second.
5. Clocks become noisy the minute lubrication fails. In these clocks there is an ample supply of lubrication and a carefully worked out lubricating system. Thus we have overcome the most troublesome defect in electric clocks. The slow speed under which these motors operate, and the lack of heat generated, permit a system of lubrication that requires no attention.

SELF STARTING MOTORS

The new Whitehead 300 R. P. M. self starting motor contains features of lubrication and design resulting from experience in building several hundred thousand clocks. It is of a unique design and fully covered by patent applications. Its slow speed insures long life and silent operation. It operates in synchronism even for wide variations in voltage, and develops ample power at the low current consumption of only $1\frac{1}{4}$ watts. Therefore, operating costs are extremely low, while long life is assured by a positive lubricating system and slow moving parts.

METHOD OF OPERATION

In these electric clocks a synchronous motor drives the time train, and in alarms, chimes and striking clocks, it also automatically winds the springs so as to store energy and operate the sounding mechanism. This system is covered by a patent and is applied to all sounding clocks with a chime, strike or alarm. It has proved to be most satisfactory, as it avoids all the troubles associated with dual motors, buzzers, shifting mechanisms and other methods. The method of winding springs is an exclusive patented feature. The springs are automatically wound by the motor, are positive in operation and infallible in action. This also means that in alarm clocks, instead of putting out a buzzer type of alarm, we use a fine sounding bell.

AUXILIARY WINDING DEVICE

The auxiliary winding device in these clocks has met with much favor, as it enables the dealer to demonstrate striking, chime and alarm clocks without current. When clocks are in operation they of course function automatically and no hand winding is necessary, but for demonstration purposes an auxiliary wind is provided on chimes and strikes. On alarms all that is necessary is to set the hands forward 12 hours which will automatically wind the alarm sounding device.

Summarizing we wish to stress:

1. LOW WATTAGE CONSUMPTION results in less operating cost, less heat, less noise and long life.
2. SLOW SPEED motors result in long life and quiet operation.
3. AMPLE POWER results in freedom from stoppages and line surges.
4. NO CONTACTS, NO BUZZERS, NO DUAL MOTORS result in silent operation.
5. AMPLE PROVISION FOR LUBRICATION results in long life and less noise.
6. ELIMINATION OF NOISE because of slow speed, low power input and ample lubrication results in a clock that will give satisfactory service for many years.

PATENT NUMBERS

1,695,484, 1,907,919, 1,935,686, 1,957,694, 1,978,973, 1,717,500, 1,437,012, 1,465,090

MOVEMENT ILLUSTRATIONS

1552

Raised Gold Numerals

Numerals of gold, raised above a frosted silver dial surface, place New Haven Chimes on an artistic level unequaled by clocks of the past. Modern cubist numerals, whose gold polished surfaces stand out in varied lights, while plainly indicating the hour; give these clocks an air of richness and charm.

1553

1920

MOVEMENT USED IN 8-DAY AUTOMATIC AND 8-DAY SIGNAL AUTOMATIC.

This Movement has Steel Pinions and Heavy Brass Plates. There is Spring Power to Run it for 14 Days. The Small Balance Wheel, Quick-Beat Escapement, and Light Train are Special Features.

IT HAS AN ALMOST SILENT BEAT.

1554

Accuracy of Movement

Though a clock may be ever so beautiful and melodious, its first mission is that of keeping reliable time. The finely trained New Haven Pendulum, swinging with a steadfast rhythm, makes this clock an accurate timekeeper.

MOVEMENT ILLUSTRATIONS

1555

Back plate stamped No. 0014 found in New Haven Regulator No. 2 weight driven Movement. {Lubertus H. Hayenga Collection}

1556

Front view of New Haven Regulator No. 2 Movement. Clock illustration #512 page 133 {Lubertus H. Hayenga Collection}

1557

Top view of New Haven No. 2 Regulator Movement. {Lubertus H. Hayenga Collection}

1558

Side view of New Haven Regulator No. 2 Movement. {Lubertus H. Hayenga Collection}

MOVEMENT ILLUSTRATIONS

1559

Front view of **8-Bell Chime No. 1** movement.
Clock illustration #343 page 96.
{ Photo courtesy of
Michael's Antique Clocks of Oklahoma City}

1560

Side view of **8-Bell Chime No. 1** movement.
Clock illustration #343 page 96.
{Photo courtesy of
Michael's Antique Clocks of Oklahoma City}

1561

Back view of **8-Bell Chime No. 1** movement.
{Photo courtesy of
Michael's Antique Clocks of Oklahoma City}

1562

Side view of **8-Bell Chime No. 1** movement.
{Photo courtesy of
Michael's Antique Clocks of Oklahoma City}

MOVEMENT PARTS

1564

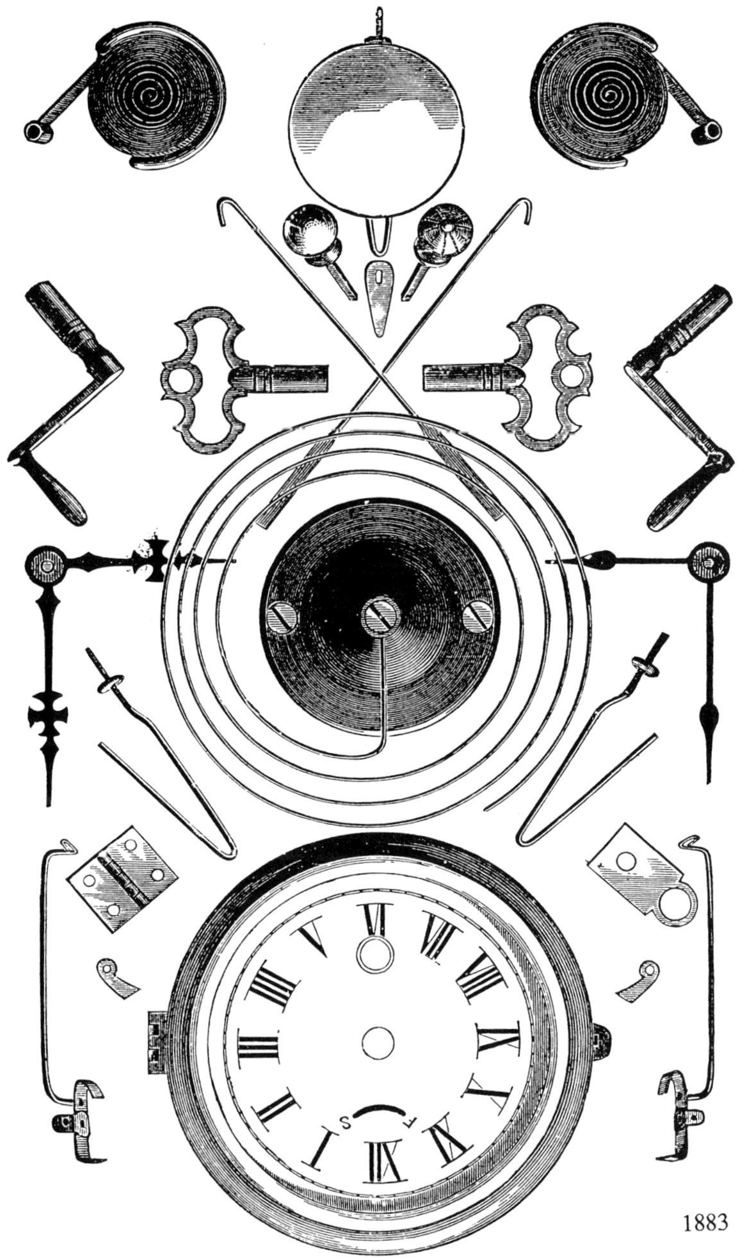

1883

1563

1883

We will supply Clock Parts and Trimmings illustrated above, and any others wanted, in quantities.

Lists and prices furnished on application.

THE PATENT DIAL PENDULUM BALL.

By means of this Ball a Clock can be quickly and closely regulated. A move of the pointer one degree to the right or left will change the Clock about one minute fast or slow in twenty-four hours. To move the pointer turn the screw or nut at the bottom of the ball to the right or left. A shorter or longer rod will show greater variation. We will fit any of our Visible Pendulum Clocks, either with this or the Imitation Mecurial Ball.

MOVEMENT PARTS

CLOCK HANDS

1565

1936

	Pair	Dozen
No. **54320.** Nugget	**$0.15**	**$0.75**

1566

1936

	Pair	Dozen
No. **58364.** Petite	**$0.15**	**$1.00**

1567

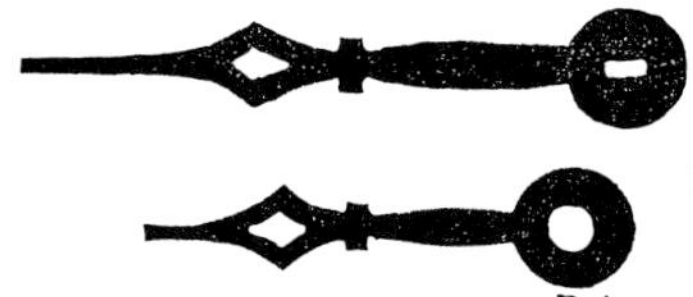

1936

	Pair	Dozen
No. **58338.** 12 Day Jeweled ..	**$0.15**	**$1.00**

1568

1936

	Pair	Dozen
No. **582153.** "Duo" 6″ Dial	**$0.20**	**$1.50**

1569

1936

	Pair	Dozen
No. **583381.** 7¼″ Bronze	**$0.25**	**$2.50**

1570

1931

No. 217
New Haven
Doz. Pairs,
$0.60

1571

1931

No. 206
New Haven
Doz. Pairs, $0.50

1572

1931

No. 222
New Haven
Doz. Pairs, $0.60

1573

1931

No. 212
New Haven Chime
Per Pair, $0.30

1574

1931

No. 213
New Haven Chime
Per Pair, $0.20

1575

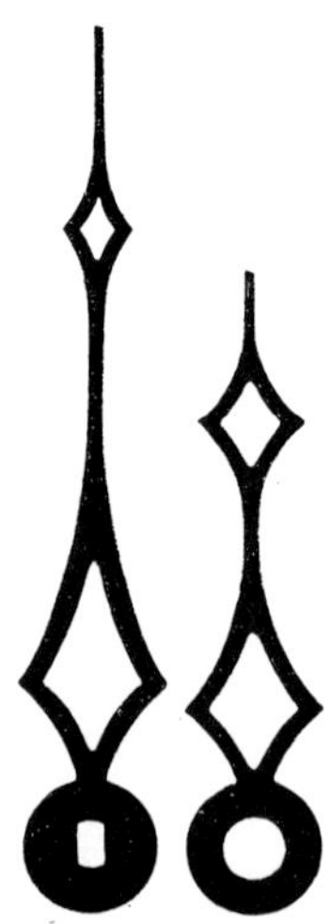

1931

No. 214
New Haven, Brass
Per Pair, $0.20

1576

1931

No. 215
New Haven
Doz. Pairs, $0.35

1577

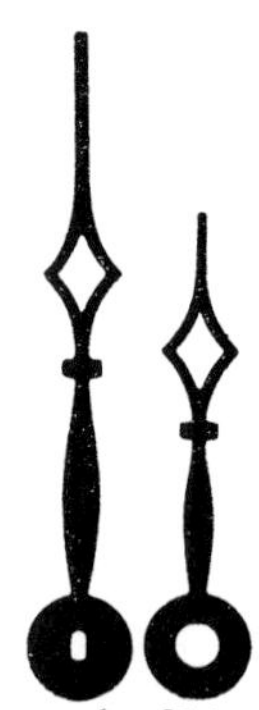

1931

No. 216
New Haven
Doz. Pairs,
$0.90

MOVEMENT PARTS
CLOCK HANDS

NEW HAVEN USA TRADE MARK

1936

1578

	Pair	Dozen
No. 583170. Tick-Tock	$0.15	$0.75

1579

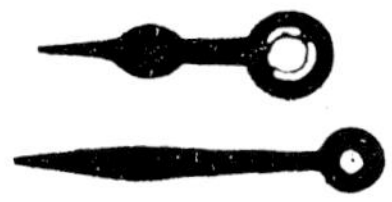

	Pair	Dozen
No. 582287. "Germ"	$0.15	$0.50

1580

	Pair	Dozen
No. 543201. Tat-Too Junior & Tidy Tot	$0.15	$0.75
No. 582091. Tidy Tot Radium.	.30	3.00

1581

	Pair	Dozen
No. 54316. Tat-Too Steel	$0.15	$0.75
No. 543206. Tat-Too Radium	.35	3.50

1582

	Pair	Dozen
No. 583180. Electric	$0.15	$0.85

1583

	Pair	Dozen
No. 543200. St. Elmo & Sting	$0.15	$0.75

1584

	Pair	Dozen
No. 543209. Radium Tick-Tock	$0.35	$3.50

1585

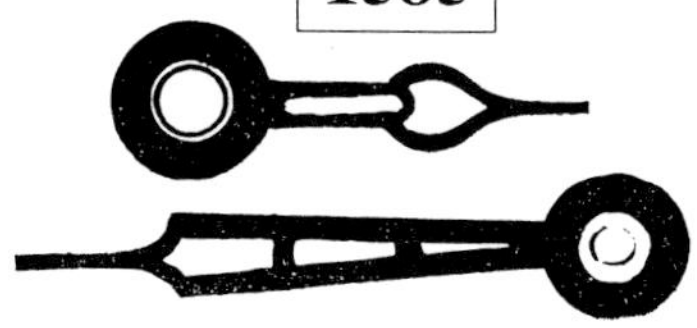

	Pair	Dozen
No. 54322. Westinghouse Radium	$0.45	$4.50

1586

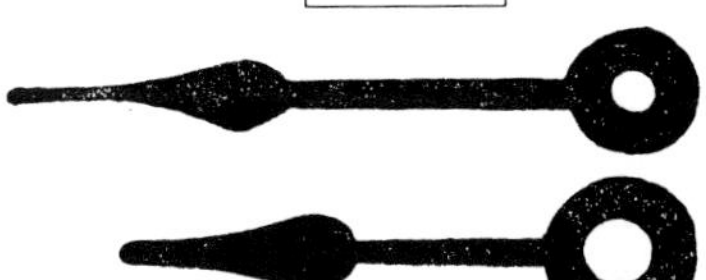

	Pair	Dozen
No. 543203. Auto	$0.15	$0.75

1587

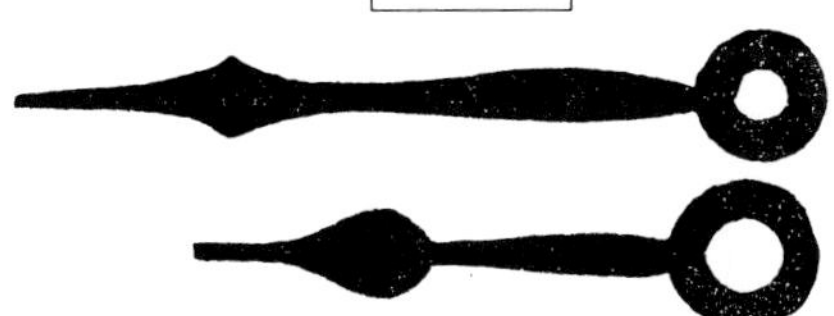

	Pair	Dozen
No. 543204. Tom-Tom Steel	$0.15	$1.00
No. 543210. Tom-Tom Radium	.35	3.50

1588

	Pair	Dozen
No. 54371. Chime Clock Hands	$0.50	$5.00
No. 543710. Kitchen Plate Clock Hands......	.35	3.50

1589

	Pair	Dozen
No. 543208. Ti-Tan Radium	$0.35	$3.50

1590

	Pair	Dozen
No. 582159. 8-Day Jeweled	$0.15	$0.75

1591

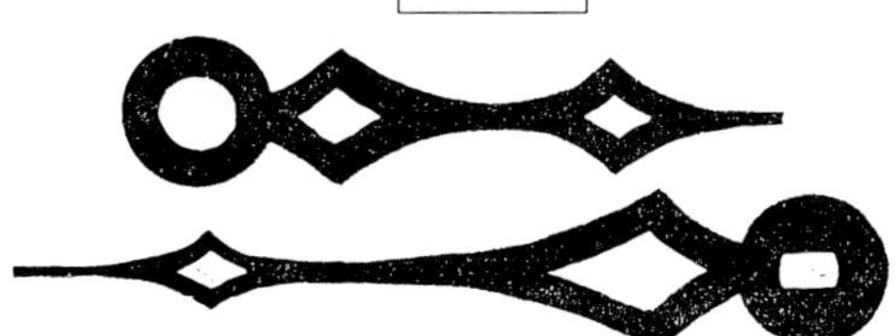

	Pair	Dozen
No. 583167. "Larina" Kitchen	$0.20	$2.40

MOVEMENT PARTS

ELECTRIC CLOCK HANDS

1931

1592

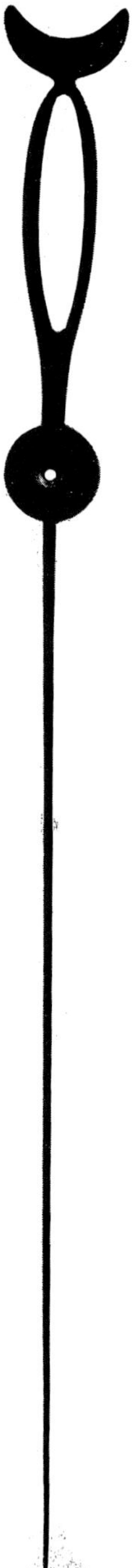

No. N.E.-1

1593

No. N.E.-5

1594

No. N.E.-6

1595

No. N.E.-7

1596

No. N.E.-8

1597

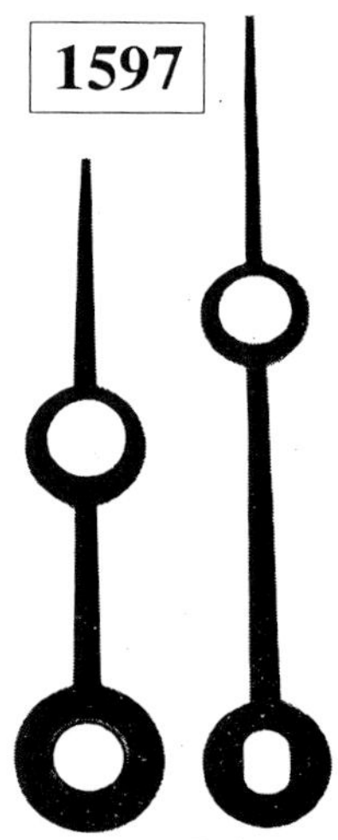

No. N.E.-9

1598

No. N.E.-10

1599

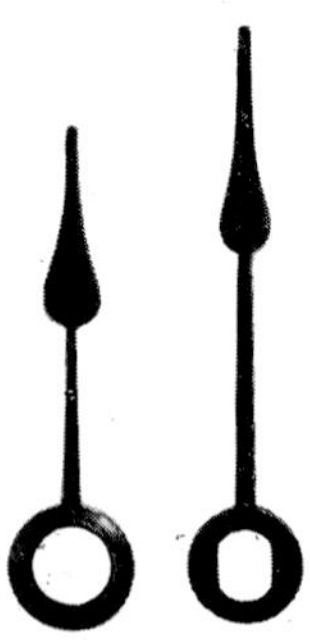

No. N.E.-11

1600

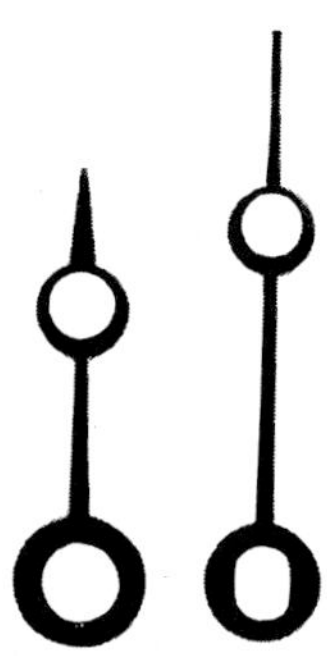

No. N.E.-12

1601

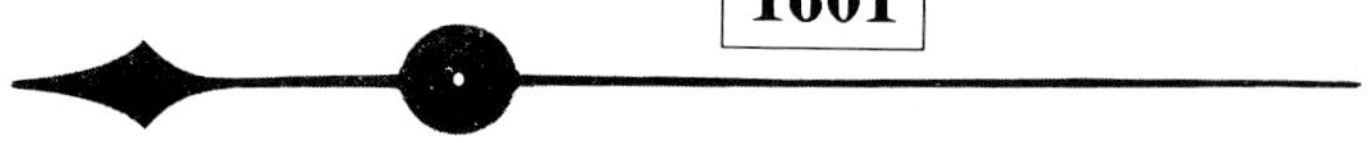

No. N.E.-2

1602

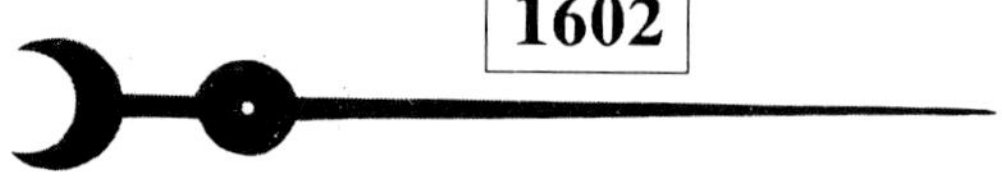

No. N.E.-3

1603

No. N.E.-4

No.	Dozen	Each
N.E.-1—Wall Sweep Hand, 12 in. Dial	$1.25	$0.15
N.E.-2—Split Sweep Hand, Mod. 965	.90	.10
N.E.-3—Split Sweep Hand, Mod. 961	.75	.10
N.E.-4—Split Sweep Hand, Mod. 962	.65	.10

No.	Doz. Pairs	Pair
N.E.-5—12 in. Wall, H. & M. Hands	$0.90	$0.10
N.E.-6—Madena, H. & M. Hands	2.00	.20
N.E.-7—Whitney, H.&M.Hand, 7 in. Dial	2.00	.20
N.E.-8—Whitney, H. & M. Hands	2.00	.20

No.	Doz. Pairs	Pair
N.E.- 9—Whalley, H .& M. Hands	$2.00	$0.20
N.E.-10—Warwick, 6 in. Dial	1.25	.15
N.E.-11—Petite, 3½ in. Dial	1.25	.15
N.E.-12—Square, 3½ in. Dial	1.50	.15

1604

MOVEMENT PARTS

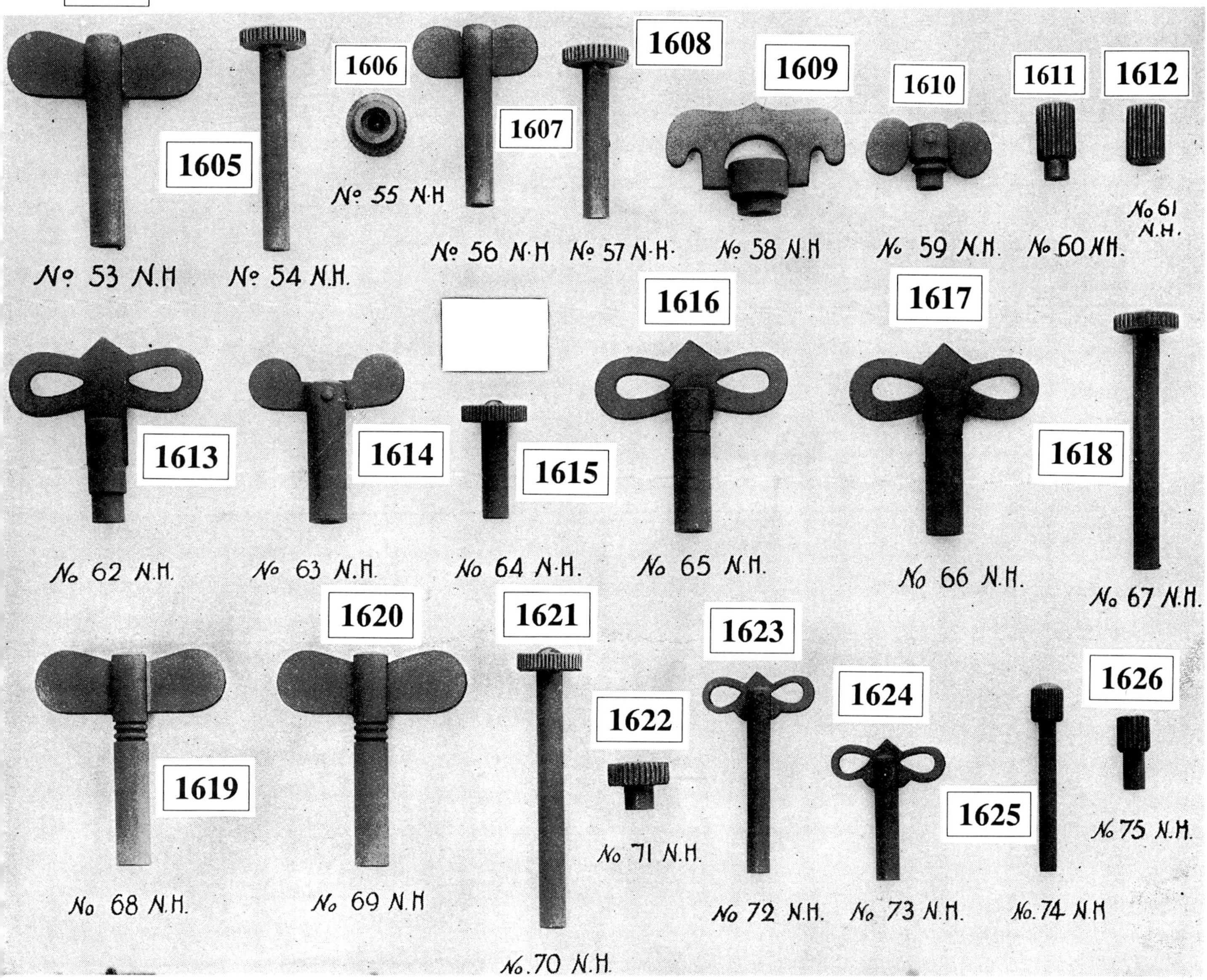

NEW HAVEN WINDING AND SETTING KEYS

No. 53NH.	Time Winding Keys	Per Doz.,	$1.25	—For Tattoo, Slumber Stopper.
No. 53NH.	Alarm Winding Keys	Per Doz.,	1.25	—For Tattoo.
No. 54NH.	Time Setting Keys	Per Doz.,	1.00	—For Tattoo.
No. 55NH.	Alarm Setting Keys	Per Doz.,	.45	—For Tattoo, Slumber Stopper, Commuter.
No. 56NH.	Alarm Winding Keys	Per Doz.,	1.00	—For Slumber Stopper.
No. 57NH.	Time Setting Keys	Per Doz.,	1.00	—For Slumber Stopper.
No. 58NH.	Time Winding Keys	Per Doz.,	1.00	—For Jump.
No. 59NH.	Alarm Winding Keys	Per Doz.,	1.00	—For Jump.
No. 60NH.	Time Setting Keys	Per Doz.,	.40	—For Jump.
No. 61NH.	Alarm Setting Keys	Per Doz.,	.40	—For Jump.
No. 62NH.	Time Winding Keys	Per Doz.,	1.25	—For Commuter
No. 63NH.	Alarm Winding Keys	Per Doz.,	1.00	—For Commuter.
No. 64NH.	Time Setting Keys	Per Doz.,	.50	—For Commuter.
No. 65NH.	Time Winding Keys	Per Doz.,	1.00	—For Signal.
No. 66NH.	Alarm Winding Keys	Per Doz.,	1.00	—For Signal.
No. 67NH.	Time Setting Keys	Per Doz.,	1.00	—For Signal.
No. 68NH.	Time Winding Keys	Per Doz.,	1.00	—For Automatic.
No. 69NH.	Alarm Winding Keys	Per Doz.,	1.00	—For Automatic.
No. 70NH.	Time Setting Keys	Per Doz.,	1.00	—For Automatic.
No. 71NH.	Alarm Setting Keys	Per Doz.,	.50	—For Automatic Signal.
No. 72NH.	Time Winding Keys	Per Doz.,	.45	—For Junior Tattoo.
No. 73NH.	Alarm Winding Keys	Per Doz.,	.40	—For Junior Tattoo.
No. 74NH.	Time Setting Keys	Per Doz.,	.35	—For Junior Tattoo.
No. 75NH.	Alarm Setting Keys	Per Doz.,	.35	—For Junior Tattoo.

NEW HAVEN USA TRADE MARK

1931

MOVEMENT PARTS

WINDING KEYS

1627

No. 544190. Time and Alarm: Tick-tock, Tell-Tale, Tidy-Tot, Tat-too Jr. Each...$0.15
Per dozen .... .65

1628

No. 544302. Time and Alarm: Titan.
Each $0.15
Per dozen .... .65

1629

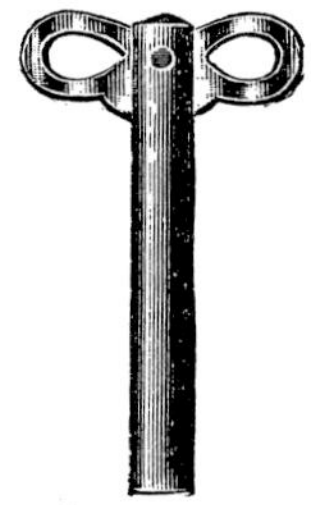

No. 54419. Time: Brownie.
Each $0.15
Per dozen .... .65

1630

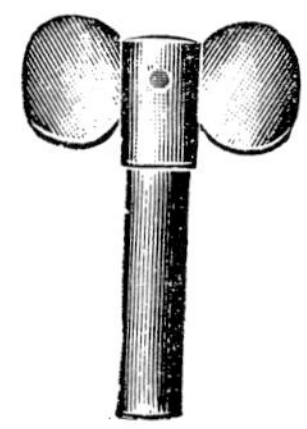

No. 54420. Time: Manola and St. Elmo. Each.$0.15
Per dozen .... .65

1631

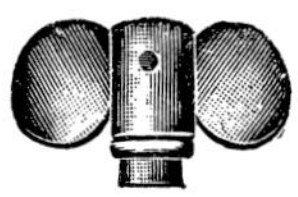

No. 54413. Alarm: Beacon.
Each $0.15
Per dozen .. .65

1632

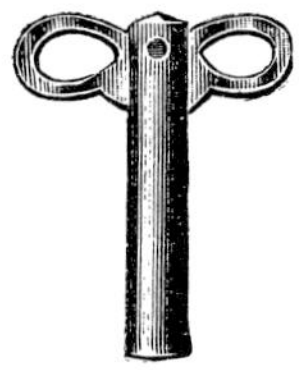

No. 54418. Alarm: Brownie.
Each $0.15
Per dozen .... .65

1633

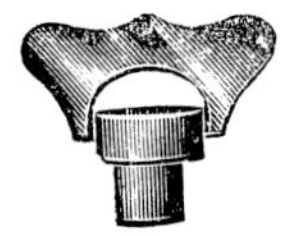

No. 54412. Time Sting. Ea..$0.15
Per dozen .. .65

NEW HAVEN
USA
TRADE MARK

1936

SETTING KEYS

1634

No. 545303. Alarm: Tick Tock.
Each $0.15
Per dozen .. .75

1635

No. 545302. Alarm: Tell-Tale and Tidy-Tot.
Each $0.15
Per dozen .. .75

1636

No. 545305. Time: Tell-Tale & Tick-Tock.
Each $0.15
Per dozen 1.25

1637

No. 54523. Alarm: Automatic.
Each ...$0.15
Per dozen. .75

1638

No. 54521. Alarm: Prompter.
Each .$0.15
Per dozen .75

1639

No. 545306. Time: Tat-Too, Jr. and Tidy-Tot. Each $0.15
Per dozen .... .75

1640

No. 54530. Time: Automatic and Ti-Tan. Each $0.15
Per dozen .... .75

1641

No. 54516. Time: Trilby.
Each $0.15
Per dozen .. .75

1642

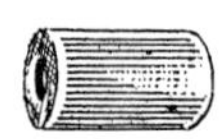

No. 58337. Tat-Too Alarm O. S.
Each $0.15
Per dozen .. 1.00

1643

No. 545307. Time: Special St. Elmo.
Each $0.15
Per dozen .... .75

1644

No. 54519. Time: Sting.
Each ...$0.15
Per dozen. .75

1645

No. 54528. Time: Tat-Too and Tom-Tom. Ea..$0.15
Per dozen .... .75

1646

No. 54529. Time: Prompter. Each $0.15
Per dozen .... .75

MOVEMENT PARTS

CLICK SPRINGS

1647

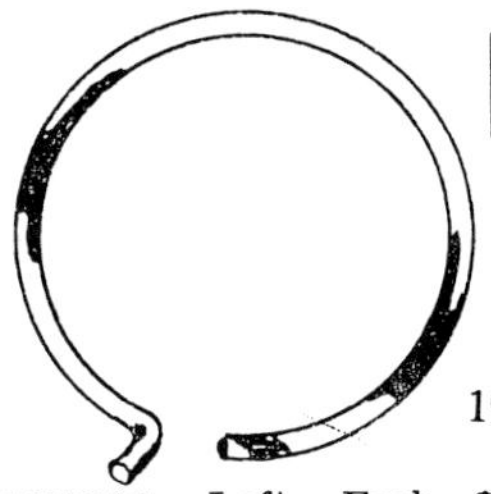

1936

No. **547041.** Left. Each..**$0.15**
Per dozen .... **.40**

1648

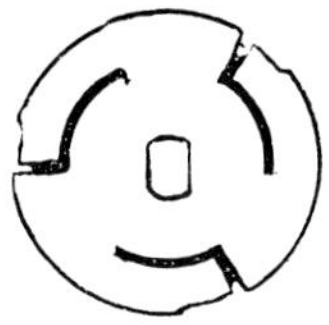

1936

No. **582016.** Tatoo, Jr.; Time.
Each **$0.15**
Per dozen **1.25**

1649

1936

NEW HAVEN CLICK SPRINGS

No. **54704.** New Haven. Left Side. Each.**$0.15**
Per dozen .... **.40**
No. **547041.** New Haven. Right Side. Each.**$0.15**
Per dozen .... **.40**

1650

1936

No. **582055.** Tatoo; Time.
Each **$0.15**
Per dozen **1.00**

1651

1936

No. **5820480.** Click and Rivet.
Each **$0.15**
Per dozen .... **.40**

1652

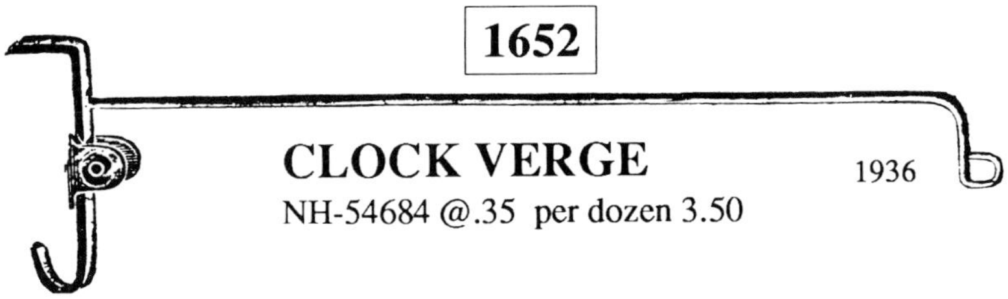

CLOCK VERGE

NH-54684 @.35 per dozen 3.50

1936

1653

1936

NEW HAVEN PENDULUM RODS

			Each	Per dozen
No. **545941.**	3¼"x1 9/16"	blade	**$0.15**	**$1.20**
No. **545942.**	3⅜"x1 9/16"	blade	**.15**	**1.20**
No. **545943.**	3½"x1 9/16"	blade	**.15**	**1.20**
No. **545945.**	4½"x1 9/16"	blade	**.15**	**1.20**
No. **5459450.**	4¾"x1 9/16"	blade	**.15**	**.60**
No. **545946.**	4½"x1½"	blade	**.15**	**1.20**

1654

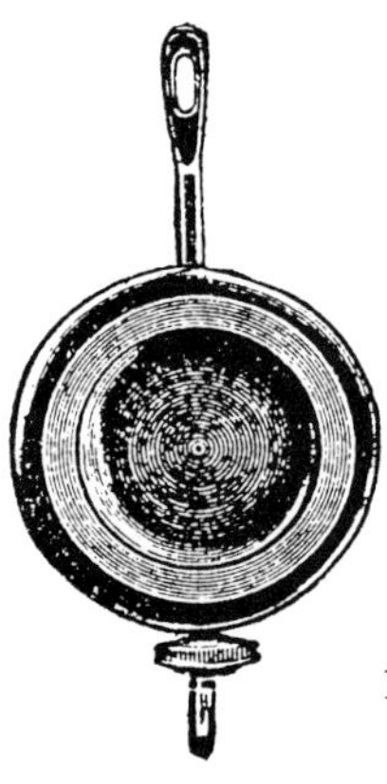

1936

PENDULUM BALL

NH-54580.....@$0.45

PENDULUM BALLS

1655

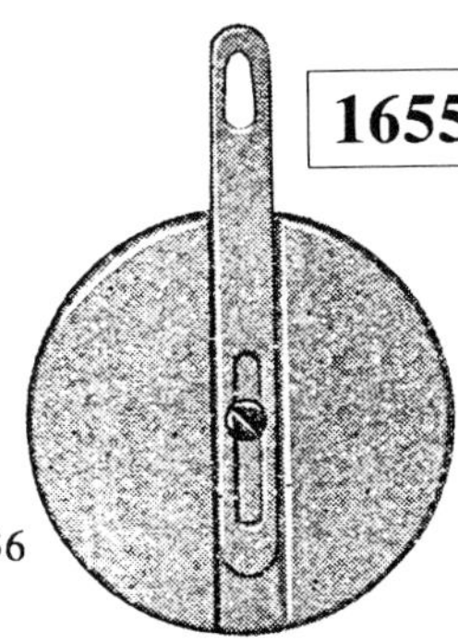

1936

No. 97
New Haven Chime
Each, $0.50

1656

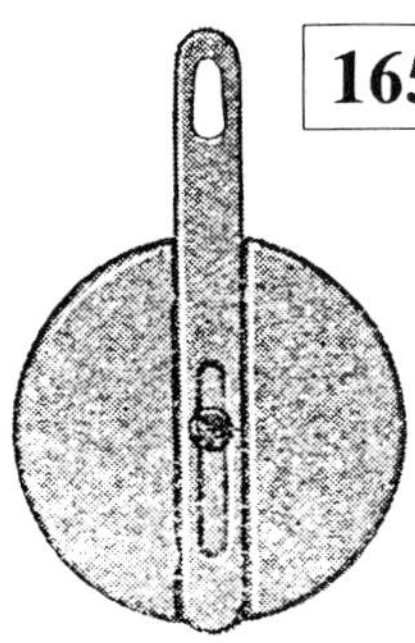

1936

No. 98
New Haven Mantel
Each, $0.50

NOVELTY CLOCKS

1657

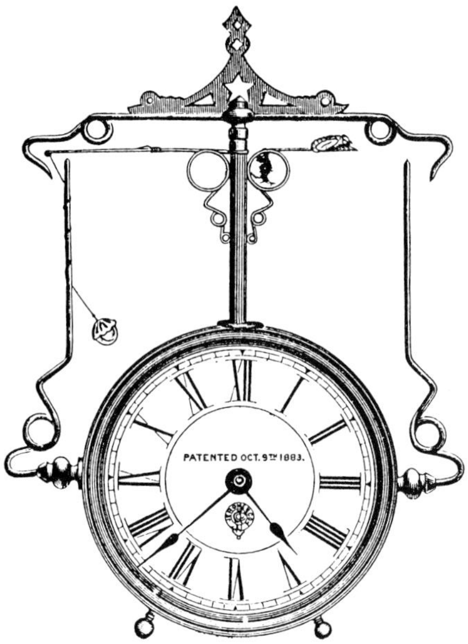

1885

FLYING PENDULUM No. 3

Nickel.
Height, 9 ³/₄ inches. Dial, 4 inches.
1 Day Time, Nickel.............. $6.75

1658

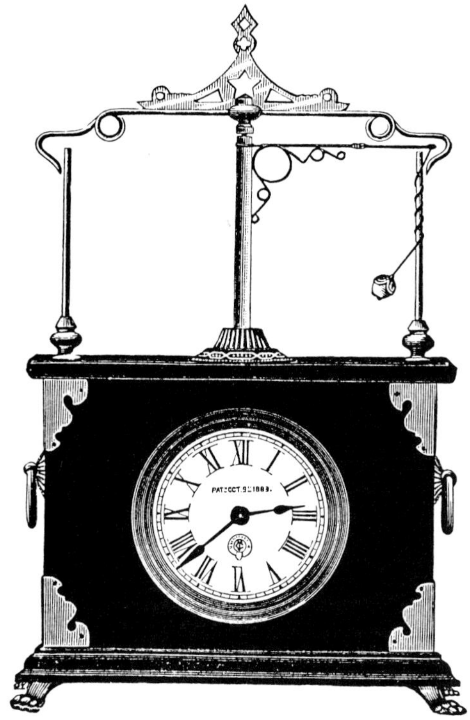

1885

FLYING PENDULUM No. 1

MAHOGANY, OAK OR EBONY.
WITH GILT ORNAMENTS.
Height, 10 ¾ inches. Dial, 3 inches.
1 Day Time, Mahogany...... $8.20
1 Day Time, Oak................ $8.20
1 Day Time, Ebony........... $8.20

1659

1884

No. 5107

1 Day, Time.......$5.18

FLYING PENDULUM CLOCK.

Cases are finished in Oak, Mahogany and Ebony.
The Latest Novelty in Clock Escapements. The Flying Ball takes the place of the Pendulum, and keeps good time.
The best Show Window Attraction ever made. Will draw a crowd wherever exhibited.

1660

1885

FLYING PENDULUM No. 2

Nickel.
Height, 9 ½ inches. Dial, 3 inches.
1 Day Time, Nickel.............. $7.55

NOVELTY CLOCKS

1661

1889

RACQUET.

FINISHED IN GOLD.

1 Day Time......$4.10

1662

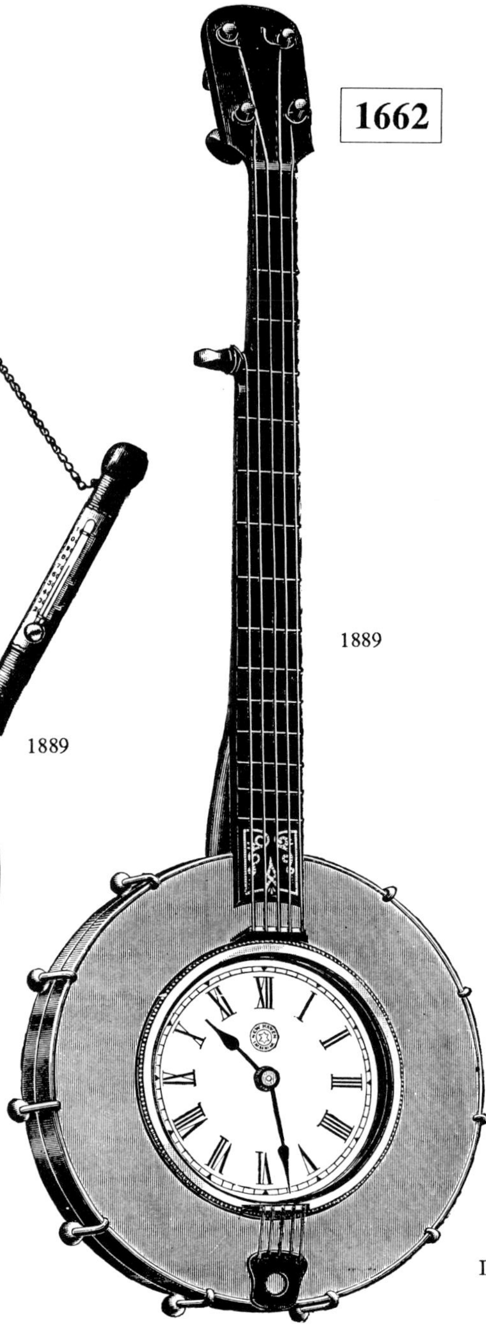

1889

BANJO.

SILVER PLATED AND
OXIDIZED.

Height, 7 inches. Dial, 2 inches.

Cut Pinion Movement.

(Liegeman) 1 Day Time..$8.75

This Movement has Screw Pillars, Cut and Polished Steel Pinions, a Straight Line Lever with Double Roller Club Tooth Escapement and Polished Springs.

1663

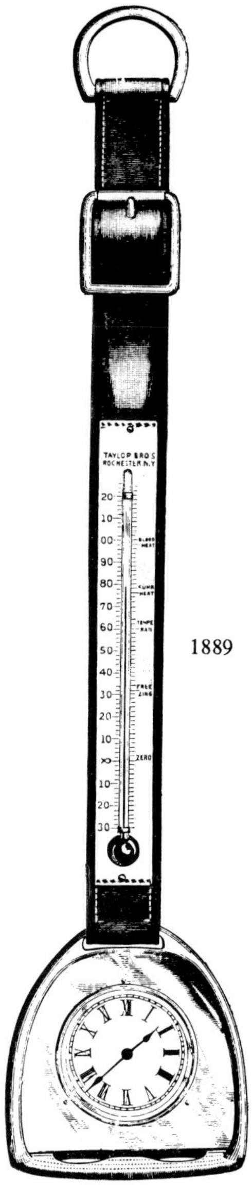

1889

STIRRUP.

FINISHED IN GOLD AND SILVER.

Leather Strap, with Gilded Stirrup and Plate.

Height, 17 inches. Dial, 2 inches.

1 Day Time......$8.00

NOVELTY CLOCKS

1664

1886

ARABESQUE.

FINISHED IN PLAIN SILVER.

1 Day Time $6.25

1665

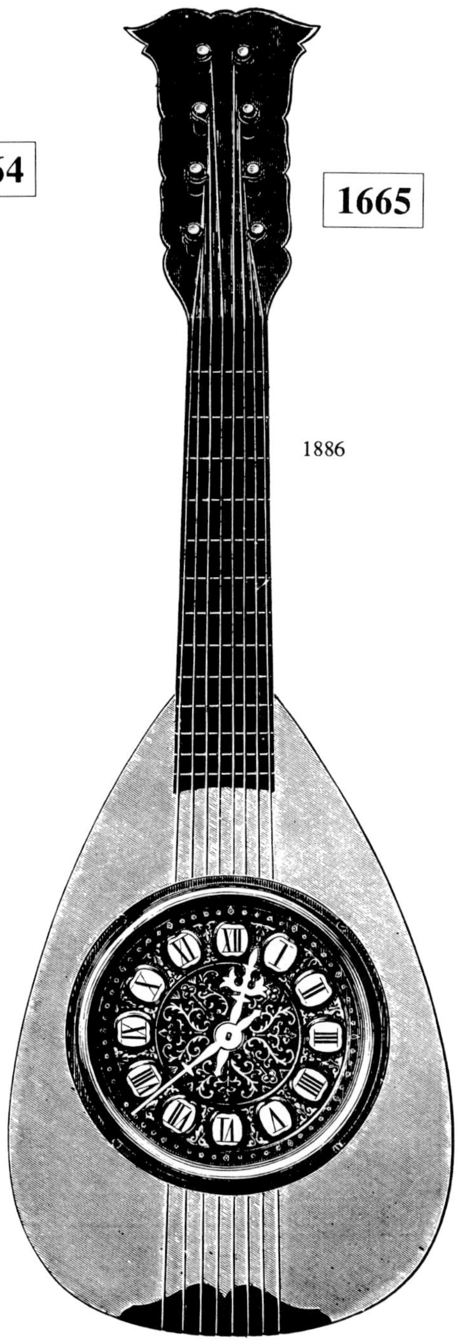

1886

MANDOLIN.

FINISHED IN BLACK AND SILVER.

Length, 23 inches. Dial, 5 inches.

Cut Pinion Movement.

8 Day Time $10.00

1666

1889

PARASOL.

PARASOL.

FINISHED IN BLACK AND GOLD.

Height, 15 inches.

15 Day Time...........$12.00

The handle can be turned to be used as a hanging or table clock.

NOVELTY CLOCKS

1667

1911

NIGHT CLOCK, No. 5, GAS
HEIGHT ....6⅛ INCHES
WIDTH ....5¾ INCHES
DIAL, TRANSLUCENT GLASS ......5 INCHES
FINISH, GOLD PLATED
MOVEMENT, 1 DAY TIME PIECE
PRICE ....LIST, $5.50

1668

1886

NIGHT CLOCK.

FURNISHED WITH HOSE AND BURNER CONNECTING WITH GAS JET.

1669

FOR GAS.

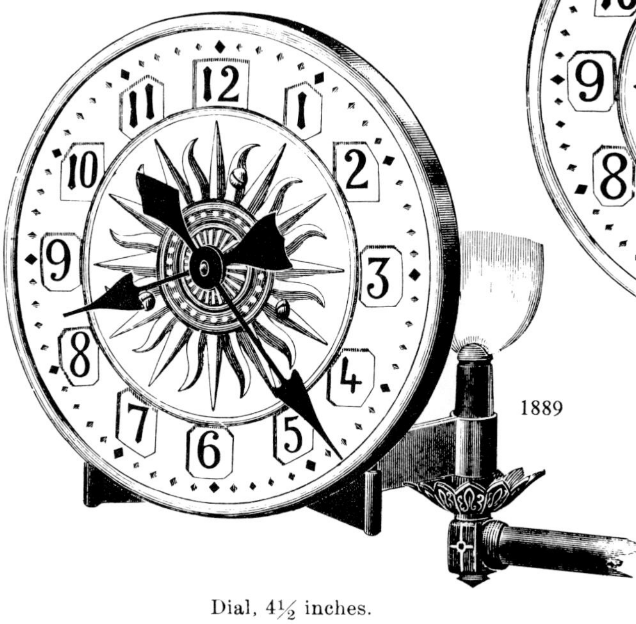

1889

Dial, 4½ inches.

FOR LAMP.

1670

1889

Dial, 4½ inches.

NIGHT CLOCK.

BERLIN IRON LAMP BASE. TRANSPARENT PORCELAIN DIAL, WITH GILT FIGURES AND GILDED RIM.

1 Day Time for Lamp................$5.20
1 Day Time for Gas.................. 4.50

This Movement has Screw Pillars, Cut and Polished Steel Pinions, a Straight Line Lever with Double Roller Club Tooth Escapement and Polished Springs.

NOVELTY CLOCKS

1671

1911

NIGHT CLOCK, No. 5, LAMP

HEIGHT ....6½ INCHES
WIDTH ....5¾ INCHES
DIAL, TRANSLUCENT GLASS ......5 INCHES
FINISH, GOLD PLATED
MOVEMENT, 1 DAY TIME PIECE
PRICE ....LIST, $5.80

1672

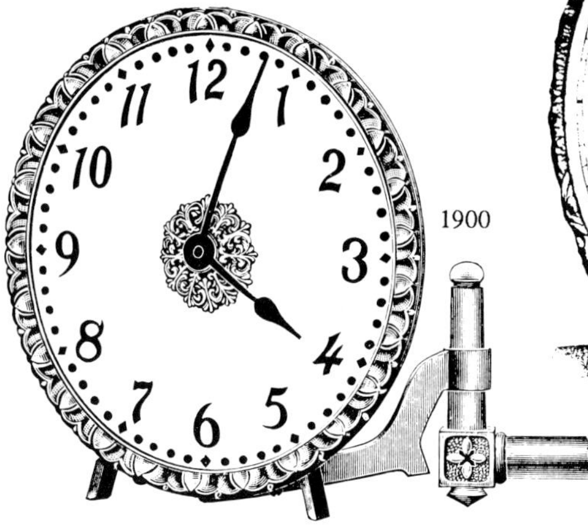

1900

NIGHT CLOCK No. 3—For Gas.

Cast Brass Sash.

Best French Satin Gold Finish.
Height, 6 inches. Dial, 5½ inches.
Translucent Porcelain Dial, Gold Center.
One-day Time.

List price, $4.75

1673

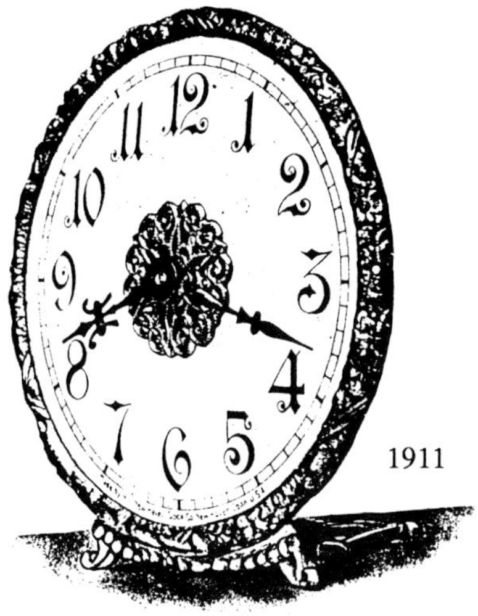

1911

CHISWICK

HEIGHT ....5⅛ INCHES
WIDTH ....5 INCHES
DIAL, PORCELAIN.......4 INCHES
FINISH, ORMOLU GOLD PLATE
MOVEMENT, 1 DAY TIME PIECE
PRICE....................................LIST, $5.10

1674

1886

NIGHT CLOCK
IN TRAVELING CASE.

1 Day Time.......................................$8.00
Leather Case, $1.25 net extra.

1675

1900

NIGHT CLOCK No. 3—With Lamp.

Cast Brass Sash.

Best French Satin Gold Finish.
Height, 6¼ inches. Dial, 5½ inches.
Translucent Porcelain Dial, Gold Center.
One-day Time.

List price, $5.15

NOVELTY CLOCKS

1676

1900

Jeweled Sash
fitted, as desired, with
Brilliant, Ruby,
Emerald, Amethyst,
Sapphire,
or
alternate stones.

COREA

French Satin Gold Finish.
Jeweled or Gold Rope Sash.
Height, 4½ inches. Width, 4¼ inches.
One-day Time.
Three-inch Ivorized Porcelain Dial with Gilt Center.

List price, with Jeweled Sash, $7.00
" " with plain Rope Sash, 5.00

1677

1900

CAMERON

French Satin Gold Finish.
Jeweled or Plain Gold Rope Sash.
Height, 6 inches. Width, 4¼ inches.
One-day Time.
Three-inch Ivorized Porcelain Dial with Gilt Center.

List price, with Jeweled Sash, $7.00
" " with plain Rope Sash, 5.00

1678

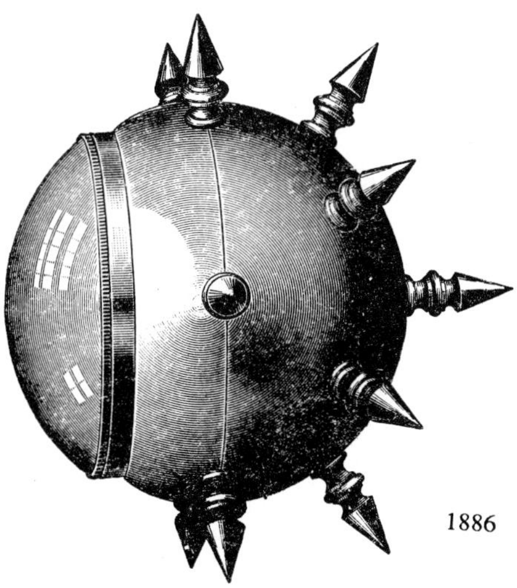

1886

MACE T. P.

SIDE VIEW.

The lens used in this clock is the best crystal, of large magnifying power, producing very novel effect.

1679

1886

MACE T. P.

Finished in Black and Gold, and Silver and Gold.

1 Day Time........................$5.75

1680

1885

HUMMING BIRD.

NICKEL.

Height, 9 inches. Dial, 3½ inches.

Code Word.		Price.
(Mother)	1 Day Time	$4 85
(Moth)	1 " " Alarm	5 75

Fitted with Fine Cut Pinion and Nickel Plated Movement.

NOVELTY CLOCKS

1681

1913

CRESCENT STAND
(WITH JASPER, BACK VIEW.)

STAND ONLY . . . . LIST, $3.65
STAND AND JASPER CLOCK. LIST, 13.10
HEIGHT, 7½ INCHES. WIDTH, 6¼ INCHES.
FINISH, GOLD PLATE.

1683

1913

IVAN
LIST, $10.20

HEIGHT, 4¼ INCHES. WIDTH, 4⅜ INCHES.
DIAL, PORCELAIN, 1½ INCHES.
CASE, SOLID BRASS, WITH MAGNIFYING LENSES OVER DIAL AND BACK OF MOVEMENT.
FINISH, GOLD PLATE.

1682

1913

JASPER FIGURE
(WITH JASPER.)

FIGURE ONLY . . . . LIST, $14.40
FIGURE AND JASPER CLOCK. LIST, 23.85
HEIGHT, 14½ INCHES. WIDTH, 7¼ INCHES.
FINISH, ORMOLU GOLD PLATE, WITH IMPORTED BRAZILIAN ONYX BASE.

1684

1913

FRONT VIEW

1685

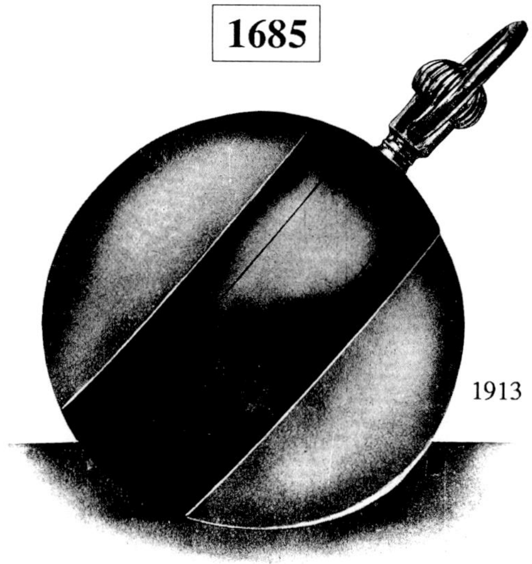

1913

(SHOWING SIDE VIEW)

JASPER
LIST, $9.45

HEIGHT, 3½ INCHES. WIDTH, 2½ INCHES.
DIAL, PORCELAIN, 1½ INCHES, WITH MAGNIFYING LENSES OVER DIAL AND BACK OF MOVEMENT.
FINISH, GOLD PLATE.

NOVELTY CLOCKS

1686

1900

ASTORIA (Jeweled Watch)

Finest 14-Carat Gold Plate Finish.
Stem Wind and Stem Set. Jeweled Front.
Fitted with Folding Back as Desk Ornament.
Watch Height, 3¼ inches. Width, 2¾ inches.
One-day Time. 1¾-inch White Porcelain Dial.

List price, Watch, $7.00

List price, Watch, with Dolphin Stand, $10.50

1687

1900

CRESCENT STAND

Finished in Bronze, Nickel or Gilt
Height, 7¾ inches. Width, 6½ inches.
Base, 3½ inches.
Either the Jasper or Astoria will fit this.

List price, $3.50

1688

1900

DOLPHIN STAND

Best French Satin Gold Finish. Malachite Base.
Height, 8 inches. Base, 3½ inches.
Either the Astoria or Jasper will fit this.

List price, $3.50

Fitted with Solid Brazilian Onyx Base, extra list, $1.50

1689

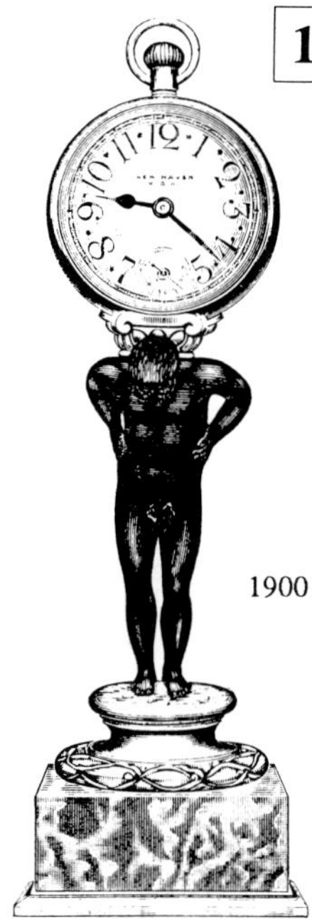

1900

ATLAS

Height, 9¼ inches. Lens, 2½ inches.

List price, $15.30

Extra list price for Brazilian
Onyx Base, $2.10

1690

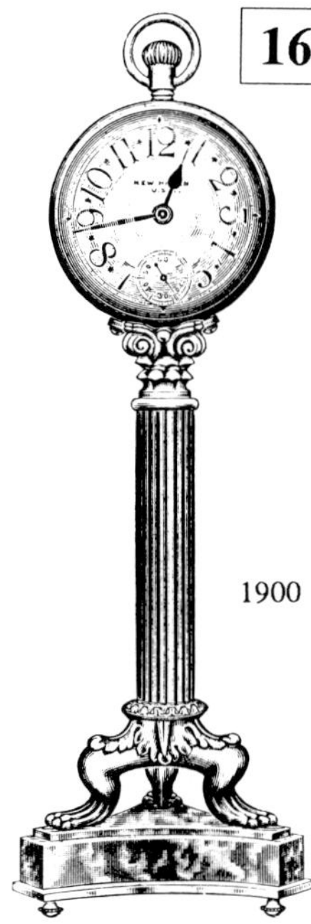

1900

MISTRAL

Height, 10¼ inches.

Lens, 2½ inches.

List price, $14.00

Dark Green and French Gilt Finish with
Malachite Base and best French Magnifying Lens Back and Front.
One-day Time Piece.

Stem Wind and Stem Set with 1¾-inch
White Porcelain Dial,
Blue Numerals.

NOVELTY CLOCKS

1691

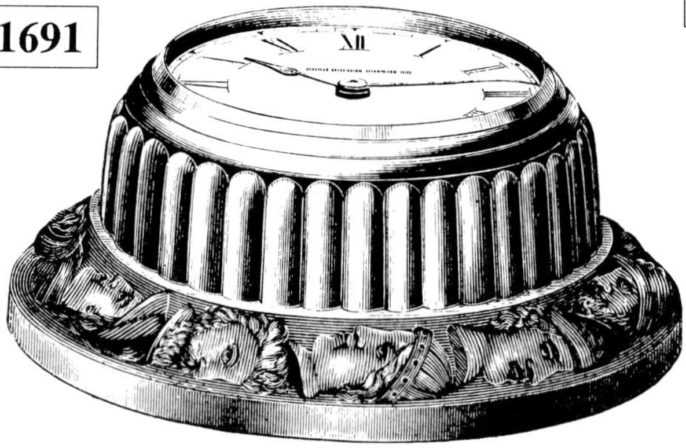

1889

1692

1889

PAPER WEIGHT.

WATCH MOVEMENT.

1 Day Time.............$9 00

1693

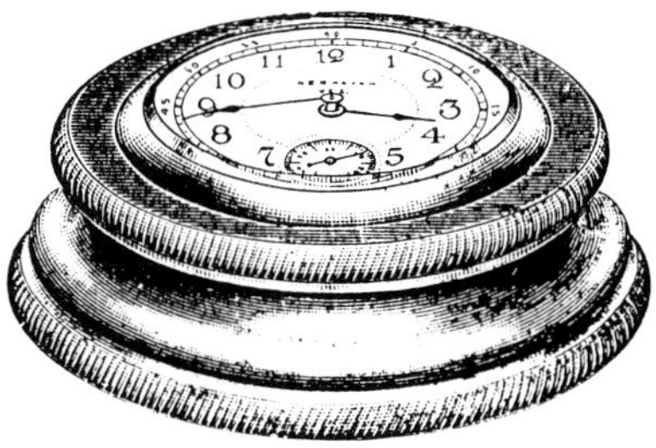

PAPER WEIGHT, No. 4

DEPTH...............1 INCH
WIDTH...............3 INCHES
DIAL, WHITE...1¾" INCHES
FINISH, OXIDIZED BRONZE
MOVEMENT, 1 DAY TIME PIECE
PRICE................................LIST. $2.50

1911

For more information on Low Tile clocks see pages 292-293

1694

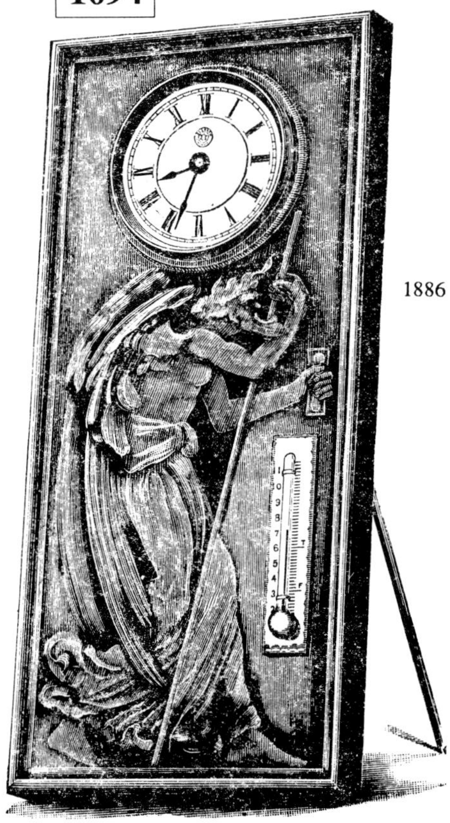

1886

TIME.

LOW TILE ENCASED IN BRASS.

Height, 9 inches.

Code Word.	Price.
(Theorem) 1 Day Time	$7.80

1695

1895

PAPER WEIGHT

NICKEL.

Calendar on Bottom of Clock.

List, $2.50   1 Day Time,

1696

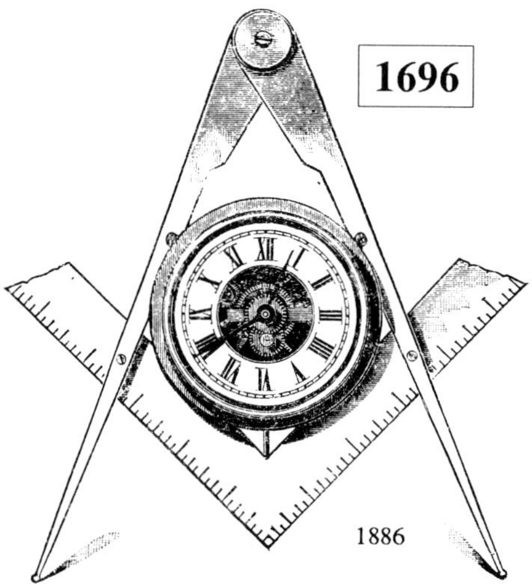

1886

MASONIC.

METAL FRAME, NICKEL PLATED.

Height, $8\frac{1}{2}$ inches.   Dial, $2\frac{1}{2}$ inches.

1 Day Time $7.00
1 " " Alarm 8.00

1697

1895

Side View.

NOVELTY CLOCKS

VOLUNTEER.

1889

1698

VOLUNTEER

FINISHED IN FROSTED

SILVER AND GILT.

2 inch Dial. Cut Pinion Movement.

1 Day Time.........$3.75

1699

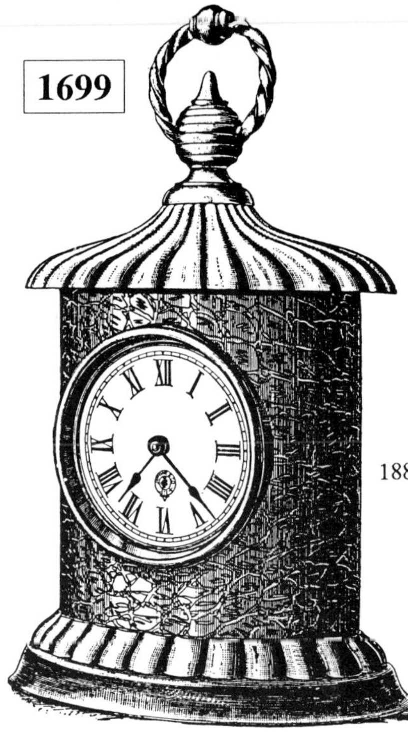

1885

PUCK

Height, 9 inches. Dial 3 inches.
1 Day Strike, Gilt.......... List, $6.00
1 Day Strike, Alarm, Gilt. List, $6.75

1700

"AIGRETTE"

Height, 3½ inches. 40-hour Movement, 2 inch Dial. Raised Numerals. Three-Ply Pentagon Case of 3 Colors, furnished in Pearl, Blue, Pink or Green. Bowed Crystal. $3.44

1701

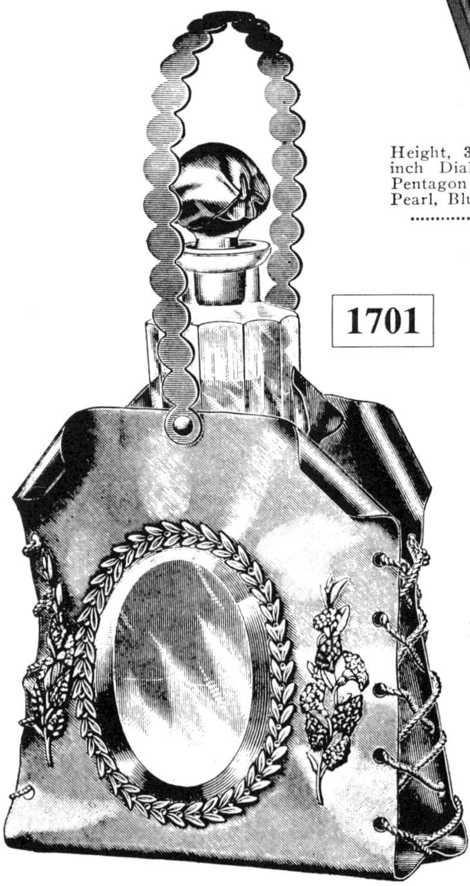

1889

1702

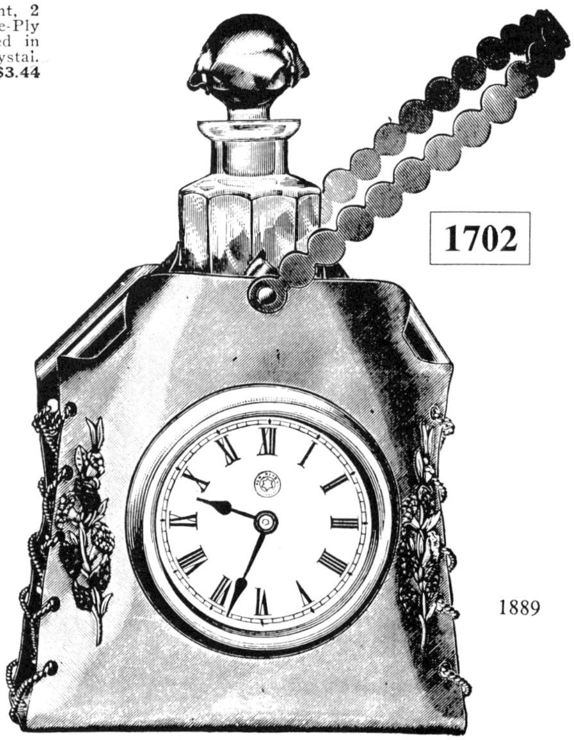

1889

BOUQUET.

GOLD OR SILVER PLATE, BEVELED GLASS MIRROR, AND CUT GLASS SCENT BOTTLE.

Code Word.	Height, 8 Inches.	Price.
(Desire) 1 Day Time	....	$7.00

This Movement has Screw Pillars, Cut and Polished Steel Pinions, a Straight Line Lever with Double Roller Club Tooth Escapement and Polished Springs.

NOVELTY CLOCKS

1703

1886

BOATMAN.

FINISHED IN GOLD AND SILVER.

1 Day Time.......... $7.80
1 " " Alarm...... 8.80

1705

1886

ROBIN.

FINISHED IN GOLD AND SILVER.

1 Day Time.............. $7.30
1 " " Alarm.............. 8.30

1706

1886

SENTINEL.

FINISHED IN GOLD AND SILVER.

1 Day Time............$7.80
1 " " Alarm...... 8.80

1707

1886

RIDER.

FINISHED IN GOLD AND SILVER.

1 Day Time..........$7.80
1 " " Alarm...... 8.80

1704

1889

LIGHTHOUSE.

OXIDIZED SILVER

AND GILT.

Height, 12½ inches. Dial, 2 inches.

Cut Pinion Movement. Revolving Lantern.

Code Word.		Price.
(Lief)	1 Day Time..	$18.00

This Movement has Screw Pillars, Cut and Polished Steel Pinions, a Straight Line Lever with Double Roller Club Tooth Escapement and Polished Springs.

NOVELTY CLOCKS

1708

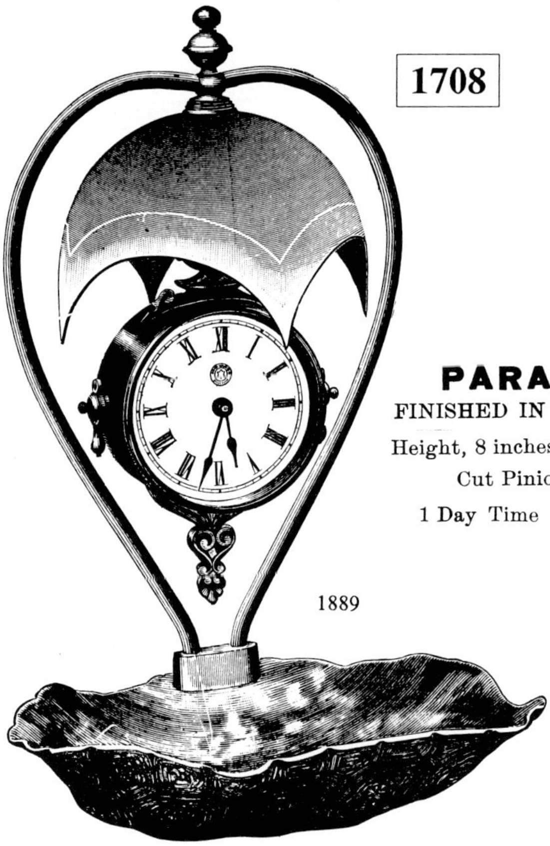

1889

PARACHUTE.

FINISHED IN ANTIQUE BRASS.

Height, 8 inches. Dial, 2 inches.

Cut Pinion Movement.

1 Day Time $6.00

PARACHUTE

1709

1885

No. 1008.

SOLID CAST BRASS ANTIQUE FINISH.

Height, 7¾ inches. Dial, 3 inches.

1 Day Time....................... $12 00
1 " " Alarm................ 13 10

1710

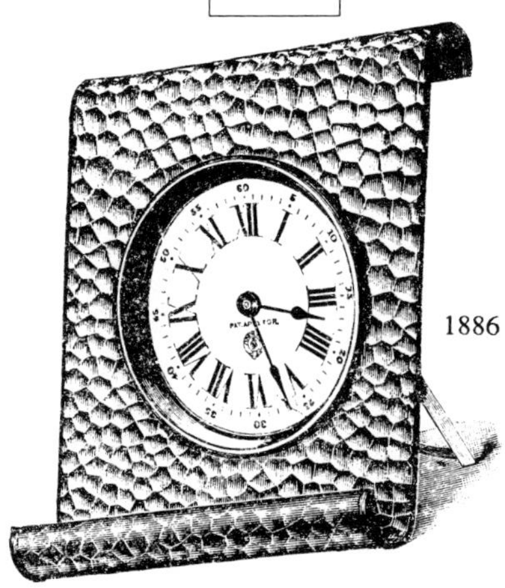

1886

SCROLL.

HAMMERED BRASS.

Height, 6½ inches. Dial, 3½ inches.

1 Day Time...................$3.00
1 " " Alarm 4.00

1711

1885

LINNET.

NICKEL.

Height, 9 inches. Dial, 4 inches.

1 Day Time...$6 00
1 " " Alarm.. .... 6 90
1 " " Calendar.... .... 6 35
1 " " " Alarm.. 7 45

This Clock is made with a combination of silver and gold, highly finished.

1712

1885

BOBBY.

NICKEL OR BRASS.

Height, 6½ inches. Dial, 3 inches.

1 Day Time, Alarm, Nickel.........$5 75
1 " Strike, Nickel............ ... 5 75
1 " " Polished Brass....... 5 75
1 " Time Alarm, " 5 75

NOVELTY CLOCKS

1713

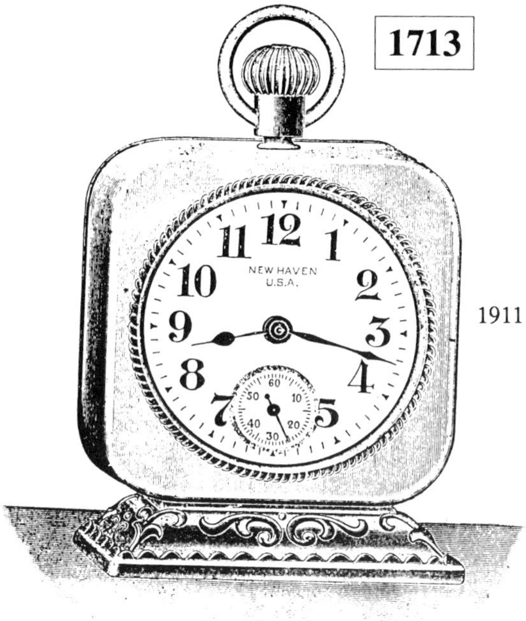

1911

DION

HEIGHT ....3 1/4 INCHES
WIDTH ....2 INCHES
DIAL, WHITE ....1 1/2 INCHES
CASE, HEAVY METAL, HOLDING A COMPLETE WATCH
FINISH, SILVER OR OXIDIZED BRONZE
MOVEMENT, 1 DAY TIME PIECE
PRICE..LIST, $2.90

1714

1911

DATEO

HEIGHT ....3 1/2 INCHES
WIDTH ....5 3/4 INCHES
DIAL, PORCELAIN ......2 INCHES
CASE, HEAVY BRASS, WITH CALENDAR HOLDER
FINISH, OLD BRASS
MOVEMENT, 1 DAY TIME PIECE
PRICE...LIST, $2.90

1715

1911

DISQUE

DISQUE

HEIGHT ....3 INCHES
WIDTH ....3 1/2 INCHES
DIAL, PORCELAIN ......2 INCHES
CASE, HEAVY BRASS
FINISH, OLD BRASS
MOVEMENT, 1 DAY TIME PIECE
PRICE, PLAIN ....LIST, $2.50
PRICE, LETTERED ....LIST, 2.65

1716

1911

TRAVELER

HEIGHT ....2 7/8 INCHES
WIDTH ....2 INCHES
DIAL, WHITE ....1 1/2 INCHES
CASE, HEAVY BRASS, HOLDING A COMPLETE WATCH
FINISH, GUN METAL, OLD BRASS, NICKEL, COPPER, OR POLISHED BRASS
MOVEMENT, 1 DAY TIME PIECE
PRICE..LIST, $2.20

1717

1911

TABARD

HEIGHT ....3 INCHES
WIDTH ....2 1/4 INCHES
DIAL, WHITE ....1 1/2 INCHES
CASE, HEAVY METAL FRAME, SILVER FINISH, HOLDING A COMPLETE WATCH
MOVEMENT, 1 DAY TIME PIECE
PRICE...LIST, $2.65

NOVELTY CLOCKS

1718

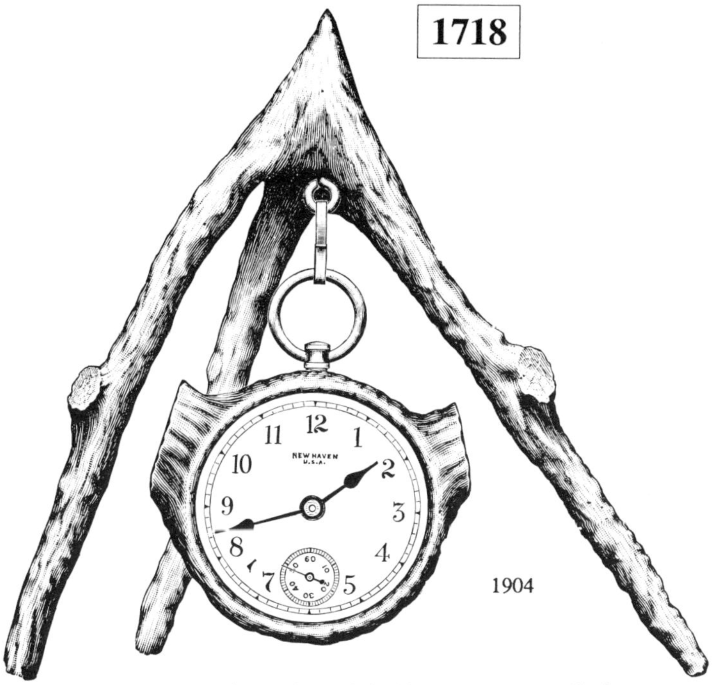

1904

Rustic Tripod. 1-day time, 2-inch dial, Sting movement. Clock and stand covered with an exact imitation old Hickory wood bark **$2.40**

1719

1886

TRAY.

ROYAL COPPER, WITH SILVER ORNAMENTS.

1 Day Time $10.50
1 " " Alarm 11.50

1720

1889

THE BELL.

BRASS FINISH WITH ORNAMENTS IN RELIEF.

Height, 12½ inches. Dial, 5 inches.

8 Day Strike on a Cathedral Gong............$19.50

1721

1906

BELLONA

Height, 9¾ inches. Width, 6½ inches.
Two-inch Porcelain Dial.
Ormolu Gold Plated Case.

One-day Time .... List Price, $7.65
One-day Time, Alarm ... " " 9.15

NOVELTY CLOCKS

1722

1900

MAINTENON

French Satin Gold or Bronze Finish.
Height, 8 inches. Width, 3½ inches.
One or Eight-day Time with 2-inch Ivorized Porcelain Dial.

List price, One-day Time, Satin Gold,		$4.25
"	" One-day Time, Bronze,	3.85
"	" Eight-day Time, Satin Gold,	5.75
"	" Eight-day Time, Bronze,	5.35

Fitted with Jeweled Sash, in Brilliant, Ruby, Emerald, Amethyst, Sapphire, Pearl and Turquoise, or alternate stones. Extra list, $2.10

1723

1906

LOIRE

Height, 6½ inches. Width, 7 inches.
Two-inch Porcelain Dial.
Ormolu Gold Plated Case.

One-day Time . . . . List Price, $6.25
One-day Time, Alarm . . . . " " 7.75

1724

1900

ELDORA

French Satin Gold or Bronze Finish.
Height, 6¾ inches. Width, 4¾ inches.
One-day Time with Two-inch Ivorized Porcelain Dial.

List price, Satin Gold, $5.00
" " Bronze, 4.50

1726

1900

STING

Dial, 2 inches.

List price, One-day Time, $1.75

1727

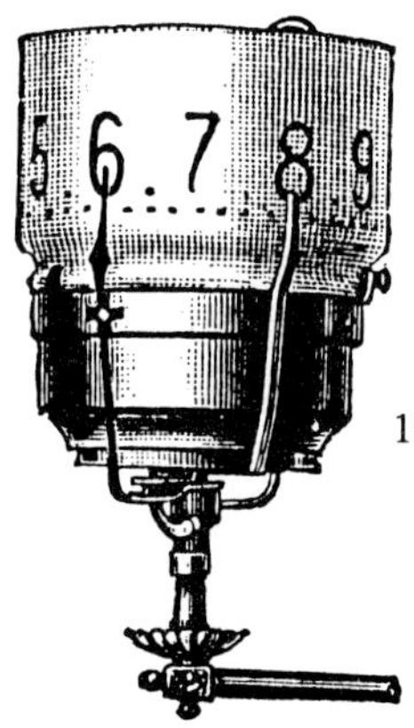

1885

AMERICAN
ILLUMINATED NIGHT CLOCK.

Patented May 14th, 1872.

Diameter of Globe, 6 inches.

IS ARRANGED
AS AN
ATTACHMENT
TO ANY
GAS FIXTURE.

1 Day Lever, Time, for Gas. - -	$7 20
1 Day Lever, Time, Alarm, for Gas, -	9 45
1 Day Lever, Time, for Lamp. - -	8 70
1 Day Lever, Time, Alarm, for Lamp -	10 95

DIRECTIONS TO SET THE CLOCK RUNNING.—Take off the old burner and screw on the one furnished with the clock, then set the clock on the burner so as to have the pointer opposite the point you wish to see the time from; to wind, lift it off the burner, by the thimble at the bottom, grasping the clock in the left hand, and wind with the right. It can be set in the same manner by holding the clock and turning the thimble to the right or left, as may be required, taking care to press the thimble up while setting the pointer to the right time.

NOVELTY CLOCKS

1728

1885

No. 1024.
CAST BRASS, HAND FINISHED.
Height, 9¼ inches. Dial, 3 inches.

8 Day Time..........$16 00

1729

Ca. 1931

CLOISONNE CLOCKS

2¾ inches Square Cloisonne Case—Colors, Pink, Blue or Yellow—2 inch Silvered Dial with Raised Numerals, Convex Crystal—Nickel Plated Back with small easel rest—Dependable 40 Hour Movement.

1730

1885

No. 1025.
CAST BRASS, SILVERED PANELS.
Height, 10 inches. Dial, 1¾ inches.

1 Day Time $14 00

1731

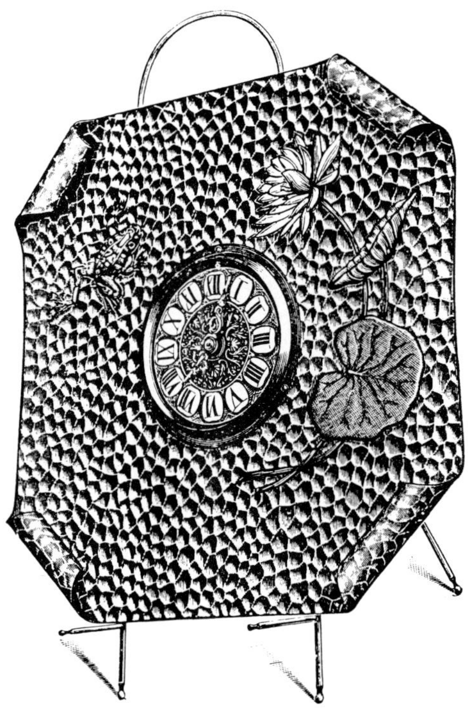

1885

No. 1004 EXTRA.
ROYAL COPPER WITH SOLID SILVER ORNAMENTS.

Dial, 3½ inches. Height, 14 inches.

8 Day Time............$27 00

NOVELTY CLOCKS

1732

1885

No. 1015.

SOLID CAST BRASS ANTIQUE FINISH.

Height, 7 inches. Dial, 3 inches.

1 Day Time..........................$11 40
1 " " Alarm................ 12 50

1733

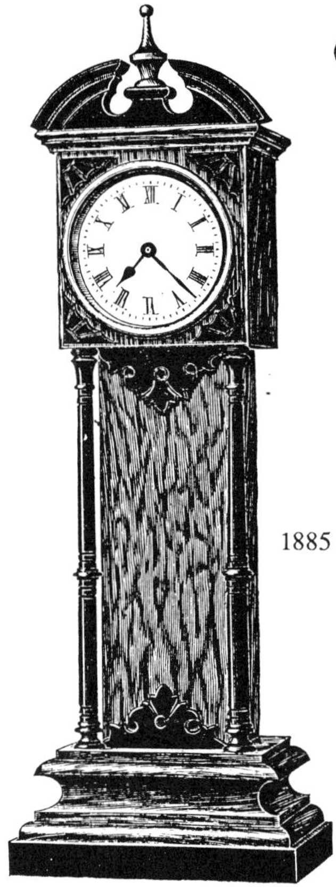

1885

GRANDFATHER.

OAK, TRIMMED WITH EBONY.

Height, 14 inches.

Dial, 2 inches.

8 Day Time....$12 00

1734

1885

No. 1005.

SOLID CAST BRASS ANTIQUE FINISH.

Height, 8 inches. Dial, 3 inches.

1 Day Time......................$12 00
1 " " Alarm...... .... 13 10

1735

1885

No. 1020.

CAST BRASS, RICHLY CHASED.

Height, 11 inches. Dial, 1¾ inches.

1 Day Time........$18 50

The above is a most artistic combination of Candle-stick and Clock.

1736

1906

TOLTEC

Height, 6 ¼ inches. Width, 5 ¼ inches. Two-inch White Dial. Ormolu Gold Plated Case. One Day Time. List price, $2.10.

1737

1906

KAZAN

Height, 6 ¼ inches. Width, 5 ½ inches. Two-inch White Dial. Ormolu Gold Plated Case. One Day Time. List price, $2.10.

NOVELTY CLOCKS

1738

1886

STANDARD.

FINISHED IN GOLD AND SILVER.

1 Day Time $8.00

1739

1880

LITTLE DUKE.

Nickel.

Height, 6¼ inches.

1 Day Time.

1 Day Time, Alarm.

1740

1895

No. 60

Height, 5 ½ inches.

1 Day Time...$4.20

1741

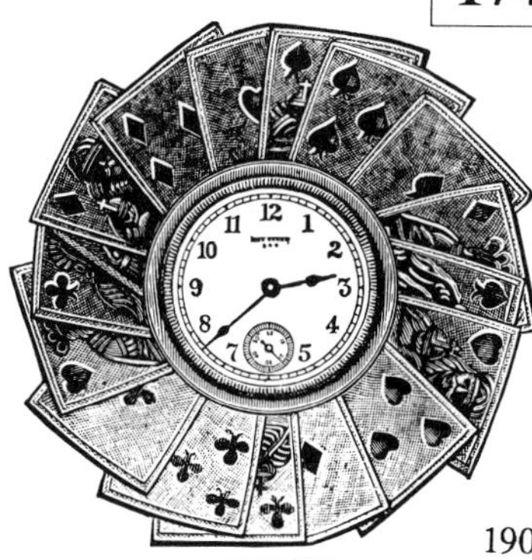

1906

FLUSH

Height, 5 inches. Width, 5 inches.

Two-inch White Dial.

Ormolu Gold Plated Case.

One-day Time ....List Price, $2.50

1742

1895

No. 63

Height, 4 ½ inches.

1 Day Time...$4.30

1743

1885

SNIPE.

NICKEL.

Height, 4½ inches. Dial, 4 inches.

Code Word.		Price.
(Muff)	1 Day Time..........................	$4 30

1744

1895

No. 64

QUADRUPLE PLATE, LACQUERED AND NON-TARNISHABLE.

Height, 5 ½ inches.

1 Day Time.............$4.20

NOVELTY CLOCKS

1745

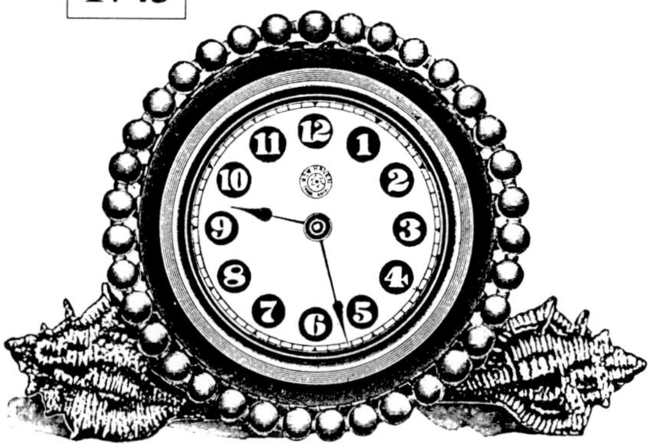

1895

No. 65.

Quadruple Silver Plate, Lacquered and Non-tarnishable.

Height, 3½ inches; Dial, 2 inches.

1 Day Time, List, $4.00

1746

1900

CREMORNE

French Satin Gold Finish.

Height, 5¼ inches.
Four-inch Ivorized Porcelain Dial with Gilt Center.
One-day Time.

Jeweled Sash, fitted as desired, in Brilliant, Ruby, Emerald, Amethyst, Sapphire, or alternate stones.

List price, with Jeweled Sash, $7.00

List price, with plain Rope Sash, $5.00

1747

1906

AIRE

Height, 6 inches. Width, 5 inches. Two-inch White Dial. Ormolu Gold Plated Case. One Day Time. List price, $2.10.

1748

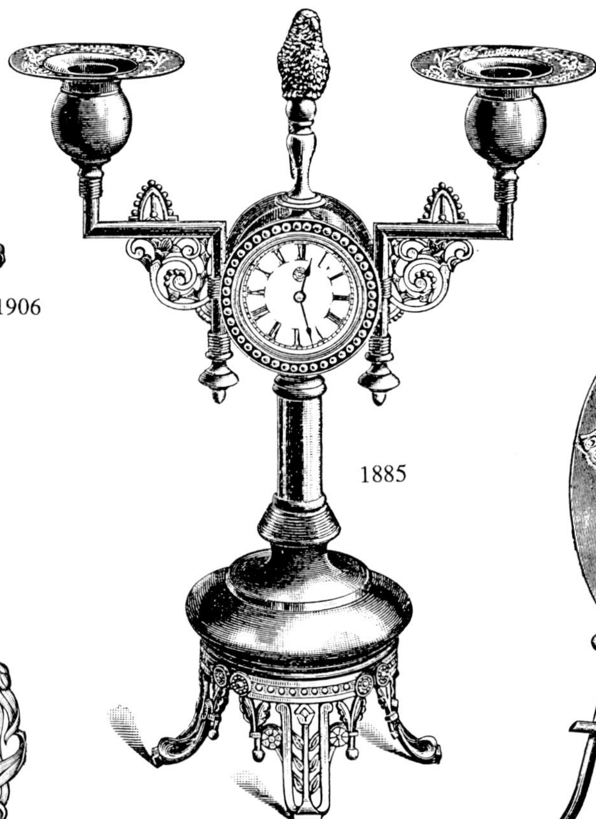

1885

No. 1031.

CANDELABRA (SILVER PLATED).

Height, 12¾ inches.
Dial, 1¾ inches.

1 Day Time...$32 00

1749

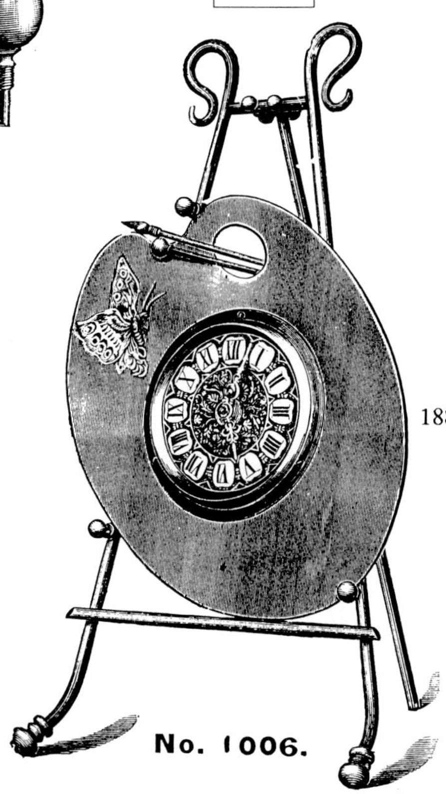

1885

No. 1006.

ROYAL COPPER WITH SOLID SILVER ORNAMENTS.

Dial, 3 inches.

1 Day Time..............$15 75
1 " " Alarm.......$16 85

1750

1906

YUMA

Height, 6 ¼ inches. Width, 5 ¼ inches. Two-inch White Dial. Ormolu Gold Plated Case. One Day Time. List price, $2.10.

NOVELTY CLOCKS

1752

1886

FAN.

ROYAL COPPER, WITH SILVER ORNAMENTS.

1 Day Time $11.75
1 " " Alarm 12.75

1751

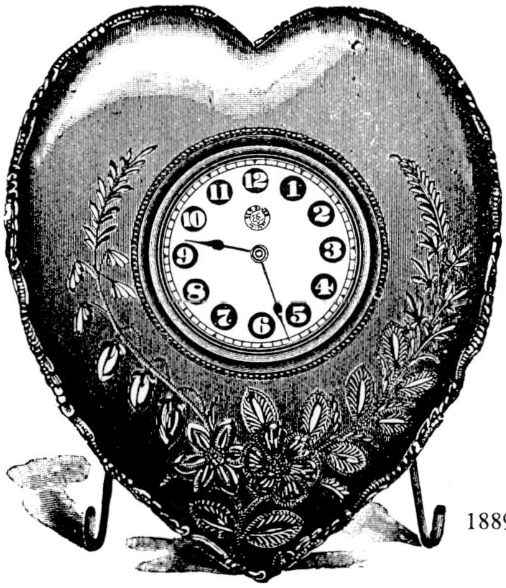

1889

No. 59

QUADRUPLE PLATE, LACQUERED
AND NON-TARNISHABLE.

Height, 6 inches.

1 Day Time.............$5.30

1754

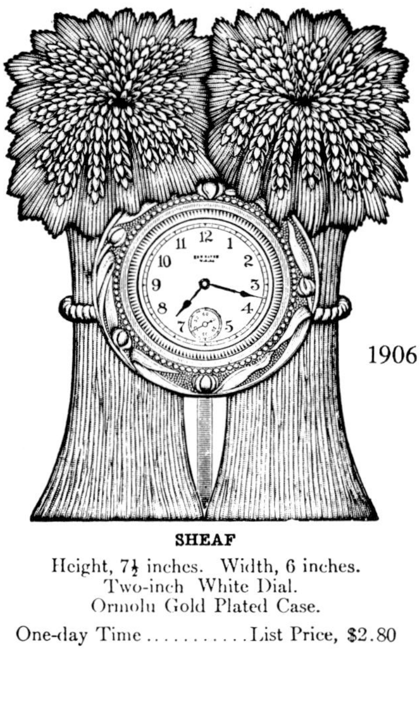

1906

SHEAF

Height, 7½ inches. Width, 6 inches.
Two-inch White Dial.
Ormolu Gold Plated Case.

One-day Time .... List Price, $2.80

1753

1889

LANTERN NIGHT CLOCK.

BRASS HALL LANTERN, STUDDED WITH JEWELS, CONTAINING TRANSLUCENT DIAL CLOCK.

FINISHED IN ANTIQUE BRASS AND ANTIQUE SILVER.
FIVE-INCH OPAL DIAL.

1 Day Time, Antique Brass............... $12.00
1 Day Time, Antique Silver.............. 13.00

NOVELTY CLOCKS

1755

1906

DIXON

Height, 9 inches. Width, 6¾ inches.
Two-inch Porcelain Dial.
Ormolu Gold Plated Case.

One-day Time . . . . .List Price, $9.75

1756

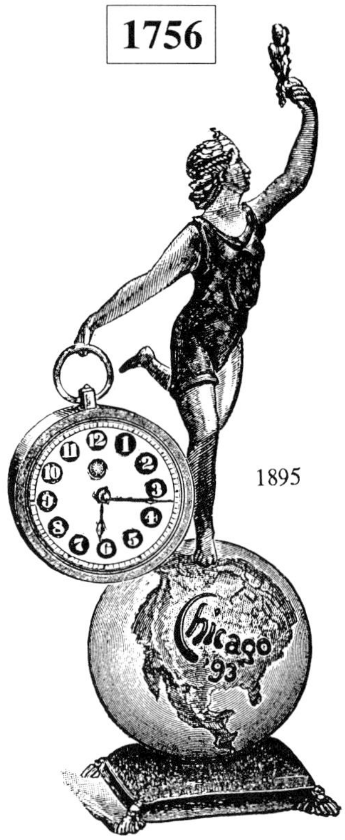

1895

No. 76

QUADRUPLE PLATE, LACQUERED AND NON-TARNISHABLE.

Height, 10 inches. Dial, 2 inches.
1 Day Time, Silver..............$6.00
1 Day Time, Silver, Gold....$7.00

1757

1895

No. 74

Height, 6 ½ inches.
1 Day Time...$4.20

1758

1895

No. 72

QUADRUPLE PLATE, LACQUERED AND NON-TARNISHABLE.

Height, 8 inches.
1 Day Time..............$7.00

1759

1895

No. 61

Height, 5 ¾ inches.
1 Day Time...$4.20

1760

1906

MINOT

Height, 9¾ inches. Width, 6¾ inches.
Two-inch Porcelain Dial.
Ormolu Gold Plated Case.

One-day Time . . . . List Price, $7.00
One-day Time, Alarm . . . . " " 8.50

NOVELTY CLOCKS

1761

1906

LANTIER

Height, 6½ inches. Width, 5¾ inches.
Two-inch Porcelain Dial.
Ormolu Gold Plated Case.
One-day Time.. ....List Price, $4.25

1762

MANGROVE

Height, 10¼ inches. Width, 5¼ inches.
Ivorized Dial, 2 inches.
French Satin Gold Finish.

List price, One-day Time, $7.00
" " Eight-day Time, 8.50

Fitted with Jeweled Sash in Brilliant, Ruby, Emerald, Amethyst, Sapphire, Pearl and Turquoise, or alternate stones.

Extra list, $2.10

1900

MANGROVE

1763

1900

ARDEN

French Satin Gold or Bronze Finish.
Height, 8½ inches. Width, 4¾ inches.
One-day Time, with 2-inch Ivorized Porcelain Dial.

List price, Satin Gold, $7.00
List price, Bronze, plain Sash, $6.30

Fitted to order only with Jeweled Sash in Brilliant, Ruby, Emerald, Amethyst, Sapphire, Pearl and Turquoise or alternate stones.

Extra list, $2.10

1764

1886

FALCON HUNTER.

FINISHED IN SILVER AND GILT.
1 Day Time.........$6.25

1765

1900

DALTON

French Satin Gold or Bronze Finish.
Height, 10 inches. Width, 3½ inches.
One-day Time with Two-inch Ivorized Porcelain Dial.

List price, Satin Gold, $4.25
" " Bronze, 3.85

Fitted with Jeweled Sash, in Brilliant, Ruby, Emerald, Amethyst, Pearl and Turquoise, Sapphire, or alternate stones. Extra list, $2.10

NOVELTY CLOCKS

ONE-DAY NOVELTY CLOCKS. TIMEPIECE OR INTERMITTENT ALARM.
2-INCH PORCELAIN DIAL. FRENCH BRONZE, OLD BRASS,
GILT, SILVER, OR GUN METAL FINISH

FITTED WITH 1-DAY INTERMITTENT ALARM, EXTRA LIST, $1.50

1766

1913

ZEBINA
LIST, $3.65
HEIGHT, 3⅞ INCHES. WIDTH, 3½ INCHES.

1767

1913

ZERBA
LIST, $3.65
HEIGHT, 5¾ INCHES.
WIDTH, 3¼ INCHES.

1768

1913

PADUA
LIST, $2.90
HEIGHT, 4 INCHES. WIDTH, 3¼ INCHES.

1769

ZETUS 1913
LIST, $2.90
HEIGHT, 3¼ INCHES. WIDTH, 3½ INCHES.

1770

1913

PEQUOT
LIST, $2.90
HEIGHT, 4⅛ INCHES. WIDTH, 3 INCHES.

1771

1913

ZETA
LIST, $3.65
HEIGHT, 3⅞ INCHES. WIDTH, 3¾ INCHES.

1772

1913

ZENDA
LIST, $3.65
HEIGHT, 5½ INCHES.
WIDTH, 2⅞ INCHES.

1773

1913

ZELDA
LIST, $3.65
HEIGHT, 3⅞ INCHES. WIDTH, 3¾ INCHES.

NOVELTY CLOCKS

METAL NOVELTIES

French Bronze, Old Brass, Gilt, Silver, or Gun Metal Finish

1775

1917

1774

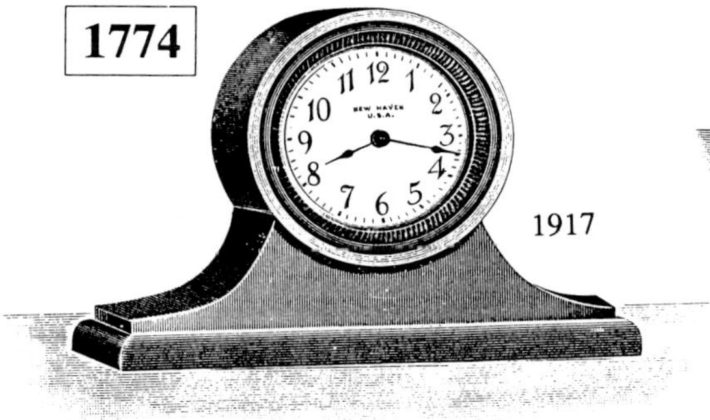

1917

RAWSON
8-Day Time
Height, 4½ Inches Width, 8¼ Inches
Dial, Porcelain, 2¼ Inches
Arabic or Roman
Beveled Bowed Glass

RODNEY
1-Day Time, or 1-Day Intermittent Alarm
Height, 3⅝ Inches Width, 5½ Inches
Dial, Porcelain, 2 Inches

1777

1913

ZILPHA
LIST, $4.35
HEIGHT, 4 INCHES.
WIDTH, 3¾ INCHES.

1776

1917

INDEE
1-Day Time
Height, 3⅞ Inches Width, 3⅜ Inches
Dial, 2 Inches
Case, Solid Brass, Finished in Nickel, Old Brass, Polished Gilt or Copper

1778

1906

FENWICK

Height, 6¼ inches. Width, 6 inches.
Two-inch Porcelain Dial.
Ormolu Gold Plated Case.
One-day Time .... List Price, $4.90

1779

1913

CHATHAM
LIST, $3.65
HEIGHT, 4⅜ INCHES. WIDTH, 3¼ INCHES.

1780

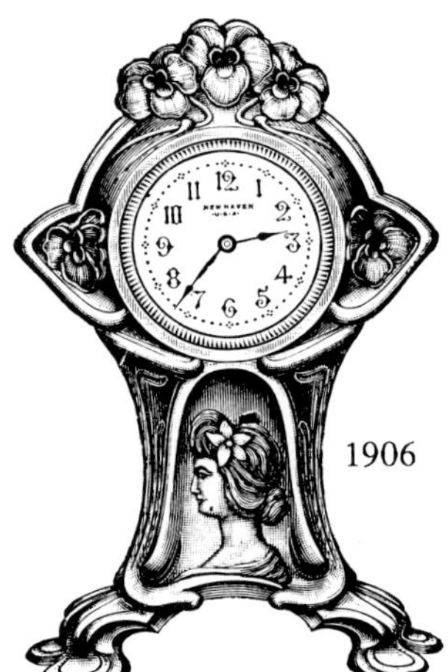

1906

PELHAM
Height, 6½ inches. Width, 4½ inches.
Two-inch Porcelain Dial.
Ormolu Gold Plated Case.
One-day Time .... List Price, $3.75
One-day Time, Alarm ... " " 5.25

1781

1913

ZIRKEL
LIST, $4.35
HEIGHT, 4 INCHES. WIDTH, 3⅛ INCHES.

NOVELTY CLOCKS

1782

1917

BELLAIRE

Height,	- - - - -	8 Inches
Width,	- - - - -	5½ Inches

1783

1900

CLOVELLY

Height, 4¼ inches. Width, 4¼ inches.
Dial, 2 inches.
French Satin Gold Finish.
Ivorized Porcelain Dial. One-day Time.
List price, $3.05

1784

1917

FENBY

Height,	- - - - -	5½ Inches
Width,	- - - - -	4½ Inches

Dial, Porcelain, 2 Inches

1787

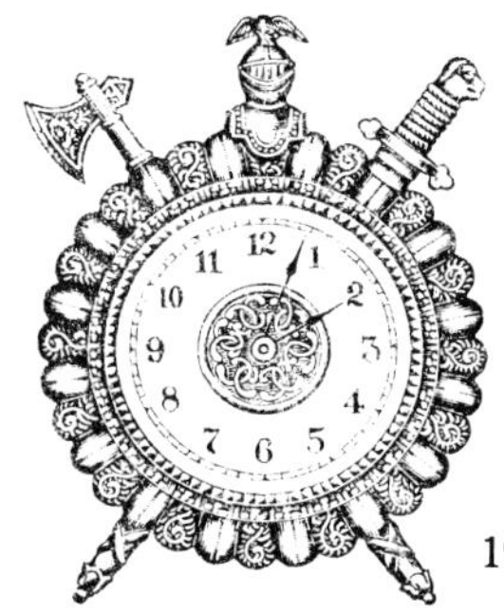

1900

PHILLIPINA

Height, 4¼ inches. Width, 3¼ inches.
Dial, 2 inches.
French Satin Gold Finish.
Ivorized Porcelain Dial. One-day Time.

1786

1900

MANILA

Height, 4 inches. Width, 3¼ inches.
Dial, 2 inches.
French Satin Gold Finish.
Ivorized Porcelain Dial. One-day Time.

1785

1900

TRIANON

French Satin Gold or Bronze Finish.
Height, 7½ inches. Width, 3¾ inches.
One or Eight-day Time, with 2-inch Ivorized Porcelain Dial.

List price,	One-day Time,	Satin Gold,	$4.25
" "	One-day Time,	Bronze,	3.85
" "	Eight-day Time,	Satin Gold,	5.75
" "	Eight-day Time,	Bronze,	5.35

Fitted with Jeweled Sash, in Brilliant, Ruby, Emerald, Amethyst, Sapphire, Pearl and Turquoise, or alternate stones. Extra list, $2.10

1788

1917

FERRO

Height,	- - - - -	8¾ Inches
Width,	- - - - -	6¼ Inches

NOVELTY CLOCKS

1789

1917

1790

1917

1791

1917

BINGHAM

Height, - - - - - 8 Inches
Width, - - - - - 5½ Inches

FALLIS

Height, - - - - - 6 Inches
Width, - - - - - 4½ Inches

NAPIER

Height, - - - - - 7 Inches
Width, - - - - - 4½ Inches

METAL NOVELTIES

1-Day Time. 2-inch White Dial. Ormolu Gold Plate Finish

1792

1900

FLEURY

Height, 5 inches. Width, 4 inches.
Dial, 2 inches.
French Satin Gold Finish.
Ivorized Porcelain Dial. One-day Time.
List price, $3.35

1793

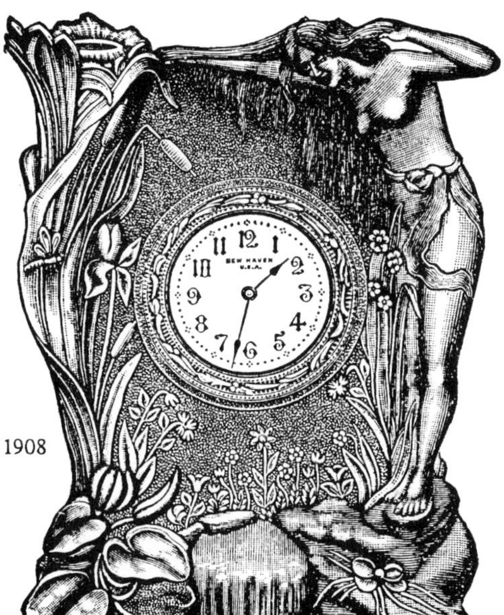

1908

7063
Rich Ormolu Gold Plated, Eight-Day Time,
Ivorized Porcelain Dial,
Height 7½ inches, Width 5½ inches,
$12 20

1794

1908

7062
Ormolu Gold Plated, One-Day Time,
Height 9 inches, Width 6 inches,
Dial 2 inches,
$8 30

1795

1906

VIETA

Height, 6 inches. Width, 5½ inches.
Two-inch Porcelain Dial.
Ormolu Gold Plated Case.
One-day Time . . . . List Price, $4.25

NOVELTY CLOCKS

METAL NOVELTIES

1-Day Time. 2-inch Porcelain Dials. Ormolu Gold Plate Finish

1796

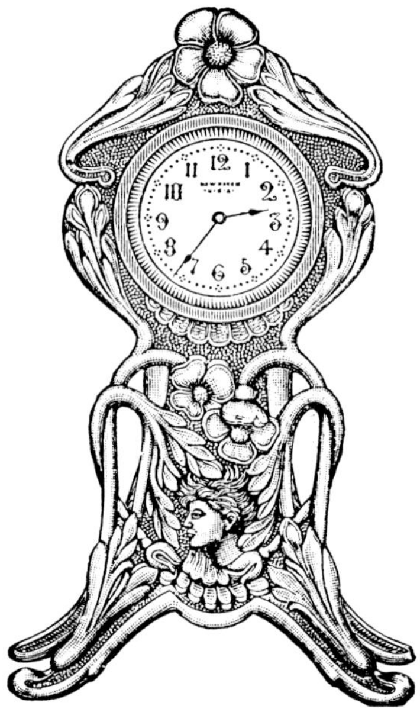

7061
Ormolu Gold Plated, Ivorized Porcelain Dial,
One-Day Time, Height 8¼ inches,
Width 5 inches, Dial 2 inches,
$6 55

1797

1906

FERNEY

Height, 5¾ inches. Width, 5¾ inches.
Two-inch Porcelain Dial.
Ormolu Gold Plated Case.

One-day Time . . . List Price, $5.15

1798

1917

NACHUSA

Height, - - - - - 7 Inches
Width, - - - - - 5½ Inches

1800

1917

DARIEN

Height, - - - - - 8½ Inches
Width, - - - - - 6¾ Inches

1801

1917

CHARIOT

Height, - - - - - 9 Inches
Width, - - - - - 9½ Inches

1802

1917

CORFU

Height, - - - - - 10 Inches
Width, - - - - - 6 Inches

NOVELTY CLOCKS

1803

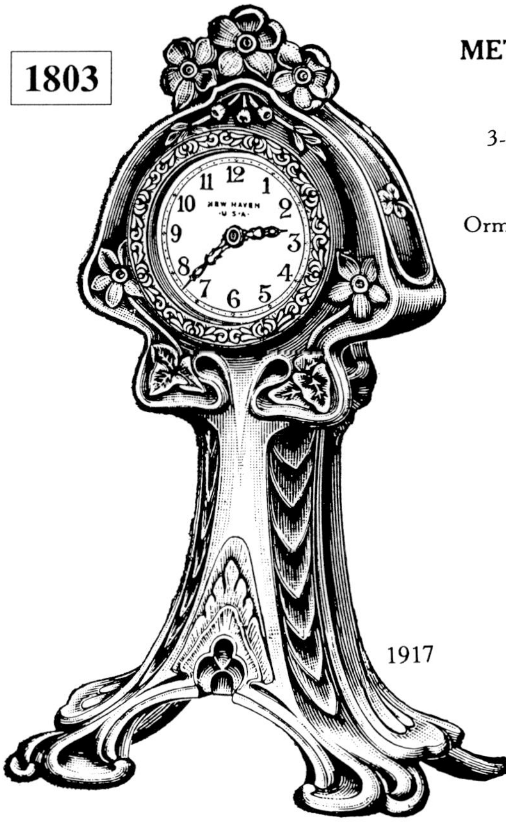

1917

FOLLEN

Height, - - - - - 11½ Inches
Width, - - - - - 7 Inches

METAL NOVELTIES

8-Day Time.

3-inch Porcelain Dial.

Bevel Glass.

Ormolu Gold Plate Finish

ARVERNE

Fitted with Jeweled Sash in Brilliant, Ruby, Emerald, Amethyst, Sapphire, Pearl and Turquoise, or alternate stones. Extra list, $2.10

1804

1900

ARVERNE

Height, 10½ inches. Width, 7½ inches.
Ivorized Dial, 2 inches.
French Satin Gold Finish.
List price, One-day Time, $10.50
" " Eight-day Time, 12.00

1805

1900

TYRONE

TYRONE

Height, 5¼ inches. Width, 3¾ inches.
Dial, 2 inches.
French Satin Gold Finish.
Ivorized Porcelain Dial. One-day Time.
List price, $3.05

1806

1913

MANTON

LIST, $9.45

HEIGHT, 5 INCHES. WIDTH, 4⅛ INCHES.
DIAL, 3 INCHES.
CASE, SOLID BRASS.
FINISH, GILT, SILVER, OR GUN METAL.

1807

1900

LASCAR

Height, 5½ inches. Width, 4½ inches. Dial, 2 inches.
French Satin Gold Finish.
Ivorized Porcelain Dial. One-day Time.
List price, $3.35

NOVELTY CLOCKS

ONE-DAY NOVELTY CLOCKS IN SOLID BRASS CASES.
TIMEPIECE OR INTERMITTENT ALARM. 2-INCH PORCELAIN DIAL.
GILT, OLD BRASS, SILVER OR GUN METAL FINISH

FITTED WITH INTERMITTENT ALARM, EXTRA LIST, $1.50

1808

1913

PANAY
LIST, $5.25
HEIGHT, 3½ INCHES. WIDTH, 3⅜ INCHES.

1809

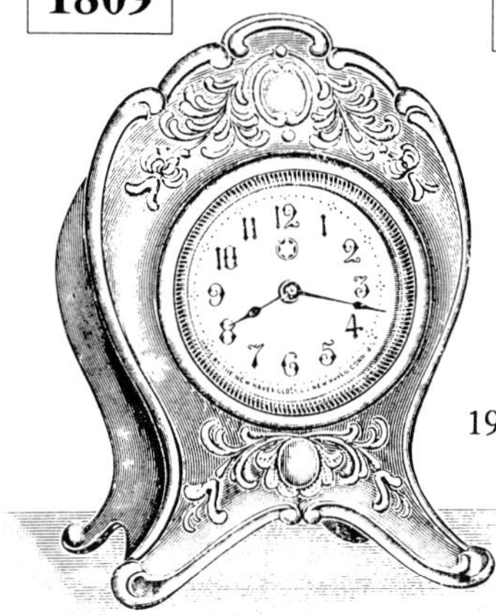

1913

CASTINE
LIST, $3.65
HEIGHT, 5⅛ INCHES.
WIDTH, 3⅛ INCHES.

1810

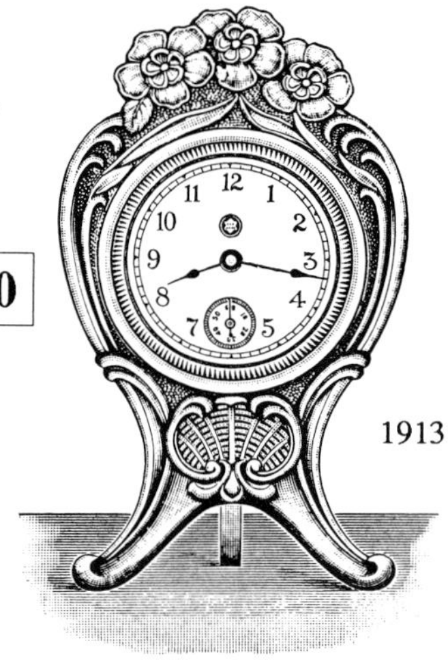

1913

DURAND
LIST, $1.90
HEIGHT, 5½ INCHES.
WIDTH, 3¼ INCHES.

1811

1913

IRMA
LIST, $3.95
HEIGHT, 3½ INCHES. WIDTH, 4⅛ INCHES.

1812

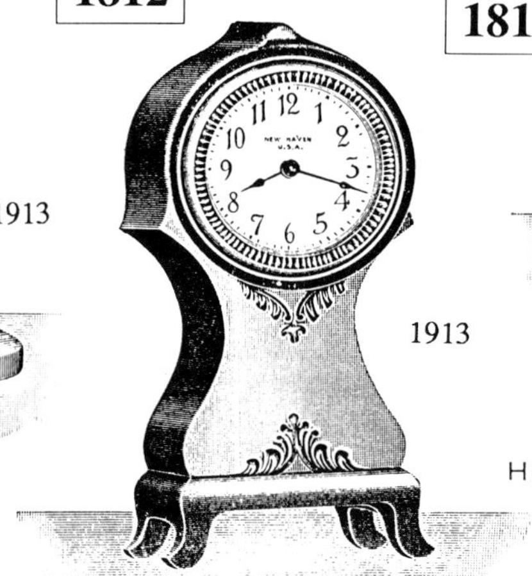

1913

ZANTE
LIST, $3.65
HEIGHT, 5⅞ INCHES.
WIDTH, 3 INCHES.

1813

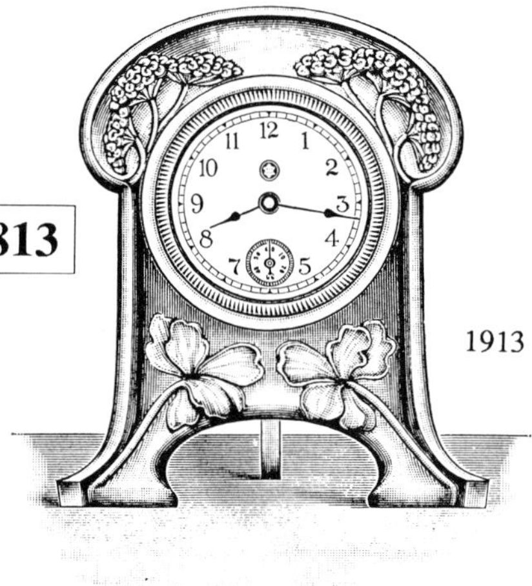

1913

DRUID
LIST, $2.20
HEIGHT, 5 INCHES. WIDTH, 4¼ INCHES.

1814

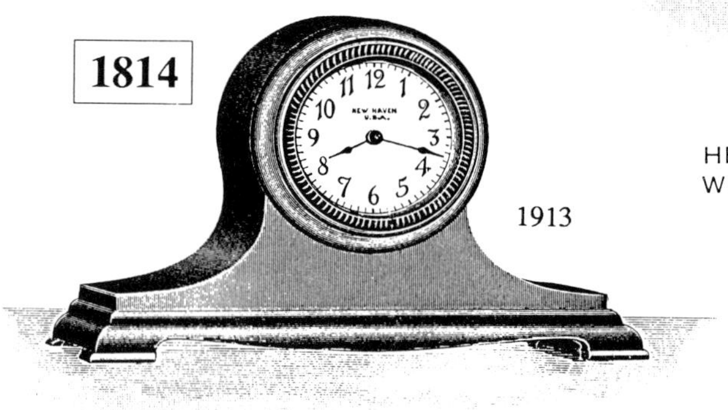

1913

RODEN
LIST, $4.35
HEIGHT, 4 INCHES. WIDTH, 6¾ INCHES.

1815

1913

ZARBINE
LIST, $4.35
HEIGHT, 3½ INCHES. WIDTH, 6⅝ INCHES.

NOVELTY CLOCKS

METAL NOVELTIES

8-Day Time. 3-inch Porcelain Dials.

Bevel Glass. Ormolu Gold Plate Finish

1816

1913

RODMAN

LIST, $4.35

HEIGHT, 3⅝ INCHES. WIDTH, 7⅜ INCHES.

1817

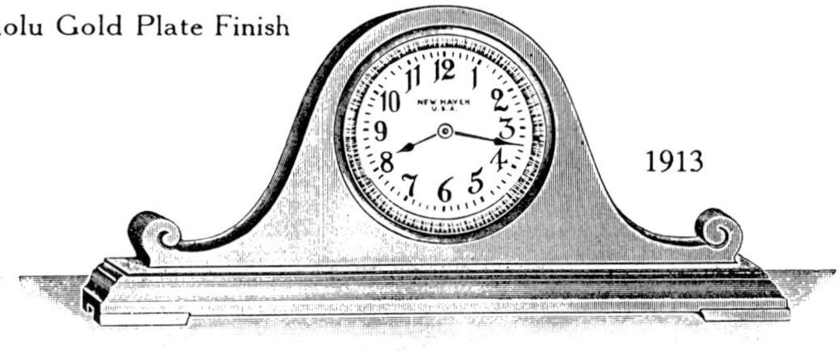

1913

RODELL

LIST, $5.10

HEIGHT, 3½ INCHES. WIDTH, 7½ INCHES.

1818

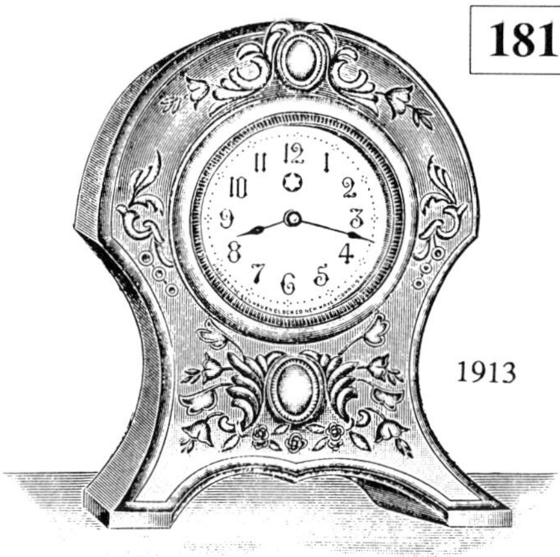

1913

CARLETON

LIST, $3.65

HEIGHT, 5 INCHES. WIDTH, 4 INCHES.

1819

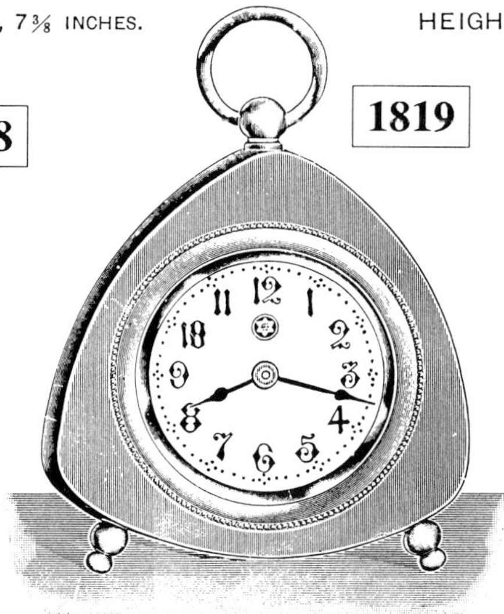

MACON

LIST, $3.65

HEIGHT, 4¼ INCHES.

WIDTH, 3⅛ INCHES.

1913

1820

1913

CALAIS

LIST, $3.65

HEIGHT, 5⅝ INCHES. WIDTH, 3⅝ INCHES.

1821

1913

CANNES

LIST, $3.95

HEIGHT, 5⅝ INCHES. WIDTH, 4½ INCHES.

1822

1913

CALIPH

LIST, $3.95

HEIGHT, 5 INCHES. WIDTH, 4 INCHES.

1823

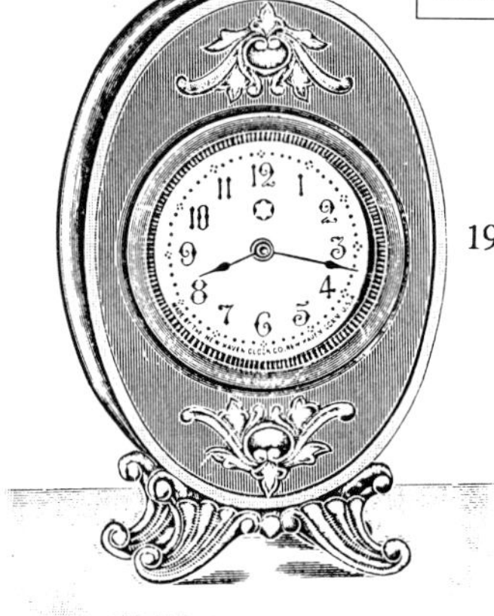

1913

COLON

LIST, $3.95

HEIGHT, 5¼ INCHES. WIDTH, 3¼ INCHES.

NOVELTY CLOCKS

ONE-DAY NOVELTY CLOCKS. TIMEPIECE.
2-INCH PORCELAIN DIAL. FRENCH BRONZE, OLD BRASS, GILT, SILVER, OR GUN METAL FINISH

1824

1913

PINDAR
LIST, $5.50
HEIGHT, 4 3/8 INCHES. WIDTH, 3 3/8 INCHES

1825

EBERLE
1913
LIST, $3.65
HEIGHT, 7 3/4 INCHES. WIDTH, 4 1/2 INCHES.

1826

1913

COLIMA
LIST, $3.95
HEIGHT, 6 INCHES. WIDTH, 4 1/8 INCHES.

FITTED WITH 1-DAY INTERMITTENT ALARM. EXTRA LIST, $1.50

1827

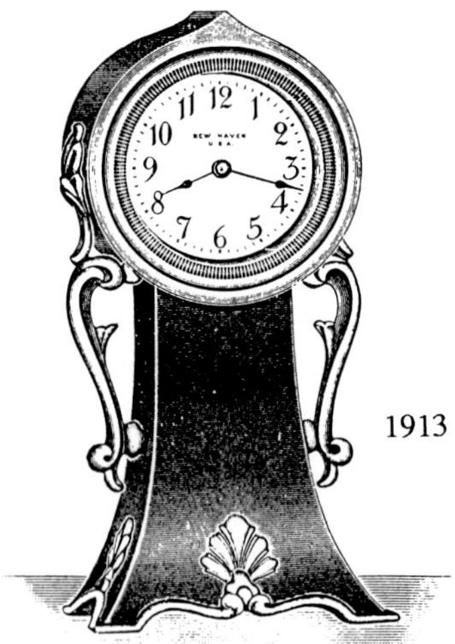

1913

ZINNIA
LIST, $4.35
HEIGHT, 5 7/8 INCHES. WIDTH, 3 INCHES.

1828

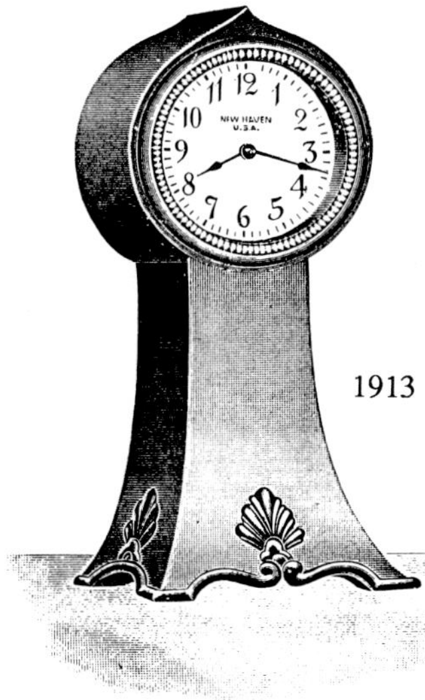

1913

ZALIA
LIST, $3.65
HEIGHT, 6 INCHES.
WIDTH, 3 INCHES.

1829

1913

PONCE
LIST, $5.50
HEIGHT, 4 3/8 INCHES. WIDTH, 3 3/8 INCHES.

NOVELTY CLOCKS

ONE-DAY TIMEPIECE NOVELTY CLOCKS. 2-INCH PORCELAIN DIALS.
ORMOLU GOLD PLATE FINISH. BEVELED GLASS

1830

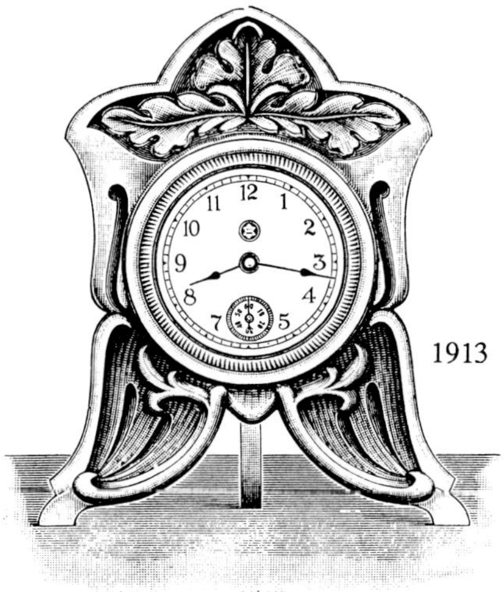

1913

DELONG
LIST, $2.20
HEIGHT, 5½ INCHES. WIDTH, 4½ INCHES.

1831

HAZEL 1913
LIST, $1.90
HEIGHT, 4⅝ INCHES. WIDTH, 3¼ INCHES.

1832

1913

ALSACE
LIST, $2.65
HEIGHT, 6 INCHES. WIDTH, 5¼ INCHES.

1833

1913

BERMUDA
LIST, $5.80
HEIGHT, 9¾ INCHES. WIDTH, 6½ INCHES.

1834

1913

NAVAN
LIST, $1.90
HEIGHT, 4¼ INCHES. WIDTH, 3¼ INCHES.

1835

1913

BARODA
LIST, $5.80
HEIGHT, 9¾ INCHES. WIDTH, 7 INCHES.

NOVELTY CLOCKS

ONE-DAY NOVELTY CLOCKS.
TIMEPIECE OR INTERMITTENT ALARM. 2-INCH PORCELAIN DIALS.
ORMOLU GOLD PLATE OR SATIN SILVER FINISH

WITH INTERMITTENT ALARM. EXTRA LIST, $1.50

1836

1913

RACHEL
LIST, $5.80
HEIGHT, 6¾ INCHES. WIDTH, 5¾ INCHES.

1837

HURON 1913
LIST, $1.90
HEIGHT, 4 INCHES. WIDTH, 4 INCHES.

1838

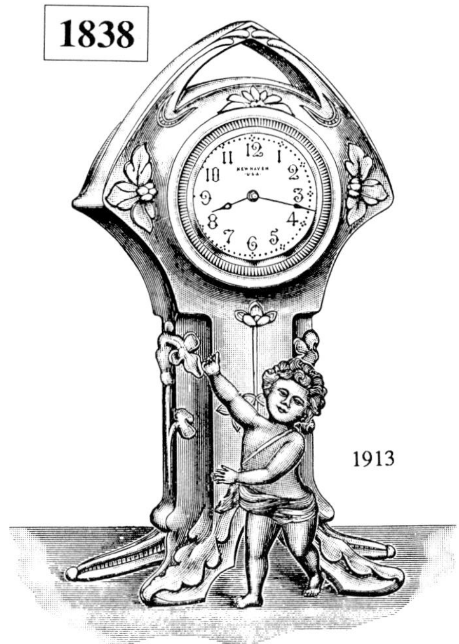

1913

DELANO
LIST, $4.35
HEIGHT, 8 INCHES. WIDTH, 4¾ INCHES.

1839

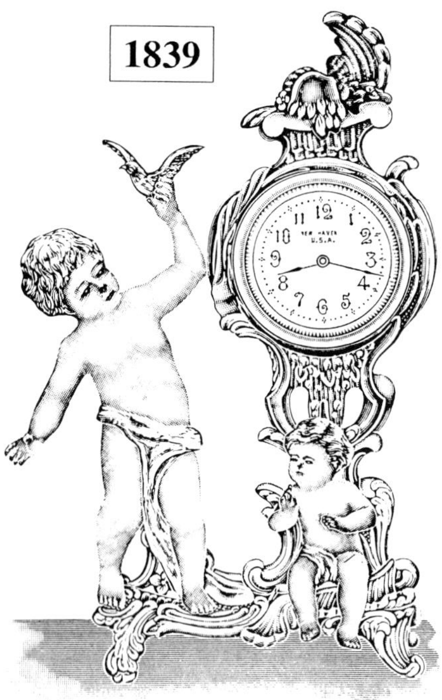

1914

YESNER
LIST, $5.40
HEIGHT, 9¼ INCHES. WIDTH, 6 INCHES.

1840

1906

VAUVERT

Height, 7¾ inches. Width, 4¾ inches.
Dial, 2 inches.
Heavy Imported Crystal Case, with Satin Gold Ornaments.
Porcelain Dial.

One-day Time . . . . List Price, $22.90

1841

1913

CARNOT
LIST, $5.10
HEIGHT, 10 INCHES. WIDTH, 6½ INCHES.

NOVELTY CLOCKS

ONE-DAY TIMEPIECE NOVELTY CLOCKS.
2-INCH PORCELAIN DIALS. ORMOLU GOLD PLATE FINISH.
BEVELED GLASS.

1842

1913

NATOMA
LIST, $4.35
HEIGHT, 6¾ INCHES. WIDTH, 5¾ INCHES.

1843

1913

DUPONT
LIST, $1.90
HEIGHT, 5⅞ INCHES.
WIDTH, 3⅜ INCHES.

1844

1913

DREW
LIST, $4.35
HEIGHT, 7 INCHES. WIDTH, 5 INCHES.

1845

1913

MINDEN
LIST, $5.10
HEIGHT, 8½ INCHES. WIDTH, 7½ INCHES.

1846

1913

NATHALIA
LIST, $1.90
HEIGHT, 5½ INCHES. WIDTH, 3½ INCHES.

1847

1913

DUNSTAN
LIST, $4.80
HEIGHT, 10½ INCHES. WIDTH, 5¾ INCHES.

NOVELTY CLOCKS

1848

1913

YANCOPIN

LIST, $3.95

WEIGHT, 6 INCHES. WIDTH, 6 INCHES.

1849

1913

YANTUS

LIST, $4.35

HEIGHT, 6½ INCHES. WIDTH, 6¾ INCHES.

Most white metal and brass ornamentals and clock cases are gilded or silver plated. Great care must be taken when cleaning these surfaces. The thin layer of gilding or silver plating will be easily removed by cleaning solutions and strong rubbing. If you believe cleaning is necessary, hire the services of a professional and avoid the expense of re-gilding or re-plating.

1850

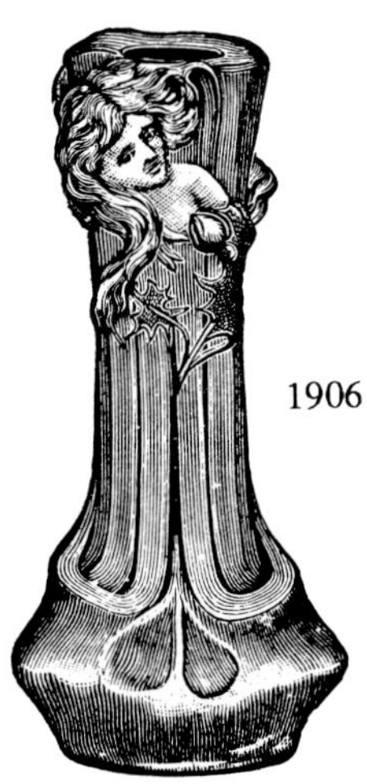

1906

1851

1906

1852

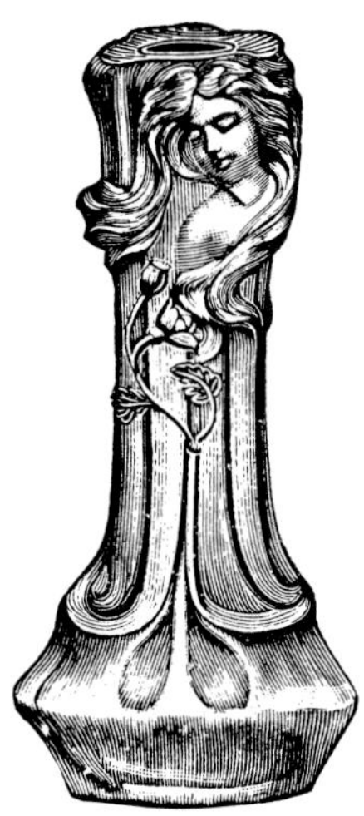

1906

VALETTA SET

Clock, Height, 10¾ inches. Width, 4¾ inches. Vases, Height, 7¾ inches.
2-inch Porcelain Dial, with beveled Glass. Ormolu Gold Plated Case, Cast Sash.
1-Day Time, Clock, List Price, $8.35. Vases per pair, List Price, 4.90. Set Complete, List Price, $13.25

1853

1900

TRALEE—Showing Brilliant Sash.
Height, 4½ inches. Width, 3 inches.
Dial, 2 inches.
French Satin Gold Finish.
Ivorized Porcelain Dial. One-day Time.

NOVELTY CLOCKS

ONE-DAY TIMEPIECE NOVELTY CLOCKS. 2-INCH PORCELAIN DIALS.
ORMOLU GOLD PLATE FINISH. BEVELED GLASS

1854

DAUPHIN 1913

LIST, $4.80

HEIGHT, 8¼ INCHES. WIDTH, 6¾ INCHES.

1855

1913

BENTON

LIST, $4.80

HEIGHT, 7 INCHES. WIDTH, 5 INCHES.

1856

1880

ACORN.

Nickel or Brass.

$2\frac{1}{2}$ inch Dial.
1 Day Time.
1 Day Time, Alarm.

1857

Ca. 1937

MONKEY

A novelty clock with 30-hour movement and cast metal case in silvered finish with black and silvered dial; or bronze finish with black and polished gilt dial. Height, 3¼ in. Width, 4 in.

1858

1907

FORLI

Height, 7½ inches. Width, 6½ inches.
Two-inch Porcelain Dial.
Ormolu Gold Plated Case.

One-day Time . . List Price, $4.90

1859

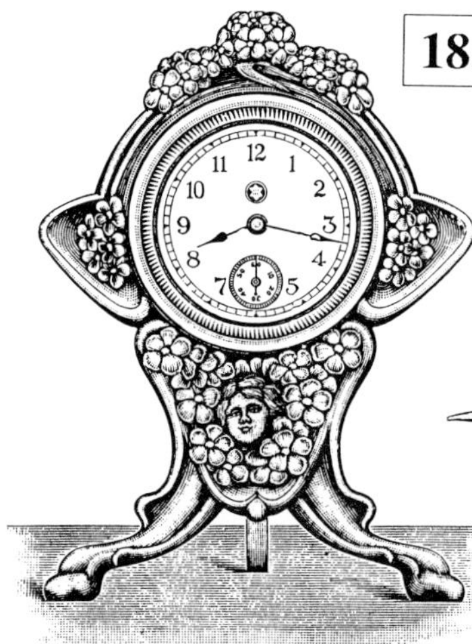

DIGBY 1913

LIST, $1.90

HEIGHT, 6⅛ INCHES. WIDTH, 4¼ INCHES.

1860

1913

DELAVAN

LIST, $4.80

HEIGHT, 7 INCHES. WIDTH, 7½ INCHES.

NOVELTY CLOCKS

ONE-DAY TIMEPIECE NOVELTY CLOCKS. 2-INCH PORCELAIN DIALS.
ORMOLU GOLD PLATE FINISH

1863

1913

NEVA
LIST, $2.65
HEIGHT, 5⅞ INCHES. WIDTH, 4⅝ INCHES.

1861

1913

AKRON
LIST, $4.80
HEIGHT, 8½ INCHES. WIDTH, 6 INCHES.

1862

1913

YAKIMA
LIST, $3.95
HEIGHT, 6¼ INCHES. WIDTH, 5¾ INCHES.

1864

NILE 1913
LIST, $2.65
HEIGHT, 5 INCHES. WIDTH, 4¼ INCHES.

1865

1886

TOBOGGAN.

SILVER PLATED AND OXIDIZED.

Height, 7 inches. Dial, 2 inches.

1 day Time.

Cut Pinion Movement.

1866

PIETY 1913
LIST, $5.80
HEIGHT, 5⅝ INCHES. WIDTH, 6 INCHES.

NOVELTY CLOCKS

1867

DINSMORE 1913
LIST, $4.35
HEIGHT 8 INCHES. WIDTH, 7 INCHES.

1868

NOME 1913
LIST, $2.65
HEIGHT, 5½ INCHES. WIDTH, 5 INCHES.

1869

1913

DORIS
LIST, $3.65
HEIGHT, 5¾ INCHES. WIDTH, 5 INCHES.

1870

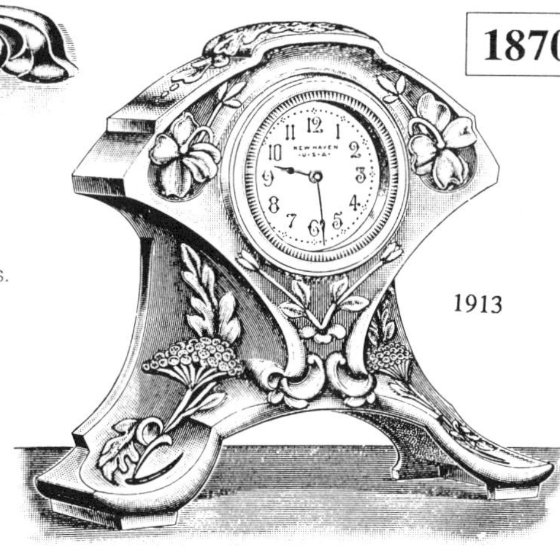

1913

ROSITA
LIST, $4.65
HEIGHT, 6 INCHES. WIDTH, 6¼ INCHES.

1871

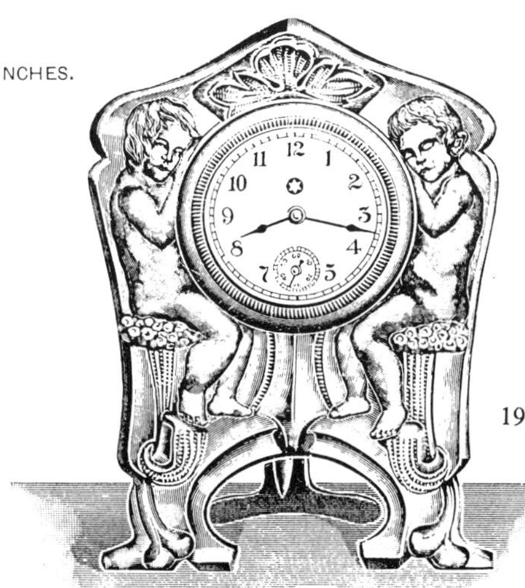

1913

NAVINA
LIST, $2.20
HEIGHT, 5¾ INCHES. WIDTH, 4¼ INCHES.

1872

DRESDEN 1913
LIST, $3.65
HEIGHT, 5¾ INCHES. WIDTH, 4½ INCHES.

1873

NARUNA 1913
LIST, $4.35
HEIGHT, 6¾ INCHES. WIDTH, 5¾ INCHES.

NOVELTY CLOCKS

1874

1913

EDISTO
LIST, $3.65
HEIGHT, 6 INCHES. WIDTH, 4¾ INCHES.

1875

1913

NASHOBA
LIST, $2.20
HEIGHT, 7 INCHES. WIDTH, 4½ INCHES.

1876

1913

NAOLA
LIST, $3.65
HEIGHT, 5¾ INCHES. WIDTH, 5½ INCHES.

1877

1913

YAMASKA
LIST, $3.65
HEIGHT, 6¼ INCHES. WIDTH, 5½ INCHES.

1878

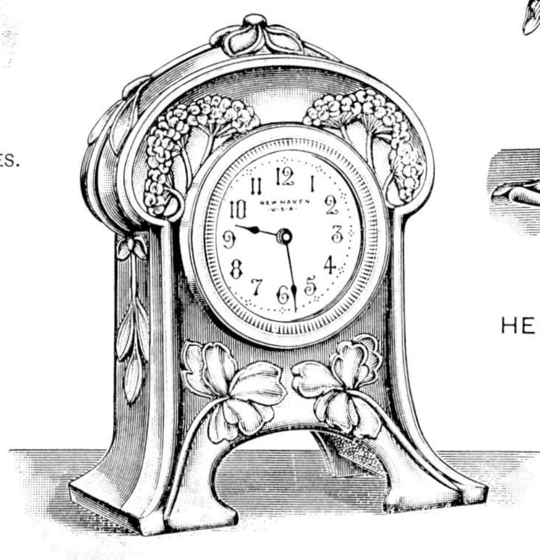

ROLAND 1913
LIST, $3.95
HEIGHT, 5¼ INCHES. WIDTH, 4¼ INCHES.

1879

DULUTH 1913
LIST, $3.65
HEIGHT, 5¾ INCHES. WIDTH, 4½ INCHES.

1880

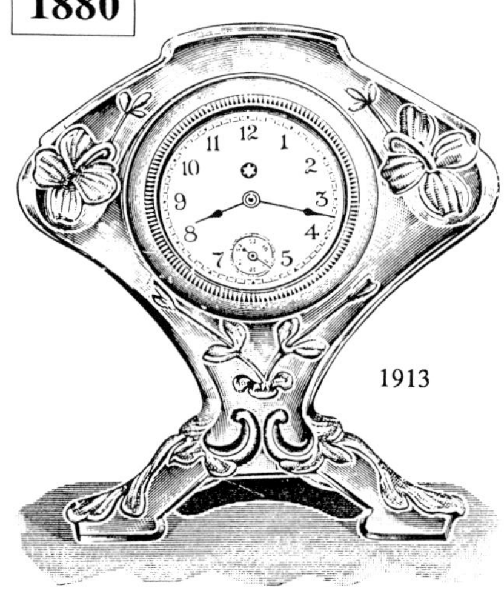

1913

YARNABY
LIST, $2.20
HEIGHT, 5¼ INCHES. WIDTH, 4¾ INCHES.

NOVELTY CLOCKS

EIGHT-DAY NOVELTY CLOCKS.
FINE QUICK-BEAT TIMEPIECES. 3-INCH PORCELAIN DIALS.
ORMOLU GOLD PLATE FINISH. BEVELED GLASS

1881

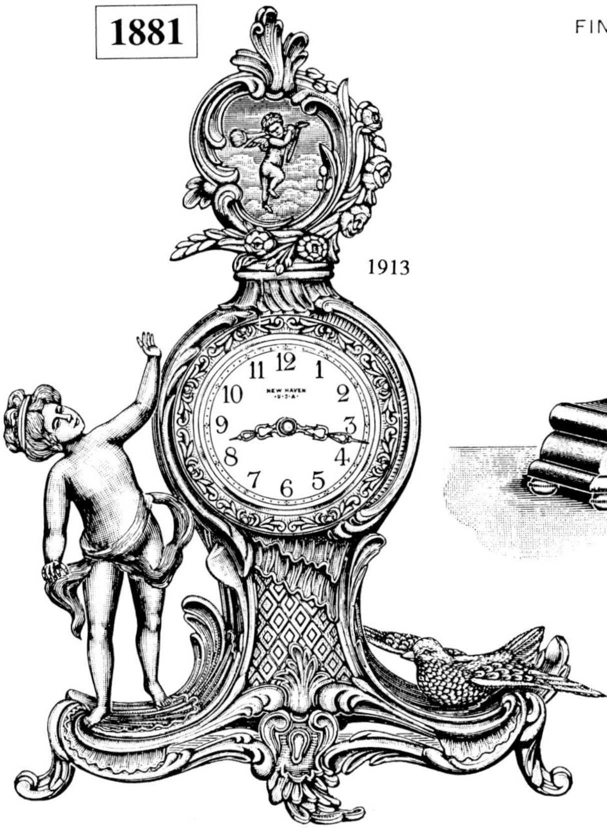

1913

HENLEY
LIST, $10.90
HEIGHT, 11¾ INCHES. WIDTH, 8¾ INCHES.
DECORATED PORCELAIN PANEL.

1882

ROSS
LIST, $9.45
HEIGHT, 4½ INCHES. WIDTH, 8½ INCHES.

1913

1883

1886

ARCHER.
SILVER PLATED AND
OXIDIZED.
1 day Repeating Strike.

1884

1913

HARLEM
LIST, $10.90
HEIGHT, 10¾ INCHES. WIDTH, 6¼ INCHES.

NOVELTY CLOCKS

EIGHT-DAY NOVELTY CLOCKS.
FINE QUICK-BEAT TIMEPIECES. 3-INCH PORCELAIN DIALS.
ORMOLU GOLD PLATE FINISH. BEVELED GLASS

1885

ROVENA 1913

LIST, $6.50

HEIGHT, 4 INCHES. WIDTH, 3½ INCHES.

1886

1913

BERBICE

LIST, $10.15

HEIGHT, 11 INCHES. WIDTH, 6½ INCHES.

1887

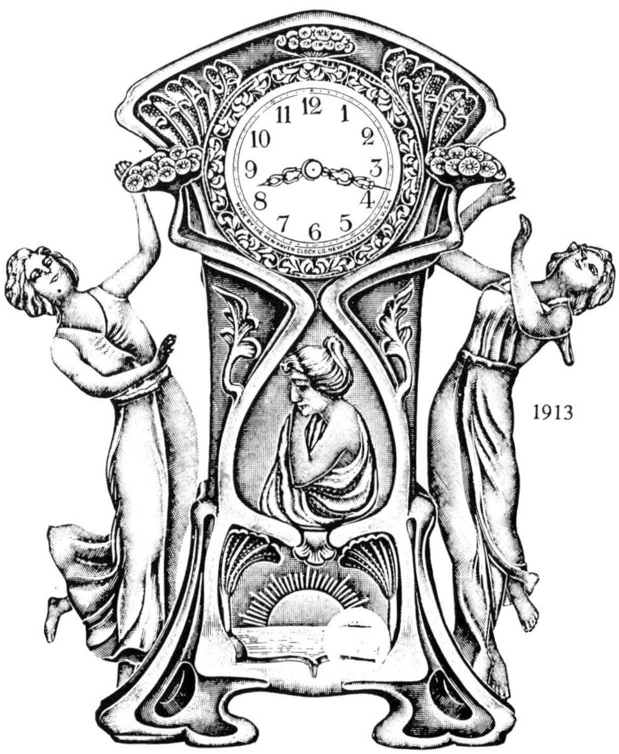

1913

HARWICH

LIST, $11.60

HEIGHT, 10¾ INCHES. WIDTH, 9½ INCHES.

1888

1913

HALSEY

LIST, $10.15

HEIGHT, 9⅝ INCHES. WIDTH, 7⅝ INCHES.

NOVELTY CLOCKS

1889

1913

UTICA

LIST, $7.25

HEIGHT, 6 1/4 INCHES. WIDTH, 4 7/8 INCHES.
DIAL, 3 INCHES.
CASE, SOLID BRASS, FANCY FIGURED.
FINISH, GILT OR SILVER.

1890

UNCAS

LIST, $10.15

HEIGHT, 6 INCHES. WIDTH, 5 7/8 INCHES.
DIAL, 3 1/2 INCHES.
FINISHES, FRENCH BRONZE, OLD BRASS, GILT, SILVER, OR GUN METAL.

1891

1913

ROMITA

LIST, $7.25

HEIGHT, 4 INCHES.

WIDTH, 3 1/2 INCHES.

1892

1913

BEDFORD

LIST, $10.15

HEIGHT, 10 INCHES. WIDTH, 8 3/8 INCHES.

1893

1913

HADLEY

LIST, $10.15

HEIGHT, 9 7/8 INCHES. WIDTH, 9 3/8 INCHES.

NOVELTY CLOCKS

1894

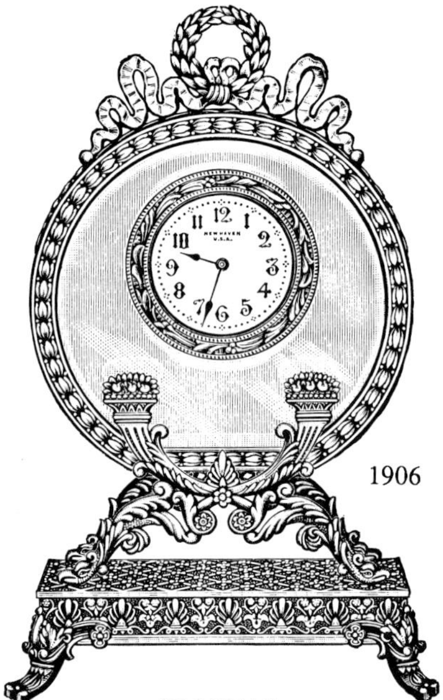

1906

VALREAS

Height, 9 ¼ inches. Width, 5 ¾ inches. Dial, 2 inches. Heavy Imported Crystal Case, with Satin Gold Ornaments. Porcelain Dial. One Day Time. List price, $18.75.

1895

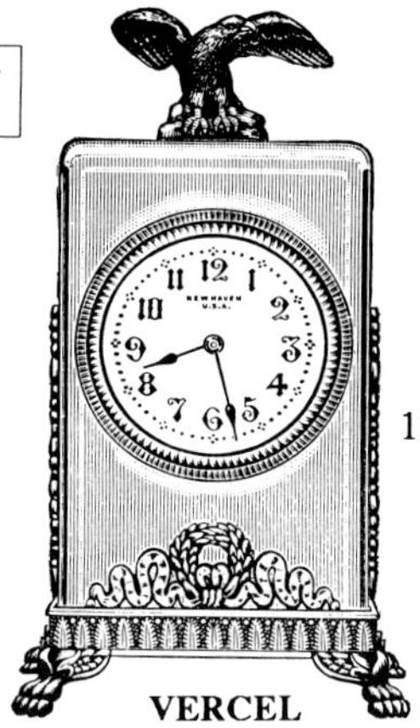

1906

VERCEL

Height, 6 ¼ inches. Width, 3 ½ inches. Dial, 2 inches. Heavy Imported Crystal Case, with Satin Gold Ornaments. Porcelain Dial. One Day Time. List price, $13.90.

1896

1907

ANDORRA

Height, 10 ¼ inches. Width, 3 ¾ inches. Two-inch Porcelain Dial. Ormolu Gold Plated Case. One Day Time. List price, $4.25

1897

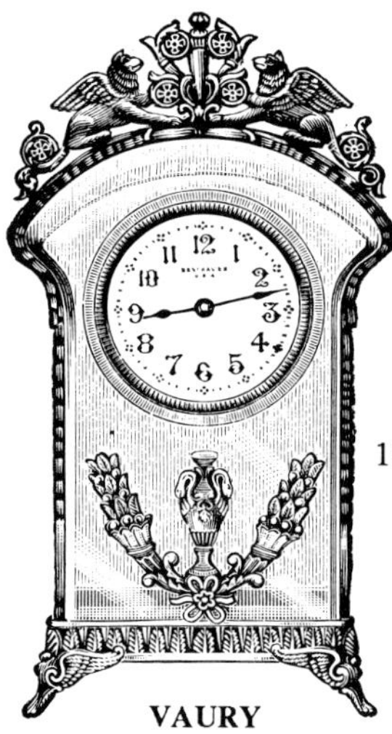

1906

VAURY

Height, 7 ⅜ inches. Width, 3 ¾ inches. Dial, 2 inches. Heavy Imported Crystal Case, with Satin Gold Ornaments. Porcelain Dial. One Day Time. List price, $18.10.

1898

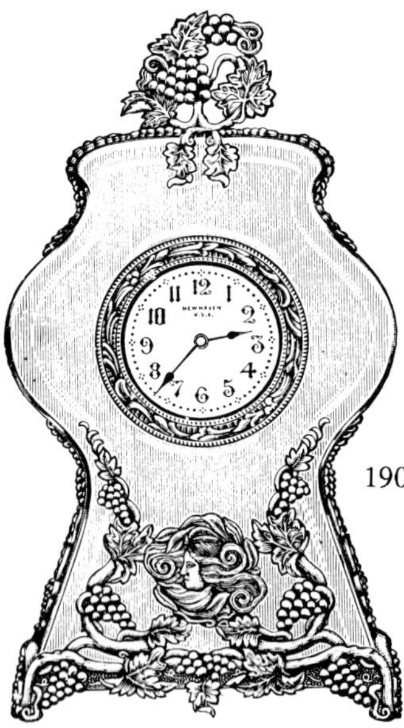

1906

VAYRAC

Height, 8 ½ inches. Width, 4 ¾ inches. Dial, 2 inches. Heavy Imported Crystal Case, with Satin Gold Ornaments. Porcelain Dial. One Day Time. List price, $26.40.

1899

1907

LISBON

Height, 6 inches. Width, 6 ½ inches. Two-inch White Dial. Ormolu Gold Plated Case. One Day Time. List price, $2.50.

1900

1907

ARDENNES

Height, 11 ½ inches. Width, 4 ¼ inches. Two-inch Porcelain Dial. Ormolu Gold Plated Case. One Day Time. List price, $4.25

NOVELTY CLOCKS

1901

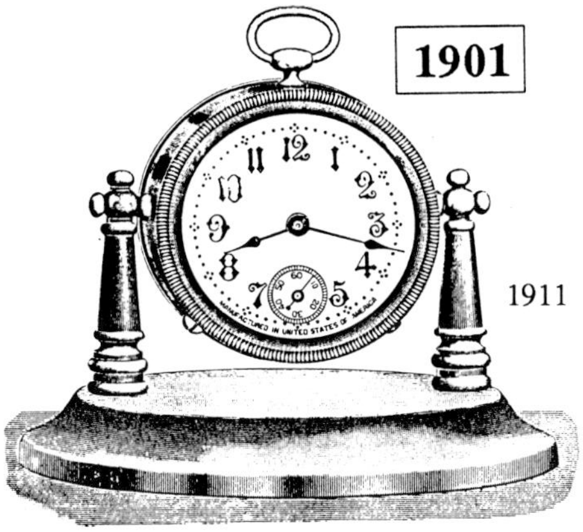

1911

INCA

HEIGHT ....3 ⅞ INCHES
WIDTH ....4 ⅜ INCHES
DIAL, IVORINE ....2 INCHES
CASE, SOLID BRASS
FINISH, POLISHED GILT
MOVEMENT, 1 DAY TIME PIECE
PRICE ....LIST, $3.C5

1902

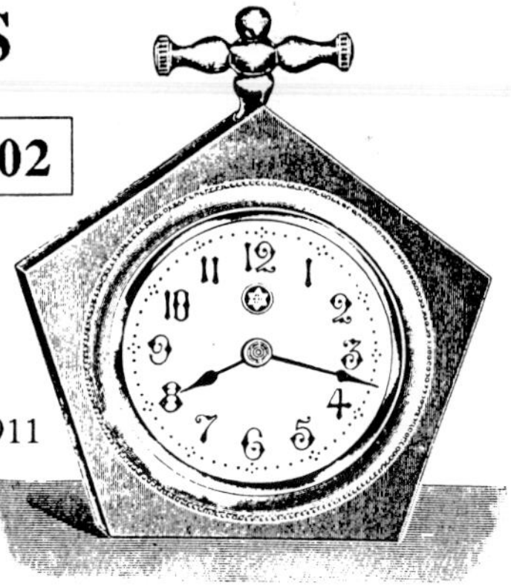

1911

MARLIN

MARLIN

HEIGHT ....3 ⅝ INCHES
WIDTH ....3 ⅛ INCHES
DIAL, PORCELAIN ....2 INCHES
CASE, SOLID BRASS
FINISH, GILT, SILVER OR GUN METAL
PRICE, 1 DAY TIME PIECE ...LIST, $3.65
PRICE, WITH 1 DAY INTERMITTENT ALARM ....LIST EXTRA, 1.50

1903

Ca. 1937

GNOME

A novelty clock with 30-hour movement and cast metal case in silvered finish with black and silvered dial; or bronze finish with black and polished gilt dial. **Height,** 4 inches. Width, 4⅛ inches.

1904

1913

ARIEL

LIST, $5.80

HEIGHT, 8 INCHES. WIDTH, 4½ INCHES.

1905

1900

ARIEL

French Satin Gold or Bronze Finish.
Height, 8 inches. Width, 4½ inches.
One-day Time, with 2-inch Ivorized Porcelain Dial.

List price, Satin Gold, $7.00
" " Bronze, 6.30

1906

1907

AMORET

Height, 6¾ inches. Width, 3½ inches.
2-in. Porcelain Dial.
Ormolu Gold Plated Case.
One-day Time . . . . List Price, $3.75

1907

1878

" WEE WAG."

No. 100.

Black or Nickel-plated, with Burnished Brass Trimmings.
1 Day. Time. Spring.
Hight, 5 3-4 inches.

1908

1878

" WEE WAG."

No. 101.

Black or Nickel-plated, with Real Bronze Trimmings.
1 Day. Time. Spring.
Hight, 6 1-4 inches.

Above also furnished with Lever Movements, with French Method of Winding, Setting, and Regulating at the Back.

NOVELTY CLOCKS

MECHANICAL CLOCKS
MOVABLE DIALS.

1909

1890

MANDOLIN - Nickel
Dial, 4 inches.
1 Day Time...$2.30
1 Day Time, Alarm..$2.80

1910

1890

SHAVER - Nickel
Dial, 4 inches.
1 Day Time...$2.30
1 Day Time, Alarm..$2.80

1911

1890

BULL FIGHT - Nickel
Dial, 4 inches.
1 Day Time...$2.30
1 Day Time, Alarm..$2.80

1912

1890

PET - Nickel
1 Day Time...$2.60

1913

1890

PUFF - Nickel
1 Day Time...$2.60

NOVELTY CLOCKS - WOOD CASES

The year entered beside each clock is the year of the catalog from which the illustration was taken. Some clocks may have been produced a few years earlier.

1915

1885

No. 1011, No. 2.

OAK, WITH BRASS HOOPS.

(Mud) 1 Day, Time.........$5 35

1914

1886

CIGAR LIGHTER.

1 Day Time.........$9.50

1916

1885

No. 1011.

OAK WITH BRASS HOOPS.

Height, 6 inches. Dial, 2½ inches.

Cut pinion movement.

1 Day Time $4 80

NOVELTY CLOCKS - WOOD CASES

SOLID MAHOGANY CLOCKS.

8 Day Lever Time Movements. 2 ¼ inch Porcelain Dials. Beveled Bowed Glass. Can be furnished in Mahogany, Brown Mahogany, Circassian Walnut Finish, Golden and Fumed Oak, at same list.

1917

1917

HARBIN
Height, 4¼ Inches Width, 3¼ Inches
Also fitted 1-Day Intermittent Alarm
at same list

1918

1917

HARVEY
Height, 4 Inches Width, 6⅞ Inches
Also fitted 1-Day Intermittent Alarm
at same list

1919

1917

HARPER
Height, 4½ Inches Width, 8 Inches
Also fitted 1-Day Intermittent Alarm
at same list

1920

1917

HALSTEAD
Height, 4 Inches Width, 8¼ Inches
Also fitted 1-Day Intermittent Alarm
at same list

1921

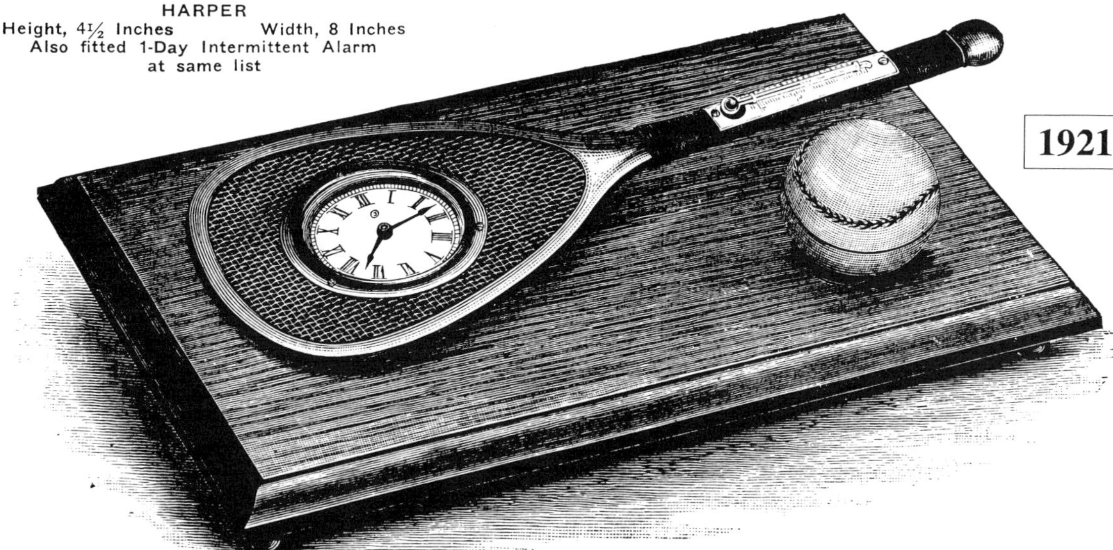

1889

THE TENNIS INK SET.

GILT MOUNTED ON ANTIQUE OAK BASE.

Racquet containing Clock and Thermometer, and forming a Paper Clip.
Tennis Ball containing Porcelain Ink Well. Size of base, 9¾x6¼ inches.

1 Day Time.................. $9.50

NOVELTY CLOCKS - WOOD CASES

The year entered beside each clock is the year of the catalog from which the illustration was taken. Some clocks may have been produced a few years earlier.

Mahogany Finish

1922

1917

BARTO
1-Day Time, White Dial
Height, 4¼ Inches Width, 3¼ Inches
Dial, 2 Inches

1923

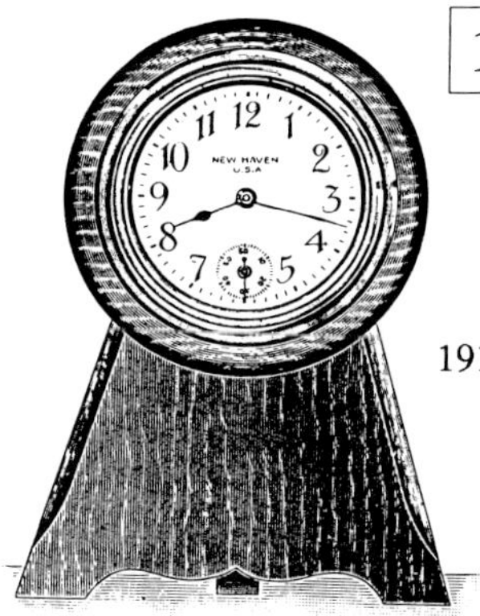

1917

BAGLEY
1-Day Time, White Dial
Height, 4⅝ Inches Width, 3½ Inches
Dial, 2 Inches

1924

1917

BATON
1-Day Time, White Dial
Height, 4¼ Inches Width, 3⅛ Inches
Dial, 2 Inches

1925

1917

BOYCE
1-Day Time
Height, 3¾ Inches Width, 6⅞ Inches
Dial, 2 Inches

1926

1917

ADRIAN
1-Day Time
Height, 3¾ Inches Width, 6½ Inches
Dial, 1¾ Inches

1927

1917

TAPIS
1-Day Time
Height, 4¼ Inches Width, 3⅛ Inches
Dial, 1½ Inches
For Desk or Wall Paper Clip

1928

1917

DATUS
1-Day Time
Height, 3⅝ Inches Width, 5¾ Inches
Dial, Porcelain, 2 Inches

1929

1917

BAXTER
1-Day Time
Height, 4⅛ Inches Width, 3¾ Inches
Dial, 2 Inches

NOVELTY CLOCKS - WOOD CASES

SOLID MAHOGANY CLOCKS

8-Day Lever Time Movements. 2¼-inch Porcelain Dials. Beveled Bowed Glass.

Silver Dials furnished on 8-Day Time only, at Extra List

1930

1917

HARLEE
Height, 4¾ Inches Width, 3¼ Inches
Inlaid Case
Also fitted 1-Day Intermittent Alarm
at same list

1931

1917

HODGES
Height, 4 Inches Width, 7½ Inches
Also fitted 1-Day Intermittent Alarm
at same list

1932

1917

HALE
Height, 4¾ Inches Width, 8⅜ Inches
Also fitted 1-Day Intermittent Alarm
at same list

1933

1917

RADWAY
Height, 4½ Inches Width, 6½ Inches
Also fitted 1-Day Intermittent Alarm
at same list

1934

1917

HARDY
Height, 4 Inches Width, 3¾ Inches
Also fitted 1-Day Intermittent Alarm
at same list

1935

1917

QUAKER
Height, - - - - - - 6¼ Inches
Width, - - - - - - 11 Inches

NOVELTY CLOCKS - WOOD CASES

SOLID MAHOGANY CLOCKS

8-Day Lever Time Movements. 3½-inch Oval Porcelain Dials Beveled Glass

1936

1917

OVAL No. 53
Height, 6¼ Inches Width, 12 Inches

1937

1917

OVAL No. 52
Height, 6¾ Inches Width, 7¼ Inches

1938

1917

QUEDA
Height, - - - - - - 6½ Inches
Width, - - - - - - 11¾ Inches

1939

1917

JACOBEAN No. 55
Height, - - - - - - 6⅛ Inches
Width, - - - - - - 13½ Inches

1940

1917

ADAM No. 52
Height, - - - - - - 6½ Inches
Width, - - - - - - 4¾ Inches
Case, - - - - - Marquetry Inlaid

1941

1917

JACOBEAN No. 53
Height, - - - - - - 7¼ Inches
Width, - - - - - - 5½ Inches

1942

1917

WESLEY
Height, - - - - - - 6¾ Inches
Width, - - - - - - 5 Inches

NOVELTY CLOCKS - WOOD CASES

1943

Ca. 1937

PEQUOT

An attractive desk clock with a 30-hour movement visible through a transparent case. Available in Clear or Green with black and silvered dial; and in Amber with black and polished gilt dial. Height, 2⅞ inches.

1944

1907

VAUX

Height, 4¾ inches. Width, 4¾ inches.
Two-inch Porcelain Dial, with Beveled Glass.
Solid Mahogany Case, with Brass Feet.

One-day Time . . . . List Price, $2.80

1945

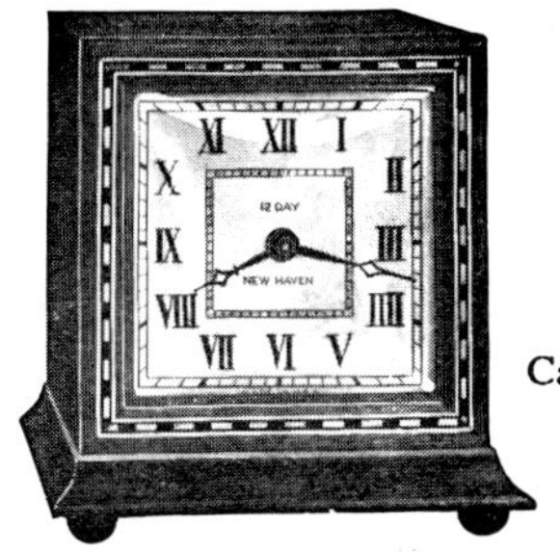

Ca. 1937

HOBBY

8 Day movement, 2½-inch silvered dial, mahogany case, with decorative face. Height, 3¼ inches. Width, 3½ inches.

EIGHT-DAY NOVELTY CLOCKS.
FINE, QUICK-BEAT TIMEPIECES. 2¼-INCH PORCELAIN DIAL.
FRENCH CAST SASH WITH BEVELED, BOWED GLASS.
SOLID MAHOGANY CASES

1946

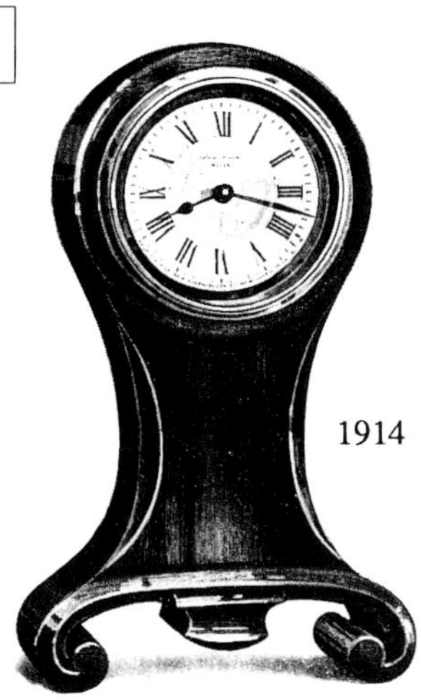

1914

HORTON
LIST, $7.25
HEIGHT, 6½ INCHES. WIDTH, 3¾ INCHES.

1947

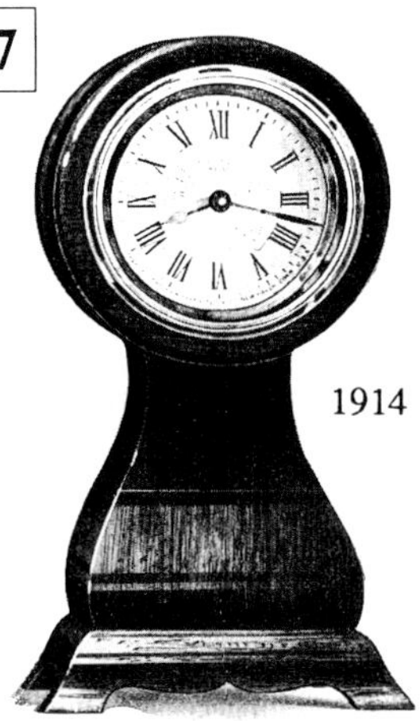

1914

HILTON
LIST, $7.25
HEIGHT, 6¼ INCHES. WIDTH, 3⅜ INCHES.

1948

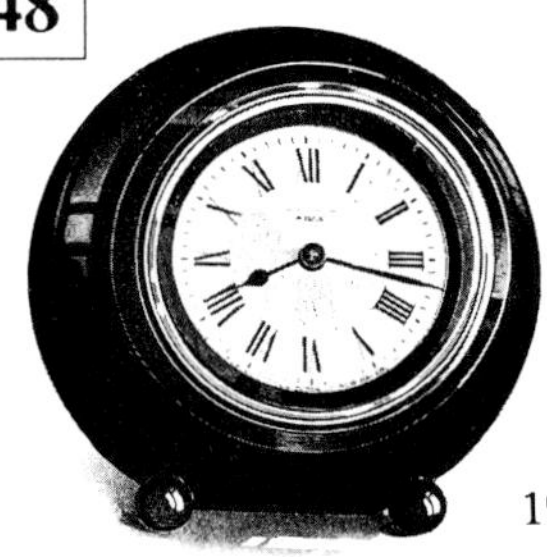

1914

QUIZ
LIST, $6.50
HEIGHT, 3¾ INCHES. WIDTH, 4 INCHES.

1949

1914

EDINA
LIST, $5.80
HEIGHT 6¼ INCHES. WIDTH, 4½ INCHES.
INLAID MARQUETRY DECORATION.

1950

1914

HOBSON
LIST, $7.25
HEIGHT 6½ INCHES. WIDTH, 3¾ INCHES.

1951

1914

HELMAR
LIST, $7.25
HEIGHT, 6⅞ INCHES. WIDTH, 3½ INCHES.

NOVELTY CLOCKS - WOOD CASES

Movements Are One Day Alarm. Rings Intermittently Every Twenty Seconds for Eight Minutes.
Has Silent Switch. Cases are Beautifully Finished in Mahogany.

1952

1917

ISON
1-Day Time
1-Day Intermittent Alarm
Height, 4 Inches Width, 8 Inches
Dial, Porcelain, 2 Inches

1953

1917

IVES
1-Day Time
1-Day Intermittent Alarm
Height, 3¼ Inches Width, 6⅛ Inches
Dial, Porcelain, 2 Inches

Above Clocks also furnished in Old Ivory Enamel Finish at Extra List

1954

1917

GREEN

Height,	- - - - - -	6¾ Inches
Width,	- - - - - -	7¾ Inches
Dial,	- - - - - - -	4⅛ Inches

1955

1917

GLADSTONE

Height,	- - - - - -	6¾ Inches
Width,	- - - - - -	13¾ Inches
Dial,	- - - - - - -	4⅛ Inches

1956

1917

GARFIELD

Height,	- - - - - -	6½ Inches
Width,	- - - - - -	13 Inches
Dial,	- - - - - - -	4⅛ Inches

1957

1917

GARCIA

Height,	- - - - - -	6¼ Inches
Width,	- - - - - -	13 Inches
Dial,	- - - - - - -	4⅛ Inches

NOVELTY CLOCKS - WOOD CASES

1958

IVY 1917
1-Day Time
1-Day Intermittent Alarm
Height, 4¼ Inches Width, 3½ Inches
Dial, Porcelain, 2 Inches

1959

IDEAL 1917
1-Day Time
1-Day Intermittent Alarm
Height, 4¼ Inches Width, 3¼ Inches
Dial, Porcelain, 2 Inches

1960

IONIA 1917
1-Day Time
1-Day Intermittent Alarm
Height, 3¼ Inches Width, 3¼ Inches
Dial, Porcelain, 2 Inches

1961

SELDEN 1911
HEIGHT ....4½ INCHES
WIDTH ....3½ INCHES
DIAL, PORCELAIN 2 INCHES
CASE, SOLID MAHOGANY
MOVEMENT, 1 DAY TIME PIECE
PRICE ....LIST, $2.90

1962

CASTRO 1913
ONE-DAY ALARM. LIST, $2.65
HEIGHT, 6½ INCHES. WIDTH, 6½ INCHES.
DIAL, FANCY, TWO-COLOR, 3¾ INCHES.

1963

1913

NEMO
ONE-DAY INTERMITTENT ALARM. LIST, $3.65
HEIGHT, 4¾ INCHES. WIDTH, 4¼ INCHES.
DIAL, WHITE, 2 INCHES.

1964

1911

EUNICE
HEIGHT ....6½ INCHES
WIDTH..................4½ INCHES
DIAL, PORCELAIN ......2 INCHES
CASE, SOLID MAHOGANY WITH INLAID MARQUETRY DECORATIONS
PRICE, 1 DAY TIME PIECE..........................LIST, $4.35
PRICE, 8 DAY TIME PIECE ....LIST EXTRA, 1.50
PRICE, WITH 1 DAY INTERMITTENT ALARM MOVEMENT ....LIST EXTRA, 1.50

1965

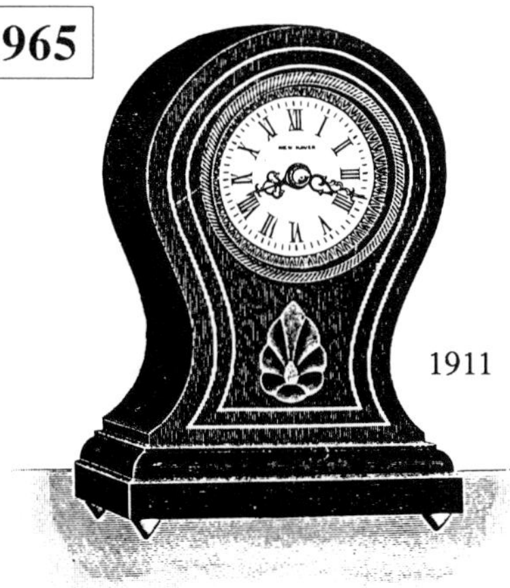

1911

EUNICE
(SHOWING 8 DAY FITTING)
HAS PORCELAIN DIAL, ROMAN NUMERALS, FANCY BRASS HANDS, FRENCH CAST SASH WITH HEAVY BEVELED BOWED GLASS

1966

NEGUS 1911
HEIGHT ....4¾ INCHES
WIDTH ....4½ INCHES
DIAL, WHITE ....2 INCHES
CASE, SOLID OAK
FINISH, FLEMISH
MOVEMENT, 1 DAY TIME PIECE
PRICE ....LIST, $2.20

PLUSH CLOCKS

1967

1886

SUN FLOWER, Plush.

PLUSH AND BRONZE, ASSORTED COLORS.

1 Day Time.............$5.50

1969

1885

SNUG.

PLUSH, LEATHER, MAHOGANY

OR OAK.

Height, 5 inches. Dial, 3 inches.

Code Word.		Price.
(Rug)	1 Day Time, Plush.............	$4 00
(Ruby)	1 " " Leather............	3 65
(Sandwich)	1 " " Wood.............	3 65

1968

1885

OWL.

Height, 13 inches. Dial, 3 inches.

Code Word.		Price.
(Gymnast)	1 Day Time.............	$29 95

Beautifully Carved in Wood, with Plush Base.

1970

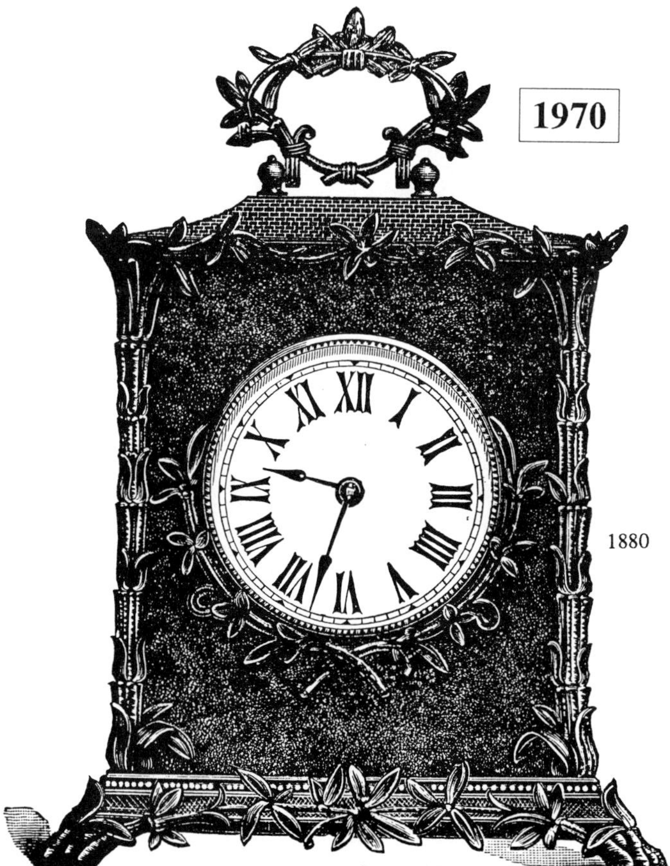

1880

ROB ROY.

BRONZE AND PLUSH CARRIAGE.

Height, 9½ inches. Dial, 3½ inches.
1 Day Time, Alarm. 8 Day Time. 1 Day Half-hour Strike.

This Clock is bronze frame, finished in the handsomest manner, with silk plush (any color) front, sides, and back.

PLUSH CLOCKS

1971

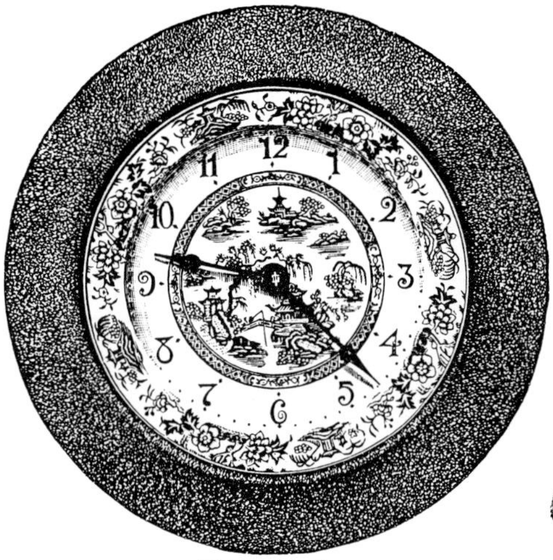

1885

Height, 10 inches.

Front View.

1972

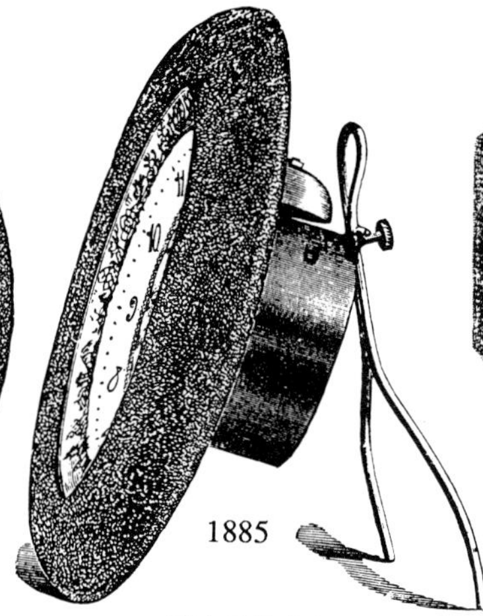

1885

Dial, 5½ inches.

Side View.

ZEPHYR.

PLAQUE, WITH PLUSH BORDER.

1 Day Time, Alarm.... ....$8 25

1 " Strike, Cut Pinion Movement............ 9 00

1973

1885

FLASH.

ASSORTED PLUSH CASES.

Height, 6 inches. Dial, 3 inches.

Code Word.		Price.
(Rule)	1 Day Time..................	$4 15
(Rum)	1 " " Alarm.....	4 95

1974

1886

CHALET.

PLUSH, ASSORTED COLORS.

1 Day Time, Alarm..... ....$5.25

1975

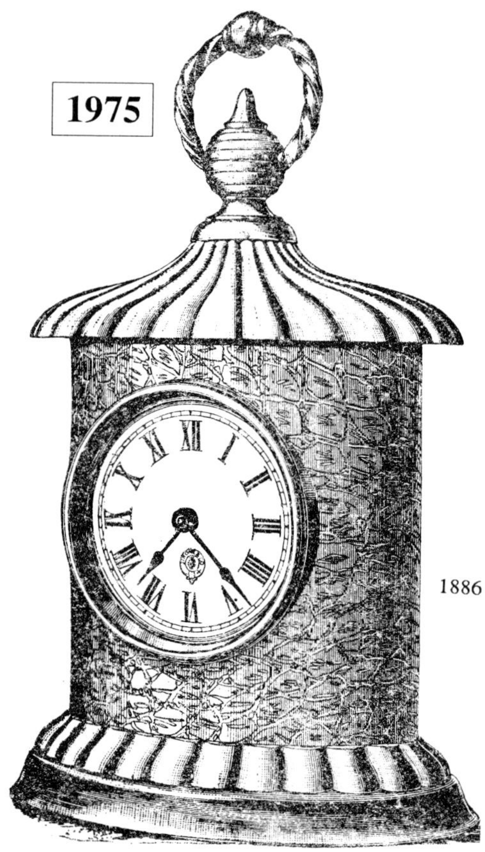

1886

PUCK

Height, 9 inches. Dial 3 inches.

1 Day Strike, Plush........... List, $6.40

1 Day Strike, Alarm, Plush. List, $6.75

PLUSH CLOCKS

1976

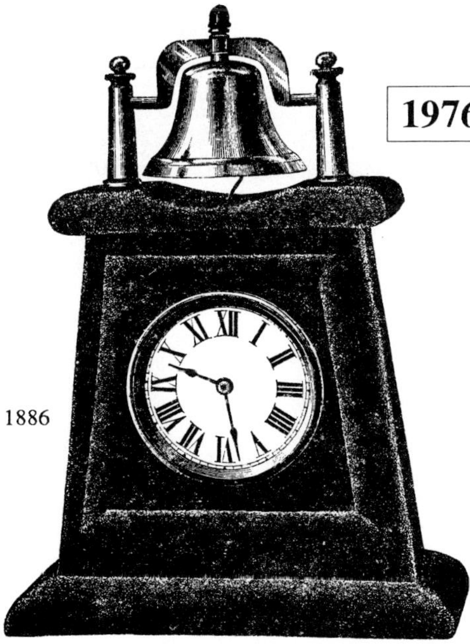

1886

TYRRILL.

PLUSH, ASSORTED COLORS.

Code Word.	Price.
(Abbott) 1 Day Time, Alarm	$6.00
(Abdal) 1 " Strike	6.00

1977

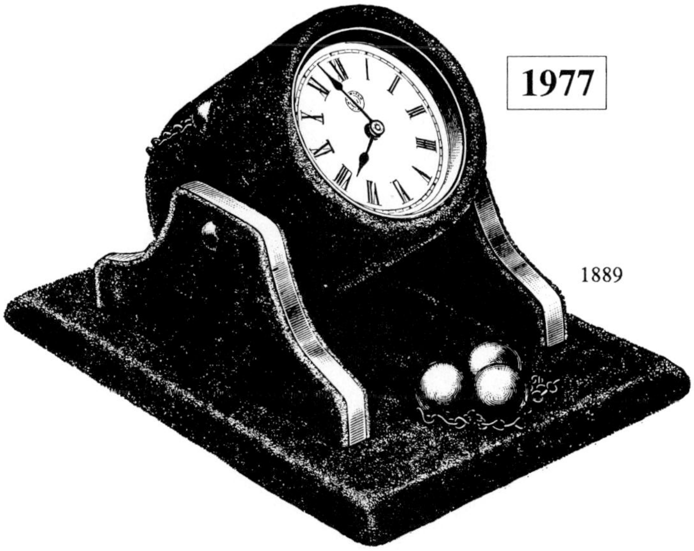

1889

MORTAR.

PLUSH, ASSORTED COLORS.

Code Word.	Price.
(Abasis) 1 Day Time	$5.00

1978

1886

PLUSH TRAVELING CASE.

STING STRIKE AND ALARM.

Code Word.	Price.
(Airy) 1 Day Repeating Strike	$6.00
(Alar) 1 " Alarm	5.00

1979

1889

STING (Octagon).

PLUSH, ASSORTED COLORS.

Code Word.	Price.
(Abased) 1 Day Time	$4.50

PLUSH CLOCKS

To receive top prices, plush clocks must have plush that still retains its nap and vivid color.

1980

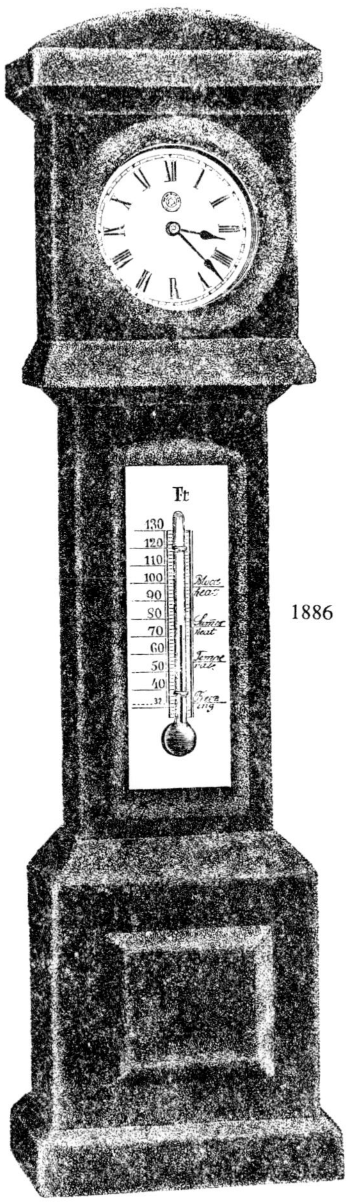

1886

GRANDFATHER.

1 Day Time $10.00

1981

1885

THE LARK. (Alarm.)

PLUSH AND FANCY BRASS.

1 Day Time, Alarm, Extra Brass.....$6 05
1 " " " Plush 6 55

1982

1885

THE LARK.

PORCELAIN, PLUSH AND FANCY BRASS.

1 Day Time, Plush.... .. ..$5 45
1 " " Porcelain.......... 4 95
1 " " Extra Brass........ 6 45

PLUSH CLOCKS

This Clock is made in the showiest possible manner, consisting of solid Gilt Frame, French Sash and Porcelain Dial, Silk Plush sides and front, of any color, and fitted with the same movement and Cathedral Gong as the Ivanhoe.

1983

1883

To receive top prices, plush clocks must have plush that still retains its nap and vivid color.

The year entered beside each clock is the year of the catalog from which the illustration was taken. Some clocks may have been produced a few years earlier.

LAMMERMOOR.

BRONZE AND PLUSH.

Height, 12½ inches. Dial, 5 inches.
8 Day Half-hour Strike on Cathedral Gong, Plain Dial.
8 Day Half-hour Strike on Cathedral Gong, Visible Escapement.

1985

1883

ALBUM,

BRONZE, METAL OR PLUSH.

Height, 5¼ inches.

1 Day Time, Bronze................$7 55
1 " " Plush...... .... 7 55

1984

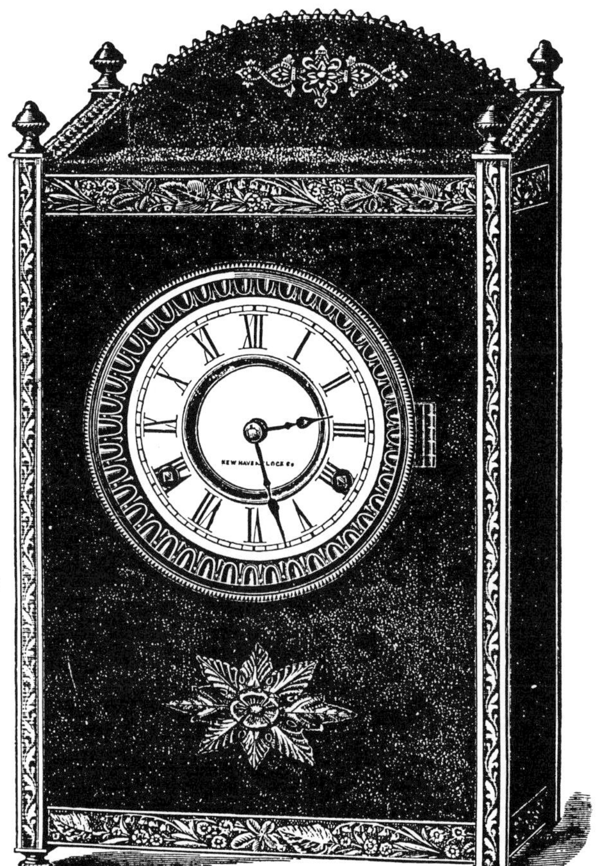

1883

MIDLOTHIAN.

GILT AND PLUSH.

Height, 15 inches. Dial, 5 inches.
8 Day Half-hour Strike on Cathedral Gong.

This Clock is made in the showiest possible manner, consisting of solid Gilt Frame, French Sash and Porcelain Dial, Silk Plush sides and front, of any color, and fitted with the same movement and Cathedral Gong as the Ivanhoe.

PORCELAIN CLOCKS

Other New Haven models with imported majolica, jasperware and porcelain cases are known which were not illustrated in catalogs.

1986

1895

WINDSOR

Height, 8 ¼ inches. Width, 7 ½ inches.
Fancy Etched Silver. 3 inch Dial.
1 Day Time, Alarm....$7.00

1987

1895

SAN REMO

Height, 10 ½ inches without Ribbon. Width, 9 inches.
1 Day Time....$6.50

1988

1895

DELTA

DELTA

Height, 8 ½ inches without Ribbon. Width, 7 inches.
1 Day Time....$6.00

1989

1895

WALDORF

Height, 8 ¼ inches. Width, 7 ½ inches.
Fancy Etched Silver. 3 inch Dial.
1 Day Time, Alarm....$7.00

PORCELAIN CLOCKS

In the case of Porcelain clocks, color is the most important factor in estimating value. Bright reds, blues and frosted pinks are usually considered to be the most desirable. In order to receive top prices, clocks must be in ORIGINAL MINT condition and in good running order. Damaged dial, chipped porcelain will substantially reduce prices.

1990

1895

ST. LOUIS
Height, 7 inches. Width, 9 inches.
1 Day Time....$6.00

1991

1895

POMPADOUR
Height, 14 ½ inches.
Width, 8 ½ inches.
8 Day, Half-Hour Strike..$20.00

1992

1895

MELROSE
Height, 8 ¼ inches. Width, 7 ½ inches.
Fancy Etched Silver. 3 inch Dial.
1 Day Time, Alarm....$7.00

1993

1895

ST. CLOUD
Height, 7 ½ inches.Width, 9 inches.
1 Day Time....$6.00

PORCELAIN CLOCKS

1994

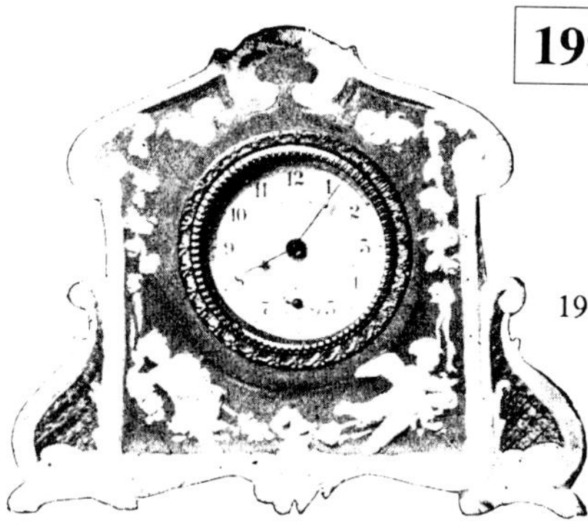

1900

LINWOOD

Imported Wedgwood Case.
Blue and Gray with White Decoration.
Height, 5 inches. Width, 6 inches.
One-day Time Piece, Two-inch
Dial, Gilt Knurled Sash.

List price, $3.05

1996

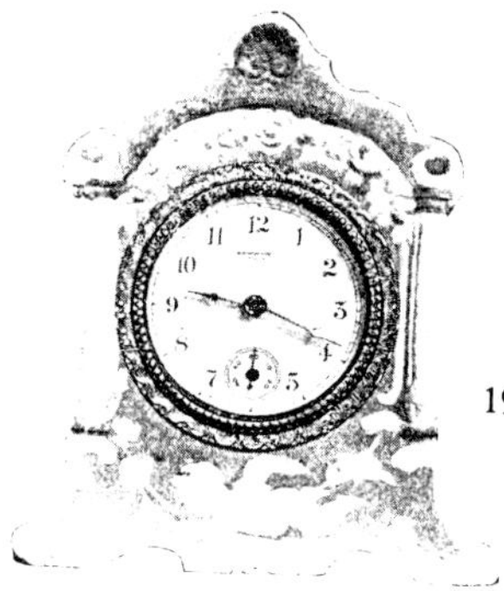

1900

LEE

Imported Wedgwood Case.
Blue and Gray with White Decoration.
Height, 5 inches. Width, 4¼ inches.
One-day Time Piece, Two-inch
Dial, Gilt Knurled Sash.

List price, $2.35

1995

1900

MOLINO

Blue or Green Wedgwood Case,
White Decoration.
Height, 10½ inches. Width, 9½ inches.
Dial, 4 inches.
Eight-day, Half-hour Strike,
Cathedral Gong.
Ivorized Porcelain Dial with Gilt Center,
Empire Sash, Heavy Beveled Glass.

List price, $18.75

Fitted with Jeweled Sash in Brilliant, Ruby,
Emerald, Amethyst, Sapphire or alternate stones, extra list, $2.80

1997

1900

MACRI

Blue or Green Wedgwood Case,
White Decoration.
Height, 11¾ inches. Width, 12 inches.
Dial, 4 inches.
Eight-day, Half-hour Strike,
Cathedral Gong.
Ivorized Porcelain Dial with Gilt Center
Empire Sash, Heavy Beveled Glass.

List price, $23.00

Fitted with Jeweled Sash in Brilliant, Ruby,
Emerald, Amethyst, Sapphire, or alternate stones, extra list, $2.80

1998

1900

MINTO

Blue or Green Wedgwood Case,
White Decoration.
Height, 11¼ inches. Width, 11 inches.
Dial, 4 inches.
Eight-day, Half-hour Strike,
Cathedral Gong.
Ivorized Porcelain Dial with Gilt Center,
Empire Sash, Heavy Beveled Glass.

List price, $21.50

Fitted with Jeweled Sash in Brilliant, Ruby,
Emerald, Amethyst, Sapphire, or alternate stones, extra list, $2.80

PORCELAIN CLOCKS

1999

1900

LATIMER

Imported Fancy Decorated Case.
Height, 11 inches. Width, 7¼ inches.
Eight-day, Half-hour Strike, Cathedral Gong, Dial, 4 inches.
Ivorized Porcelain with Gilt Center, Empire
Sash, Heavy Beveled Glass.

List Price, $14.00

Fitted with Jeweled Sash in Brilliant, Ruby, Emerald, Amethyst, Sapphire, or alternate stones, extra list, $2.80

2000

1900

NELSON

Fancy Porcelain Case.
Height, 13 inches. Width, 12¼ inches. Dial, 4 inches.
Ivorized Porcelain Dial, Gilt Center, Eight-day,
Half-hour Strike, Cathedral Gong.

List price, $11.15

Fitted with Jeweled Sash in Brilliant, Ruby, Emerald, Amethyst, Sapphire, or alternate stones, extra list, $2.80

In the case of Porcelain clocks, color is the most important factor in estimating value. Bright reds, blues and frosted pinks are usually considered to be the most desirable. In order to receive top prices, clocks must be in ORIGINAL MINT condition and in good running order. Damaged dial, chipped porcelain will substantially reduce prices.

2001

1900

2002

1900

2003

1900

TRIAD SET

Dark Blue or Dark Green Porcelain Bowl with best Gold Finish Trimmings.

Height of Clock, 21½ inches.
Width, 9 inches.
Height of Candelabra, 18 inches.

Fitted with Eight-day, Half-hour Strike, Cathedral Gong. Four-inch Ivorized Porcelain Dial, Fancy Gilt Center, or Decorated Porcelain Dial, Empire Sash and Heavy Beveled Glass.

List price, $47.25

Fitted with Jeweled Sash, in Brilliant, Ruby, Emerald, Amethyst, Sapphire or alternate stones, at extra list of $2.80.

PORCELAIN CLOCKS

Other New Haven models with imported majolica, jasperware and porcelain cases are known which were not illustrated in catalogs.

2004

1900

PENOLA

Height, 8¼ inches. Width, 8½ inches.
Dial, 4 inches.
Imported Decorated Porcelain Case,
Gilt Knurled Sash.

List price, One-day Time, $4.25

2006

1900

PARKMAN

Height, 10¼ inches. Width, 8¼ inches.
Dial, 4 inches.
Imported Decorated Porcelain Case,
Gilt Knurled Sash.

List price, One-day Time, $4.25

2005

1900

THISTLE

Decorated Porcelain Body,
Gilt Ornaments.
Eight-day Time Piece, with Empire Sash,
Beveled Glass, Five-inch Ivorine
Dial, with Fancy Gilt Center.
Height, over all, 25¾ inches.
Dial, 5 inches.

List price, $18.10

2007

1900

VOLNEY

Imported Porcelain Case.
Height, 5¼ inches. Width, 4½ inches.
Dial, 2 inches.
One-day Time Piece. Gilt Knurled Sash.
List price, $2.10

2008

1900

MARION

Imported Decorated Porcelain Case.
Height, 6 inches. Width, 4 inches.
Dial, 2 inches.
One-day Time Piece. Gilt Knurled Sash.
List price, $2.35

PORCELAIN CLOCKS

In the case of Porcelain clocks, color is the most important factor in estimating value. Bright reds, blues and frosted pinks are usually considered to be the most desirable. In order to receive top prices, clocks must be in ORIGINAL MINT condition and in good running order. Damaged dial, chipped porcelain will substantially reduce prices.

2009

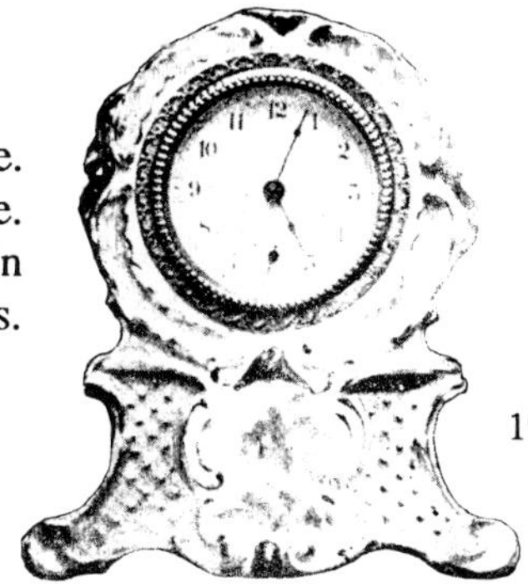

1900

VERONA

Imported Porcelain Case.
Height, 5 inches. Width, 4½ inches.
Dial, 2 inches.
One-day Time Piece. Gilt Knurled Sash.
List price, $2.10

2011

1900

WILSON

Imported Porcelain Case.
Height, 5 inches. Width, 4½ inches.
Dial, 2 inches.
One-day Time Piece. Gilt Knurled Sash.
List price, $2.10

2010

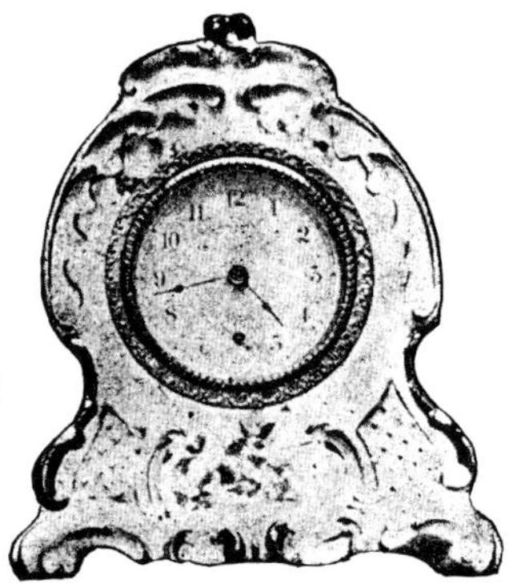

1900

WINDHAM

Imported Porcelain Case.
Height, 5¼ inches. Width, 4½ inches.
Dial, 2 inches.
One-day Time Piece. Gilt Knurled Sash.
List price, $2.10

2012

1900

WILMOT

Imported Porcelain Case.
Height, 5 inches. Width, 4½ inches.
Dial, 2 inches.
One-day Time Piece. Gilt Knurled Sash.
List price, $2.10

2014

1900

AVAR

Imported Decorated Porcelain Case.
Height, 3¾ inches. Width, 5¼ inches.
Dial, 2 inches.
One-day Time Piece. Gilt Knurled Sash.
List price, $2.10

2013

1900

AMBROSE

Imported Decorated Porcelain Case.
Height, 3¾ inches. Width, 5¼ inches.
Dial, 2 inches.
One-day Time Piece. Gilt Knurled Sash.
List price, $2.10

2015

1900

ANNABEL

Imported Decorated Porcelain Case.
Height, 3¾ inches. Width, 5¼ inches.
Dial, 2 inches.
One-day Time Piece. Gilt Knurled Sash.
List price, $2.10

PORCELAIN CLOCKS

2016

1885

No. 1066 A.

FAIENCE WARE,—WITH FLORAL ORNAMENTATION.

1 Day Time,	$5.60
1 Day Time, Alarm,	6.70

Visible Movement.

2017

1885

No. 1066 C.

FAIENCE WARE,—WITH FLOWERS IN RELIEF.

1 Day Time,	$6.60
1 Day Time, Alarm,	7.70

Visible Movement.

2018

1885

No. 1066 B.

FAIENCE WARE,—WITH FLORAL ORNAMENTATION.

1 Day Time,	$6.10
1 Day Time, Alarm,	7.20

Visible Movement.

2019

1885

No. 1066 D.

FAIENCE WARE,—WITH FLOWERS IN RELIEF.

1 Day Time,	$7.15
1 Day Time, Alarm,	8.25

Visible Movement.

PORCELAIN CLOCKS

2020

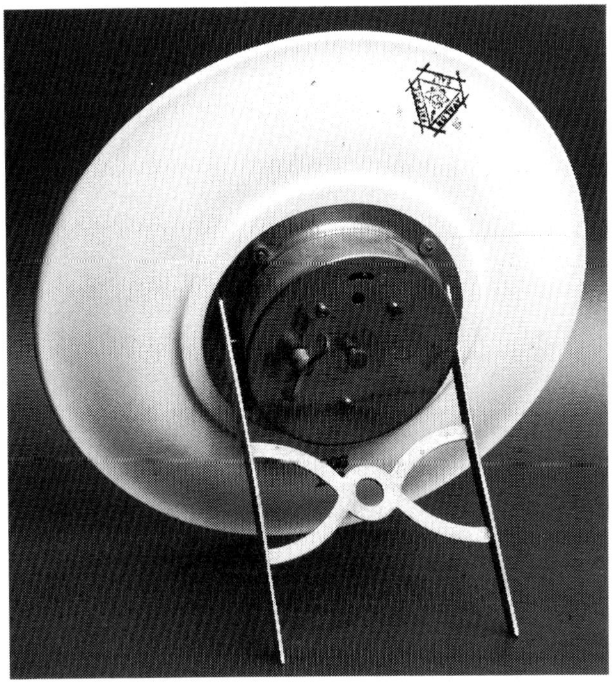

Rear view of **No. 1066** B.
{Robert E. Reichel Collection}

2021

No. 1066 B.
FAIENCE WARE - WITH FLORAL ORNAMENTATION.
Height, 10 inches. 1 Day Time. Visible Movement.
{Robert E. Reichel Collection}

2022

Close-up of **No. 1066** B. 1 Day Time Movement.
{Robert E. Reichel Collection}

2023

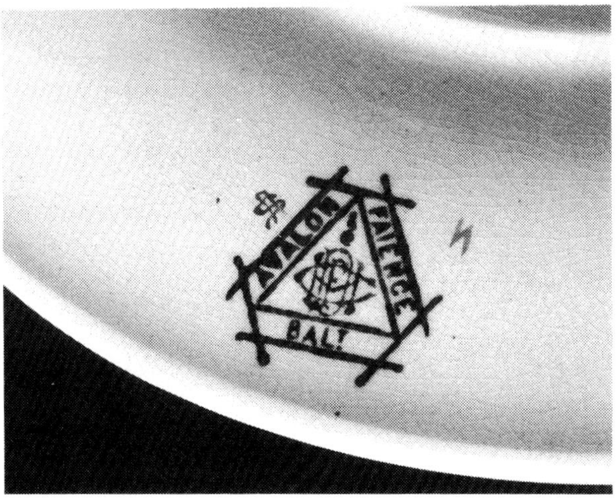

Close-up of **No. 1066** B. Monogram.
{Robert E. Reichel Collection}

PORCELAIN CLOCKS

2024

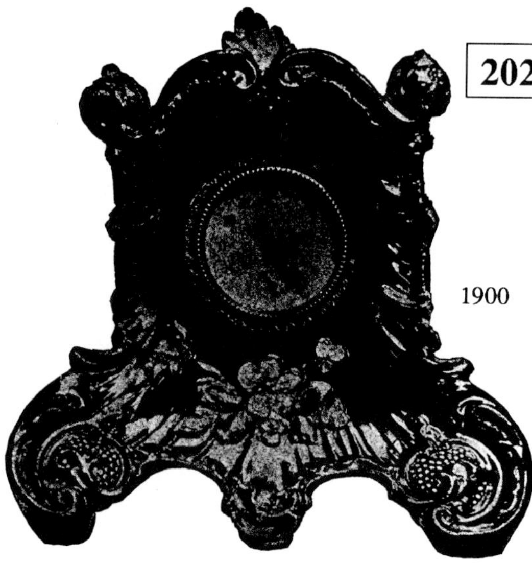

1900

ADDISON

Height, 7¼ inches. Width, 7 inches. Dial, 2 inches.
Imported Fancy Majolica Porcelain Case.
Gilt Knurled Sash.

List price, One-day Time, $3.05

2025

1900

VINCENT

Height, 9⅜ inches. Width, 6⅜ inches.
Two-inch Ivorized Porcelain Dial.
Imported French Porcelain Case.
Gilt Knurled Sash.

List price, One-day Time, $4.25

2026

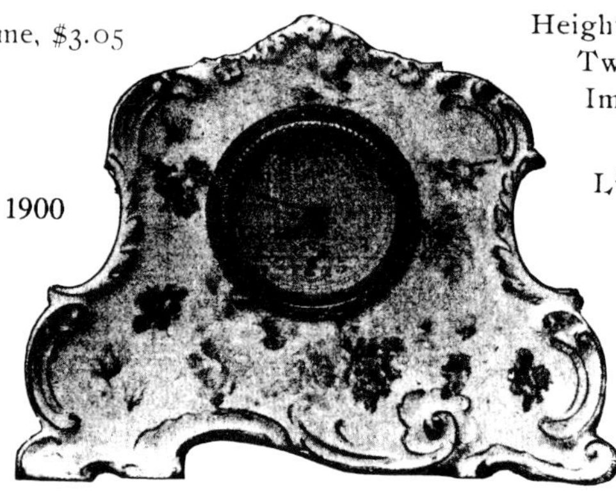

1900

CREIGHTON

Height, 5½ inches. Width, 7 inches.
Dial, 2 inches.

Imported Decorated Porcelain Case.
Gilt Knurled Sash.

List price, One-day Time, $2.80

2027

1900

ABNER

Height, 7 inches. Width, 6¼ inches. Dial, 2 inches.
Imported Fancy Majolica Porcelain Case.
Gilt Knurled Sash.

List price, One-day Time, $3.05

2028

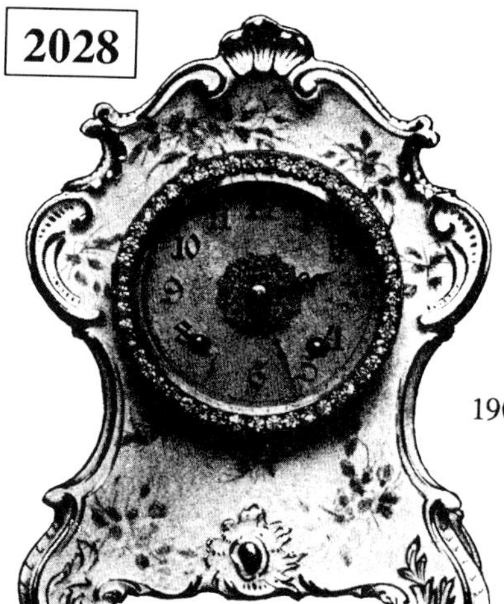

1900

AUDUBON—Showing Jeweled Sash.

Height, 10½ inches. Width, 7⅝ inches.
Imported Decorated Porcelain Case.
Eight-day, Half-hour Strike, Cathedral Gong.
Four-inch Dial, Pearl or Ivorized Porcelain with Gilt Center, Empire Sash,
Heavy Beveled Glass.

List price, $9.75

Fitted with Jeweled Sash in Brilliant, Ruby, Emerald, Amethyst, Sapphire, or alternate stones, extra list $2.80

STATUETTES AND CANDLESTICKS

2029

1911

HORSE TAMER

HEIGHT ....7 INCHES
WIDTH ....7 INCHES
PRICE, BRONZE FINISH ..LIST, $2.05
PRICE, GILT FINISH ....LIST, 2.25

2030

1911

HORSE

HEIGHT ....8½ INCHES
WIDTH ....9¼ INCHES
PRICE, BRONZE FINISH ....LIST, $1.45
PRICE, GILT FINISH ....LIST, 1.65

2031

1911

TRUMPETER

HEIGHT ....7½ INCHES
WIDTH ....6 INCHES
PRICE, BRONZE FINISH ....LIST, $1.35
PRICE, GILT FINISH ....LIST, 1.55

2032

1911

BONNER

HEIGHT ....7½ INCHES
WIDTH ....8½ INCHES
PRICE, BRONZE FINISH ....LIST, $1.45
PRICE, GILT FINISH ....LIST, 1.65

2033

1911

DIANA

HEIGHT ....6¼ INCHES
WIDTH ....5 INCHES
PRICE, BRONZE FINISH ....LIST, $ .70
PRICE, GILT FINISH ....LIST, .90

2034

1911

STAG

HEIGHT ....9 INCHES
WIDTH ....6 INCHES
PRICE, BRONZE FINISH ....LIST, $1.45
PRICE, GILT FINISH ....LIST, 1.65

2035

1911

LION

HEIGHT ....5¾ INCHES
WIDTH ....8½ INCHES
PRICE, BRONZE FINISH ....LIST, $1.75
PRICE, GILT FINISH ....LIST, 1.95

STATUETTES AND CANDLESTICKS

2036

1911

DOG

HEIGHT ....6½ INCHES
WIDTH ....8 INCHES
PRICE, BRONZE FINISH ....LIST, $1.35
PRICE, GILT FINISH ....LIST, 1.55

2037

1911

SCIPIO

HEIGHT ....9 INCHES
WIDTH ....8¼ INCHES
PRICE, BRONZE FINISH ....LIST, $2.35
PRICE, GILT FINISH ....LIST, 2.55

2038

1913

MUSIC
LIST, $1.60
HEIGHT, 8½ INCHES. WIDTH, 9 INCHES.

2039

1911

SCOUT

HEIGHT ....9¾ INCHES
WIDTH ....9½ INCHES
PRICE, BRONZE FINISH ....LIST, $1.90
PRICE, GILT FINISH ....LIST, 2.10

2040

1911

HANNIBAL

HEIGHT ....7¼ INCHES
WIDTH ....8¼ INCHES
PRICE, BRONZE FINISH .LIST, $2.35
PRICE, GILT FINISH ......LIST, 2.55

2041

1911

PANDORA

HEIGHT ....6¾ INCHES
WIDTH ....7¼ INCHES
PRICE, BRONZE FINISH ....LIST, $1.35
PRICE, GILT FINISH ....LIST, 1.55

2042

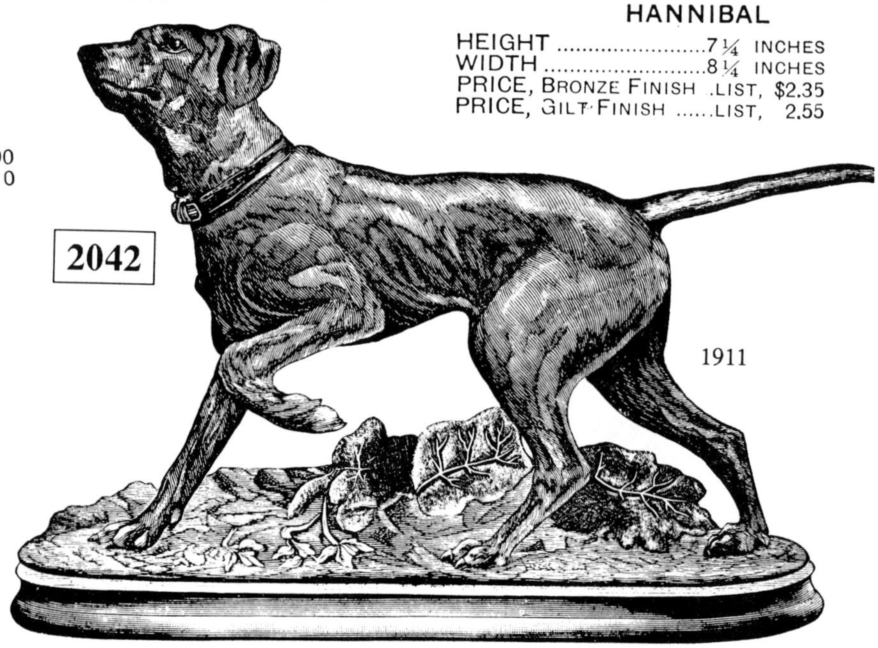

1911

POINTER

HEIGHT ....8 INCHES
WIDTH ....12 INCHES
PRICE, BRONZE FINISH ....LIST, $2.50
PRICE, GILT FINISH .... LIST, 2.70

STATUETTES AND CANDLESTICKS

2043

2044

HERNDON CANDLESTICKS
HEIGHT ....8¼ INCHES
WIDTH ....3½ INCHES
METAL, ORMOLU GOLD PLATE FINISH
PRICE, PER PAIR ....LIST, $5.10

HERNDON CANDLESTICKS

2047

2048

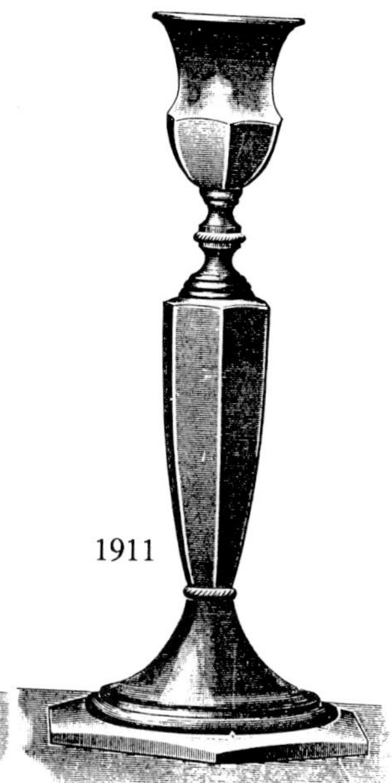

2045

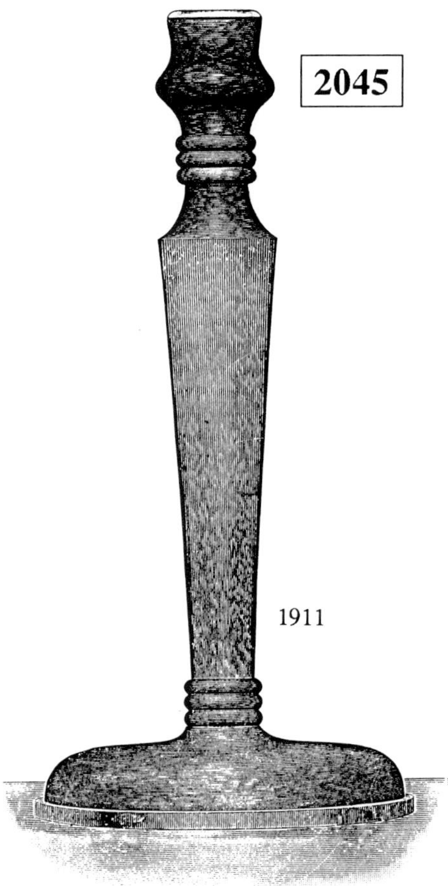

2046

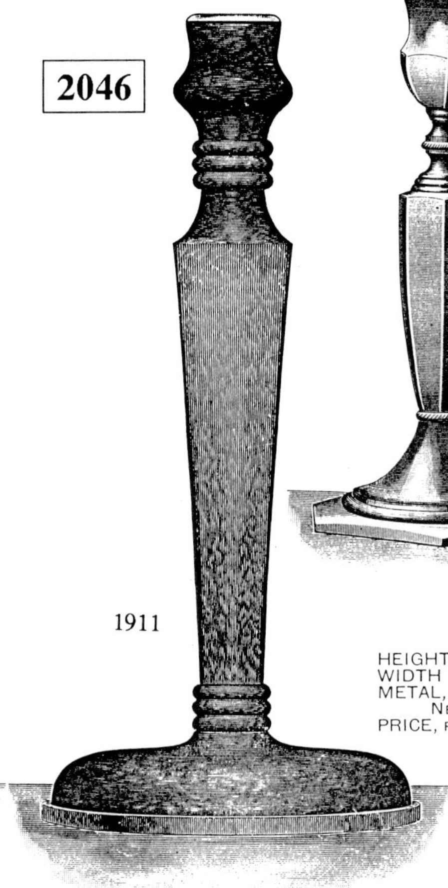

No. 387. CANDLESTICKS
HEIGHT ....8½ INCHES
WIDTH ....4 INCHES
METAL, ORMOLU GOLD PLATE, FRENCH BRONZE,
NEW RITZ, OR SILVER
PRICE, PER PAIR ....LIST, $4.65

No. 3197. CANDLESTICKS
HEIGHT ....12½ INCHES
WIDTH ....5¾ INCHES
SOLID MAHOGANY, WITH BRASS CANDLE CUPS
PRICE, PER PAIR ....LIST, $5.80

STATUETTES AND CANDLESTICKS

2049

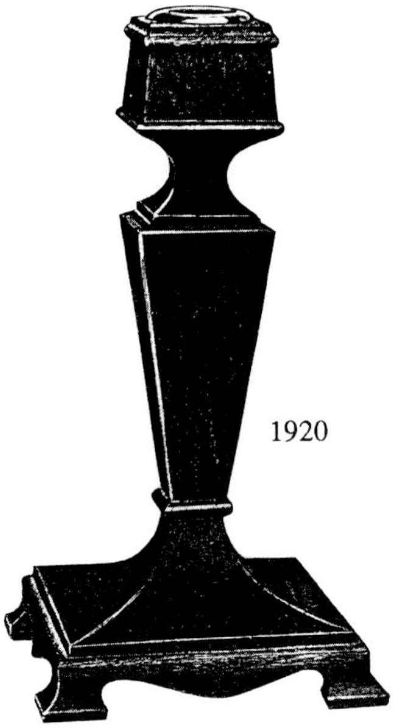

1920

No. 3817
CANDLE STICKS
Height, 9 Inches Width, 4½ Inches
Solid Mahogany
With Brass Candle Cups

2050

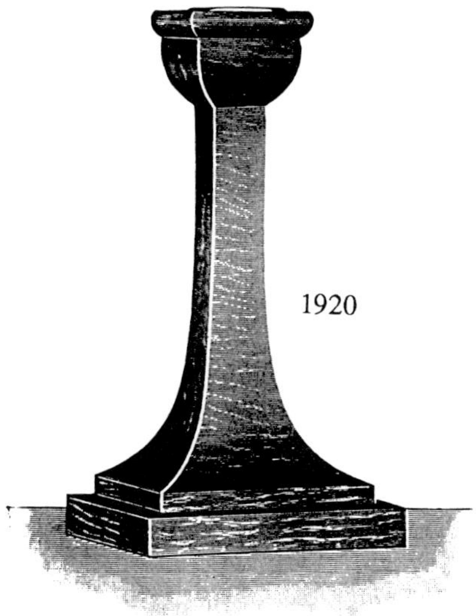

1920

No. 3345
CANDLE STICKS
Height, - - - - - - 6¼ Inches
Width, - - - - - - 3 Inches
Solid Mahogany, with Brass Cups

2051

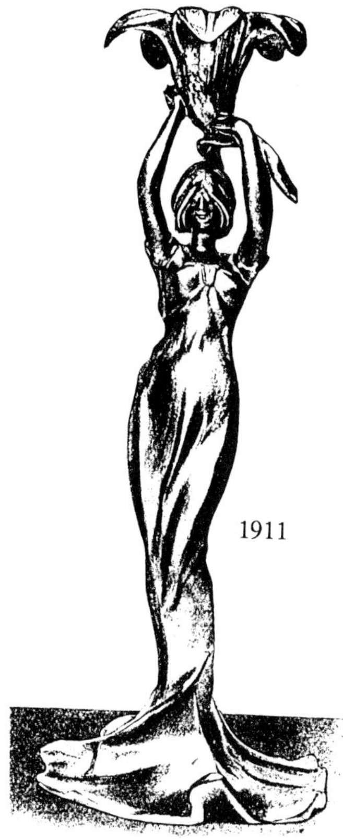

1911

TROUVILLE
CANDLESTICKS. PER PAIR, LIST, $11.60
HEIGHT, 12 INCHES. WIDTH, 4¾ INCHES.

2052

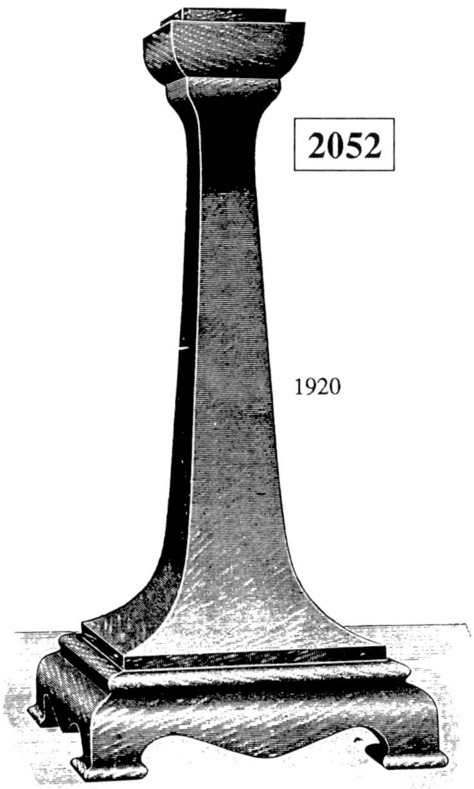

1920

2053

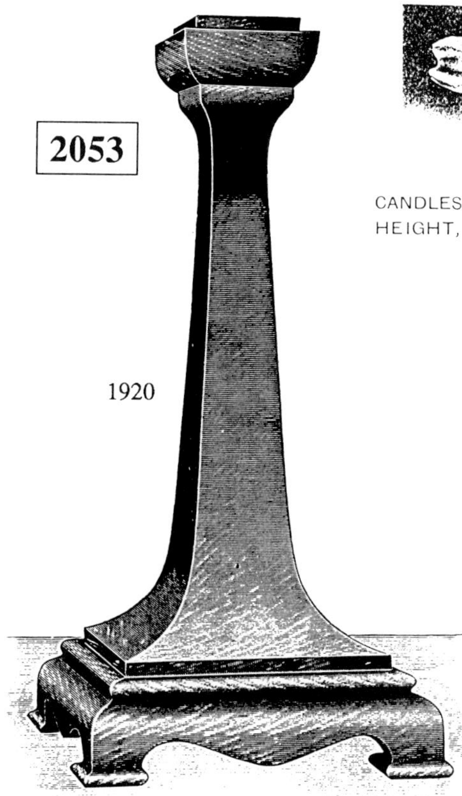

1920

PARAGON CANDLE STICKS
Height, 9 Inches Width, 4¼ Inches
Solid Mahogany
With Brass Candle Cups

2054

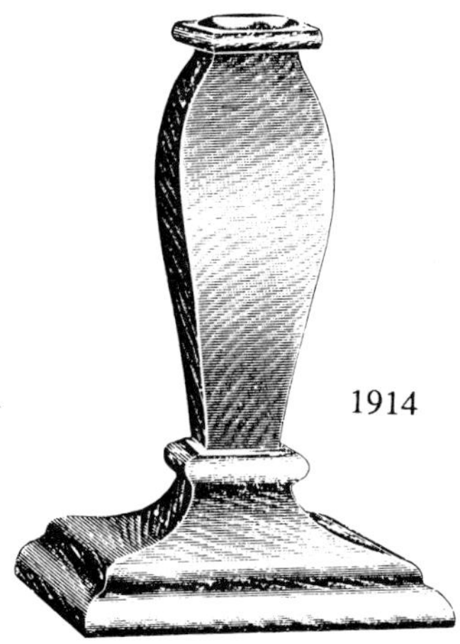

1914

No. 3320
PER PAIR. LIST, $6.50
HEIGHT, 6½ INCHES. WIDTH, 4 INCHES.
SOLID MAHOGANY, WITH BRASS CANDLE CUPS.

WATCHMAN'S CLOCKS

2055

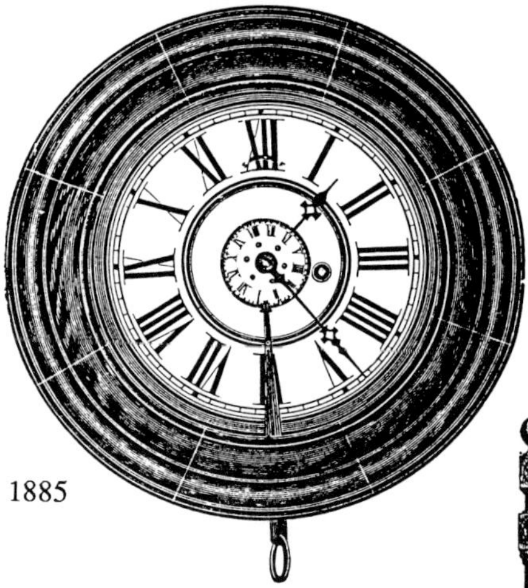

1885

THE WATCHMAN'S DETECTOR

THE WATCHMAN'S DETECTOR

Walnut. Dial 8 inches. 1 Day Time $12.00

This is the cheapest and simplest Watchman's Clock made. The movement is a thirty-hour lever time-piece, and the watchman's dial is attached to the hour hand of the clock. The figures on this dial are reversed from the time dial of the clock, and by means of the slides from the centre dial, which are operated by the pull from the outside of the case, the faithfulness of the watchman is seen at once. The sash is fastened by a patent lock which cannot be tampered with, and by using a thirty-hour movement the Clock requires winding every day, and the slides are pushed back in place at the same time. There is nothing about this to get out of order, and the movement cannot be stopped by any tampering with the pull. We commend this to the attention of our customers wishing a cheap and reliable watchman's clock.

2056

1885

2057

1883

ELECTRIC WATCHMAN'S CLOCK

BLACK WALNUT CASE.

Height, 64 inches. Dial 12 inches. 8 inch Register.

8 Day Timepiece.

ELECTRIC WATCHMAN'S CLOCK

BLACK WALNUT.

Height, 64 inches. Dial 12 inches. 8 inch Register.

8 Day Time.... Net $150.00

The above without Time Dial.... Net $140.00

WATCHMAN'S CLOCKS

The Newman Portable Watchman's Clock.

THE ONLY ONE THAT CANNOT BE TAMPERED WITH.

2058

1900

Front View.

Portable watches have been made for many years, but owing to the fact that watchmen could tamper with them by the use of one duplicating key for registering and by other means, manufacturers and other users of watchmen's clocks as well as insurance underwriters have generally accepted them ***only as a makeshift.***

In many institutions a Portable watch is the only system of Recorder that can be used, such, for instance, as lumber yards, railroad yards, mines, factories that are not enclosed, chemical and packing houses where the brine or chemicals destroy the efficacy of other systems, and for general outdoor watching.

Portable watches, therefore, being a necessity, it remained for some one to discover a method to insure ***absolutely correct results.***

Electrical watchmen's clocks have been and are a great source of annoyance in many places, and in such institutions as enumerated above they are absolutely worthless. There are too many contingencies on account of the many wires and their delicate construction as well as the high cost for repairs.

For these reasons this invention will, we predict, create ***a revolution in the future systems of controlling watchmen.*** The public want simplicity and correctness; they also want a watchman's clock that will produce a ***perfect record at a low price and that will not get out of order.***

The average cost for repairs of an electrical watchman's clock in three years is much more than the first cost of our sixteen-key watch.

The watch needs cleaning and oiling about once a year at a cost of about $1.50 at any watchmaker's. It can be returned the same day, thus no waiting for electricians or specialists.

The following are the principal features of improvements:

1st. To duplicate the key, our dies and matrix *must* be used. Keys cannot be purchased without application by the proper authorities.

2d. The case is made of aluminum for lightness, and to prevent tarnishing or injury from acids and grease.

3d. The dial is arranged for twenty-four hours. This gives a record for the Sunday day registering when necessary.

(A 24 or 48-hour dial should be used on every other watchman's clock that has a safety lock, although they all use a 12-hour dial, which the watchman can duplicate, mark it up and go to sleep. The record made in locking and unlocking on a 24-hour dial denotes whether it has been locked or unlocked in the *night* or *day.*)

4th. The mechanism that does the marking is ***on the key,*** which runs through a little house of steel bronze, so made that the key fits exactly and has but one motion. The house is separated from the movement and tightly closed so that no dirt or dust can affect the works. Dirt or dust getting into the house would not affect the marking, as there are no springs, cams or levers within.

2059

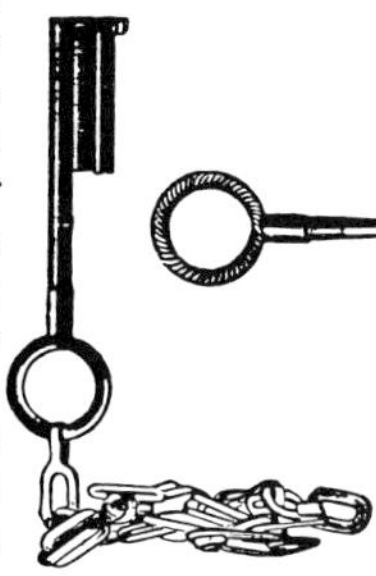

2060

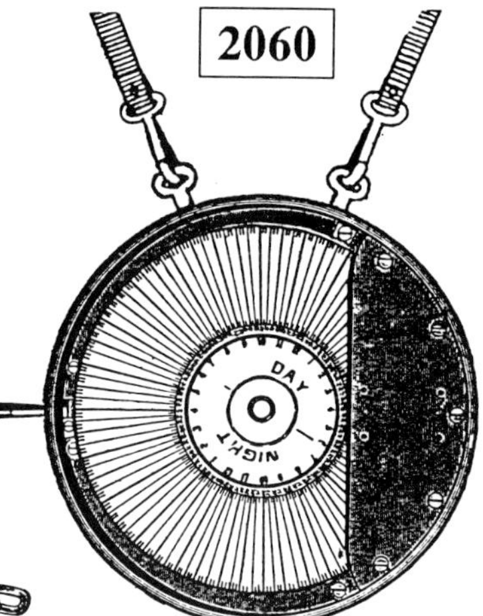

1900

Back view with lid off.

5th. This, among other things, does away with a ***clumsy leather pouch.*** The watchman throws a strap over his shoulder, carrying the watch to the station where the key is fastened.

6th. To operate it he need not lay down his lantern.

7th. The registering is made by ***embossing the numbers of the station into the dial*** whenever he turns the keys.

Cut of box in which key chain is riveted. The box is fastened from within with two screws which are sealed so that it cannot be removed without detection. Specially adapted to outside stations. Size of an ordinary match safe. Price, 80c. each.

8th. It has a capacity for sixteen stations in one line from center. More can be added if necessary.

9th. A ***timepiece dial*** shows the watchman the time to start on his rounds, and the time he should take in making such rounds, which prevents his rushing to the stations without seeing everything along his route.

Where it is desired or necessary to use a twelve-hour watchman's dial they will be furnished.

The movement is of the finest quality and jeweled, and is far enough away from the case to prevent its injury by sudden jarring.

The watch weighs less than one and one-half pounds, which is less than one-third as heavy as other watches on the market.

The watches are warranted and sent out on approval.

THE PRICE...

Of 8 key Watch is $ 90.00

Of 16 key Watch is $100.00

Including one year's supply of paper dials. Extra keys $2.00 each.

Extra dials $5.00 per box.

WATCHMAN'S CLOCKS

The Newman System of Watchman's Clocks.

2061

2062

THE MECHANICAL OR STATIONARY CLOCK.

The best System of Watch Clocks, especially where from one to six stations are necessary.

MORE THAN 40,000 IN USE ALL OVER THE WORLD.

This clock has the following advantages over other systems:

1st. It is ***absolutely*** tamper proof.

2nd. It serves as a large, ***fine-looking, eight-day time-piece.***

3rd. It is simplicity itself, therefore there is nothing to get out of order more than there is in an ordinary clock, having only two more wheels, in addition to which is a button with a needle at its end, which pricks the dial as the watchman presses it.

4th. It is bound to report every such operation and gives the watchman no loophole to crawl out.

5th. Each station is independent of the other, so that if one clock should get out of order, the others will still record, while you can take it to any good jeweler and get it repaired in an hour or two.

6th. It can be placed anywhere and can be moved at will by anyone, as it is fastened up with only two screws, which, however, are inside the case, so that the watchman cannot take the clocks down and carry them to some convenient place.

7th. It is the only Watch Clock that will record the night watchman's record independent of the day watchman's on Sundays or holidays, the records being made at night on a separate part of the paper.

8th. It is the only clock in reach of all requiring Watchmen's Clocks, as it costs no more than any ordinary timepiece and less to keep it in order.

The cases are made of finely finished Oak or Cherry, so that they are nice enough for the most elaborately finished office.

WHY THE CLOCK CANNOT BE TAMPERED WITH, EVEN THOUGH THE WATCHMAN HAD A DUPLICATE KEY AND DIAL.

The locking device is one of the best features. It registers on the inside of the paper dial by puncturing a large hole the very time the clock is locked and unlocked, consequently, ***by the use of our 48-hour dial,*** it would be impossible for the watchman to tamper with the clock, even though he had a key and duplicate paper dials. For instance, suppose you lock your clock at 10 o'clock in the morning; before you do so, you set the correct time of "1st day" under the pricking devices; when locking a hole is punched, registering the time you locked it; should the watchman have a key and a duplicate paper dial which he replaces for the one you put in, he must replace it to correspond with the time on the face of the clock. The quarter circle of the first night being under the pricking devices (as it is night when he opens it and puts on the new dial, you having done so in the day time), when he locks the clock he is obliged to leave a record showing conclusively that he has displaced your dial by another. He cannot leave it unlocked, as the case will remain wide open when not locked.

FIGURE A. FRONT VIEW OF CLOCK. SIZE 14 x 14 IN.

FIGURE D.

TO PULL FROM OUTSIDE.

Where street watchmen or any outside watchmen are employed we can fasten an attachment on the case on either the square Regulator Nos. 1 or 2, to pull from outside.

It can also be arranged to press a button by electric attachment if the clock should be hung far from the watchman's station. When ordering for outside pulling, state which is needed.

2063

EASY TO DO SO ON A 12-HOUR DIAL.

All other Watchmen's Clocks ***use a 12-hour dial,*** and even by the use of an indicating lock, ***are worthless so far as tampering is concerned,*** because it is easy for watchmen to produce duplicate paper dials and set them as you did yours, when using a 12 or 24-hour dial; this is one of our strongest points and makes this clock absolutely positive.

Should there be more than one of these clocks and one should get out of order in the night, there is a record from the other stations, and the next morning simply unscrew it from its hanging place, repair it and have it back in time to use at night. The movement has no intricate wheels or mechanism; there are in addition to the regular gearing of an ordinary clock just two wheels; otherwise it is exactly like any other clock movement in its construction.

1900

THE LATEST IMPROVEMENTS.

1st. ***A 48-hour dial;*** other clocks only have a 12-hour dial *and this is not satisfactory.* The advantage of this 48-hour dial over the 12, is that the record of the watchman on Saturday night and Sunday night are independent of each other, hence you need not come down on Sunday to change your dial; the clock, therefore, can be operated for Saturday night, Sunday and Sunday night, all records being separate. ***The dial needs to be removed every other day only.***

2nd. ***The Arrangement for Pricking the Paper Dial*** when opened or closed, which is now done on the inside toward the center, making a large hole that cannot escape your notice to show tampering.

3rd. The different designs on the dial, showing cut of the first day, which means the day that the dial is put on, and the following night, day and night, so that should the watchman attempt to tamper with the clock by duplicate dial, he would necessarily be exposed very quickly by the record he would make when locking the door.

The clock also has an ***Improved Pricking Device*** which cannot be broken by the watchman without using undue force.

The Square Clock as represented in Fig A, is operated by an 8-day lever movement, so that it will run in any position. The movement is a quick train, and if forgotten, will run fourteen days without winding. They should, however, be wound every week, to insure keeping time. The movement is enclosed in an extra case to keep out moisture and dust.

Fig. C.—Dial 12 in., size: 42¾x14½ inches, oak case, 7-day watchman's dial, purely mechanical.

Fig. C is made of the same material as Fig. B, length of case 42¾ in., width 14½ in., dial 12 in. The movement is operated in the same manner as Regulator No. 1, but the watchman's dial makes a revolution only once in seven days, so that it needs attention but once a week, at the same time when clock is wound. It is always in sight, however.

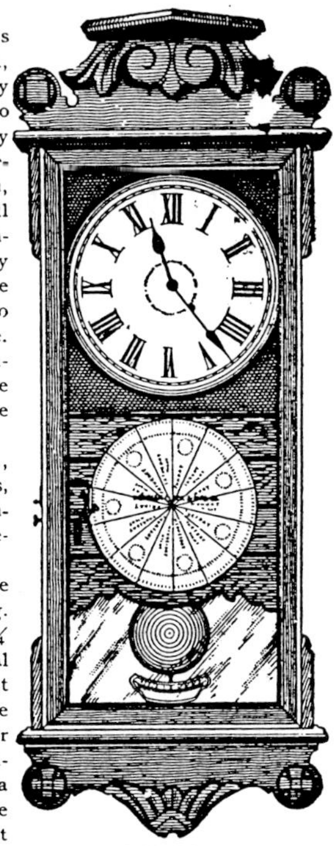

FIGURE C.

Fig. B.—Oak or Cherry. Regulator No. 1 as represented in Fig. B, is operated by a single movement running 8 days. Length of case 33¾ in., width 11¼ in.; dial 8 in. The movement is heavy and well built, and is especially recommended where close time is required. This clock will keep good time.

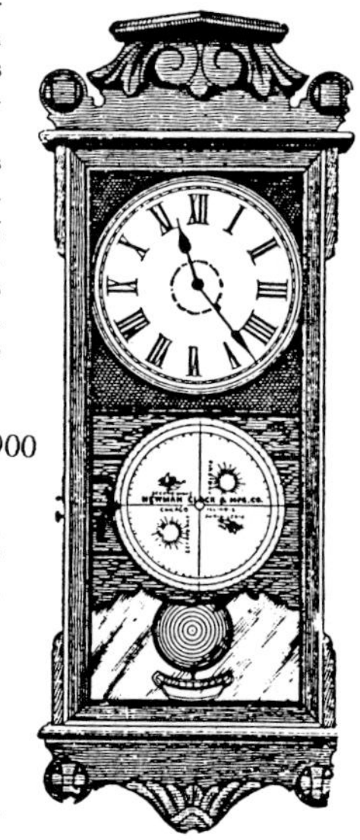

FIGURE B.

Figure A,	$20.00
Figure B,	23.00
Figure C,	28.00
Attachment on clock for pulling from outside of building,	$1.10

2064

WATCHES

2065

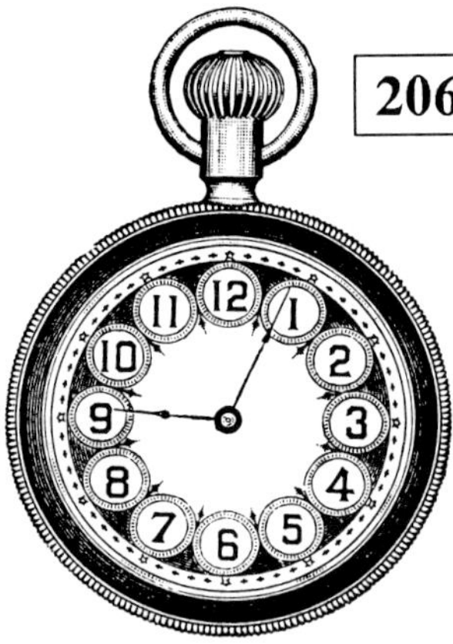

1911

2066

1911

Back, showing Engraved Initial

2067

1911

LADY NELSON

SIZE, No. 6
FINISH, Gun Metal Bezel and Back with Gilt Body, Pendant and Bow
CRYSTAL, Heavy and Beveled
DIAL, Black Arabic Numerals on White Circle with Gilt Border and Background of Dark Red
BACK, with Hand Engraved Initial (any Letter)
MOVEMENT, 1 day Stem Wind and Stem Set
PRICE ....List, $3.20

NELSON

SIZE, No. 16
FINISH, Gun Metal Bezel and Back with Gilt Body, Pendant and Bow
CRYSTAL, Heavy and Beveled
DIAL, Black Arabic Numerals on White Circle with Gilt Border and Background of Dark Red
BACK, with Hand Engraved Initial (any Letter)
MOVEMENT, 1 day Stem Wind and Stem Set
PRICE ....List, $2.20

2068

1889

2069

1889

FULL SIZE. SILVER.

STING WATCH.

FULL SIZE. NICKEL.

1 Day Time, Nickel.......................................$3.00
1 Day Time, Silver.......................................3.20

This Movement has Screw Pillars, Cut and Polished Steel Pinions, a Straight Line Lever with Double Roller Club Tooth Escapement and Polished Springs.

WATCHES

2070

1900

2071

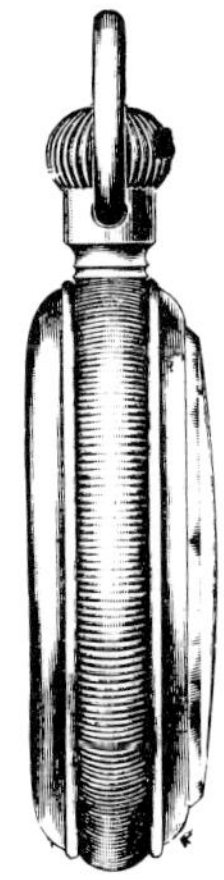

1900

2072

1900

NEW HAVEN WATCH, No. 8—STEM WIND, STEM SET

Nickel Finished Case with Knurled Edge and Heavy Beveled Crystal.
This Watch is an 18 size, has a Sunk Second, and is the thinnest and smallest made by any Clock Company.

List price, Nickel,	$2.35
" " Gilt,	2.65
" " 14-Carat Gold Plate,	3.50

2073

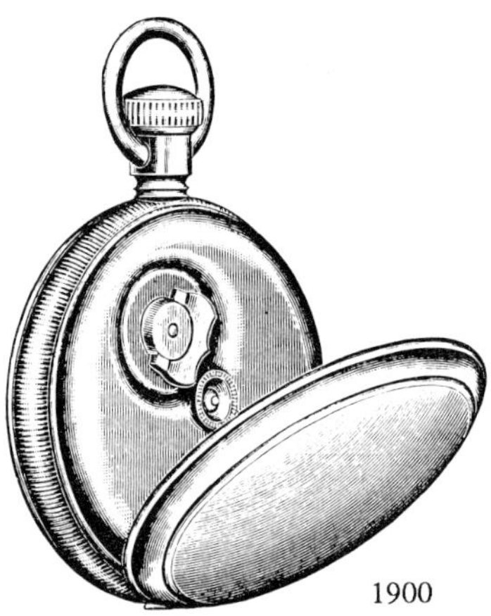

1900

2074

1900

2075

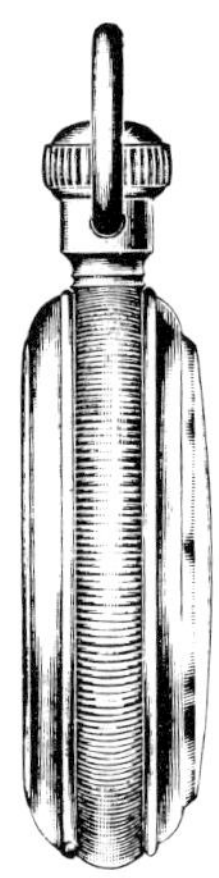

1900

2076

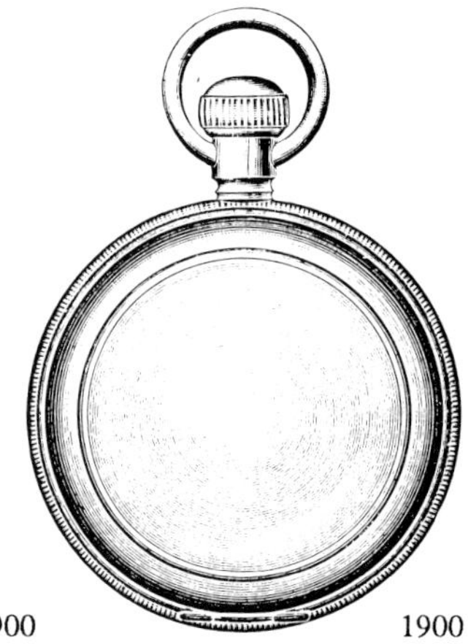

1900

THE HERO

Nickel Finished Case with Knurled Edge, Hinged Back and Heavy Beveled Crystal. Back Wind and Back Set. Sunk Second.
The Thinnest Back Wind Watch in the market—18 size.

List price, Nickel,	$1.80
" " Gilt,	2.05
" " Bicycle Holder,	.40

WATCHES

2077

1911

2078

1911

"C" Fancy Back for Lady Clare only

LADY CLARE

SIZE, No. 6
CRYSTAL, Heavy and Beveled
DIAL, White Arabic
CASE, Heavy Metal
MOVEMENT, 1 day Stem Wind and Stem Set
PRICE, Nickel Finish ....list, $3.05
PRICE, Gilt Finish ....list, 3.20
PRICE, Gun Metal Finish ....list, 3.20
PRICE, 14 Karat Gold Plate ....list, 3.80
PLAIN Back and Bezel furnished unless otherwise specified. Can supply fancy Back without extra charge

2079

1911

YALE

SIZE, No. 16 "Thin Model"
CRYSTAL, Heavy and Beveled
DIAL, White Arabic
CASE, Heavy Metal with Knurled Edge
MOVEMENT, 1 day Stem Wind and Stem Set
PRICE, Nickel Finish ....list, $1.85
PRICE, Gilt Finish ....list, 2.00
PRICE, Gun Metal Finish ....list, 2.00
PRICE, 14 Karat Gold Plate ....list, 2.60

2080

1911

JEROME

SIZE, No. 16
CRYSTAL, Heavy and Beveled
DIAL, White Arabic
CASE, Heavy Metal
MOVEMENT, 1 day Stem Wind and Stem Set
PRICE, Nickel Finish ....list, $1.70
PRICE, Gilt Finish ....list, 1.85
PRICE, Gun Metal Finish ....list, 1.85
PRICE, 14 Karat Gold Plate ....list, 2.45

2081

1911

RANGER

SIZE, No. 16
CRYSTAL, Heavy and Beveled
DIAL, White Arabic
CASE, Heavy Metal with Knurled Edge
MOVEMENT, 1 day Stem Wind and Stem Set
PRICE, Nickel Finish ....list, $1.70
PRICE, Gilt Finish ....list, 1.85
PRICE, Gun Metal Finish ....list, 1.85
PRICE, 14 Karat Gold Plate ....list, 2.45

2082

1911

YALE FANCY BEZEL

Jerome, Ranger, and Yale, furnished with Plain Back and Bezel unless otherwise specified. Can supply with Fancy Backs either design, or with Fancy Back and Fancy Bezel, without extra charge

WATCHES

EXTRA FOR GILT OR GUN METAL FINISH . . . . LIST, $.15
EXTRA FOR 14 KARAT GOLD PLATE FINISH . . . . LIST, .75
THESE WATCHES ARE REGULARLY FITTED WITH PLAIN BACKS AND BEZELS. FANCY BACKS AS ILLUSTRATED ON PAGE 7 WILL BE FURNISHED AT NO EXTRA CHARGE, WHEN ORDERS SPECIFY, EXCEPT ON THE ELITE WATCH.

2083

1913

THE ELM CITY WATCH

SIZE No. 16. "THIN MODEL."
NICKEL FINISH . . . . . . LIST, $1.85
THIS MODEL IS NOT MADE IN GILT OR 14 KARAT GOLD PLATE FINISH.

2084

1913

THE RANGER WATCH

SIZE No. 16.
NICKEL FINISH . . . . . . LIST, $1.70

2085

1913

THE JEROME WATCH

SIZE No. 16.
NICKEL FINISH . . . . . . LIST, $1.60

2086

1913

THE ELITE WATCH

SIZE No. 12. "THIN MODEL."
SOLID PLAIN GERMAN SILVER CASE, HEAVILY NICKELED . . . . . LIST, $3.05
SOLID GERMAN SILVER CASE, NICKEL PLATED, FANCY ENGRAVED BACK AND BEZEL . . . . LIST, 3.20
SOLID COMPOSITION METAL CASE, RESEMBLING GOLD, OR GUN METAL PLAIN OR FANCY ENGRAVED BACK, LIST, 3.20

2087

1913

THE LADY CLARE WATCH

SIZE No. 6.
NICKEL FINISH . . . . . . LIST, $3.05

2088

1913

THE GEM WATCH

SIZE No. 16. "THIN MODEL."
NICKEL FINISH . . . . . . LIST, $2.00
THIS MODEL IS NOT MADE IN GILT OR 14 KARAT GOLD PLATE FINISH.

WATCHES

2089

1917

THE SERVICE WATCH

With Radium Dial

The material used on the dial and hands is of the highest quality and is guaranteed to last several years.

2090

1917

THE SERVICE WATCH

(Size No. 6)

Made in nickel, gilt or gun metal finish. Stem wind and stem set. A guaranteed timepiece. Made with either a plain or fancy back.

2091

1917

THE SERVICE
Convertible Wrist Watch
Radium Dial

2094

1917

N
THE KERMIT
Fancy Back

2092

1917

THE ELITE

Representing the highest development in the manufacture of a low-priced watch

This is an open face, 12 size extra thin model, German Silver case, plain or fancy snap back and bezel with patented stem wind, stem set, antique pendant and gilt crown, sunk second dial and heavy beveled glass.

Case is finished in Nickel, Gilt or Gun Metal.

2093

1917

THE KERMIT

The thinnest model in a low priced Gentleman's 16 size watch on the market

This is an open face, 16 size thin model, German Silver case, plain or fancy snap back and bezel with patented stem wind, stem set, antique pendant and gilt crown, sunk second dial and heavy beveled glass.

Case is finished in Nickel, Gilt or Gun Metal.

When orders so specify fancy backs and bezels (see illustrations, page 5) will be supplied, but fitted regularly with plain backs and bezels.

2095

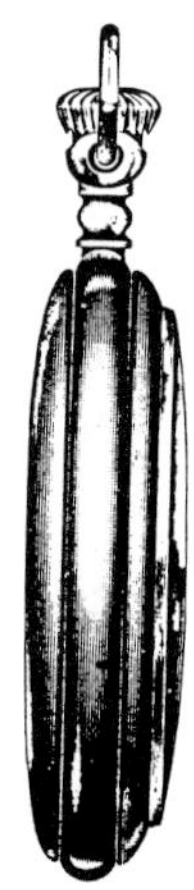

1917

KERMIT
Side View

WATCHES

2096

1937

LOCOMOTIVE

(Nickeled or Gold-color case—plain or engraved)

Nickeled or gold-color case with locomotive back. Railroad type dial. Black hour numerals and red minutes. Unbreakable crystal. Sunk second dial.

2097

Ca. 1932

ELM CITY POCKET

A smartly designed, low priced pocket watch with nickel plated, dust-proof case; attractive black and silvered dial; and sunk second dial.

2098

1917

THE GEM

Just what its name implies—built to please the eye

This is an open face, 16 size plain or fancy snap back and bezel with patented stem wind, stem set, antique pendant bow and crown, sunk second dial and heavy beveled glass.

Case is finished in Nickel or Gun Metal. (This pattern is not made in other finishes.)

2099

Ca. 1932

POPEYE POCKET

Nickel-plated case; silvered dial with Popeye brightly illustrated in colors thereon, and his constantly moving arms indicating the correct time. Unbreakable crystal; and a second dial with an animated " Wimpy " in full charge.

2100

1937

"V" Back "Q" Back

MODEL "B" BACKS

Illustrated above are the two types of backs which are also available for the Model B pocket watch. Specify style when ordering.

2101

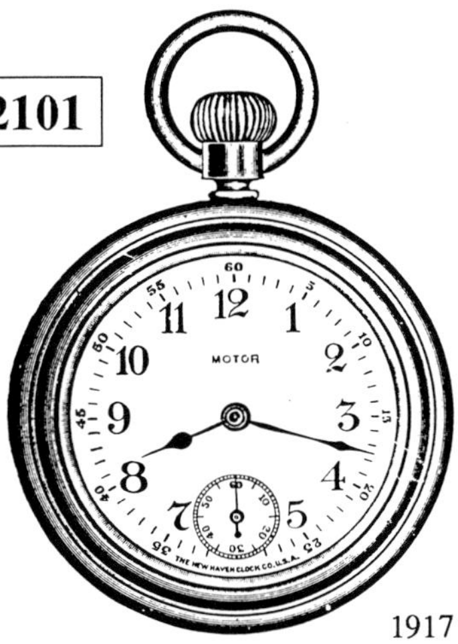

1917

THE MOTOR

Practical and Serviceable

This is an open face, 16 size plain or fancy snap back and bezel with patented stem wind, stem set, sunk second dial and heavy beveled glass.

Case is finished in Nickel or Gun Metal. (This pattern is not made in other finishes.)

2102

1937

NEW HAVEN Streamline

Nickel plated, dust-proof case with fixed top ring. Two-tone silvered dial. Sunk second dial. Unbreakable crystal.

RETAIL $1.50

2103

EARL (7-Jewel; "Bek" Gold Case)

Exceptionally thin, "BEK" YELLOW GOLD plated case. 7-Jewel movement. Silvered dial with etched numerals. Flush second hand. Unbreakable crystal.

RETAIL $6.95

EARL (2-Jewel; Chrome Case)

Similar to the above but with 2-jewel movement and chromium plated case.

RETAIL $3.50

WATCHES

2104

1937

MODEL "A"

Nickel plated, dust-proof case. Two-tone etched dial. Sunk second dial. Unbreakable crystal.

RETAIL $1.25

2105

1937

MODEL "B"

(Gold-color case—engraved)

Engraved, gold-color case with locomotive ("V") back or linear-design ("Ϙ") back. Two-tone, gold-color dial with an unbreakable crystal. Sunk second dial.

RETAIL $1.50

2106

1937

MODEL "B" Radium

Nickel-plated case. Black and white dial with RADIUM numerals and hands. Unbreakable crystal. Sunk second dial.

2107

1937

MODEL "B"

(Nickeled case—plain or engraved)

Plain, nickel plated case with plain back or locomotive ("V") back. Also engraved, nickel plated case with locomotive ("V") back or linear-design ("Ϙ") back. Metal dial with etched numerals and an unbreakable crystal. Sunk second dial.

RETAIL $1.25

2108

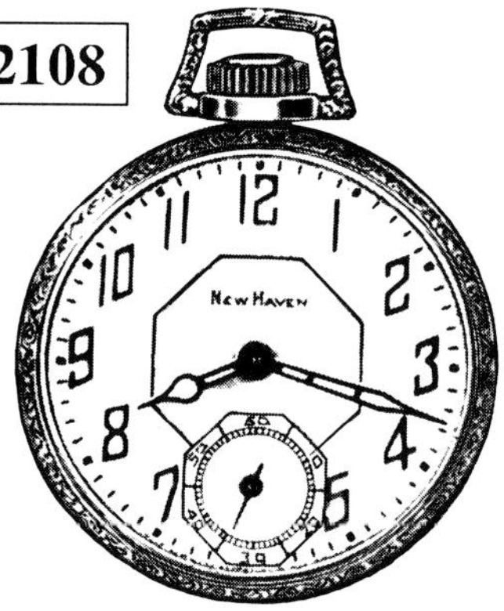

1937

NIFTY

Gold-plated case with engraved back, bezel and top ring. Silvered dial with raised numerals and Radium hands. Sunk second dial. Unbreakable crystal.

2109

1937

MODEL "D"

Chromium plated case. Black track with chrome numerals. Silvered dial; unbreakable crystal.

2110

1937

MODEL "E"

Thin, chromium plated case with engraved back, bezel and top ring. Silvered dial with black numerals; second hand. Unbreakable crystal.

WATCHES

2111

1937

100-MILE PEDOMETER

Nickel finish case. Hand may be set forward or backward, as desired, by removing back of case and turning hand set. Registers to 100 miles. May be hooked to belt or pocket. Adjustable to any length stride. Packed in lots of 48. Shipping weight: 7½ lbs.

RETAIL $2.50

2112

NURSES' SPECIAL

An attractive little pendant watch designed especially for nurses' use. This watch is unusually small in size and is suspended from a chromium-plated chatelaine. Chromium-plated case; etched numeral dial; RED SECOND HAND; and an unbreakable crystal. Packed in one dozen lots. Shipping weight: 2½ lbs.

RETAIL $3.50

2113

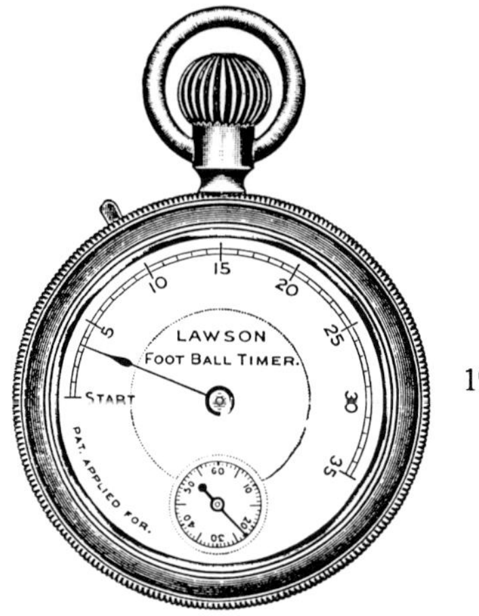

1917

LAWSON FOOTBALL TIMER

Size, No. 16

Registers minutes from 1 to 60, and the seconds. Has lever for starting and stopping.

Case, nickel plated.

Crystal, heavy and beveled.

2114

1937

TWO-TIMER

Combination watch and timer. Revolving second dial with stationary pointer. Side slide for stop-watch control. Chromium-plated case. Black numerals on a chrome band. Metal dial; unbreakable crystal. Packed in lots of 48. Shipping weight: 9 lbs.

2115

1937

Timers

SPORTS TIMER

Octagon case, nickel-plated. Indicates playing or working time from one second to the hour. Packed in lots of 48. Shipping weight: 9 lbs.

RETAIL $2.50

2116

1937

FOOTBALL TIMER

Octagon case, nickel-plated. Indicates playing or working time from one second to one hour. Packed in lots of 48. Shipping weight: 8¼ lbs.

RETAIL $4.00

WATCHES

2117

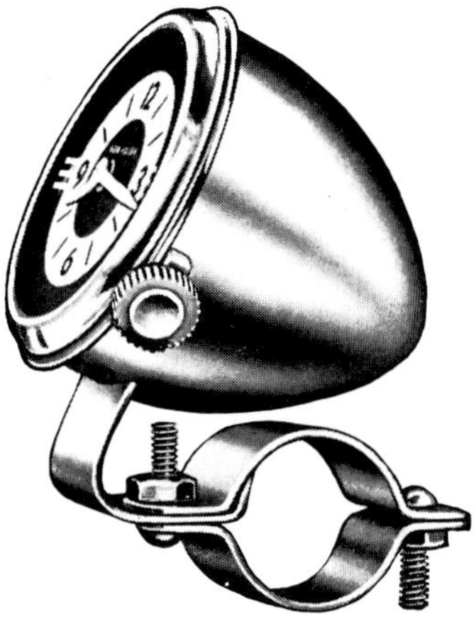

1937

2118

1911

E BACK

2119

1929

2120

1929

Novelty Watches

BICYCLE CLOCK

Chrome, streamlined case. Visible aviation type dial. Unbreakable crystal. Stem wind, stem set, 30-hour movement. Size: Width, 2 inches. Height, $2\frac{1}{2}$ inches. Depth, $2\frac{1}{8}$ inches. Bolts to handle bars. Packed in lots of 24. Shipping weight: 8 lbs.

TIP-TOP SERVICE

Silvered Dial, 6 Size, Thin, Nickel Silver Crown, Streamline Bow, Stem Wind, Pull-out Stem Set, Cut Steel Pinions, Removable Main Spring, Sunk Second Dial, Nickel-plated Dust-proof Case.

Silvered Dial, each.......$4.44 Silvered Radium, each......$5.10

2121

1937

2122

1937

2123

1937

THE MANHATTAN WATCH.

Nickel Plated	$ 7 50	Gold Plated	$ 8 70
" " Sweep Second	8 80	" " Sweep Second	10 00

We herewith show lithographs of the Manhattan Watch, for which we are the sole agents for the trade.

This Watch has been entirely reconstructed with a straight line lever escapement; winds at the stem, has a new hand-setting attachment, which is direct acting and the simplest and most effective ever invented. They are made with and without a sweep second hand, have enameled dials and are cased in either a combination silver metal case, or a gold plate on French oroide metal, having every appearance, as well as wearing qualities of a solid silver or gold cased watch. They are packed in boxes of six each, and if Jobbers prefer, when purchasing gross lots, can have them packed singly in wooden boxes that will bear transit by mail. Sample watches will be forwarded to any of our customers by mail on application.

These watches have our guarantee in every particular and can be regulated closer than any imported medium grade watch.

WATCHES

2124

1929

2125

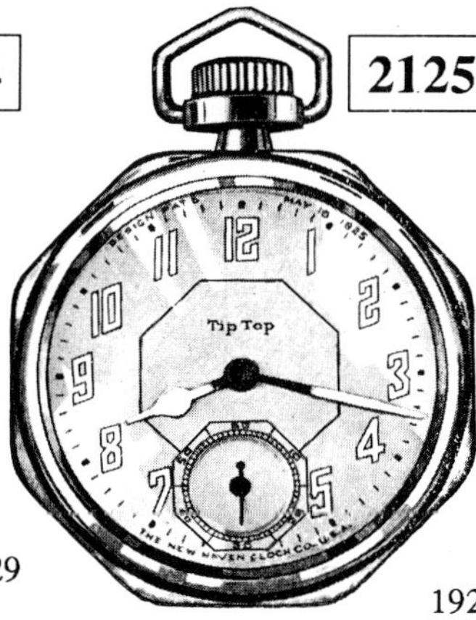

1929

TIP-TOP POCKET

Silvered Dial, 14 Size, Thin, Nickel Silver Crown, Streamline Bow, Stem Wind Pull-out Stem Set, Cut Steel Pinions, Removable Main Spring, Sunk Second Dial. Nickel-plated. Dust-proof Case.

Silvered Dial, each..........$2.08 Silvered Radium, each........$3.08

2126

Ca. 1932

NEWHAVEN POCKET

(Radium Dial)

The latest in pocket watches - a streamline case with fixed top ring; unbreakable crystal, attractive black and white dial with radium numerals and hands, and sunk second dial.

2127

1937

TIP TOP

Nickel plated, dust-proof case with engraved back and bezel. Fixed top ring. Two-tone silvered dial. Sunk second dial. Unbreakable crystal.

RETAIL $1.50

2128

1929

2129

1929

TOMMY-TICKER POCKET, ENGRAVED

Silvered Dial, 14 Size, Cushion Shape, Nickel-plated Case, Thin Model. Pull-out Set, Stem Wind, Sunk Second Dial. Engraved Back and Bezel, Nickel Silver Crown, Semi-octagon Bow.

Silvered Dial, each........$2.08 Silvered Radium, each.....$3.08

2130

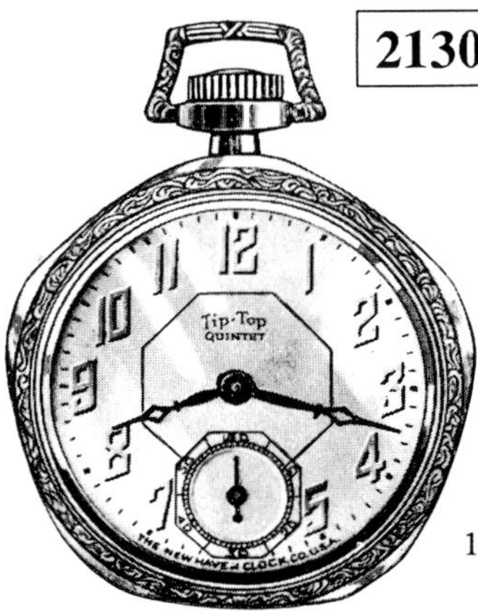

1929

2131

1929

TIP-TOP QUINTET

Chromium Plated Case. Engraved Back, Bezel and Bow

Quintet Shaped Chromium Plated Case, 14 Size, Silvered Dials, Pull-out Stem Set, Stem Wind, Cut Steel Pinions, Removable Main Spring, Sunk Second Dial.

Raised Numerals, each..........$2.74 Silvered Radium, each...........$3.46

2132

Ca. 1932

NEW HAVEN POCKET

(Silvered Dial)

The latest in pocket watches - a streamline case with fixed top ring; attractive two-tone silvered dial, black numerals, unbreakable crystal, and sunk second dial

WATCHES

Lapel Watches

2133

1937

PENTAGON LAPEL No. 1

Enameled case in a color selection of Red, Tan or Black with nickel plated sides, bezel and top ring. Two-tone silvered dial with an unbreakable crystal. Braided cord of black or tan leather. Packed in gross lots. Shipping weight: 33 lbs.

RETAIL $1.50

2134

1937

PENTAGON LAPEL No. 2

Enameled case in a color selection of Red, Tan or Black. Nickel plated sides, bezel and top ring. Black numerals on a chrome band. Metal dial with an unbreakable crystal. Braided cord of black or tan leather. Packed in gross lots. Shipping weight: 33 lbs.

RETAIL $1.50

2135

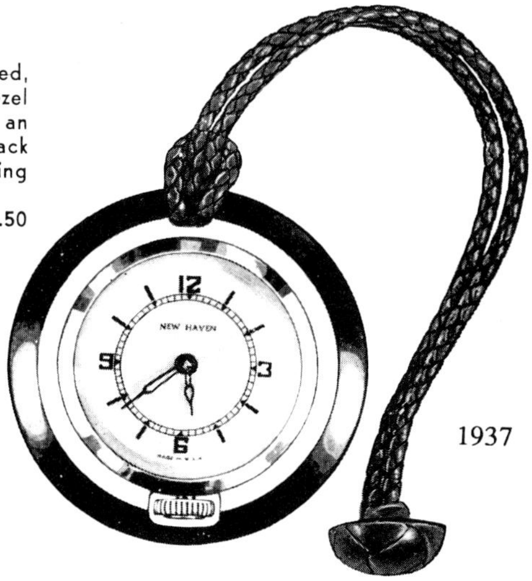

1937

LAPEL Jr.

Case of plastic composition—exceptionally thin and light in weight—in color combinations of black and red or black and white. Transparent back. Metal dial with modernistic indicators and numerals. Unbreakable crystal; chromium plated bezel. Braided cord of black leather. Packed in lots of 24. Shipping weight: 6¼ lbs.

RETAIL $3.50

2136

1937

DON WINSLOW Lapel Watch

Don Winslow colorfully reproduced on the back of a black enameled case. Nickel plated sides, bezel and top ring. Two-tone silvered dial; unbreakable crystal. Braided cord of black leather. Packed in lots of 36. Shipping weight: 7 lbs. 10 oz.

RETAIL $1.50

2137

1937

CLYDE BEATTY Lapel Watch

Clyde Beatty colorfully reproduced on the back of a black enameled case. Nickel plated sides, bezel and top ring. Two-tone silvered dial; unbreakable crystal. Braided cord of black leather. Packed in lots of 36. Shipping weight: 7 lbs. 10 oz.

RETAIL $1.50

Lapel Watches

2138

1937

LAPEL TIMER

Black, "crackle-finish" case. Nickel-plated sides, bezel and top ring. Black numerals on a chrome band. Metal dial; unbreakable crystal. Moving second dial with stationary hand. Braided cord of black leather. Packed in gross lots. Shipping weight: 33 lbs.

2139

1937

LONE RANGER Lapel Watch

Lone Ranger colorfully reproduced on the back of a black enameled case. Nickel plated sides, bezel and top ring. Two-tone silvered dial; unbreakable crystal. Braided cord of black leather. Packed in lots of 36. Shipping weight: 8 lbs. 5 oz.

2140

1937

LONE RANGER Lapel Watch

Lone Ranger colorfully reproduced on the back of a black enameled case. Nickel plated sides, bezel and top ring. Two-tone silvered dial; unbreakable crystal. Braided cord of black leather. Packed in lots of 36. Shipping weight: 8 lbs. 5 oz.

RETAIL $1.50

2141

1937

LONE RANGER Fob Watch

Lone Ranger colorfully reproduced on the back of a black enameled case. Nickel-plated sides, bezel and top ring. Two-tone silvered dial; unbreakable crystal. Equipped with a leather holster and a gun. Packed in lots of 36. Shipping weight: 10 lbs.

WATCHES

2142

A

2143

B

2144

D

2145

E

2146

F

2147

G

FOR ELM CITY, GEM AND JEROME

2148

I

2149

K

2150

L

2151

M

2152

N

2153

R

FOR ELM CITY AND GEM

2154

E. A.

2155
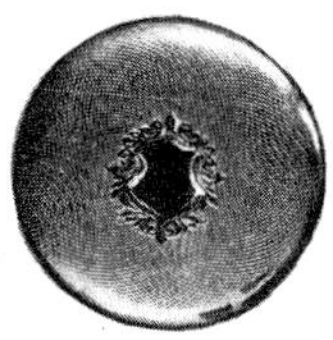
E. B.

FOR ELITE

2156

L. A.

2157
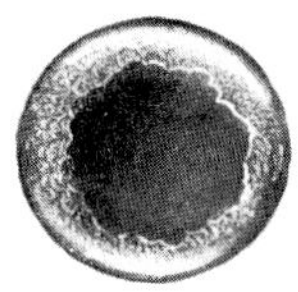
L. B.

2158

L. C.

2159

L. D.

FOR LADY CLARE

2160

R. A.

2161

R. B.

2162
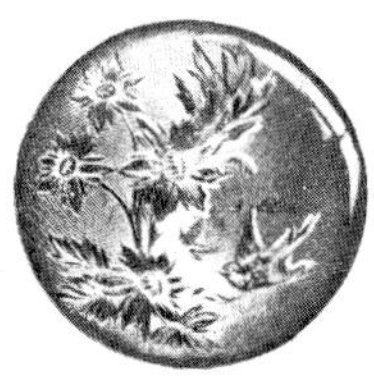
R. C.

2163

R. D.

2164

R. E.

2165

R. F.

FOR RANGER

FANCY BACKS FURNISHED WITHOUT EXTRA CHARGE ON ALL WATCHES (EXCEPT THE ELITE WATCH)
PLAIN BACKS REGULARLY FURNISHED.

THE NEW HAVEN SPORTS TIMER
FOOTBALL
BASKETBALL
BOAT RACING
AUTO RACING
HOCKEY
WRESTLING
RUNNING
BOXING
Stop Lever for "Time Out"
Red Numerals for Timing.
Start Lever.
Standard Time.
THE NEW HAVEN SPORTSTIMER
PATENT PENDING
-A- Watch
$2.50
-A- Timer
Boys!
THE NEW HAVEN SPORTS TIMER
is a
Perfect Timekeeper
and
Sports Timer
Combined.
Better for Games than a Stop Watch.
EST. THE NEW HAVEN CLOCK CO. NEW HAVEN CONN. 1817
THE NEW HAVEN SPORTS TIMER
Watch
Timer
Packed 12 Watches in Carton
THE NEW HAVEN CLOCK CO. NEW HAVEN CONN.
Two Minutes to Play!
By the
STOP LEVER

WATCHES - ADVERTISING

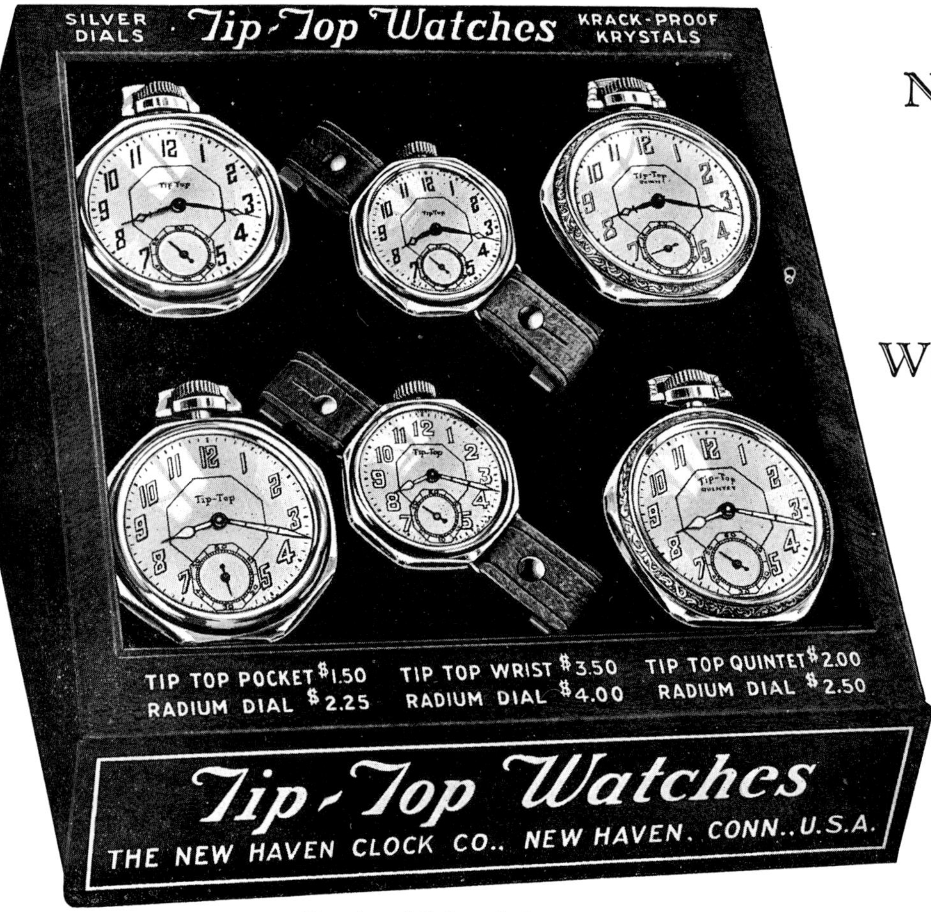

New Haven Tip-Top Pocket and Wrist Watches

Dimensions of display and sales case, 6¾ inches long, 5¼ inches high, 7 inches deep, made of metal, of strong, durable construction attractively finished in dull mahogany, full-vision glass front, storage space in back for extra stock; removable display plaque which displays the watches to the best advantage, creating an unusual eye-catching sales appeal.

No. 12 NEW HAVEN TIP-TOP POCKET AND WRIST WATCH SAMPLER, with handsome mahogany finished metal display and sales case.

This watch sampler consists of twelve (12) popular priced pocket and wrist watches with silver dials. All Nationally advertised.

Every watch is given a four-day test run and carefully inspected for accuracy before leaving the factory. They are really *True Time Tellers.*

Summary of No. 12 *Tip-Top Watch Sampler:*	List Price Each	Total List Price	Suggested Consumer's Price	Total Retail Value
3—*only,* TIP-TOP POCKET WATCHES (14 Size), Silver Dial, *Ornate Hands, Black Cubist Numerals*	$2.08	$6.24	$1.50	$4.50
1—*only,* TIP-TOP POCKET WATCH (14 Size), Silver Dial, *Radium Hands and Numerals*	3.08	3.08	2.25	2.25
3—*only,* TIP-TOP QUINTET, Chromium Plated Case, Quintet Shaped, Engraved Back, Bezel and Bow. Silver Dial, *Raised Numerals*	2.74	8.22	2.00	6.00
1—*only,* TIP-TOP QUINTET, Chromium Plated Case, Quintet Shaped, Engraved Back, Bezel and Bow. Silver Dial, *Radium Numerals and Hands*	3.46	3.46	2.50	2.50
2—*only,* TIP-TOP WRIST WATCHES (6 Size), Silver Dial, *Ornate Hands, Black Cubist Numerals*	4.76	9.52	3.50	7.00
2—*only,* TIP-TOP WRIST WATCHES (6 Size), Silver Dial, *Radium Hands and Numerals*	5.44	10.88	4.00	8.00
1—*only,* Handsome Display Case		1.60		
		$43.00		$30.25

2167

2166

2168

2169

1883

SMITH'S ✣ WATCH ✣ SIGNS.

(SEE CUT.)

ILLUSTRATIONS AND PRICE LIST

— OF —

Smith's American Shield, Watch, Clock, and Jewelry

SIGNS

AND THE AMERICAN GEM SPECTACLE SIGNS.

PERFECTED.

NOTICE.—Letters Patent have just been granted us for valuable improvements on these Signs, and also on our SHIELD SIGNS. We believe the Trade will admit that we have at last produced Signs that for Beauty, Durability, and Cheapness have not been, and cannot be excelled.

Each Sign is packed securely in a strong wooden box, and can be forwarded to any part of the world with perfect safety. Larger sizes manufactured to order.

Regular sizes 36 and 47 inches. Special cut. See Price List.

GEM SIGN DESIGN No. 2.

1883

2170

WRIST WATCHES

2171

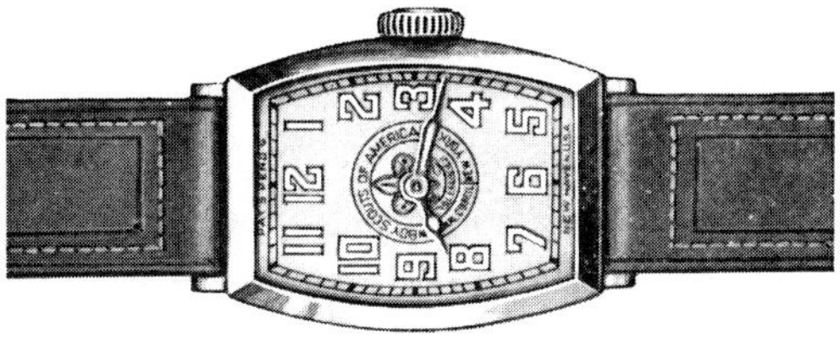

Ca. 1926

BOY SCOUT (7-Jewel)—6/0 size, 7-jewel movement. "BEK" YELLOW GOLD plated case. Stainless, chromium plated back—curved to fit the wrist. Boy Scout seal on dial. Radium hands and numerals. Unbreakable crystal. Pigskin strap with gold plated buckle. RETAIL $6.50

2172

Ca. 1926

BROWNIE—8/0 size. Chromium plated case with stainless back—curved to fit the wrist. Two-tone silvered dial with an unbreakable crystal. Choice of black leather strap or link bracelet.
RETAIL $3.50

2173

Ca. 1926

COUNT (7-Jewel)—6/0 size, 7-jewel movement. "BEK" YELLOW GOLD plated case. Stainless, chromium plated back—curved to fit the wrist. Metal dial with etched numerals and an unbreakable crystal. Pigskin strap with gold plated buckle.
RETAIL $4.95

2174

Ca. 1926

COUNTESS (7-Jewel)—8/0 size, 7-jewel movement. "BEK" YELLOW GOLD plated case. Stainless, chromium plated back—curved to fit the wrist. Metal dial with etched numerals and an unbreakable crystal. Pigskin strap with gold plated buckle. RETAIL $5.95

2175

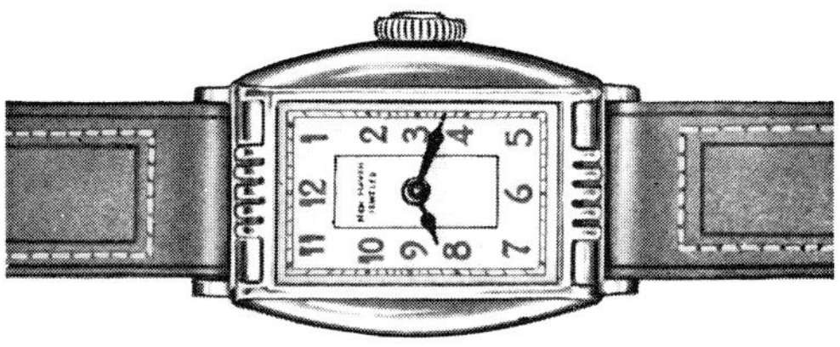

Ca. 1926

EARL (2-Jewel) — 6/0 size, 2-jewel movement. "BEK" YELLOW GOLD plated case. Stainless, chromium plated back—curved to fit the wrist. Metal dial with etched numerals and an unbreakable crystal. Pigskin strap with gold plated buckle.
RETAIL $4.75

2176

Ca. 1926

ELF (2-Jewel)—8/0 size, 2-jewel movement. Chromium plated case with stainless back—curved to fit the wrist. Metal dial with black numerals and an unbreakable crystal. Choice of black leather strap or link bracelet. RETAIL $3.95

2177

Ca. 1926

DUCHESS (7-Jewel)—8/0 size, 7-jewel movement. "BEK" YELLOW GOLD plated case. Stainless, chromium plated back—curved to fit the wrist. Metal dial with etched numerals and an unbreakable crystal. Two-tone brown "checker-board" strap with chromium plated ratchet clasp.
RETAIL $6.75

2178

Ca. 1926

DUKE (7-Jewel)—8/0 size, 7-jewel movement. "BEK" YELLOW GOLD plated case. Stainless, chromium plated back—curved to fit the wrist. Metal dial with etched numerals and an unbreakable crystal. Pigskin strap with gold plated buckle.
RETAIL $5.95

WRIST WATCHES

Compensated for Temperature Changes.

"BEK" Gold plated cases. Curved to fit the wrist.

2179

Ca. 1926

WALES (7-Jewel)—6/0 size, 7-jewel movement. "BEK" YELLOW GOLD plated case. Stainless, chromium plated back—curved to fit the wrist. Metal dial with etched numerals and an unbreakable crystal. Pigskin strap with gold plated buckle.
RETAIL $5.95

2180

Ca. 1926

YALE—6/0 size. "BEK" YELLOW GOLD plated case. Stainless, chromium plated back—curved to fit the wrist. Two-tone silvered dial with an unbreakable crystal. Pigskin strap with gold plated buckle.
RETAIL $3.95

2181

Ca. 1926

JEROME (2-Jewel)—6/0 size, 2-jewel movement. Chromium plated case with stainless back—curved to fit the wrist. Metal dial with black numerals and an unbreakable crystal. Choice of black leather strap or link bracelet.
RETAIL $3.95

2182

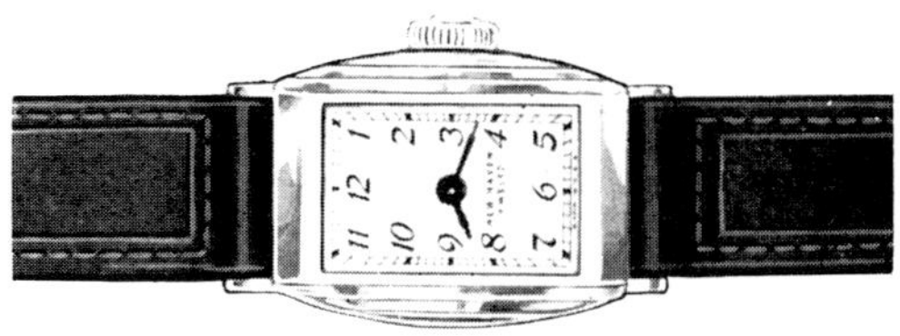

Ca. 1926

JEROME JR. (2-Jewel)—8/0 size, 2-jewel movement. Chromium plated case with stainless back—curved to fit the wrist. Metal dial with black numerals and an unbreakable crystal. Choice of black leather strap or link bracelet.
RETAIL $3.95

2183

Ca. 1926

JEROME (7-Jewel)—6/0 size, 7-jewel movement. Chromium plated case with stainless back—curved to fit the wrist. Metal dial with etched numerals and an unbreakable crystal. Choice of black leather strap or link bracelet.
RETAIL $5.00

2184

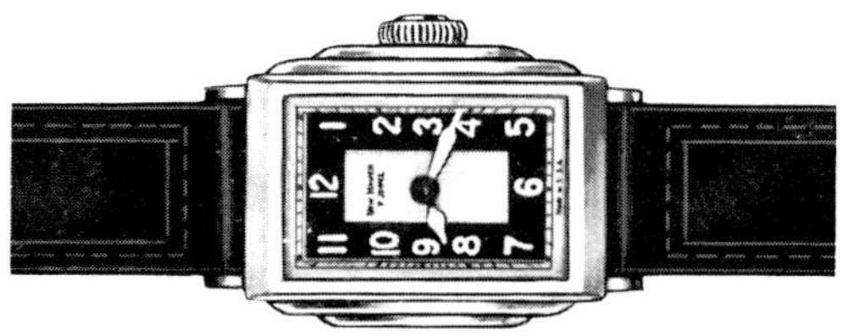

Ca. 1926

JEROME (7-Jewel)—6/0 size, 7-jewel movement. Chromium plated case with stainless back—curved to fit the wrist. Black and silvered dial with radium numerals and hands. Unbreakable crystal. Choice of black leather strap or link bracelet.
RETAIL $5.50

2185

Ca. 1926

INCA (2-Jewel)—6/0 size, 2-jewel movement. Chromium plated case with stainless back—curved to fit the wrist. Metal dial with black numerals and an unbreakable crystal. Choice of black leather strap or link bracelet.
RETAIL $3.95

2186

Ca. 1926

LADY JOAN—8/0 size. Chromium plated case with stainless back—curved to fit the wrist. Metal dial with black numerals and an unbreakable crystal. Choice of black leather strap or link bracelet.
RETAIL $3.50

WRIST WATCHES

2187 2188

Ca. 1926

TIP-TOP WRIST

Silvered Dial, 6 Size, Thin, Stem Wind, Pull-out Stem Set, Cut Steel Pinions, Removable Main Spring, Sunk Second Dial, Detachable Genuine Pig Skin Strap. Non-corrosive Chromium Plated Back that hugs the wrist snugly.

Silvered Dial, each..........$1.76 Silvered Radium, each $5.44

2189

Ca. 1926

DICK TRACY—6/0 size. Chromium plated case with stainless back—curved to fit the wrist. Dick Tracy reproduced in full colors on a silvered dial. Unbreakable crystal. Blue leather strap.

RETAIL $3.50

2190

Ca. 1926

FAD—6/0 size. Chromium plated case with stainless back—curved to fit the wrist. Two-tone silvered dial with an unbreakable crystal. Choice of black leather strap or link bracelet. RETAIL $2.75

Chromium plated cases.
Curved to fit the wrist.

2191 2192

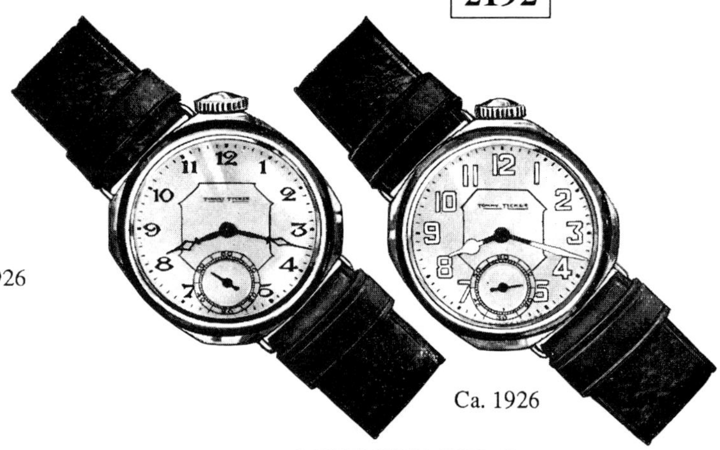

Ca. 1926

TOMMY-TICKER WRIST

Silvered Dial, 6 Size, Cushion Shape, Nickel-plated Case. Thin Model, Pull-out Set. Stem Wind. Offset, easy-to-read angle, Sunk Second Dial, Detachable Genuine Pig Skin Strap, Non-corrosive Chromium Plated Back that hugs the wrist snugly.

Silvered Dial, each ....$5.10 Silvered Radium, each...........$5.78

2193

Ca. 1926

BOY SCOUT—6/0 size. Chromium plated case with stainless back—curved to fit the wrist. Boy Scout seal on dial. Radium numerals and hands. Unbreakable crystal. Black leather strap.

RETAIL $4.00

2194

Ca. 1926

GEM—6/0 size. Chromium plated case with stainless back. Two-tone silvered dial with an unbreakable crystal. Choice of black leather strap or link bracelet. RETAIL $2.75

2195

Ca. 1926

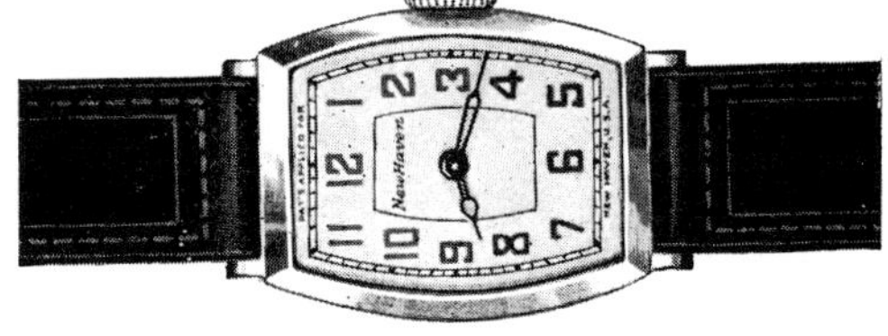

ELITE—6/0 size. Chromium plated case with stainless back—curved to fit the wrist. Metal dial with etched numerals and an unbreakable crystal. Choice of black leather strap or link bracelet.

RETAIL $3.25

WRIST WATCHES

2196

Ca. 1936

PAR ROUND—6/0 size. "BEK" YELLOW GOLD plated case. Stainless steel back. Two-tone silvered dial with etched numerals and an unbreakable crystal. SUNK SECOND DIAL. Pigskin strap with gold plated buckle.

2197

Ca. 1937

2198

Ca. 1937

CW905—"Countess"—$8.90
(Advertised Consumer's Price $6.50)
Filled Yellow Gold Plated case, 7 Jewel movement, Leather Strap, Silvered dial with black numerals.

2199

Ca. 1937

CW907—"Duchess"—$10.20
(Advertised Consumer's Price $7.50)
Case in Filled Yellow Gold Plate, two-tone brown leather strap in checkerboard design. Ratchet clasp, 7 Jewelled movement.

2200

Ca. 1937

CW908—"Fad"—$3.80
(Advertised Consumer's Price $2.90)
White metal case, black and silvered dial, smartly styled, leather strap.

2201

Ca. 1937

"Orphan Annie"
CW902—$4.60
(Advertised Consumer's Price $3.50)
American made guaranteed timepiece, Chromium finished case, curved back. Character in color on silver dial.

2202

Ca. 193

CW901—"Count"—$8.20
(Advertised Consumer's Price $5.95)
This new watch has a dependable 7 Jewel Movement. Case of Filled Yellow Gold Plate; silvered dial, raised gilt figures.

WRIST WATCHES

Krack Proof Krystals

2203

1931

2204

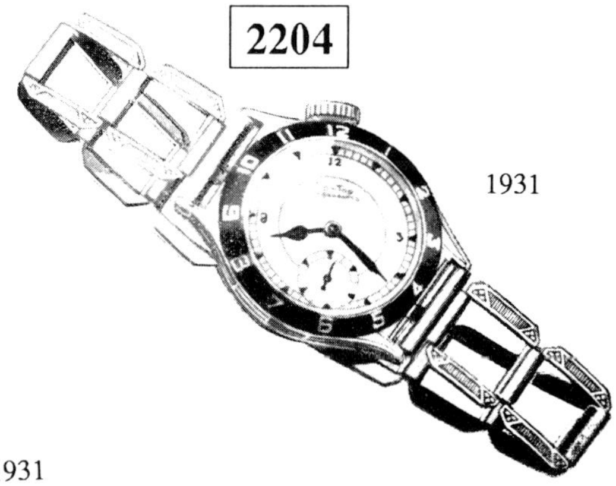

1931

TIP TOP PAR Compensated for Temperature

A distinctive chromium plated wrist watch with silvered dial and raised numerals on black enamel. Diamond cut escape wheel, polished and hardened pinions, main spring arbors and balance staff. Available with leather strap, metal mesh band or link bracelet and cased in a handsome black and red box.

2205

Ca. 1926

POPEYE—6/0 size. Chromium plated case with stainless back—curved to fit the wrist. Popeye characters colorfully reproduced on a silvered dial. Unbreakable crystal. Blue leather strap.
RETAIL $3.50

2206

Ca. 1926

ORPHAN ANNIE—8/0 size. Chromium plated case with stainless back—curved to fit the wrist. Orphan Annie colorfully reproduced on a silvered dial. Unbreakable crystal. Leather straps in Red, Blue or Tan.
RETAIL $3.50

2207

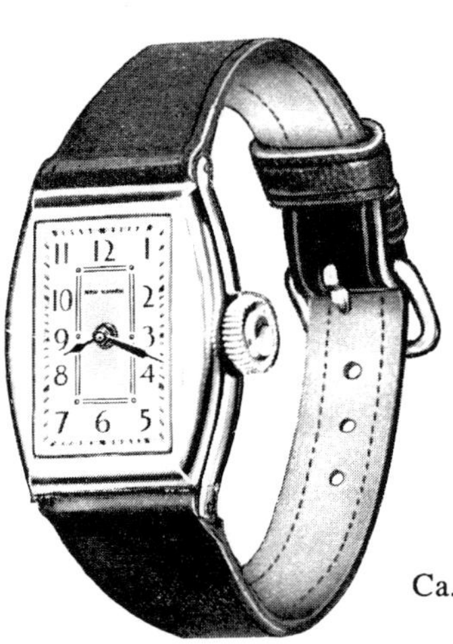

Ca. 1927

New FAD

A smartly styled wrist watch with new and thinner case in chromium plate; and new, attractively designed two-tone silvered dial. Available with detachable leather strap or link bracelet.

2208

Ca. 1927

New NEW HAVEN

A handsomely styled wrist watch with new and thinner case in chromium plate; and newly designed black and silvered dial. Choice of link bracelet or leather strap.

2209

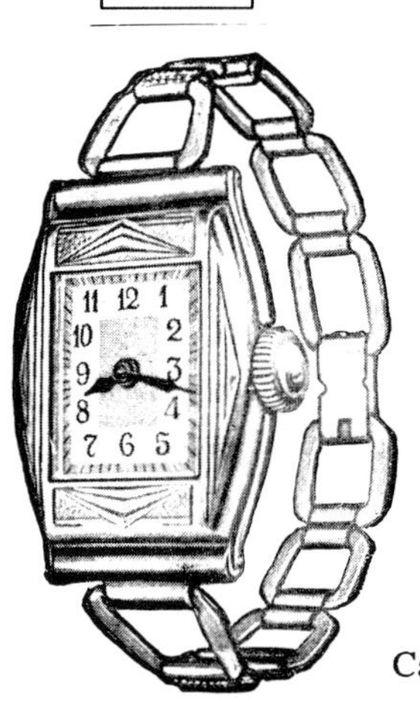

Ca. 1927

New GEM

A wrist watch of smart and distinctive appearance with new and thinner case, engraved and chromium plated. New, attractive two-tone silvered dial. Equipped with detachable leather strap or link bracelet.

WRIST WATCHES

2210

BROWNIE—8/0 size. Chromium plated case with stainless steel back—curved to fit the wrist. Silvered dial with black numerals and an unbreakable crystal. Choice of black leather strap or link bracelet.

2211

BROWNIE

A wrist watch of unusual value and distinction. Chromium plated case, thin and small; and an attractive two-tone silvered dial. Available with detachable leather strap, or link bracelet with ratchet clasp.

2212

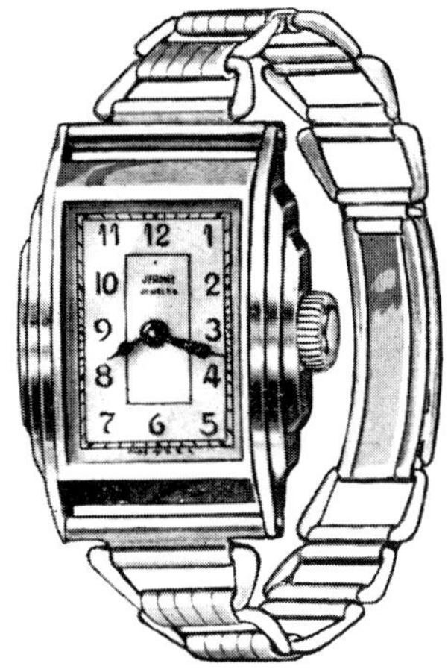

JEROME (Jeweled)
(Silvered Dial)

A beautifully designed Wadsworth case in chromium finish, thin and small; with a dependable 6\0 JEWELED movement; and silvered dial with etched gold numerals and black hands. Choice of link bracelet with ratchet clasp; or leather strap.

NEW HAVEN USA TRADE MARK

Ca. 1927

2213

JEROME Jr. (Jeweled)

A small, 8\0 size wrist watch with genuine Wadsworth case in chromium plate; JEWELED New Haven movement; and raised numerals on a silvered dial. A choice of leather strap; three-strand corded leather strap with ratchet clasp; or link bracelet with ratchet clasp.

2214

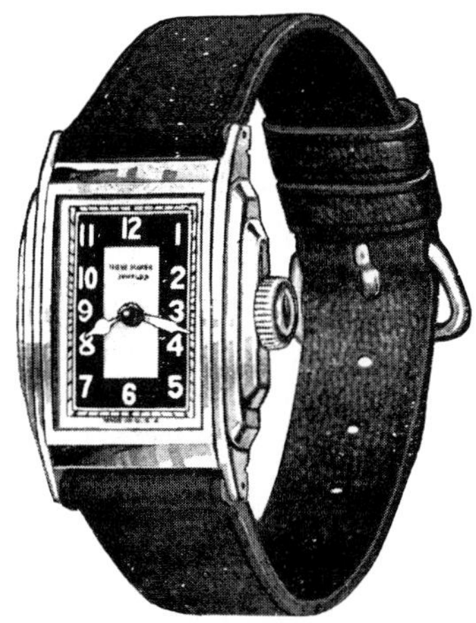

JEROME (Jeweled)
(Radium Dial)

A beautifully designed Wadsworth case in chromium finish, thin and small; with a dependable 6\0 JEWELED movement; and black and silvered dial with radium numerals and hands. Choice of link bracelet with ratchet clasp; or leather strap.

2215

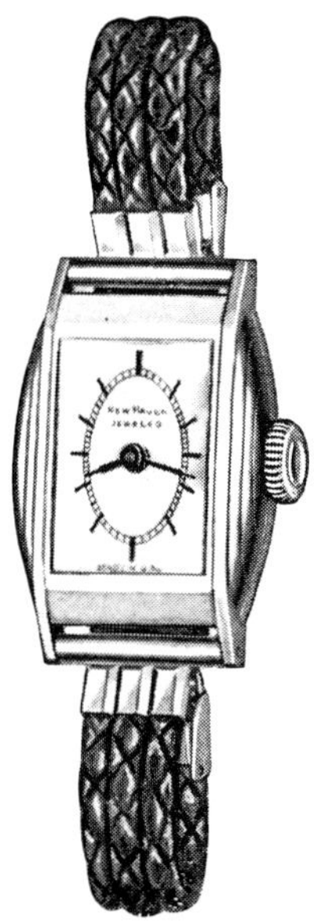

JEROME Jr. (Jeweled)

A small 8\0 size wrist watch with genuine Wadsworth case in chromium plate; JEWELED New Haven movement; and oval index dial in silver finish. A choice of leather strap; three-strand corded leather strap with ratchet clasp; or link bracelet with ratchet clasp.

WRIST WATCHES

Chromium-plated cases.
Curved to fit the wrist.

2216

Ca. 1937

CW900—"Duke"—$8.90
(Advertised Consumer's Price $6.50)
A simple designed wrist watch, case of Filled Yellow Gold Plate; Chrome Plated back, 7 Jewels.

2218

Ca. 1936

TROPHY—6/0 size. Engraved, chromium plated case with stainless steel back—curved to fit the wrist. Metal dial with black numerals and an unbreakable crystal. SUNK SECOND DIAL. Choice of black leather strap or link bracelet.

2219

Ca. 1937

CW903—"Elf"—$6.80
(Advertised Consumer's Price $5.00)
Jewelled movement with a silver dial. White metal case with a silk cord.

2221

Ca. 1936

COMMUTER—6/0 size. Chromium plated case with stainless steel back—curved to fit the wrist. Metal dial with black numerals and an unbreakable crystal. Choice of black leather strap or link bracelet.

2217

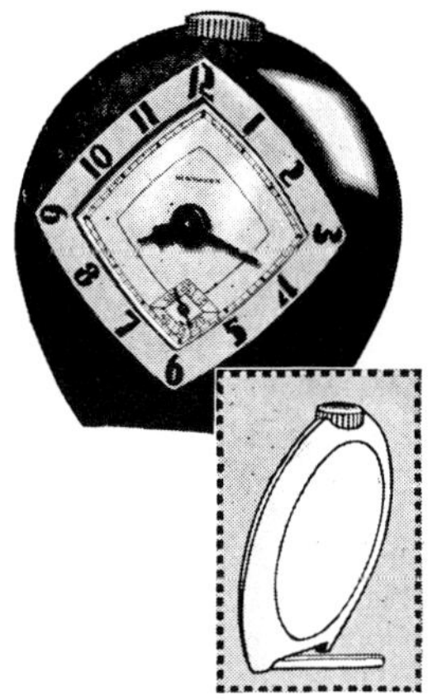

Ca. 1937

Hand Bag Watch
CW911—$4.00
(Advertised Consumer's Price $2.95)
Combination Ladies' Hand Bag and Dresser Timepiece, Black Enamel Case. The supporting legs fold into case and become invisible.
Size 2¼" x 2¼".

2220

Ca. 1937

SERVICE

A popularly priced, thin model wrist watch that hugs the wrist. Case in chromium plate, designed and built by Wadsworth; silvered, index dial; and a sturdy, dependable movement of proven merit. Available with pigskin strap or link bracelet with detachable lugs.

INDEX

ADVERTISING CLOCKS 25-27

Balcony Electric 27
Bragg, Cummings & Co. 25
Duffy's Pure Malt Whiskey 26
Harvard Brewing Co. 27
Lucky Strike 25
Movements 8-day time..25,26
R. A. Patterson Tob. Co. 25
Referee Electric 27
Saturn Electric 27
Snowball Baking Powder
Pure Wholesome 27

ALARM CLOCKS 28-48

Abbey Jr. Artlarm 47
Acorn 35
Alcove Tattoo 41
Alert 28
Alert - Extra Case. 28
Alert -T. P. 28
Artlarm, Abbey Jr. 47
Artlarm, Doris 47
Artlarm, Easel 43
Artlarm, Square 47
Artlarm, Tat-Too 30
Artlarm, Tat-Too, Radium..30
Artlarm, Tat-Too Jr. 46
Auto 38
Bamboo 28
Beacon 31, 33
Beacon - Luminous 33
Boudoir Tattoo 45
Brummel 29
Cairo 43
Campus 31
Capri 31
Cascade 29
Castro 46
Cecil 45
Ceres 45
Ceylon 31
Champion 32, 33

ALARM CLOCKS (continued)

Chico 43
Chrome - Square 48
Circle 39
Cisco 45
Commuter 39
Day Dream 35
Doris Artlarm 47
Easel Artlarm 43
Egyptian.............................28
Eight-Day Automatic
Black Dial 40
Eight-Day Automatic
White Dial 40
Eight-Day Self-Setter 41
Elfin 39
Eunice 46
Flash - Luminous 37
Fly 37
Futurist - Musical 48
Giant 38
Globe 35
Guest Tattoo 45
Herald 37
Irma 46
Jigger 35
Junior Tattoo 41
Junior Tattoo, Leather Case 41
Kalon 42
Keating 42
Keith 42
Kelvin 42
Knox 42
Lark Repeating Strike 36
Lark Switchless 36
Lark T. P. 36
Lark Traveling Case 36
Library Tattoo 44
Macon 41
Marvel 46
Moonshine 29
Morita 43
Negus 43

ALARM CLOCKS (continued)

Nelson No. 2 48
Nemo 42
Padlock 30
Padua 41
Paral 29
Popeye 48
Prompter 40
Rapid Fire 39
Sentry 29
Shield No. 1, T. P. 30
Shield No. 2, T. P. 30
Slumber Stopper 48
Snipe 48
Soft- Tick 48
Spero 31
Sphere 40
Sprite 34
Square Artlarm 47
Square Tell-Tale 44
Start 37
Sting 34, 39
Sting Repeating Strike........34
Study Tattoo 44
Sultana 34
Tattoo, Alcove 41
Tattoo, Boudoir 45
Tattoo, Guest 45
Tattoo - Intermittent, Black
Dial 32
Tattoo - Intermittent, White
Dial 32
Tattoo, Junior 41
Tattoo, Junior
Leather Case 41
Tattoo, Library 44
Tattoo, Study 44
Tattoo, Traveler 44
Tat-Too Artlarm 30
Tat-Too Artlarm, Radium ..30
Tat-Too Jr., Artlarm 46
Tick Tock 44, 47
Tidy-Tot............................47

ALARM CLOCKS
(continued)

Tip-Top Traveler 48
Tocsin 34
Tom-Tom, Plain Dial 44
Tom-Tom, Radium 44
Tourist No. 2 31
Traveler Tattoo 44
Victory 29
Waldo 42
Wanda 41
Watson 42
Welby - Intermittent 46
Weston 42
Whistler, Fancy Dial 35
Whistler, Roman Dial 35

AUTOMOBILE CLOCKS 49-58

Ascot, Plain Dial 56
Ascot, Radium Dial 56
Auto, Plain Dial 56
Auto, Radium Dial 56
Auto Watch 55
Belmont 55
Bicycle Clock 58
Cowl 52
Daytona 51
Devere 51
Dieppe 51
Duane 51
Fiat Motor Cycle Clock 50
Jericho 51
Jerome 52, 53
Mineola 49
Mineola Sectional View 49
Mirror Clock Combinations
No. 2 AV 55
No. 17 AV 54
No. 30 AV 57
No. 31 AV 57
No. 32 AV 57
No. 33 AV 57
No. 34 AV 57
No. 55 AV 53
No. 55 AV - 6A 55
No. 55 AV - 7A 55
No. 55 AV - 8A 55
No. 61 AV 57
No. 62 AV 57

AUTOMOBILE CLOCKS
(continued)

No. 63 AV 57
No. 73 AV 54
No. 73 AV - 1 55
No. 77 AV 58
No. 79 AV 54
No. 81 AV 54
No. 90 AV 53
No. 92 AV 53
No. 93 AV 54
No. 95 AV 53
No. 96 AV 54
No. 99 AV 53
No. 165 AV 52
No. 1001 56
No. 2001 58
No. 5000 52
No. 7001 58
Moderne 52
Mounting Bracket 50
Nebo Motor Cycle Clock 50
New Haven Trademark 57
New Ormond 50
Ormond 50
Savannah 49
Westbury 49

BANJO CLOCKS 59-64

Myran 60
Richelieu 63
Waldron 61
Walton 61
Wardrop 62
Waring 64
Waring No. 9761 62
Waring No. 9762 62
Warren 62
Warwick Timepiece 64
Washington 64
Waverly 63
Wayland 64
Welton 62
Welton No. 9765 62
Whalley 60
Whitney 64
Willard 59, 60
Willard, Edw. K. 59
Willis 59, 60
Willmar 59

BANJO CLOCKS
(continued)

Wilmot 60
Wilson 63
Windrow 61
Windrow No. 2 61
Winetka 60, 63
Winner 63
Winona 63
Winsome, Strike 60
Winsome, Time 60
Winston 61
Worthington, George 59
Worthy 63

CALENDAR CLOCKS 65-86

Austrian 78
Babson 82
Barbara 79
Beacon 85
Blake 81
Braddock 78
Cabinet 68
Cabinet No. 4 66
Cabinet No. 5 66
Cabinet No. 6 69
Cabinet No. 7 68
Calendar, Plush 83
Column 74
Calendar Mechanism 74
Conroy 82
Crown 73
Calendar Mechanism 73
Curlew 67
Dneister 70
Dneister, with Base 70
Drop Octagon 81
8, 10, 12-Inch 81
Echo 86
Fashion 73
Fly Alarm 85
Fly Time 85
Gibson 82
Globe 85
Gloriana 78
Hebe 80
Heron 72
Indicator 76
Indicator, D. R. 75
Intrepid 79

CALENDAR CLOCKS
(continued)

Jerome & Company 65-72
Leader 72
Mallard 67
Movements
Column 74
Jerome Perpetual Cal. 68
Leader Calendar 72
New Haven 69
Register, Mantel 71
Regulator No. 1 77
National Cal. Clock Co. 73,74
Number 1 Regualtor 77
O. G. 74
Plover 65
Register, Hanging 71
Register, Mantel 71
Regulator 76
Regulator, D. R. 75
Regulator No. 1 77
Rokeby (Spring) 70
Round Calendar Lever 76
Round Corner Lever 83
Rutland 80
Sampson 82, 86
Sprite 84
Sprite, Alarm 84
Start (T.P. Cal. Alarm) 84
Tampa 83
Vistula 83
Whistler 86
Whistler, Alarm 86
Wood Lever 76

CARRIAGE CLOCKS 87-93

Acme 88
Adela 90
Alert - Plush 88
Badger 91
Carriage Alarm 90
Cavalier 89
Chopin 93
Edmond 92
Kenneth 92
Kenwick 92
Leroy 93
Marvel 93
Messenger 91
Nadir 93

CARRIAGE CLOCKS
(continued)

Nana 91
Nugget 92
Nugget In Leather Case 92
Oberon 89
Padlock Repeating Strike ... 90
Panza 90
Paral 93
Paris 93
Peacock 88
Pigeon 89
Pilgrim 87
Samoa 91
Starling 89
Sutherland 91
Titania 89
Truro 90

CHIME CLOCKS 94-101

Abbey 98
Amiens 101
Art Moderne 94
Chester 100
Chime No. 4 .... 95, 96
Chime No. 6 97
Chime No. 7 97
Chippendale 94
Cloister - Self Starting Electric or 8-Day Pendulum ... 98
Duna - Self Starting Electric or 8-Day Pendulum 99
Durham - Self Starting Electric or 8-Day Pendulum ... 98
Eight-Bell Chime No. 1 96
Electric 98-99
Four-Bell Westminster No. 3 96
Four-Bell Westminster No. 5 97
Gamma 101
Granville 99
Harkness 100
Inglewood 98
Ingomar 99
Lincoln 100
Movement, Chime No. 4 95
Orleans 100
Peerless - Self Starting Elec- .. tric or 8-Day Pendulum ... 99

CHIME CLOCKS
(continued)

Sheraton 94
Sigma 101
Theta 101
Toledo 100
Washington Banjo 99
Westminster Chimes 97
Whitney Banjo 94
Willard Banjo 94
Willcock Patent 95
Winsome Banjo 94

CRYSTAL REGULATORS 102-107

Taunton 106
Tecumseh 102
Thames 102
Theone 106
Theron 102
Thespia 102
Theta 106, 107
Movement 107
Thopas 104
Thoreau 103
Thrace 105
Thurston 105
Tioga 103
Tolland 103
Toluso 105
Torla 105
Tucson 104
Tuxford 106
Tycoon 104

DIALS 108-109

American Gilt 108
American White 108
Cast Sash 109
Empire Sash, Convex Glass 108
French Sash 108
Ivorine Dials 109
Ivorized Porcelain Dials ... 109
Pearl Dials 109
Porcelain Dials 108
Silver Dials 108
Struck Sash 109
Visable Escapements 108

ELECTRIC CLOCKS 110-115

8-Inch Wall 110
12-Inch Wall 110
14-Inch Wall
(EC 611-31T) 113
Alford Alarm 115
Alford Timepiece 115
Amiens 111
Banjo No. 3
(NHW 611-131T) 113
Bantry Alarm 114
Campus Alarm 113
Claremont Timepiece 115
Commuter Alarm 114
Devere Strike 114
Devon Timepiece 115
Easel 113
EC 611-30T 112
EC 611-31T 113
EC 611-37S 112
Elko Timepiece 2 TB 113
Elm City Electric Clocks .. 111
Elwood Strike 114
Eureka Alarm 113
Farmington Strike
(NHW 611-85S) 112
Futurist Alarm 113
Gallery 110
Gallery (EC 611-30T) 112
Gothic 111
Harmonious 111
Inca (NHW 611-134T) 113
Larina 110
Luray (NHW 611-53T) 112
Madena 110
Maravia 110
Mars Alarm 114
Montrose Timepiece 115
NHW 611-53T 112
NHW 611-73S 112
NHW 611-76S 112
NHW 611-85S 112
NHW 611-131T 113
NHW 611-134T 113
Octet Alarm 114
Onyx No. 2 Timepiece 114
Petite 111
Renfrew Timepiece 115
Seaman Timepiece 115
Shelley Strike

ELECTRIC CLOCKS (continued)

(NHW 611-76S) 112
Square 110
Stylis Timepiece 114
Tambour No. 8 Strike
(NHW 611-73S) 112
Tambour No. 90 111
Tattoo Alarm 114
Toledo 111
Turin Chime or Strike 112
Vilna Timepiece 115
Wall, 8-Inch 110
Wall, 12-Inch 110
Wall, 14-Inch 113
Warwick Banjo 110
Whalley Banjo 110
Whitney Banjo 110
Whitney (EC 611-37S) 112
Winslow Easel Alarm 115
Winslow Easel Timepiece 115

FIGURE CLOCKS 116-130

Benvenuto Cellini 119
Bernard Palissy 119
Brennus 126
Carthage 120
Clotho 121
Clovis 120
Don Caesar 116
Don Juan 116
Duquay Tronin 125
Fame 129
Fisher Boy 122
Flower Girl 129
Flower Girl Extra 129
Flute Player 119
Gaul 116
Gaulois and Amazon 117
Horse 130
Horse Extra 130
Hunter and Dog 120
Imogene 118
Infantry 122
Ivanhoe 124
Ivanhoe Extra 124
Jean Bart 125
Knight 127
Knight Extra 127
La Caresse 127

FIGURE CLOCKS (continued)

La Fonte 124
La Forge 125
Lulli 122
Marine 128
Mignon 121
Norman 123
Octavius 119
Philip 122
Poetry 130
Psyche 117
Raphael 118
Rebecca 128
Roman 121
Rubens 118
Ruth 125
Saxon 128
Sir Christopher 121
Spaniard 117
Van Dyke 118

HANGING CLOCKS 131-181

Acorn Lever 147
Admiral 142
Aida 180
Amoor 139
Austrian 148
Balcony, 12-Inch, 14-Inch 177
Bank 163
Barbara 138
Barometer Regulator 143
Beat Indicator (Saturn) 144
Bishop 164
Blake 164
Boston, The 136
Braddock 163
Branford 163
Brazilian Line
Hanging A, B, C 172
Cambria (Spring) 152
Camden Line
Hanging A, B, C 170-171
Canton Lever 175
Chatelaine 180
Columbia 137
Commodore 147
Cuba 173
Dneiper 148

HANGING CLOCKS
(continued)

Drop Octagon
8, 10, 12 Inch 165
10 Inch (Gilt) 165, 168
12 Inch 158, 166, 168
Drop Octagon,
Gilt, Buttons 168
Drop Octagon, No. 2 165
Drop Octagon R. C. 164
Drop Octagon, Small 167
Drop Octagon,
Small (Gilt) 167
Drop Octagon,
Small, R. C. 167
Drop Octagon
Small (Spring) 166, 168
Eagle 176
Elfrida 140
Emperor 166
Emperor, Brass Bands 166
Engine Lever 175
Enquirer 140
Erie 164
Fatinitza, Gilt 155
Faustina, Gilt 155
Gallery 174
12-Inch 174, 178
Walnut & Gilt 174
Gambia 142
Glass Dial Clocks 176
Transparent, Translucent 177
Glenor 157
Gloriana 137
Grayson 157
Grecian 141
Griffin 157
Harvard 163
Hebe 138
Hudson 150
Imperial 162
Interpid 139
Jestla 159
Julian 158
Juneau 154
Kenilworth 154
Kitchenette 178
Lacona 158
Larina 178
Lawen 178
Lawen Kitchen 139

HANGING CLOCKS
(continued)

Lever
Acorn 147
Canton 175
Engine 175
Round 139
Round Corner Octagon .. 178
Star 181
Synchronized 176
Wood 175, 178
Yacht 174
Lucerne 177
Madena 157
Madero 160
Mandarin 180
Manor 158
Maravia 157
Marino 160
Marne 159
Matanzas 172
Maywood 140
Merchants Line
Hanging A 170
Mersey Line Hanging A ... 170
Meuse 139
Mimosa 160
Mirada 160
Movements
Office No. 1 146
Regulator No. 2 133
Saturn Double Spring 144
Saxon 169
Nansen 156
Nason 156
Neafus 156
Neese 156
New York, The 136
No. 25 153
No. 25, Gilt 153
No. 26 152
No. 26, Gilt 152
No. 1012 179
No. 1013 - 1014 179
No. 1022 181
No. 3004 147
No. 3043 159
Norseman 149
Norwich Line
Hanging A, B, C 170-171
Obi 142

HANGING CLOCKS
(continued)

Octagon, Drop
12 Inch 158, 166, 168
Office No. 1 .... 145, 146
Office No. 2 148
Pan 179
Programme 161
Referee 158
Regulator 150
Regulator, Small 162
Regulator, B. B. (Spring) .. 155
Regulator, C. C. 153
Regulator, D. R. 161
Regulator No. 0 132
Regulator No. 00 131
Regulator No. 00 No. 2 131
Regulator No. 1 132
Regulator No. 2 133
Regulator No. 3 134
Regulator No. 6 134
Regulator No. 9 135
Regulator No. 10 135
Riverton 159
Round Corner Octagon 178
Round Lever 139
Salem 173
Saracen 149
Saturn 144
Saxon 169
Saxon Mosaic, 8-Inch 169
Saxon Mosaic, 10-Inch 169
Small Drop Octagon 167
Small Drop Octagon,
R. C. 167
Small Drop Octagon
(Spring) 166, 168
Small Regulator 162
Solent 143
St. Louis, The 136
Standard Time No. 1 151
Standard Time No. 2 151
Standish 173
Stanton Hanging Kitchen . 173
Stanton Weight Regulator 141
Star Lever 181
Synchronized Lever 176
Tampa 158
Tampico 143
Thornton 141
Trojan 138

HANGING CLOCKS (continued)

Vamoose 137
Vandal 162
Ventura 181
Vidette 149
Viking 150
Winnipec 154
Winnipec Regulator 154
Wood Lever 175, 178
Yacht Lever 174

INKSTAND CLOCKS 182-183

Buffalo Ink 183
Louis 182
Maintenon 182
Scribe 183
Scribe, The (Ink Stand) 183
Secretary Ink 182

LEATHER CASES 184

Courier 184
Jasper Clock 184
Junior Tattoo 184
Kenneth 184
Kenwick 184
Leathern 184
Magnus 184
Panza 184
Sting Time 184

LONGCASE CLOCKS & STANDING REGULATORS 185-190

Annapolis 189
Arlington 190
Blenheim 191
Brooksdale 190
Chapel Chime No. 1 191
Chippendale 186
Giant 186
Hall Clock No. 1 187
Hall Clock No. 1 Mvt. 191
Minneapolis 189
Regulator No. 4 187
Regulator No. 8 185
Regulator No. 10 189
Sheffield 190

LONGCASE CLOCKS & STANDING REGULATORS (continued)

Sherbrooke 188
Thornton 186
Wellington 187
Wilmington 188

MANTEL CLOCKS - CABINET CLOCKS 192-216

Adam No. 2 206
Adam No. 52 215
Amphion 204
Antony 197
Arab 197
Arrow 202
Aubert 210
Aubrey 208
Auburn 208
Aurania 196
Aurine 210
Auris 212
Aurora 208
Ausprey 200
Austell 211
Auston 212
Austro 211
Banshee 207
Black Bird 201
Brutus 205
Cabinet No. 1 .... 192, 193
Movement and Label 192
Cabinet No. 2 193
Cabinet No. 3 194
Cabinet No. 4 195
Cabinet No. 5 195
Cabinet No. 6 195
Cabinet No. 7 197
Cabinet No. 8 204
Cabinet No. 9 198
Cabinet No. 10 201
Cabinet No. 516 204
Cadmus 205
Caliban 196
Cassius 205
Cato 193
Clifton 196
Colonial 206
Corbin 196
Delhi 213
Delmar 213

MANTEL CLOCKS - CABINET CLOCKS (continued)

Edina 204
Effington 209
Eggleston 209
Etruria 201
Funston 195
Gallia 199
Hecla 200
Horace 194
Inwood 216
Irene 216
Irwin 216
Jacobean No. 1 206
Jacobean No. 53 215
Loanda 216
Lombard 216
Lopez 213
Lorne 213
Lorraine 216
Lurine 203
Malo 208
Marius 205
Martin 202
Medea 200
Melita 200
Mendon 200
Merton 208
Montross 214
Morea 203
Myra 203
Nero 193
Oberlin 210
Olean 207
Olivia 207
Olympus 215
Oporto 212
Oregon 211
Orin 215
Oriole 211
Orison 210
Orville 209
Osborne 209
Ostend 212
Pallas 214
Pedestal Clock No. 1 214
Percival 204
Petrel 194
Piedmont 207
Pompey 201
Portia 214

MANTEL CLOCKS - CABINET CLOCKS (continued)

Pyrrhus 194
Remus 205
Romulus 205
Russia 202
Scipio 193
Servia 199
Stuart 206
Sulla 197
Thrush 202
Umbria 199
Undine 203
Valkyrie 198
Virgil 194
Vreda 198
Wesley 214

MANTEL CLOCKS - ENAMELED IMITATION CASES 217-222

Favorita 221
Florence 220
Fortuna 222
Galatea 218
Genesta 220
Gladys 220
Gretchen 222
Halcyon 221
Henrietta 217
Hermit 221
Hildegarde 219
Huntress 219
Ianthe 219
Mohican 218
Mona 218
Oriva 218
Regina 222
Siren 220

MANTEL CLOCKS - ENAMELED IRON CASES 223-236

Abbottsford 223
Aberdeen 224
Aberfoyle 225
Aegean Set 235
Aragon 226
Arno 234

MANTEL CLOCKS - ENAMELED IRON CASES (continued)

Baccarat 230
Belfast 225
Bolivia 226
Charlton 234
Chateau 234
Chiselhurst 234
Colonna 226
Constance 227
Corydon 227
Cuba 228
Cumberland 224
Czarin 232
Czarina 228
De Soto 226
Dominica 226
Fairfield 230
Fairmont 230
Fayette 233
Ferncliff 233
Fremont 231
Fulton 231
Harcourt 236
Hayti 229
Honduras 226
Ivanhoe 224
Kenmore 226
Leland 236
Margarita 228
Melrose with
Coupe H 223
Monaco 236
Montrose 227
Munich 225
New Monaco 228
Newark 231
Newberne 233
Olinda 235
Pocahontas 229
Rosalind 228
Sinbad 229
Talmage 231
Tempest 235
Thetis 229
Titan 236
Trafalgar 225
Tyndall 230
Wabash 233
Wallace 232
Waltham 232

MANTEL CLOCKS - ENAMELED IRON CASES (continued)

Warwick 231
Washington 227
Waverly 224
Wexford 232
Windemere 223

MANTEL CLOCKS - ENAMELED WOOD CASES 237-246

Alexander 241
Armorel 245
Bancroft 242
Bingen 239
Brighton 244
Burlington 243
Canfield 240
Carmona 241
Castile 246
Claypool 243
Concord 238
Cordova 246
Cranston 240
Dante 246
Edgebrook 237
Edgemont 242
Edgewood 242
Elbridge 237
Elsmere 237
Elsworth 237
Joliet 239
Loring 238
Malvern 244
Mantua 244
Milan 245
Numa 241
Ottawa 241
Pembroke 243
Prescott 239
Riviera 244
Rosetti 246
Seneca 243
Seville 245
Tyndall 245
Woodbine 238
Woodbridge 242
Woodlake 239
Woodmont 240
Woodside 238

MANTEL CLOCKS - ENAMELED WOOD CASES (continued)

Woodvale 240

MANTEL CLOCKS - MARBLE CASES 247-263

Kismet 257
Lorna 257
Magic 260
Mandan (Set) 256
Marble and Bronze, No. 1027 254
Marble No. 1001 249
Marble No. 1005 249
Marble No. 1008 248
Marble No. 1010 252
Marble No. 1012 251
Marble No. 1016 247
Marble No. 1019 248
Marble No. 1026 250
Marble No. 1031 253
Marble No. 1032 253
Marble No. 1033 253
Marble No. 1034 253
Marble No. 1050 254
Marble No. 1051 248
Marble No. 1054 255
Marble No. 1065 250
Marble No. 1076 255
Marble No. 1078 251
Marble No. 1086 249
Marble No. 1100 249
Marble No. 1144 255
Marble No. 1149 250
Marble No. 1150 250
Marble No. 1151 247
Marble No. 1152 252
Marble No. 1153 255
Marble No. 1154 254
Marble Set No. 1006 248
Marble Set No. 1011 251
Marble Set No. 1035 252
Marble Set No. 1037 247
Marjorie 259
Marston (Set) 256
Mermaid 261
Minevra 261
Mistletoe 261
Modjeska 259
Monarch 258

MANTEL CLOCKS - MARBLE CASES (continued)

Mugwump 257
Mystery 262
Napolean 258
Natalie 258
Nautilus 262
Nebula 262
Nirvana 262
Nymph 263
Parole 263
Peerless 259
Phantom 263
Priscilla 260
Qui Vive 260

MANTEL CLOCKS - MARBLOID CASES 264-265

Illyria 264
Ione 265
Ivaca 265
Nubian 265
Parian 264
Patrician 265
Poligny 264

MANTEL CLOCKS - METAL CASES 266-281

Acrobat 280
Allier 277
Altair 281
Amanda 281
Annecy Set 266
Antique Brass 276
Antique Brass, No. 1 276
Antique Brass, No. 2 274
Arnold Set 270
Corolan 267
Criterion Set 268
Cupid Set 270
Falaise Set 267
Falmouth 278
Flores 279
Furness 271
Hampden 281
Hilda 278
Isere Set 273
Lakeside Set 268
Larchmont Set 272

MANTEL CLOCKS - METAL CASES (continued)

Laurel 279
Lenore Set 269
Libitina Set 273
Libourne 277
Louis 281
Lyon Set 269
No. 26 280
No. 1009 277
No. 1019 278
No. 1023 277
Orpheus 279
Pavia Set 271
Plevna 277
Rochelle Set 272
Rosemont Set 275
Serpentine 281
Sicilian Set 271
Spartan 276
Trouville Set 274
Truxton Set 275
Uno 276
Vergennes Set 266

MANTEL CLOCKS - MUSICAL CLOCKS 282

Lohengrin 282

MANTEL CLOCKS - TAMBOUR CLOCKS 283-291

Hale 291
Hanford 291
Harper 291
Lansing 288
Linton 287
Lucifer 288
Luzon 287
Quayle 291
Quebec 290
Queda 287
Questa 290
Quinby 289
Quincy 290
Quirk 290
Quito 291
Radway 291
Romero 291
Tambour No. 1 284

MANTEL CLOCKS - TAMBOUR CLOCKS (continued)

Tambour No. 2 283
Tambour No. 3 285
Tambour No. 4 289
Tambour No. 5 289
Tambour No. 6 286
Tambour No. 7 285
Tambour No. 8 290
Tambour No. 9 284
Tambour No. 11 288
Tambour No. 13 288
Tambour No. 14 286
Tambour No. 15 283
Tambour No. 16 287
Tambour No. 17 286
Tambour No. 18 285

MANTEL CLOCKS - TILE CASES 292-293

Chelsea Tiles 292, 293
Ilium 292
No. 1017 .... 292, 293
No. 1017, Movement 293

MANTEL CLOCKS - WOOD CASES 294-349

Adige 311
Agrippa 341
Alderman 339
Alps 297
Ambassador 337
Andes 294
Apollo 312
Aral 322
Argyle 309
Argyle Time 309
Arion 340
Arno 320
Auburn 303
Avon 319
Ayr 325
Bobolink 333
Bonita 333
Boreas 334
Boudoir Clocks 349
Boyne 319
Brazilian Line 345
Calumet 324

MANTEL CLOCKS - WOOD CASES (continued)

Camden Line 344
Captain 338
Carmen 327
Cecile 327
Celeste 327
Champion 313
Charmer 335
Christine 329
Cinderella 329
Clarita 329
Clochette 327
Clyde 326
Colonel 338
Column 305
Conway 309
Coquette 321
Corsair 317
Cottage 304
Cottage, Extra 303
Cottage, Gem 295, 303
Cottage, Jerome & Co. 304
Cottage, No. 2 303
Councilman 339
Countess 313
Coupon Preferred 296
Cygnet 343
Dandy Line 348
Danube 322
Dee 329, 346
Derby 300
Derby, T. P. 300
Desna 320
Diadem 300
Direct 340
Don 329, 346
Doon 346
Dora 320
Douro 319
Drave 316
Duchess 298
Duna 321
Ebro 320
Eclipse 294
Elbe 315, 316
Elbe Patent Dial Pendulum ....
Ball 315, 401
Elma 328
Enchantress 325
Euphrates 327

MANTEL CLOCKS - WOOD CASES (continued)

Ganges 318
Gem 298
Gem Cottage 295, 303
General 338
Geneva Set 336
Gingerbread 314
Gothic 301
Gothic Boudoir 349
Gothic Gem 295
Gothic Gem, D. R. 295
Gothic Gem, V. P. 295
Gothic, Round 336
Gothic, Round, Extra 306
Gothic, Round, V. P. 306
Gothic, Sharp, V. P. 302
Gothic, Small 301, 302
Gothic, Small, V. P. 302
Governor 337
Gracchus 341
Grayling 326
Guide 306
Guide, Full Gilt 306
Guide, Gilt Pillar 306
Guide, V. P. 306
Gypsy Line 344
Humber 323
Hustler Line 339
Indus 318
Irex 328
Ivy 334
Janitor 339
Jerome & Co. Cottage 304
Jerome & Co. Label 304
Juvenal 341
Kama 326
Lena 317
Lieutenant 338
Liffey 319
Liris 331
Lobbyist 337
Loire 345
Maas 319
Mail 313
Maine Line 346
Major 338
Marshal 339
Mayflower 343
Mayor 339
Merchants Line 348

MANTEL CLOCKS -
WOOD CASES (continued)

Mersey 321
Mersey Line 343
Metal Trimmed Mantels 345
Metellus 341
Military Line 338
Mirror 349
Moselle 316
Mottled Tuscan 294
Movements
Column, 30hr T & S 305
Duchess, 30hr, T & S 298
Elbe, 8-day, T, S & A 315
Gingerbread, 8-day, T, S & A 314
Gothic, 30hr, T & S 301
Indus, 30hr 318
Jerome & Co. Cottage, 30hr, T & A 304
Sigma, T, S & A 314
Small Gothic, 30hr, T & A 301
Nancy 340
Neckar 323
Nereid 325
Neva 324
No. 251 300
No. 254 299
No. 501 312
No. 502 312
No. 503 323
No. 504 326
No. 505 ....329, 335
No. 506 327
No. 508 334
No. 509 313
No. 510 316
No. 511 322
No. 512 322
No. 513 323
No. 514 323
No. 515 317
No. 517 317
No. 518 321
No. 519 324
No. 520 324
No. 521 335
No. 522 323
No. 523 335
No. 524 345

MANTEL CLOCKS -
WOOD CASES (continued)

No. 525 345
No. 529 336
No. 750 310
No. 751 308
No. 752 308
No. 753 308
No. 754 310
No Name 294
Norwich Line 347
Nutwood 340
O. B. 309
O. B. Time 309
O. G. (No. 1 or No. 2) 307
O. O. G. 307
O. O. G. Extra V. P. 307
Oakdale 312
Occidental 311
Occidental Extra 311
Octagon Gem 297
Octagon Gem, Extra 297
Octagon Gem, Extra Gilt 297
Octagon Gem, Gilt 297
Octagon Prize 298, 299
Octagon Prize, Sm. Extra 299
Octagon Prize, V. P. 298, 299
Octagon Rocket 297
Oder 312
Osage 326
Ouse 324
Oval Boudoir 349
Parisian 322
Parthian 336
Patrol Line 342
Picket Line 342
President 337
Prize 304
Prize Winner Line 340
Pruth 324
Pyramid 294
Rambler 328
Rarus 310
Rarus, T. P. 310
Rattler 303
Recorder 339
Regulus 341
Reliable 294
Rhine 316
Riga 321
Rocket 294

MANTEL CLOCKS -
WOOD CASES (continued)

Roman Line 341
Rose Gilt 303
Round Gothic 336
Round Gothic, Extra 306
Round Gothic, V. P. 306
Sabine 330
Santee 333
Secretary 337
Seine 330
Seminole 331
Senator 337
Sergeant 338
Severn 331
Shamrock 335
Shannon 330
Sharp Gothic "A" 300
Sharp Gothic "G" 300
Sharp Gothic, V. P. 302
Sigma 314
Small Gothic 301, 302
Small Gothic, V. P. 302
Small Octagon Prize, Extra 299
Small Tuscan, Extra 296
Square Boudoir 349
Stamboul 340
Statesman Line 337
Sterling 309
Sunol 340
Tagus 330
Tamar 332
Tarquin 341
Tees 332
Tennessee 326
Thames 330
Theiss 320
Tiber 330
Tigris 332
Titania 331
Tomahawk 333
Trent 321
Tuscan 296
Tuscan, D. R. 296
Tuscan, Mottled 294
Tuscan, Small, Extra 296
Tuscan, V. P. 296
Tyne 311
Vampire 328
Vindex 332

MANTEL CLOCKS - WOOD CASES (continued)

Volante 332
Volga 334
Warwick 309
Weser...............................325
Yarana 333
Zingara 334

MISSION CLOCKS 350-356

Dulce 351
Durance Hanging.............351
Dutch No. 20 350
Dutch No. 40 350
Dutch No. 45 350
Gordon.............................351
La Cruz 351
La Lanza 353
La Plata 354
Liso (No. 7216) 356
Los Alamos 353
Los Barrios 353
Los Santos 353
Lute (No. 7213) 356
Minas (7250) 356
Minch (7252)....................356
Minho (7251)....................356
Moro 355
No. B5084 355
No. B5085 (San Felipe)....355
No. 7211 356
No. 7213 (Lute) 356
No. 7216 (Liso) 356
No. 7250 (Minas) 356
No. 7251 (Minho).............356
No. 7252 (Minch) 356
Palos 355
San Benito 355
San Carlos 353
San Felipe (No. B5085)....355
San Jose 352
San Juan...........................354
San Luis 353
San Lucas 351
San Marcos 356
San Martin 352
San Mateo 352
San Pablo 354
San Pedro.........................352
San Rafael 352

MISSION CLOCKS (continued)

Sancho Line 354
Santa Clara 352
Tosca 356
Vega 355

MOVEMENT ILLUSTRATIONS 357-400

1-Day Sm. Lever Timepiece
(No. 400) 385
1-Day Sm. Slip-in Alarm
(No. 434) 387
(No. 435) 388
1-Day Marine 395
1-Day Tat-Too Alarm
(No. 505) 389
1-Day Tick-Tock Alarm
(No. 590) 391
1-Day Tidy-Tot Alarm
(No. 426) 386
1-Day Tom-Tom Alarm
(No. 506) 390
8-Bell Chime No. 1 400
8-Day Alarm 395
8-Day Automatic and 8-Day
Signal Automatic 398
8-Day Duo-Strike, Rack &
Snail (No. 700) 392
8-Day Motor Clocks 394
8-Day Pend. Chime, Rack &
Snail (No. 380) 383, 384
8-Day Pend. Time & Strike
(No. 303) 361
(No. 304) 362
(No. 307) 363
(No. 308) 364
(No. 309) 365
(No. 310) 366
(No. 311) 367
(No. 312) 368
(No. 320) 369
(No. 321) 370
(No. 335) 379
(No. 337) 381
8-Day Pend. Timepiece
(No. 325) 371
(No. 327) 372
(No. 328) 373

MOVEMENT ILLUSTRATIONS (continued)

(No. 329) 374
(No. 330) 375
(No. 332) 376
(No. 333) 377
(No. 334) 378
(No. 336) 380
(No. 338) 382
8-Day Time & Strike, Rack &
Snail, (No. 750) 393
12-Day Motor Clock 394
Automatic (8-day) 398
Chime No. 1 (8-Bell)........400
Elm City Electric 396
Junior Tattoo 395
Marine (1-day).................395
Motor Clocks...................394
No. O 357
No. 1 358
No. 1 (8-Bell, Chime)......400
No. 2 (Regulator)............399
No. 50 (Sm. Watch).........359
No. 270 (Lg. Watch) 360
No. 303 (8-day P, T,S) 361
No. 304 (8-day P, T,S) 362
No. 307 (8-day P, T,S) 363
No. 308 (8-day P, T,S) 364
No. 309 (8-day P, T, S) ...365
No. 310 (8-day P, T, S) ...366
No. 311 (8-day P, T, S) ...367
No. 312 (8-day P, T, S) ...368
No. 320 (8-day P, T, S) ...369
No. 321 (8-day P, T, S) ...370
No. 325 (8-day P, T)........371
No. 327 (8-day P, T)........372
No. 328 (8-day P, T)........373
No. 329 (8-day P, T)........374
No. 330 (8-day P, T)........375
No. 332 (8-day P, T)........376
No. 333 (8-day P, T)........377
No. 334 (8-day P, T)........378
No. 335 (8-day P, T, S) ...379
No. 336 (8-day P, T)........380
No. 337 (8-day P, T, S) ...381
No. 338 (8-day P, T)........382
No. 380 (8-day P, Chime,
R & S)...................383, 384
No. 400 (1-day Sm. L.ever
Timepiece) 385
No. 426 (1-day Tidy-Tot

MOVEMENT ILLUSTRATIONS (continued)

Alarm) 386
No. 434 (1-day Sm.Slip-in ....
Alarm) 387
No. 435 (1-day Sm. Slip-in
Alarm) 388
No. 505 (1-day Tat-Too
Alarm) 389
No. 506 (1-day Tom-Tom
Alarm) 390
No. 590 (1-day Tick-Tock
Alarm) 391
No. 700 (8-day Duo-Strike,
R & S) 392
No. 750 (8-day, T, S,
R & S) 393
Regulator No. 2 399
Saturn, Wt. Driven 395
Self Starting 397
Signal Automatic (8-day) .398
Spring (No. 1) 358
Watch, (Large) (No. 270) .360
Watch , (Small) (No. 50) .359
Weight (No. O) 357
Whitehead, 300 R.P.M.
Self Starting 397

MOVEMENT PARTS 401-407

Click Springs 407
Clock Verge 407
Electric Clock Hands 404
Hands 402, 403, 404
Keys, Winding and
Setting 405, 406
Parts & Trimmings 401
Patent Dial Pendulum
Ball 401
Pendulum Ball,
Patent Dial 401
Pendulum Balls 407
Pendulum Rods 407
Springs, Click 407
Verge, Clock 407

NOVELTY CLOCKS 408-452

Acorn 443
Aigrette 417
Aire 426

NOVELTY CLOCKS (continued)

Akron 444
Alsace 439
American Illuminated Night
Clock 422
Amoret 451
Andorra 450
Arabesque 410
Archer 447
Arden 429
Ardennes 450
Ariel 451
Arverne 435
Astoria (Jeweled Watch) ..415
Atlas 415
Banjo 409
Baroda 439
Bedford 449
Bell, The 421
Bellaire 432
Bellona 421
Benton 443
Berbice 448
Bermuda 439
Bingham 433
Boatman 418
Bobby 419
Bouquet 417
Bull Fight (Mechanical) ...452
Calais 437
Caliph 437
Cameron 413
Cannes 437
Carleton 437
Carnot 440
Castine 436
Chariot 434
Chatham 431
Chiswick 412
Cloisonne 423
Clovelly 432
Colima 438
Colon 437
Corea 413
Corfu 434
Cremorne 426
Crescent Stand 415
Crescent Stand
With Jasper 414
Dalton 429

NOVELTY CLOCKS (continued)

Darien 434
Dateo 420
Dauphin 443
Delano 440
Delavan 443
Delong 439
Digby 443
Dinsmore 445
Dion 420
Disque 420
Dixon 428
Dolphin Stand 415
Doris 445
Dresden 445
Drew 441
Druid 436
Duluth 446
Dunstan 441
Dupont 441
Durand 436
Eberle 438
Edisto 446
Eldora 422
Falcon Hunter 429
Fallis 433
Fan 427
Fenby 432
Fenwick 431
Ferney 434
Ferro 432
Fleury 433
Flush 425
Flying Pendulum No. 1 408
Flying Pendulum No. 2 408
Flying Pendulum No. 3 408
Flying Pendulum
(No. 5107) 408
Follen 435
Forli 443
Gnome 451
Grandfather 424
Hadley 449
Halsey 448
Harlem 447
Harwich 448
Henley 447
Hazel 439
Humming Bird 413
Huron 440

NOVELTY CLOCKS
(continued)

Inca 451
Indee 431
Irma 436
Ivan 414
Jasper 414
Jasper Figure 414
Kazan 424
Lantern Night Clock 427
Lantier 429
Lascar 435
Lighthouse 418
Linnet 419
Lisbon 450
Little Duke 425
Loire 422
Mace T. P. 413
Macon 437
Maintenon 422
Mandolin 410
Mandolin (Mechanical) 452
Mangrove 429
Manila 432
Manton 435
Marlin 452
Masonic 416
Mechanical,
  Movable Dials 452
    Bull Fight 452
    Mandolin 452
    Pet 452
    Puff 452
    Shaver 452
Minden 441
Minot 428
Mistral 415
Monkey 443
Nachusa 434
Naola 446
Napier 433
Naruna 445
Nashoba 446
Nathalia 441
Natoma 441
Navan 439
Navina 445
Neva 444
Night Clock 411
Night Clock Berlin Base ..411
Night Clock No. 3 Gas 412

NOVELTY CLOCKS
(continued)

Night Clock No. 3 Lamp ..412
Night Clock No. 5 Gas 411
Night Clock No. 5 Lamp ..412
Night Clock in Case 412
Nile 444
No. 59 427
No. 60 425
No. 61 428
No. 63 425
No. 64 425
No. 65 426
No. 72 428
No. 74 428
No. 76 428
No. 1004 Extra 423
No. 1005 424
No. 1006 426
No. 1008 419
No. 1015 424
No. 1020 424
No. 1024 423
No. 1025 423
No. 1031 426
No. 5107
  (Flying Pendulum) 408
No. 7061 434
No. 7062 433
No. 7063 433
Nome 445
Padua 430
Panay 436
Paper Weight 416
Paper Weight No. 4 416
Parachute 419
Parasol 410
Pelham 431
Pequot 430
Pet (Mechanical) 452
Phillipina 432
Piety 444
Pindar 438
Ponce 438
Puck 417
Puff (Mechanical) 452
Rachel 440
Racquet 409
Rawson 431
Rider 418
Robin 418

NOVELTY CLOCKS
(continued)

Rodell 437
Roden 436
Rodman 437
Rodney 431
Roland 446
Romita 449
Rosita 445
Ross 447
Rovena 448
Rustic Tripod 421
Scroll 419
Sentinel 418
Shaver (Mechanical) 452
Sheaf 427
Snipe 425
Standard 425
Sting 422
Stirrup 409
Tabard 420
Time 416
Toboggan 444
Toltec 424
Tralee 442
Traveler 420
Tray 421
Trianon 432
Tyrone 435
Uncas 449
Utica 449
Valetta Set 442
Valreas 450
Vaury 450
Vauvert 440
Vayrac 450
Vercel 450
Vieta 433
Volunteer 417
Wee Wag No. 100 451
Wee Wag No. 101 451
Yakima 444
Yamaska 446
Yancopin 442
Yantus 442
Yarnaby 446
Yesner 440
Yuma 426
Zalia 438
Zante 436
Zarbine 436

NOVELTY CLOCKS (continued)

Zebina 430
Zelda 430
Zenda 430
Zerba 430
Zeta 430
Zetus 430
Zilpha 431
Zinnia 438
Zirkel 431

NOVELTY CLOCKS - WOOD CASES 453-460

Adrian 455
Adam No. 52 457
Bagley 455
Barto 455
Baton 455
Baxter 455
Boyce 455
Castro 460
Cigar Lighter 453
Datus 455
Edina 458
Eunice 460
Garcia 459
Garfield 459
Gladstone 459
Green 459
Hale 456
Halstead 454
Harbin 454
Hardy 456
Harlee 456
Harper 454
Harvey 454
Helmar 458
Hilton 458
Hobby 458
Hobson 458
Hodges 456
Horton 458
Ideal 460
Ionia 460
Ison 459
Ives 459
Ivy 460
Jacobean No. 53 457
Jacobean No. 55 457

NOVELTY CLOCKS - WOOD CASES (continued)

Negus 460
Nemo 460
No. 1011 453
No. 1011, No. 2 453
Oval No. 52 457
Oval No. 53 457
Pequot 458
Quaker 456
Queda 457
Quiz 458
Radway 456
Selden 460
Tapis 455
Tennis Ink Set, The 454
Vaux 458
Wesley 457

PLUSH CLOCKS 461-465

Album 465
Chalet 462
Flash 462
Grandfather 464
Lammermoor 465
Lark, The 464
Lark, Alarm ,The 464
Midlothian 465
Mortar 463
Owl 461
Puck 462
Rob Roy 461
Snug 461
Sting (Octagon) 463
Sun Flower 461
Traveling Case 463
Tyrrill 463
Zephyr 462

PORCELAIN CLOCK 466-474

Abner 474
Addison 474
Ambrose 471
Annabel 471
Audubon 474
Avalon, Faience Monogram 473
Avar 471
Creighton 474

PORCELAIN CLOCKS (continued)

Delta 466
Faience Ware 472, 473
Faience Ware, Monogram 473
Latimer 469
Lee 468
Linwood 468
Macri 468
Marion 470
Melrose 467
Minto 468
Molino 468
Nelson 469
No. 1066 A 472
No. 1066 B 472
No. 1066 B Movement 473
No. 1066 C 472
No. 1066 D 472
Parkman 470
Penola 470
Pompadour 467
San Remo 466
St. Cloud 467
St. Louis 467
Thistle 470
Triad Set 469
Verona 471
Vincent 474
Volney 470
Waldorf 466
Wilmot 471
Wilson 471
Windham 471
Windsor 466

STATUETTES AND CANDLESTICKS 475-478

Bonner 475
Diana 475
Dog 476
Hannibal 476
Herndon Candlesticks 477
Horse 475
Horse Tamer 475
Lion 475
Music 476
No. 387, Candlesticks 477
No. 3197, Candlesticks 477
No. 3320, Candlesticks 478

STATUETTES AND CANDLESTICKS (continued)

No. 3345, Candlesticks....478
No. 3817, Candlesticks....478
Pandora....476
Paragon Candlesticks......478
Pointer....476
Scipio....476
Scout....476
Stag....475
Trouville....478
Trumpeter....475

WATCHMAN'S CLOCKS 479-481

Electric Watchman's Clock....479
Key, Chain, Box....480
Leather Pouch....480
Newman Portable....480
Newman System....481
Regulator No. 1....481
Square Clock....481
Square Regulator No. 1....481
Square Regulator No. 2....481
Watchman's Detector....479

WATCHES 482-494

100 Mile Pedometer....489
Bicycle Clock, Novelty....490
Case Backs.... 484, 486,487,490, 494
"C" Fancy....484
"E"....490
"N" Fancy....486
"Q"....487
"V"....487
Clyde Beatty Lapel....492
Don Winslow Lapel....492
Earl....487
Elite....485, 486
Elm City....485, 487
Gem....485, 487
Football Timer....489
Hero....483
Jerome....484, 485
Kermit....486
Lady Clare....484, 485
Lady Nelson....482

WATCHES (continued)

Lapel Jr.....492
Lapel Timer....493
Lapel Watches....492, 493
Lawson Football Timer....489
Locomotive....487
Lone Ranger....493
Lone Ranger Fob Watch....493
Manhattan Watch, The....490
Model "A"....488
Model "B"....488
Model "B" Backs....487
Model "D"....488
Model "E"....488
Motor, The....487
Nelson....482
New Haven, No. 8,....483
New Haven Pocket....491
Nifty....488
Nurses Special....489
Pedometer, 100 Mile....489
Pentagon Lapel No. 1....492
Pentagon Lapel No. 2....492
Popeye Pocket....487
Ranger....484, 485
Service Convertible Wrist Watch....486
Service Watch, The....486
Sports Timer....489
Sting Watch....482
Streamline, New Haven....487
Timers....489, 493
Tip-Top....491
Tip-Top Pocket....491
Tip-Top Quintet....491
Tip-Top Service....490
Tommy-Ticker....491
Two-Timer....489
Yale....484
Yale Fancy Bezel....484

WATCHES - ADVERTISING 495-498

Gem Sign, Design No. 2...498
Smith's American Shield Watch, Clock, Jewelry Sign....497, 498
Sports Timer....495

WATCHES - ADVERTISING (continued)

Tip-Top....496

WRIST WATCHES 499-505

Boy Scout....499, 501
Brownie....499, 504
Commuter....505
Count....499, 502
Countess....499, 502
CW900 (Duke)....505
CW901 (Count)....502
CW902 (Orphan Annie)...502
CW903 (Elf)....505
CW905 (Countess)....502
CW907 (Duchess)....502
CW908 (Fad)....502
CW911 (Hand Bag)....505
Dick Tracy....501
Duchess....499, 502
Duke....499, 505
Earl....499
Elf....499, 505
Elite....501
Fad....501, 502
Fad, New....503
Gem....501
Gem, New....503
Hand Bag Watch....505
Inca....500
Jerome....500, 504
Jerome Jr.....500, 504
Lady Joan....500
New Fad....503
New Gem....503
New New Haven....503
Orphan Annie....502, 503
Popeye....503
Par Round....502
Service....505
Tip-Top....501
Tip-Top Jr.....502
Tip-Top Par and Box....503
Tommy-Ticker....501
Trophy....505
Wales....500
Yale....500